CHEMICAL SIGNALS
IN VERTEBRATES 6

Edited by

Richard L. Doty
University of Pennsylvania
Philadelphia, Pennsylvania

and

Dietland Müller-Schwarze
College of Environmental Science and Forestry
State University of New York
Syracuse, New York

Plenum Press · New York and London

Library of Congress Cataloging-in-Publication Data

Chemical signals in vertebrates 6 / edited by Richard L. Doty and
Dietland Müller-Schwarze.
        p.   cm.
    "Proceedings of the Sixth International Conference on Chemical
Signals in Vertebrates, held June 16-22, 1991 in Philadelphia,
Pennsylvania"--T.p. verso.
    Includes bibliographical references and index.
    ISBN 0-306-44250-7
    1. Chemical senses--Congresses. 2. Vertebrates--Physiology-
-Congresses. 3. Physiology, Comparative--Congresses.   I. Doty,
Richard L.   II. Müller-Schwarze, Dietland.   III. International
Conference on Chemical Signals in Vertebrates (6th : 1991 :
Philadelphia, Pa.)
QP455.C473   1992
596'.01'826--dc20   92-22006
                                                               CIP

QP
455
.C473
1992

Proceedings of the Sixth International Conference on
Chemical Signals in Vertebrates, held June 16–22, 1991,
in Philadelphia, Pennsylvania

ISBN 0-306-44250-7

© 1992 Plenum Press, New York
A Division of Plenum Publishing Corporation
233 Spring Street, New York, N.Y. 10013

Printed in the United States of America

CHEMICAL SIGNALS
IN VERTEBRATES   6

This volume was made possible, in part, as the result of
contributions from the following corporations:

Block Drug Company, Inc.
Campbell Soup Company
Coca-Cola Company
Denali Corporation
Erox Corporation
First Brands Corporation
International Flavors and Fragrances
McNeil Specialty Products Company
PepsiCo Inc.
Ralston Purina Company
R.J. Reynolds Tobacco Company
Rohm and Haas Company
Sensonics, Inc.

The corporations listed below are given special recognition for
having provided additional funds to insure the
success of CSV VI:

Erox Corporation
Rohm and Haas Corporation

PREFACE

This volume is an up-to-date treatise of chemosensory vertebrate research performed by over 200 scientists from 22 countries. Importantly, data from over 25 taxa of vertebrates are presented, including those from human beings. Unlike other volumes on this topic, a significant number of the contributions come from leading workers in the former Soviet Union and reflect studies within a wide variety of disciplines, including behavior, biochemistry, ecology, endocrinology, genetics, psychophysics, and morphology.

Most of the studies described in this volume were presented at the Chemical Signals in Vertbrates VI (CSV VI) symposium held at the University of Pennsylvania in the summer of 1991. This international symposium was the largest and the most recent of a series of six such symposia, the first of which was held in Saratoga Springs, New York (June 6-9, 1976) and the last in Oxford, England (August 8-10, 1988). Unlike the previous symposia, Chemical Signals in Vertebrates VI lasted a full week, reflecting the increased number of participants and the desire of many to present their research findings orally to the group as a whole.

A number of the papers of the present contribution reflect, explicitly or implicitly, advances that have occurred in the chemical senses field in the last five years, including breakthroughs in understanding the basic mechanisms underlying the olfactory, gustatory, vomeronasal, trigeminal, nervus terminalis, and other oral and nasal sensory systems. Notable examples of such advances include (a) the implication of the vomeronasal system in a wide variety of behavioral and endocrine responses, (b) the better understanding of the biochemical mechanisms involved in olfactory and gustatory sensory transduction (e.g., the important role of G proteins and cyclic nucleotide second messengers in such transduction for some stimuli), (c) the realization of the complexity of chemical communication within most vertebrate classes, including fish and birds, and the elucidation of the influences of semiochemicals, particularly hormones, on such processes, and (d) advances in the measurement of human olfactory function, including the recording of odor evoked potentials.

As can be gleaned from the contents of the volume, there continues to be widespread interest in the chemical senses which is fostered, in a number of countries, by increased research funding from both governmental and private sources. Indeed, the present book would not have been possible without the generous support of the following organizations: Block Drug Company, Inc., Campbell Soup Company, Coca-Cola Company, Denali Corporation, Erox Corporation, First Brands Corporation, International Flavors and Fragrances, McNeil Specialty Products Company, PepsiCo Inc., Ralston Purina Company, R.J. Reynolds Tobacco Company, Rohm and Haas Company, and Sensonics, Inc. Special thanks are due to David Berliner of the Erox Corporation, and to Phillip G. Lewis of the Rohm and Haas Corporation, who contributed funds even after the symposium was over to insure its fiscal viability.

We wish to acknowledge the efforts of Catherine Beinhauer, Michele Kaminsky, and Jennifer Loren of Conference Management Associates, who organized and collected many of the initial manuscripts and who played a primary role in the organization and managment of the CSV VI conference. Other persons to whom we are indebted include the members of the CSV VI steering committee and the staff and students of the Smell and Taste Center. We are particularly grateful to Amanda Merwin, Donah Crawford, and Donald McKeown, who volunteered their services to insure the success of the meeting.

Richard L. Doty
Dietland Muller-Schwarze

Philadelphia, Pennsylvania
Syracuse, New York

March, 1992

CONTENTS

SECTION SIX: CHEMICAL REPELLENTS AND CHEMOSENSORY AVERSIONS

SECTION SEVEN: BEHAVIOR AND CHEMICALLY MEDIATED SOCIAL COMMUNICATION

Part 1: Fish

Part 6: Mammals -- Humans

SECTION ONE

ANATOMY AND PHYSIOLOGY OF CHEMOSENSORY SYSTEMS

ACTION OF SUCROSE ON THE SALTY TASTE RESPONSE

Keiichi Tonosaki

Department of Oral Physiology, School of Dentistry
Asahi University, 1851 Hozumi, Hozumi-cho, Motosu-gun
Gifu 501-02 Japan

INTRODUCTION

Taste stimulus adsorption is believed to occur at the taste cell microvillous membrane (Beidler, 1954; Beidler and Tonosaki, 1985). However, little is known about the mechanisms of taste transduction, due mainly to technical difficulties of inserting glass electrodes into mammalian taste cells (Tonosaki and Funakoshi, 1984,1988). We previously reported that the mouse taste cell response to sucrose is a membrane depolarization accompanied by an increase in membrane resistance. The sucrose response increases in amplitude as the membrane is depolarized and decreases in amplitude as the membrane is hyperpolarized. The same taste cell responds to NaCl with membrane depolarization accompanied by a decrease in membrane resistance. The NaCl response increases in amplitude as the membrane is hyperpolarized and decreases in amplitude as the membrane is depolarized. These results suggest that sucrose and NaCl have quite different response generation mechanisms (Tonosaki and Funakoshi, 1988; Tonosaki, 1988).

I now present evidence that sucrose pre-adaptation suppresses the cross-adaptation responses to NaCl and KCl. Intracellular recordings from receptor cells are performed in order to study the adaptation mechanism.

MATERIALS and METHODS

Adult male mice (Slc:ICR)(30-50g) were anesthetized with sodium pentobarbital and tracheotomized. Tongues were continuously superfused with distilled water at a constant rate of 0.08ml/sec. Test solutions (sucrose: 0.5M, 1.0M; NaCl: 0.05M, 0.1M; KCl: 0.05M, 0.1M) were periodically introduced by replacing the distilled water by switch without interrupting the fluid flow. In the pre-adapting experiments, the tongues were first adapted to distilled water. Then the distilled water flow was stopped, and the adapting solution flow was immediately switched on. After a few minutes, the adapting solution was replaced with the test solution followed by reintroduction of the adapting solution for about five seconds. After about one minute, the adapting solution was turned off and distilled water was immediately turned on. These experimental paradigms are indicated in the Figures. The dissection of the preparation, the recording techniques, and the taste delivery methodology are thoroughly described elsewhere (Tonosaki, 1988).

RESULTS

Figure 1 shows a typical example of mouse taste cell responses. All records (Fig 1A and B) were obtained from one taste cell. Figure 1A shows responses to sucrose, NaCl and KCl. The resting membrane potential was -45mV. The sucrose response shows a membrane depolarization which is accompanied by an increase in membrane resistance. The NaCl and KCl responses show membrane depolarization which is apparently caused by a decrease in membrane resistance. As shown in Figure 1B, pre-adapting the tongue with sucrose suppressed the NaCl and KCl responses. Similar results were obtained from thirteen experiments. In Figure 1B, the NaCl response was nearly completely suppressed and the membrane resistance slightly increased. The KCl response was also almost completely suppressed and the membrane resistance initially increased then decreased. It is not identical with that of NaCl. These results indicate that the generation mechanisms of NaCl and KCl responses after adaptation are different. Sucrose response magnitudes are much smaller than those to NaCl and KCl. The pre-adaptation to 0.5M sucrose solution also suppressed both of these responses, but the effect was weaker than that to 1.0M sucrose.

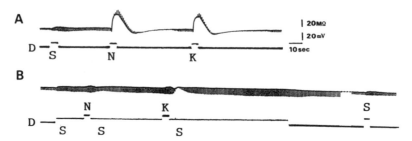

Fig. 1. All the records were obtained from a taste cell. See description in text. The time shown by the broken lines was two minutes. Note that, in Figure 1A during the NaCl and KCl responses, the upward deflected electrotonic potentials represent a decrease in membrane resistance. Abbreviations; D: distilled water; S: 1.0M sucrose; N: 0.1M sodium chloride; K: 0.1M potassium chloride.

Figure 2 is an example from another taste cell. In Fig. 2A, the NaCl and sucrose+NaCl stimulus produced similar response profiles. When the taste cell was pre-adapted to sucrose, the cross-adapted NaCl response was suppressed and the rising phase was less steep than the response of the control. When the taste cell was pre-adapted to NaCl, the cross-adapted sucrose response showed a hyperpolarization potential change which was accompanied by an increase in membrane resistance. In Fig. 2B, the KCl and KCl+sucrose stimuli produced similar response profiles. When the taste cell was pre-adapted to sucrose, suppression of the KCl cross-adaptation response was observed. When the cell was pre-adapted to KCl, the cross-adapted sucrose response showed hyperpolarization accompanied by an increase in membrane resistance. When the membrane potential was depolarized, the sucrose response magnitude was increased and the NaCl response magnitude decreased. While the membrane potential was hyperpolarized, the sucrose response was decreased and NaCl response was increased. It should be noted that the sucrose+NaCl and NaCl stimuli produced similar responses whenever the membrane potential was depolarized or hyperpolarized. Figure 3 shows that pre-adapting the taste cell to sucrose (or NaCl) affected the taste cell responses to NaCl (or sucrose). When the cell was pre-adapted to sucrose, NaCl produced a

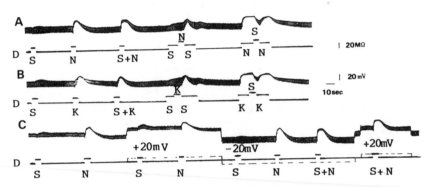

Fig. 2. All the records were obtained from one taste cell. See description in text. The symbols and the abbreviations are the same as described in Fig. 1.

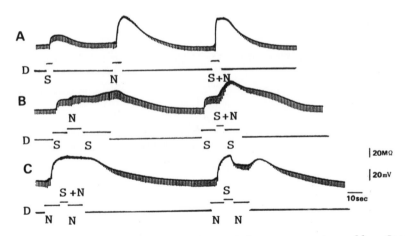

Fig. 3. All the records were obtained from one taste cell. See description in text. The symbols and the abbreviations are the same as described in Fig. 1.

different response profile to that observed for NaCl+sucrose (Fig. 3B). When the cell was pre-adapted to NaCl, sucrose+NaCl produced a different response profile than observed when it was cross-adapted to sucrose (first response in Fig. 3C). The NaCl pre-adapted taste cell does not show the membrane potential change to the sucrose+NaCl stimulus. In contrast, when the sucrose stimulus was applied to a cell adapted to NaCl (second response in Fig. 3C), the cell hyperpolarized and evidenced an increase in membrane resistance. After the application of the stimulus, however, the hyperpolarization continued for a while, even after the solution was switched back to NaCl. The potential gradually returned to a depolarized level.

DISCUSSION

Numerous psychophysical taste cross-adaptation studies have been published (Bartoshuk, et. al., 1978; Kroeze, 1978, 1979; Gilian, 1982; Lawless, 1982; Tonosaki and Funakoshi, 1989). However, the exact nature of adaptation is not known. The present data suggest that adaptation occurs in the taste receptor cell itself. There are two major hypotheses for taste reception mechanisms. One is the idea of multiple sensitivity and the other is the idea of across fiber pattern. These hypotheses have been examined only by recording nerve responses and by psychophysical experiments. Since a taste nerve fiber innervates many taste cells, it is extremely difficult to determine whether different taste receptor sites are distributed equally among individual taste cells. The mechanism of cross-adaptation has never been studied with single taste cells. The present study examined cross-adaptation with intracellular electrodes. The results suggest that activation of NaCl and sucrose sensitive sites interferes with each other's transduction process. We previously advanced the hypothesis that when sucrose adsorbs to the receptor site of the taste cell, the internal c-GMP concentration increases, leading to a decrease in potassium conductance (Tonosaki and Funakoshi, 1988). We also proposed, based on voltage-clamp experiments (Tonosaki, 1988), an alternative hypothesis: that when NaCl adsorbs to the receptor site, a NaCl-related second messenger in the cell triggers the opening of the sodium channels. From these results, it might be hypothesized that when each taste stimulus binds to certain receptor sites on the taste cell, each stimulus activates internal second messengers. The intracellular chemical processes (including second messengers) activated by different taste stimuli may interact with each other, thereby leading to the response shown in this study (Fig. 1B).

ACKNOWLEDGEMENTS

This research was supported, in part, by Grant-in-Aid to K.T. for Scientific Research from the Ministry of Education, Science and Culture of Japan.

REFERENCES

Bartoshuk, L.M., Murphy, C. and Cleveland, C.T., 1978, Sweet taste of dilute NaCl: Psychophysical evidence for a sweet stimulus, Physiol. Behav., 21:609-613.

Beidler, L.M., 1954, A theory of taste stimulation, J. Gen. Physiol., 38:133-139.

Beidler, L.M. and Tonosaki, K., 1985, Multiple sweet receptor sites and taste theory, in: "Taste, Olfaction and the Central Nervous System.", D.W. Pfaff, ed., Rockefeller University Press, New York, pp. 47-64.

Gilian, D.J., 1982, Mixture suppression: The effect of spatial separation between sucrose and NaCl, Percept. Psychophys., 32:504-510.

Kroeze, J.H.A., 1978, The taste of sodium chloride: Masking and adaptation, Chem. Senses Flavour, 3:443-449.

Kroeze, J.H.A., 1979, Masking and adaptation of sugar sweetness intensity, Physiol. Behav., 22:347-351.

Lawless, H., 1982, Adapting efficiency of salt-sucrose mixtures, <u>Percept</u>.
    <u>Psychophys</u>., 32:410-422.
Tonosaki, K., 1988, Generation mechanisms of mouse taste cell responses.
    In: "The Beidler Symposium on Taste and Smell"., I.J. Miller, Jr.,
    ed., Book Service Associates, Winston-Salem, North Carolina, U.S.A.,
    pp. 93-102.
Tonosaki, K. and Funakoshi, M., 1984, Effect of polarization of mouse
    taste cells, <u>Chem</u>. <u>Senses</u>, 9:381-387.
Tonosaki, K. and Funakoshi, M., 1988, Cyclic nucleotides may mediate
    taste transduction, <u>Nature</u> (Lond), 331:354-356.
Tonosaki, K. and Funakoshi, M., 1989, Cross-adapted sugar responses in
    the mouse taste cell, <u>Comp</u>. <u>Biochem</u>. <u>Physiol</u>., 92A:181-183.

α GUSTDUCIN: A TASTE CELL SPECIFIC G PROTEIN SUBUNIT

CLOSELY RELATED TO THE α TRANSDUCINS

Susan K. McLaughlin, Peter J. McKinnon and
Robert F. Margolskee

Roche Research Center
Roche Institute of Molecular Biology
Nutley, New Jersey   07110

INTRODUCTION

The guanine nucleotide binding proteins (G proteins) mediate signal transduction in olfactory, visual, hormonal and neurotransmitter systems (see Birnbaumer, 1990). The G proteins comprise a gene family of proteins which transduce an extracellular signal into an intracellular second messenger (e.g. cAMP, cGMP, IP3). Many different cell surface receptors couple to and interact with specific subtypes of G proteins. All G proteins are heterotrimers ($\alpha\beta\gamma$) wherein the α subunit binds guanine nucleotides and contains most, if not all, of the specificity for both receptor and effector interactions.

In the vertebrate taste cell, electrophysiological, biochemical and histological evidence suggest that G proteins are involved in the taste transduction process (Avenet and Lindemann, 1989; Roper, 1989; Kinnamon, 1988). Sweet compounds cause a GTP-dependent generation of cAMP in rat tongue membranes (Striem, Pace, Zehavi, Naim and Lancet, 1989). External application or microinjection of cAMP inactivates K channels in vertebrate taste cells and leads to depolarization of these cells (Tonasaki and Funakoski, 1988; Avenet, Hoffmann, and Lindemann, 1988). High levels of adenylyl cyclase and cAMP phosphodiesterase are found in taste tissue (Kurihara and Koyama, 1972; Law and Henkin, 1982). These results suggest a transduction pathway in which the sweet receptor leads to cell depolarization via a G protein mediated rise in cAMP. Bitter may also be transduced via a G-coupled receptor: bitter compounds such as denatonium lead to Ca^{++} release from internal stores (likely by way of G protein-coupled IP3 generation) (Akabas, Dodd and Al-Awqati, 1988). To characterize the taste transduction process and to identify the specific proteins involved, we have cloned G protein α subunit cDNAs from rat taste tissue.

MATERIALS AND METHODS

The circumvallate and foliate papillae from one hundred Sprague-Dawley rats were harvested using the method of Spielman, Brand and Wysocki (1989). The harvested papillae were immediately frozen in 100% ETOH at -70° C. An equivalent amount of non-taste lingual epithelium (devoid of taste buds) was likewise harvested. PolyA⁺ mRNA was harvested from taste and non-taste tissue using the Quick Prep kit from Pharmacia: 7.9 μg of mRNA was recovered from the taste tissue and 2.4 μg of mRNA was recovered from the control non-taste tissue. The BRL pSPORT vector and the BRL superscript kit were used to make two cDNA libraries from 1 μg of taste and 1 μg of non-taste mRNA. The taste library contained 2.6 x 10⁶ independent clones (average insert size of 1.1 kb). The non-taste library contained 4.8 x 10⁶ independent clones (average insert size of 1.0 kb). RNA for RNase protection assays was generated in vitro by T3 or T7 polymerase transcription of linearized subclone DNAs. RNase protection assays used a kit from Ambion.

PCR primers corresponding to conserved amino acids of G proteins were made on an Applied Biosystems DNA synthesizer. The following 5' primers were used: DVGGQR; KWIHCF; HLFNSIC. The following 3' primers were used: FLNKKD and VFDAVTD. Three of these primers (DVGGQR, KWIHCF and FLNKKD) have been previously used by Strathman, Wilkie, and Simon (1989) in an earlier G protein Screen. PCR reactions contained 250 pmol of each primer and 10 ng of taste cell library cDNA in a 50 μl reaction volume. The PCR reaction followed this program: 94° x 1'; 37° to 72° with a rise time of 1° per 4 seconds; 72° x 3'; for three cycles followed by: 94° x 1'; 43° x 2'; 72° x 3'; for a total of 35 additional cycles. The PCR reaction was digested with Bam HI and Eco RI, electrophoresed in a 1% agarose gel, then bands of expected size were excised, purified, and cloned into the pBluescript vector. After transformation into E. coli, individual colonies were picked, and their DNAs sequenced using T7 DNA polymerase.

RESULTS

To identify G proteins specifically expressed in mammalian taste cells, we set out to clone G protein α subunit cDNAs from a rat taste cell cDNA library. Degenerate primers corresponding to the most highly conserved regions of the α subunits were synthesized and used in the polymerase chain reaction (PCR) with taste tissue cDNA. PCR amplified products of the expected size were cloned and the DNA inserts of individual clones were sequenced. Clones were categorized according to α subtype specificity based on their inferred amino acid sequence. The predicted amino acid sequences for four different taste α subunits derived from taste tissue are shown in Figure 1. A comparison of their sequences to known α subunits demonstrates that they are clearly members of the G protein α subunit family. Sequence analysis suggests that two of the taste tissue isolates are of α_i type ($\alpha_{i-1,3}$ and α_{i-2}), one is rod transducin (α_{t-rod}) and one is a novel subunit, α gustducin (α_{gust}) differing from previously known isolates. The amino acid sequence of α gustducin is most closely related to cone and rod transducins (see below). Altogether, seven different types of α subunits were isolated (see Table 1): two types of α_i, two types of α_q, two types of transducins (rod and cone forms), and α gustducin. Different PCR primer pairs yielded different types of α-subunit clones, α_i, α_{t-rod} and α_{gust} were the most frequent isolates.

Expression of the taste tissue α subunit clones was assayed by RNase protection with RNAs from taste tissue, retina and control non-taste lingual tissue (Figure 2). Highly elevated levels of α_{i-2} and α_{q-1} were detected in taste vs. non-taste RNA. α_{t-rod} and α_{gust} were expressed in taste tissue but were not present in the control non-taste RNA. Furthermore. α_{gust} is not detectable in retina, whereas α_{t-rod} is much more highly expressed in retina than in taste tissue. Previously, rod and cone transducins had only been identified in the retina in their respective

A.

```
              ───────→  ───────→                                                    ←───────
              DVGGQR    KWIHCF                                                       FLNKKD
αs:           DVGGQRDERRKWIQCFNDVTAIIFVVASSSYNMVIREDNQTNRLQEALNLFKSIWNNRWLRTISVILFLNKQD
αolf:         DVGGQRDERRKWIQCFNDVTAIIYVAACSSYNMVIREDNNTNRLRESLDLFESIWNNRWLRTISIILFLNKQD
αi-1,3:       DVGGQRSERKKWIHCFEGVTAIIFCVALSDYDLVLAEDEEMNRMHESMKLFDSICNNKWFTDTSIILFLNKKD
αi-2:         DVGGQRSERKKWIHCFEGVTAIIFCVALSAYDLVLAEDEEMNRMHESMKLFDSICNNKWFTDTSIILFLNKKD
αo:           DVGGQRSERKKWIHCFEDVTAIIFCVALSGYDQVLHEDETTNRMHESLMLFDSICNNKFFIDTSIILFLNKKD
αz:           DVGGQRSERKKWIHCFEGVTAIIFCVELSGYDLKLYEDNQTSRMAESLRLFDSICNNNWFINTSLILFLNKKD
αq            DVGGQRSERRKWIHCFENVTSIMFLVALSEYDQVLVESDNENRMEESKALFRTIITYPWFQNSSVILFLNKKD
αt-rod:       DVGGQRSERKKWIHCFEGVTCIIFIAALSAYDMVLVEDDEVNRMHESLHLFNSICNHRYFATTSIVLFLNKKD
αt-cone:      DVGGQRSERKKWIHCFEGVTCIIFCAALSAYDMVLVEDDEVNRMHESLHLFNSICNHKFFAATSIVLFLNKKD
consensus:    DVGGQRSER KWIHCFE VT IIF ALS YD VL ED   NRM ES  LF SI N    F  SIILFLNKKD
```

B.

```
αi-1,3:       SERFLWIHCFEGVTAIIFCVALSDYDLVLAEDEEMNRMHESMKLFDSICNNKWFTDTSIIL
αi-2:         SERKKWIHCFEGVTAIIFCVALSAYDLVLAEDEEMNRMHESMKLFDSICNNKWFTDTSIIL
αt-rod:       SERKKWIHCFEGVTCIIFIAALSAYDMVLVEDDEVNRMHESLHLFNSICNHRYFATTSIVL
αgust:        SERKKWIHCFEGVTCIIFCAALSAYDMVLVEDEEVNRMHESLHLFNSICNHKYFATTSIVL
```

Fig. 1. (A) The amino acid sequences of nine previously cloned G protein α subunits are depicted. Above the α sequences are three conserved regions chosen for use as PCR primers (DVGGQR, KWIHCF and FLNKKD). A consensus sequence was determined when there was identity in 7 of 9 sequences. (B) The inferred amino acid sequence of four α subunit cDNAs obtained by PCR amplification of taste tissue cDNA. The assignment of a particular taste clone to subtype was according to amino acid match to previously known sequences. αgust does not match any previously known G protein sequences.

Table 1. Isolates of α-subunit clones from PCR amplified taste tissue cDNA

	5' KWIHCF x 3' FLNKKD	5' DVGGQR x 3' FLNKKD	5' HLFNSIC x 3' VFDAVTD
αi-2	4	-	-
αi-1,3	5	-	-
αq-1	-	1	-
αq-2	-	1	-
αt-rod	3	-	2
αt-cone	-	-	2
αgust	5	-	4

photoreceptor cells where they mediate photo-transduction (Lochrie, Hurley and Simon, 1985; Yatsunami and Khorana, 1985; Medynski, Sullivan, Smith, VanDop, Chang, Fung, Seeburg and Bourne, 1985; Tanabe, Nakada, Nishikawa, Sugimoto, Suzuki, Takahashi, Noda, Naga, Ichiyamo, Kangawa, Minamino, Matsuo, and Numa, 1985). The αt-rod isolated from rat taste tissue is virtually identical to bovine retina αt-rod (60/61 amino acids are identical in this region, suggesting a species difference rather than a subtype difference). Furthermore, RNase protection using the rat taste tissue isolate of $\alpha_{t\text{-rod}}$ as probe against rat retinal mRNA indicates that the taste tissue clone matches perfectly to the retinal mRNA (i.e., the taste and retinal α_t mRNAs have identical nucleotide sequence). Since the main difference between taste and non-taste tissue is the presence or absence of taste buds, these results are highly suggestive of taste cell expression of α_i, α_q, $\alpha_{t\text{-rod}}$ and α_{gust}. We have recently determined by in situ hybridization that α gustducin is expressed specifically within taste buds of foliate, circumvallate and fungiform papillae (data not shown).

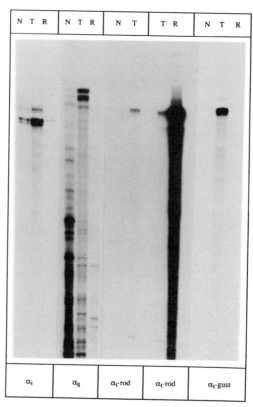

Fig. 2. α subunit expression analyzed by RNase protection: the RNAs used are indicated at the top of the figure (N=non-taste, T=taste, R=retina), the probes used are indicated at the botton of the figure. The α subunit probes correspond to the region shown in Figure 2 ($\alpha_i=\alpha_{i-2}$; $\alpha_q=\alpha_{q-1}$). Full-length protected band sizes are ~230 for α_q and ~200 bp for the other probes.

For the region displayed in Figure 1, rat taste tissue α_{gust} matches bovine $\alpha_{t\text{-rod}}$ at 58 of 61 amino acids and bovine $\alpha_{t\text{-cone}}$ also at 58 of 61 amino acids. However, this is one of the most highly conserved regions of all G proteins. Longer $\alpha_{t\text{-rod}}$ and α_{gust} clones were isolated by PCR with other primers and by screening the taste cell cDNA library with specific probes. Comparing the DNA sequence and inferred amino acid sequence of these longer clones to that of known α subunits confirms that α gustducin is most closely related to the α transducins; it is equally related to rod and cone α transducins (see Table 2). α_{gust} is much more distantly related to other α subunits (e.g. α_i, α_o, α_s). The rat taste tissue isolate of α_{t-rod} is clearly most closely related to bovine rod transducin (98% amino acid identity), and less closely related to bovine cone transducin (79% amino acid identity). This relatedness and the RNase protection data suggest that the rat taste isolate of α_{t-rod} is identical to the retinal form of rat α_{t-rod}. The bovine α_{t-cone} gene is closely related to the taste tissue isolate of rat α_{t-rod} (79% amino acid identity) and to rat α_{gust} (79% identity). These results suggest that α_{t-rod}, α_{t-cone} and α_{gust} comprise a gene family of three closely related G protein subunits, all of which are expressed in taste tissue. Presumably α_{t-rod} and α_{t-cone} are also expressed in the taste buds themselves.

Table 2. Percent sequence identify (amino acid/nucleic acid) between taste tissue α isolates and other α-subtypes

	Rat α_t-rod	Rat α gust
Bov α_t-rod	98/86	80/71
Bov α_t-cone	79/71	79/73
Rat α_i	66/68	66/65
Rat α_o	61/63	62/63
Rat α_s	45/59	46/59

DISCUSSION

G proteins play a key role in sensory transduction in vision, olfaction and taste. Specific biochemical pathways for signal transduction have been identified in both vision and olfaction. Many of the proteins involved in the visual and olfactory pathways have been cloned and characterized. The gustatory pathways have not been as well characterized as those of the visual or the olfactory systems. However, biochemical and electrophysiological evidence suggest a role for G protein coupled receptors in the transduction of both sweet and bitter tastants. Candidate second messengers include cAMP (sweet) and IP_3 (bitter); effector enzymes present in gustatory cells include cAMP phosphodiesterase and adenylyl cyclase. No direct evidence exists for a gustatory specific cyclic nucleotide responsive channel. However, gustatory K+ channels are regulated by cAMP and a cAMP dependent protein kinase.

In the present study we have identified cDNAs for seven different G protein α subunits present in gustatory tissue. α_i and α_q type mRNAs are elevated in taste vs. non-taste tissue. α_{t-rod} and α_{gust} are expressed in taste tissue but not in non-taste lingual tissue. α_{gust} mRNA is very highly expressed in taste tissue but not in retina. We have directly shown that taste buds are the site of expression of α gustducin. If indeed α_{t-rod} and α_{t-cone} are present in both visual and taste transducing cells it would suggest a remarkable commonality to these two sensory transduction pathways. Furthermore, that α gustducin is so closely related to the α transducins suggests that gustducin may be functioning analogously to the other transducins. Perhaps similarities exist between the visual and gustatory pathways in other elements (e.g. receptors, phosphodiesterases, kinases, second messengers, and ion channels). Ultimately molecular cloning, biochemical purification and in vitro reconstitution may allow us to understand and describe the molecular events of taste transduction with the same level of precision presently known for vision.

REFERENCES

Akabas, M.H., Dodd, J., and Al-Awqati, Q., 1988, A bitter substance induces a rise in intracellular calcium in a subpopulation of rat taste cells, Science, 242:2047.

Avenet, P. and Lindemann, B., 1989, Perspectives of taste reception, J. Membr. Biol., 112:1.

Avenet, P., Hofmann, F., and Lindemann, B., 1988, Transduction in taste receptor cells requires cAMP-dependent protein kinase, Nature, 331:352.

Birnbaumer, L., 1990, G proteins in signal transduction, Ann. Rev. Pharmacol. Toxicol., 30:675.

Kinnamon, S.C., 1988, Taste transduction: a diversity of mechanisms, Trends in Neurosci., 11:491.

Kurihara, K. and Koyama, N., 1972, High activity of adenyl cyclase in olfactory and gustatory organs, Biochem. Biophys. Res. Comm., 48:30.

Law, J.S. and Henkin, R.I., 1982, Taste bud adenosine-3',5'-monophosphate phosphodiesterase: activity, subcellular distribution and kinetic parameters, Res. Comm. in Chem. Pathology and Pharm., 38:439.

Lerea, C.L., Somers, D.E., Hurley, J.B., Klock, I.B., and Bunt-Milam, A.H., 1986, Identification of specific transducin α subunits in retinal rod and cone photoreceptors, Science, 243:77.

Lochrie, M.A., Hurley, J.B., and Simon, M.I., 1985, Sequence of the α subunit of photoreceptor G protein: homologies between transducin, ras and elongation factors, Science, 228:96.

Medynski, D.C., Sullivan, K., Smith, D., VanDop, C., Chang, F.-H., Fung, B.K.-K., Seeburg, P.H., and Bourne, H.R., 1985, Amino acid sequence of the α subunit of transducin deduced from the cDNA sequence, Proc. Natl. Acad. Sci. USA, 82:4311.

Roper, S.D., 1989, The cell biology of vertebrate taste receptors, Ann. Rev. Neurosci., 12:329.

Spielman, A.I., Brand, J.G., and Wysocki, L., 1989, A rapid method of collecting taste tissue from rats and mice, Chem. Senses. 14:841.

Strathmann, M., Wilkie, T.M., and Simon, M.I., 1989, Diversity of the G-protein family: sequences from five additional α subunits in the mouse, Proc. Natl. Acad. Sci. USA, 86:7407.

Striem, B.J., Pace, U., Zehavi, U., Naim, M., and Lancet, D., 1989, Sweet tastants stimulate adenylate cyclase coupled to GTP-binding protein in rat tongue membranes, Biochem. J., 260:121.

Tanabe, T., Nakada, T., Nishikawa, Y., Sugimoto, K., Suzuki, H., Takahashi, H., Noda, M., Haga, T., Ichiyamo, A., Kangawa, K., Minamino, N., Matsuo, H., and Numa, S., 1985, Primary structure of the α-subunit of transducin and its relationship to ras proteins, Nature, 315:242.

Tonosaki, K. and Funakoski, M., 1988, Cyclic nucleotides may mediate taste transduction, Nature, 331:354.

Yatsunami, K., and Khorana, H.G., 1985, GTPase of bovine rod outer segments: the amino acid sequence of the α subunit as derived from the cDNA sequence, Proc. Natl. Acad. Sci. USA, 82:4316.

AVIAN TASTE BUDS: TOPOGRAPHY, STRUCTURE AND FUNCTION

H. Berkhoudt

Department of Neurobehavioral Morphology
Zoological Laboratory, University of Leiden
P.O. Box 9516, 2300 RA Leiden, The Netherlands

INTRODUCTION

An overwhelming range of special foraging and feeding techniques have
evolved in birds. Generally, these adaptations are reflected in the
external anatomy of the bill, their main manipulative tool. Such
adaptations can also be seen in certain inner oropharyngeal structures,
including the tongue which, in many birds, has a key-role in the prehension
and transport of food (for a review of bill and tongue adaptations, see
McLelland, 1979). It was already clear to the first investigators of avian
taste buds that the topographic distribution of such buds shows a strong
correlation with the feeding techniques employed and, thus, with anatomical
details of the oropharynx.

There has long been a conflict in the literature between the
behavioral evidence of the ability of birds to use taste as a cue in food
finding and acceptance and the anatomical evidence for taste sensitive
structures (Berkhoudt, 1985a). Historically, we can distinguish two periods
of the morphological description of avian taste buds, separated by a clear
break coinciding with the change from German to English as the scientific
language.

Taste buds in birds were discovered relatively late, just after the
turn of the century, about 50 years after the description of taste organs
in fishes, amphibia, reptiles (excluding Crocodilia, which were described
later by Bath, 1906) and mammals. This delay may have been partly caused by
the fact that the contemporary expert on avian sense organs, Merkel, stated
in his main work (1880) that he was unable to find any taste buds in birds.
It took until 1904, when Botezat, in a provisional communication, published
the finding that taste buds are, in fact, present in several bird species.
As in mammals, the earlier search for avian taste buds concentrated on the
anterior tongue as the preferred location for their positioning. In birds,
however, the tongue functions as a prehensile organ for food intake and
transport of solid food. As a result, the anterior dorsal surface of the
tongue is often keratinised, thus preventing any further use for taste bud
positioning.

Botezat (1904, 1906, 1910) and Bath (1906), who were the first to
describe these organs in birds correctly had profound disputes over the

question as to whether the avian tongue contains taste buds. Both agreed about the main topography of taste buds, i.e. their location on the palatal and mandibular areas of the uncornified oral epithelium and the fact that few are contained on the anterior tongue. Their main point of dispute, apart from minor structural differences in the description of the taste bud itself, was what areas constitute "tongue". As pointed out by Moore and Elliott (1946), there is a difference in the interpretation of the posterior limit of the tongue. Bath excluded all areas posterior to the "tongue fold", while these were included by Botezat. As a result, Bath had the opinion that no tastebuds occur in the avian tongue, where Botezat claimed them in posterior areas in the dorsal aspect of the tongue.

A clear break in history was initiated by the work of Moore and Elliott (1946), who were interested in the evolution of taste on the tongue in vertebrates. This coincided with the switch from German to English as the scientific language at the end of the second world war. For birds, they meticulously mapped the number of tastebuds in six pigeon tongues. Since their study formed the starting point for all later publications in English, and the earlier German articles often appeared in obscure journals, the notion that avian tastebuds occur only in the tongue gained so much influence that even today this view is still held as credible by some workers. One of the goals for my contribution to this symposium is to show that the statement that "anatomical descriptions are rudimentary, especially for taste, and functional morphology has scarcely been approached" (Wenzel, 1984) needs readjustment, since it is not in line with the historical record. I will discuss the present state of the field of avian taste from a functional morphological perspective, parallel to the early German findings which are often cited, but seldom read.

MATERIALS AND METHODS

In parallel with the work of Bath (1906), the topography of taste buds was studied in a number of bird species such as mallard, chicken, pigeon, carrion crow, and grebe (cf. Berkhoudt, 1985a). Here fresh information about taste buds in the starling will be included. In contrast to Bath's procedure, the extraction of densities from serial sections, we have developed a new rapid staining procedure with Pontamine Skyblue which gives a quick general overview of the distribution. We generally follow this by scanning electron microscopy (SEM) for more detailed regional information (see Berkhoudt, 1977, for technical details). Details of taste bud structure and their innervation were obtained from serial sections of material fixed by perfusion in formalin 10%. The sections, either frozen or after paraffin embedding, were stained using the silver staining method of Sevier and Munger (1965). In the same species, functional morphological investigations of feeding were also carried out using high-speed film, X-ray cinematography, and, recently, on-line gaperecording, often in conjunction with electromyography (EMG) of the major jaw- and tongue muscles.

RESULTS

Topography of taste buds

Feeding adaptations of birds are reflected in the internal anatomy of the oropharynx. Bath had concluded that the distribution of taste buds in various birds is strongly correlated with this internal anatomy and thus with the feeding technique employed. Therefore, anatomical details such as the shape, size and structure of the tongue in relation to the lower mandible formed, next to structural details of the taste bud itself, an additional criterion for a subdivision of his avian taste material into three categories.

In Bath's first category fall birds with a relatively slender bill and a narrow tongue, which leaves a free strip of uncornified oral mucosa in the mandible on both sides of the tongue. As noted by Botezat (1904, 1906) and Bath (1906), these strips are the major location of taste buds in this group. However, in the chicken and pigeon they are also found in rostral and caudal palate areas and in the lingual base (cf. Berkhoudt, 1985a; Ganchrow and Ganchrow, 1985, Kurosawa et al., 1983). Our results show that the distribution of taste buds, as assessed by the rapid surface staining procedure is comparable to the tedious countings obtained from mapping the locations from serial sections. For the chicken, for instance, van Prooije's data (unpublished internal report, see Berkhoudt, 1985a) only show deviations for the posterior tongue regions compared with the data of Kurosawa et al. (1983). Van Prooije found proportions of 70%, 26% and 4% in the oral epithelium of palate, mandible and posterior region of the anterior tongue, respectively (N=17, total number of buds =250). Especially for this latter region it is very hard to tell the difference between the openings of the salivary glands and taste buds proper, while also combinations of the two seem to occur (Botezat, 1906; Ganchrow and Ganchrow, 1985). In hatchling chicks these figures were 69%, 29% and 2% (on an average of 316 taste buds, N=3, Ganchrow and Ganchrow, 1985). In this group of birds the pecking technique of food intake predominates, especially in the granivorous species investigated by Bath, such as pigeons, chickens, sparrows, finches and buntings. But birds employing more specialised feeding techniques, such as gaping in starlings, aerial sweeping in swallows and swifts, throwing and catching in hornbills etc. are also included in this first group. As already suggested by Bath, it is this difference in feeding technique that accounts for the variations observed in the topography of the areas for taste bud location.

Bath's second category contains birds that employ predominantly aquatic filter-feeding techniques, such as ducks and flamingoes. Straining of food from suspensions in ducks is the result of a suction-pressure pump mechanism (Zweers et al.,1977; Kooloos et al., 1989), in combination with different food retention techniques, dependent on the species (Kooloos et al. 1989). A similar mechanism is supposed for straining in flamingoes (Jenkin, 1957). In both cases the wide and fleshy tongue, which acts as a piston for the pumping mechanism, fills up the whole oral cavity, so that the oral mucosa of the mandible is no longer free on the sides and consequently taste buds no longer occur in this region. In mallards, additional rostral fields of tastebuds in the mandible under the tongue-tip and in the rostral and middle palate, not included in Bath's original description, were recently described by Berkhoudt (1977). It was demonstrated that the topographic distribution of taste buds in the mallard is in register with the pathway taken by the food, both in straining (Zweers et al., 1977) and pecking (Kooloos, 1986).

Bath's third group, formed by parrots and parakeets, is not described here since we have no sufficient material for proper illustration.

Structure of taste buds

Bath classified avian taste buds into three types. All share a composition of four types of cells: centrally in the bud neuroepithelial cells occur, surrounded by sustentacular or supporting cells, whilst in the periphery, follicular or sheath cells separate the inner sensory core from the surrounding epidermal cells. At the base of each taste bud, occasional basal cells are sometimes found. In the first electron microscopic study of avian taste buds, these four types of cells were also distinguished (Dmitrieva, 1981). Dimitrieva's study shows that, in the pigeon, each receptor cell has a large microvillar process at its apical end, probably identical to the "Sinnesstiftchen" described by Bath. The perigemmal cells, which form several layers around the taste bud, are identical to the follicular or sheath cells described by Bath.

Electron microscopic examination of chicken taste buds revealed light and dark cells in the centre of the bud, both showing slender microvilli at their apex (Kurosawa et al. 1983). These cells are probably identical to the neuroepithelial (sensory) and supporting cells described by Botezat and Bath. Similarly, in developing taste buds in the chick on embryonic day 20, basal and perigemmal cells can be recognized (Ganchrow and Ganchrow, 1989), while in the central-apical part of the bud the presence of light, dark and also intermediate cell types is again reported. At the tips of the light and dark cells numerous microvilli are found, contrasting with the single microvillar process found in the pigeon.

Innervation of taste buds and central projections

In birds, three neuronal nets exist for taste bud innervation, a sub-gemmal, perigemmal, and intra-gemmal net (Botezat 1904, 1906; cf. Berkhoudt, 1985a). Krol and Dubbeldam (1979), using the degeneration technique, elegantly proved in the mallard that nerve fibers of the medial ramus of the facial nerve, travelling in the inferior ramus of the ophthalmic branch of the trigeminal nerve, innervate the newly described rostral areas of taste buds in the palate. After interruption of the nerve, taste buds containing 20-30 axons in intact taste buds were devoid of innervating axons. For the anterior mandibular taste area in the mallard, innervation occurs by fibers of the chorda tympani nerve (Berkhoudt, 1980). Gentle (1983) has shown that the chorda tympani branch of the facial nerve in the chicken responds to chemical stimulation of taste buds in the region of the anterior mandibular salivary glands. Dubbeldam (1977, 1984a) showed that in the mallard and the pigeon a first central relay station occurs for gustatory afferents of nerves VII and IX in a specific subnucleus in the solitary complex, the nucleus ventrolateralis anterior (Vla of Dubbeldam, 1984). Recent work of Arends et al. (1988) gives evidence that this nucleus (here designated as mVal, the subnucleus medialis ventralis, pars anterolateralis of the nucleus of the solitary tract, nTS) projects, in the pigeon, to the medial and dorsal regions of the parabrachial nuclear complex, as do other medial tier subnuclei, which receive gastrointestinal input.

Feeding mechanics and taste function

Explaining structure is a main goal in functional morphology. Therefore, our studies of the anatomy and distribution of avian taste buds are embedded in a wider functional anatomical study of the feeding mechanism in birds. Understanding the neural basis of feeding can be formulated as one of our ultimate aims (cf. Dubbeldam, 1984). Our functional studies of the feeding mechanism of a wide range of bird species enables us to compare the topography of taste buds in the oropharynx with the positions along which taste information can be gathered during the process of intra-oral transport. Taste is a rather slow process, mediated by diffusion of taste substances from the food into the taste pit, which is filled with mucopolysaccharid substances, the presence of which can be detected by Periodic Acid-Schiff (PAS) and Pontamine Skyblue staining methods (Berkhoudt, 1977). Therefore we believe that sampling of taste information occurs only in the static and not in the dynamic phases of pecking (Zweers and Berkhoudt, 1990).

An analysis of the general pecking behavior in granivorous birds, characteristic for Bath's group I birds, shows that it can be broken down into a distinct series of phases (Zweers, 1982; Berkhoudt, 1985a). In order of occurrence: visual fixation, approach, peck, grasp, stationing and intra-oral transport. During head descent the eye closes so the final phase of the peck is only touch controlled. Taste control is not possible here, since cornification of the rostral aspects of the beak prevents taste bud positioning. This touch control is demonstrated in a sequence of film where plasticene imitation peas were offered to a pigeon. It is clear that discrimination and rejection occurs at beak tip levels. After the food is initially grasped between the beaktips, it is repositioned, the so-called stationing-phase (Zweers, 1982), between the tongue tip-anterior mandible and the monostomatic maxillary glands of the rostral palate. In birds employing the pecking type of food intake, the free strips of the mucosa adjoining the tongue are additionally exposed during grasping, since tongue-retraction and jaw-opening are coupled reflexively, as shown by EMGrecordings in pigeons (Berkhoudt, unpublished results) and chickens (W.F. van den Heuvel, personal communication). The stationing phase is the first opportunity where the most rostral taste areas can be brought into play, by contact with chemical substances from the food (Berkhoudt, 1985a,b). Later, during the subsequent phases of the intra-oral transport which are generally mediated by tongue actions, more caudal areas of taste buds become involved in discrimination. In non-alert birds, food dipped into acidic taste solutions is rejected only after complete ingestion is accomplished, i.e. at posterior tongue levels, unlike the case of alert birds, where rejection occurs at the most rostral taste areas.

REFERENCES

Arends, J.J.A., J. M. Wild and H. P. Zeigler, 1988, Projections of the nucleus of the tractus solitarius in the pigeon (Columba livia), J. Comp. Neurol. 278, 405-429.

Bath, W., 1906, Die Geschmacksorgane der Vögel und Krokodile. Arch. f.Biontologie 1, 5-47.

Berkhoudt, H., 1977, Taste buds in the bill of the mallard (Anas platyrhynchos L.), Their morphology, distribution and functional significance. Neth. J. Zool. 27, 310-337.

Berkhoudt, H., 1980, The morphology and distribution of cutaneous mechanoreceptors (Herbst and Grandry corpuscles) in bill and tongue of the mallard (Anas platyrhynchos L.), Neth. J. Zool. 27, 1-34.

Berkhoudt, H., 1985a, Special sense organs: Structure and function of avian taste receptors, in : Form and function of birds, Vol 3, A. S. King and J. Mc.Lelland, eds.

Berkhoudt, H., 1985b, The role of oral exteroceptive sense organs in avian feeding behaviour, Fortsch. Zool. 30, 269-272.

Botezat, E., 1904, Geschmacksorgane und andere nervöse Endapparate im Schnabel der Vögel (vorläufige Mitteilung), Biol. Zbl. 24, 722-736.

Botezat, E., 1906, Die Nervenendapparate in den Mundteilen der Vögel und die einheitliche Endigungsweise der peripheren Nerven bei den Wirbeltieren, Z. Wiss. Zool. 84, 205-360.

Botezat, E., 1910, Morphologie, Physiologie und phylogenetische Bedeutung der Geschmacksorgane der Vögel, Anat. Anz. 36, 428-461.

Dmitrieva, N. A., 1981, Fine structural peculiarities of the taste buds in the pigeon, Columba livia (In Russian.), Dokl. Akad. Nauk SSSR 23, 874-878.

Dubbeldam, J.L., 1977, "Touch" and "taste" within the brainstem of the mallard, 18th Dutch Fed. Meet., Leiden, 185 (abstract).

Dubbeldam, J.L., 1984a, Afferent connections of the nervus facialis and nervus glossopharyngeus in the pigeon (Columba livia) and their role in feeding behaviour, Brain Behav. Evol. 24, 47-57.

Dubbeldam, J.L., 1984b, Brainstem mechanisms for feeding in birds: Interaction or plasticity, Brain Behav. Evol. 25, 85-98.

Ganchrow, D. and Ganchrow, J.R., 1985, Number and distribution of taste buds in the oral cavity of hatchling chicks, Physiol. Behav. 34, 889-894.

Ganchrow, D. and Ganchrow, J.R., 1989, Gustatory ontogenesis in the chicken: an avian-mammalian comparison, Med. Sci. Res. 17, 223-228.

Gentle, M.J., 1983, The chorda tympani nerve and taste in the chicken, Experientia 39, 1002-1003.

Jenkin, P.M., 1957, The filter feeding and the food of flamingoes (Phoenicopteri), Phil. Trans. R. Soc. London, B, 240, 401-493.

Kooloos, J.G.M., 1986, A conveyer-belt model for pecking in the mallard (Anas platyrhynchos L.), Neth. J. Zool. 36, 47-87.

Kooloos, J.G.M., Kraaijeveld, A.R., Langenbach, G.E.J., Zweers, G.A., 1989, Comparative mechanics of filter feeding in Anas platyrhynchos, Anas clypeata and Aythya fuligula (Aves, Anseriformes), Zoomorphology 108, 269-290.

Krol, C.P.M. and Dubbeldam, J.L., 1979, On the innervation of taste buds by the n. facialis in the mallard (Anas platyrhynchos L.), Neth. J. Zool. 29, 267-274.

Kurosawa, T. , Niimura, S., Kusuhara, S., and Ishida, K., 1983, Morphological studies of taste buds in chickens, Jap. J. Zootech. Sci. 54, 502-510.

Merkel, F., 1880, Ueber die Endigungen der sensiblen Nerven der Wirbeltieren, Fues's Verlag, Leipzig.

McLelland, J., 1979, Digestive system, in: Form and Function in Birds, A.S. King and J. McLelland eds. Vol 1, Academic Press, London and New York.

Moore, R. A. and Elliott, R., 1946, Numerical and regional distribution of taste buds on the tongue of the bird, J. Comp. Neurol. 84, 119-131.

Sevier, A.C., and Munger, B.L., 1965, A silver method for paraffin sections of neural tissue, J. Neuropath. Exp. Neurol. 24, 130-135.

Wenzel, B.M., 1984, Chemical senses. In: Physiology and Behaviour of the pigeon, M. Abs ed., Academic Press, New York and London.

Zweers, G.A., 1982, Pecking in the pigeon (Columba livia L.), Behaviour 81, 173-230.

Zweers, G.A. and Berkhoudt, H., 1990, Recognition of food in pecking, probing and filter feeding birds, Proc. XXth. Int. Ornith. Congr., Christchurch, in press.

Zweers, G.A., Gerritsen, A.F.C., Kranenburg-Voogd, P. van, 1977, Mechanics of feeding in the mallard (Anas platyrhynchos L., Aves, Anseriformes), in: Contr. Vert. Evol. Vol 3, Hecht, Szalay, eds. (Karger, Basel).

EFFECTS OF NARIS CLOSURE ON THE OLFACTORY EPITHELIA OF ADULT MICE

Joel Maruniak, Frank Corotto, and Eric Walters

Biological Sciences
University of Missouri
Columbia, Missouri 65211

Over the past several years we have been investigating the effects of
unilateral naris closure on the olfactory system of adult mice. Unilateral
closure creates an experimental situation in which conditions within the
open-side nasal cavity are generally harsher than normal, while conditions
on the closed side are generally milder. Because rodents are obligatory
nasal breathers, unilateral closure forces all respiratory flows to pass
through the open side. This exposes the open side to more odors, xenobiotics, dry air, pathogens and particulate matter than the closed side or
normal nasal cavities. In addition, naris closure eliminates the nasal
cycle (a pattern of alternating airflow between the two sides of the nose;
see Ritter, 1970; Eccles, 1978; Cole and Haight, 1986). We believe that
these conditions create a chronic low level of trauma for the open side,
while not negatively affecting the closed side. Our studies have examined
the effects of unilateral naris closure on: 1) morphology of the olfactory
epithelium; 2) rate of mitosis of the neuronal stem cells of the epithelium; and 3) expression of the xenobiotic metabolizing enzyme, cytochrome
P-450.

EFFECTS OF CLOSURE ON THE MORPHOLOGY OF THE OLFACTORY EPITHELIUM

Unilateral naris closures were performed on adult mice (>120 days old)
by means of heat cautery and suture as previously described (Maruniak et
al., 1989; 1990). At various times after closure mice were killed and
prepared for routine histology or immunohistochemistry.

One of the most striking effects of naris closure was a dramatic loss
of olfactory receptor neurons on the open side beginning as early as one
month after closure (Figure 1). Those losses were generally confined to
the rostral regions of the epithelium--interestingly, these are the exact
areas expected to encounter the least conditioned air (Fry and Black, 1973;
Heyder and Rudolf, 1975; Hallworth and Padfield, 1986). Figure 2 shows
data from several experiments on the effects of unilateral closure on
numbers of receptor neurons in rostral regions of the open- and closed-side
olfactory epithelia. Following one month to eight months of closure,
losses could be observed in the rostral regions of the open side. As shown
in Figure 2, the four-month closure group in one study showed regeneration
of its receptor neuron populations. This is not surprising given that
mitosis of the basal cells has been shown to provide for either routine
replacement of receptor neurons (Graziadei and Metcalf, 1971; Moulton 1974;
Monti-Graziadei and Graziadei, 1979; Samanen and Forbes, 1984) or massive
replacement after experimental destruction of the receptor neuron population (Graziadei and DeHan, 1973; Matulionis, 1975; Monti-Graziadei et al.,
1980). In another experiment, in which naris closures lasted 1, 2, 3 or 4
months, we did not observe such synchronized regeneration in any group.

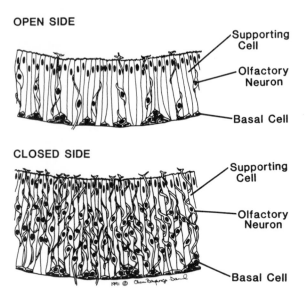

Fig. 1. Representation of the effects of unilateral naris closure on rostral regions of the olfactory epithelia. Closed sides appear to be unaffected and are indistinquishable from controls. Open sides show losses of receptor neurons and decreases in thickness.

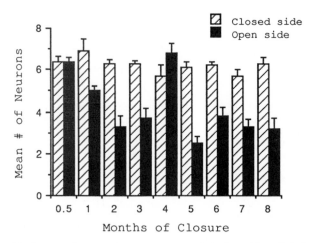

Fig. 2. Numbers of receptor neurons spanning the thickness of the rostral olfactory epithelia in naris closure mice (n=9-12/grp).

We believe, nonetheless, that regeneration had occurred, but was not synchronous among animals in any of the groups and therefore was not statistically detectable. In fact, we often observed normal-looking epithelia within groups with significant losses. We now feel that under some experimental conditions the majority of animals within a closure group regenerate their receptor neurons over a short period of time and thus appear to be synchronized. Under other conditions, regeneration may occur over a longer period: in some mice at three months, some at four, some at five or later.

Another prominent effect of closure was an increase in the number of neurons in the caudal regions of the open side beginning about two months after closure. After two months, the number of receptor neurons on the open sides averaged about 13% higher than on the closed sides. The effect was quite striking in sections viewed under light microscopy. In contrast, the closed-side olfactory epithelia always appeared to be unaffected when compared to control epithelia. The responses of middle regions of the open-side epithelia tended to be intermediate between the rostral and caudal responses, and rather variable.

EFFECTS OF CLOSURE ON THE MITOTIC RATE OF NEURONAL STEM CELLS

Previous studies have indicated that about 2-3% or so of the receptor neurons in mammals are normally replaced each day (Graziadei and Metcalf, 1971; Moulton 1974; Monti-Graziadei and Graziadei, 1979; Samanen and Forbes, 1984), but much higher rates can occur if a large fraction of the population has been experimentally destroyed (Monti-Graziadei and Graziadei, 1979; Camara and Harding, 1984). Between those two extremes there is a question of whether rates normally change in response to ordinary variations in the rate of neuronal death. Our morphological data suggested to us that unilateral naris closure might provide an experimental paradigm in which that question could be tested. Those data suggested that in rostral regions the rate of neurogenesis was insufficient to compensate for receptor neuron losses, while in caudal regions the rate over-compensated. Thus, in our next study we wanted to determine if, under conditions of low trauma, which we believe exists on the open side of naris closure animals, the rate of neurogenesis would rise in response to increased losses of receptor neurons.

Rates of neurogenesis were measured using uptake of the thymidine analog, bromodeoxyuridine (BrdU), as a marker. Animals with unilateral naris closures for 3 months were given a single injection of BrdU and then sacrificed one hour later and prepared for immunohistochemistry. Basal cells that had incorporated BrdU were visualized using a monoclonal antibody (Amersham Corp.) and the biotin-avidin-HRP technique. Counts of BrdU-positive basal cells were made in control olfactory epithelia and in open- and closed-side epithelia.

Figure 3 shows that rates of neurogenesis in the open-side epithelia were much higher than for the closed side or for control epithelia. These data clearly demonstrate that the olfactory epithelium is able to increase the production of new neurons under conditions of chronic low level trauma that shorten the lifespan of its receptor cells.

EFFECTS OF CLOSURE ON THE EXPRESSION OF CYTOCHROME P-450

The discovery of olfactory-specific isozymes of cytochrome P-450 has led to interest in the role that these enzymes play in the olfactory system (Dahl, 1988). Cytochrome P-450s generally have been implicated in metabolism of a variety of endogenous and exogenous compounds. It has been proposed that they might be involved in protecting the olfactory epithelium from xenobiotics, in metabolism of odors, and in the transduction process itself (Nef et al., 1989). In the present study we wanted to determine if the expression of P-450 immunoreactivity was different on the two sides of mice who have had a naris closed.

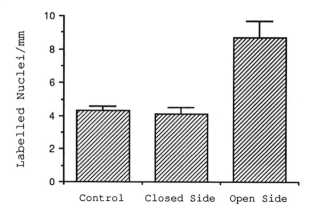

Fig. 3. Rates of neurogenesis in controls and after three months of unilateral naris closure (n=10 mice/group). The number of labelled basal cells in rostral, middle and caudal sampling areas were counted and pooled.

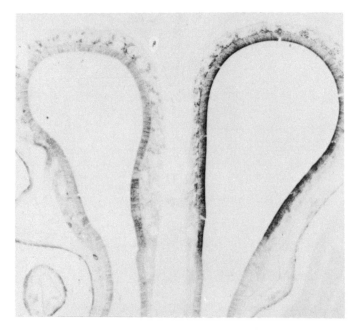

Fig. 4. Effects of unilateral naris closure on the expression of cytochrome P-450NMb immunoreactivity in the rostral olfactory epithelium. The closed side (on the right) shows normal levels of immunoreactivity while the open side shows almost none.

Adult mice with unilateral naris closure were sacrificed and their heads prepared for immunohistochemistry of paraffin sections. Antibodies for two olfactory specific forms of cytochrome P-450, NMa and NMb, were obtained from Drs. Ding and Coon at the University of Michigan. These were used to localize the distribution of P-450NMa and NMb in the olfactory epithelia of normal and naris closure animals.

We found the distribution of olfactory-specific cytochrome P-450s in normal mice to be similar to that in previous reports (Voigt et al., 1985; Baron et al., 1988). Cytochrome P-450 was present in the supporting cells of the olfactory epithelium and in the Bowman's glands of the underlying lamina propria. Both of our antibodies, anti-P-450 NMa and NMb, localized their antigens to similar regions. Figure 4 shows the distribution of P-450 NMb in a mouse after ten months of naris closure. The expression of the isozyme was dramatically diminished in the same areas of the olfactory epithelium that suffered severe losses of receptor neurons--the rostral regions--in spite of the fact that P-450 does not even appear to be present in the neurons. This suggests a direct relationship between the expression of P-450 in supporting cells and Bowman's glands, and the presence of nearby neurons. Alternatively, these data might indicate that P-450 has been "used up" in the rostral open-side epithelium.

REFERENCES

Baron, J., Burke, J., Guengerich, F., Jakoby, W. and Voigt, J., 1988, Sites for xenobiotic activation and detoxication within the respiratory tract: implications for chronically induced toxicity, Toxicol. Appl. Pharmacol. 93, 493-505.
Camara, C. and Harding, J., 1984, Thymidine incorporation in the olfactory epithelium of mice: early exponential response induced by olfactory neurectomy, Brain Res., 308, 63-68.
Cole, P. and Haight, J.S., 1986, Posture and the nasal cycle, Ann. Otol. Rhinol. Laryngol., 95, 233-237.
Dahl, A., 1988, The effect of cytochrome P-450-dependent metabolism and other enzyme activities on olfaction, in "Molecular Neurobiology of the Olfactory System," F. Margolis and T. Getchell, eds., Plenum Press, New York.
Eccles, R., 1978, The domestic pig as an experimental animal for studies on the nasal cycle, Acta Otolaryngol., 85, 431-436.
Fry, F.A. and Black, A., 1973, Regional deposition and clearance of particles in the human nose, Aerosol Sci., 4, 113.
Graziadei, P.P.C. and DeHan, R.S., 1973, Neuronal regeneration in frog olfactory system , J. Cell Biol., 59, 525-530.
Graziadei, P. and Metcalf, J., 1971, Autoradiographic and ultrastructural observations on the frog's olfactory mucosa, Z. Zellforsch., 116, 305-318.
Hallworth, G.W. and Padfield, J.M., 1986, A comparison of the regional deposition in a model nose of a drug discharged from metered aerosol and metered-pump nasal delivery systems, J. Allergy Clin. Immunol., 77, 348.
Heyder, J. and Rudolf, G., 1975, Deposition of aerosol particles in the human nose, in "Inhaled Particles IV. Part 4,"W.H. Walton, ed., Pergamon Press, Oxford.
Maruniak, J.A., Lin, P. and Henegar, J., 1989, Effects of unilateral naris closure on the olfactory epithelia of adult mice, Brain Res., 490, 212.
Maruniak, J.A., Henegar, J.R. and Sweeney, T.P., 1990, Effects of long-term unilateral naris closure on the olactory epithelia of adult mice, Brain Res., 526, 65.
Matulionis, D.H., 1975, Ultrastructural study of mouse olfactory epithelium following destruction by ZnSO4 and its subsequent regeneration, J. Anat., 142, 67-90.
Monti-Graziadei, G. and Graziadei, P.P.C., 1979, Neurogenesis and neuron regeneration in the olfactory system of mammals. II. Degeneration and reconstitution of the olfactory sensory neurons after axotomy, J. Neurocytol., 8, 197-213.

Monti-Graziadei, G. Karlan, M., Bernstein, J. and Graziadei, P., 1980,
   Reinnervation of the olfactory bulb after section of the olfactory
   nerve in monkey (*Saimiri sciureus*), *Brain Res.*, 189, 343-354.
Moulton, D.G., 1974, Dynamics of cell populations in the olfactory epitheli-
   um, *Ann. N.Y. Acad. Sci.*, 237, 52-61.
Nef, P., Heldman, J., Lazard, D., Margalit, T., Jaye, M., Hanokoglu, I. and
   Lancet, D., 1989, Olfactory-specific cytochrome P450: cDNA cloning of
   a novel neuroepithelial enzyme possibly involved in chemoreception, *J.
   Biol. Chem.*, 264, 6780.
Ritter, F.N., 1970, The vasculature of the nose, *Ann. Otol. Rhinol.
   Laryngol.*, 779, 468-474.
Samanen, D.W. and Forbes, W.B., 1984, Replication and differentiation of
   olfactory receptor neurons following axotomy in the adult hamster: A
   morphometric analysis of postnatal neurogenesis, *J. Comp. Neurol.*,
   225, 201-211.
Voigt, J., Geungerich, F. and Baron, J., 1985, Localization of a cytochrome
   P-450 isozyme (cytochrome P-450 PB-B) and NADPH-cytochrome P-450 reduc-
   tase in rat nasal mucosa, *Cancer Lett.*, 27, 241.

IS THE MOUSE VOMERONASAL ORGAN A SEX PHEROMONE RECEPTOR?

Tsuneo Hatanaka

Department of Biology, Faculty of Education
Chiba University
Chiba 260, Japan

INTRODUCTION

Many behavioral studies indicate that the accessory olfactory system of mammals contributes to sexual behavior. However, the actual nerve responses of the vomeronasal system have received little study. In rodents, movement of odorants into the vomeronasal organ is facilitated by the pumping action of cavernous vascular tissue situated along its outer surface (Meredith and O'Connell, 1979; Meredith et al., 1980). To ascertain whether the vomeronasal organ is specialized to detect chemicals from the urine of mice (which may contain a sex pheromone), single unit responses of mouse vomeronasal receptor cells and accessory olfactory bulb (AOB) neurons were recorded to several types of stimuli, including urine vapors. Next, we determined whether the vasomotor action of the organ was affected by various kinds of odor, particularly whether urine odor promotes sampling movement. As indicated this papar, the vomeronasal organ responds to a wide range of stimuli in addition to vapors from mouse urine.

MATERIALS AND METHODS

Adult mice (BALB/c strain) were anesthetized by urethane. The root of an incisor was whittled off and the vomeronasal organ exposed. Then the organ was opened to expose the receptor membrane. Subsequently, the dorsal part of the frontal bone was dissected and the olfactory bulb exposed. Unit responses were recorded extracellularly through a glass microelectrode penetrating the mucosa and AOB region. To monitor vasomotor action, a mechanotransducer probe was gently positioned on the vascular tissue.

RESULTS

Neural Responses

Fig. 1 shows examples of vomeronasal receptor responses. The unit in Fig. 1A was stimulated by urine odors thought to contain the sex pheromone and discriminated between male and female urine. However, the vapors of isoamyl acetate and trimethyl amine also induced responses. In Fig. 1B, responses of a vomeronasal receptor unit to the aqueous stimuli of glutamic acid and ammonium chloride solution are shown. The data indicate that the

Chemical Signals in Vertebrates VI, Edited by R.L. Doty and
D. Müller-Schwarze, Plenum Press, New York, 1992

27

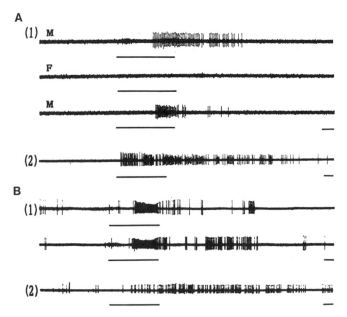

Fig. 1. Unitary responses of mouse vomeronasal
receptor cells. A: Responses to vapor
stimuli. Unit in (1) responded to male
urine (M) but not to female urine (F).
Unit in (2) showed prolonged response to
isoamyl acetate. B: Responses to aqueous
stimuli. Unit in (1) responded to L-glutamic acid solution, and unit in (2) showed
prolonged excitation to ammonium chloride
solution. Horizontal bars below each
records indicate the stimulation. Calibration (on right) is 1 sec.

mouse vomeronasal receptors respond to various type of odorants, regardless
as to whether they are presented in the vapor or aqueous phase. Unfortunately, the difficulty in holding a unit for a long time made it impossible
to obtain complete response spectra for each receptor. Only 10 of 88 units
(11.4%) responded to vapor urine odors, and only 1 of 8 units (12.5%) to
aqueous urine odors. The responsiveness to urine vapors (11.5%) was not
greater than that to other odorants (16%). AOB neurones also responded to
various type of odorants. Suppression-type responses were frequently observed in the AOB. The responsiveness of AOB neurons to urine odor (66.7%)
was greater than that of above described vomeronasal receptor units, and
that of AOB neurons to other odors (35.0%) was also greater than that of
receptors.

Odor Sampling Movement

Vasomotor movement of the mouse vomeronasal organ is controlled by the
naso-palatine nerve and is very sensitive to anesthesia. When urethane is
slightly increased (from 0.11 g to 0.22 g/ 100 g body weight) so that the
heart and breathing rates are not affected, vomeronasal movement readily
ceases. In such a case, an electrical stimulus to the naso-palatine nerve
causes vasomotor movement. A single electrical shock caused initial relaxation followed by rapid contraction and a graded relaxation to basal level
(Fig. 2A-1). Repetitive stimulation for a long period caused relaxation to

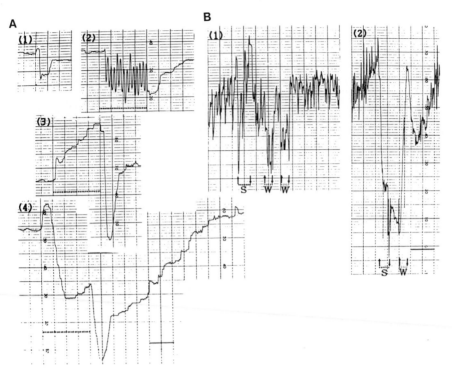

Fig. 2. Vasomotor movements of vomeronasal organ. A: When spontaneous action ceased in deep anesthetic state, single
or repetitive electrical stimuli to the naso-palatine
nerve caused vasomotor actions, (1) single shock, (2)
0.5 Hz, (3) 8 Hz, (4) 20 Hz. Downward movement corresponds to contraction. Each dotted line indicates
stimulation. Calibration is 15 sec. B: Periodical
vasomotor movement was affected by aqueous stimuli (S)
and washing (W), (1) stimulated with isoamyl acetate
solution and washed by Ringer's solution, (2) stimulated
with female urine and washed by distilled water. Calibration is 1 minute.

be separated from contraction. At a low frequency (< 0.8 Hz), large contraction and relatively rapid relaxation resulted (Fig. 2A-2), as has been
noted for spontaneous vasomotor movement. At higher frequencies (1 - 8 Hz),
relaxation continued during stimulation and contraction started following
termination of stimulation (Fig. 2A-3). When the stimulus frequency was
high (> 10 Hz), irregular contraction began during stimulation (Fig. 2A-4).

When anesthesia is weak, vascular tissue contacts periodically, but
its amplitude and frequency vary somewhat in contrast to heart beat (Fig.
2B). When vapor odors were applied, increase or decrease in amplitude and
frequency of movement could sometimes be observed. But the effects varied
according to the condition of the animal. Odorants such as acetic acid and
isoamyl acetate suppressed periodical movement or caused a continued contraction. Aqueous solution of formic acid or saturated isoamyl acetate
solution applied to the vomeronasal mucosa had little effect on movement
(Fig. 2B-1), whereas application of female urine caused a large contraction
(Fig. 2B-2).

DISCUSSION

The mouse vomeronasal system responds not only to urine odors but various general odors as well. Complete response spectra could not be obtained, but the vomeronasal receptors may be considered to be of the generalist type. In rats, a deficit behavior due to a lesion of the accessory system is partially compensated for by the main olfactory system by the reconstruction of a central neural networks (Ichikawa, 1989). Thus, some of the receptor properties of the two olfactory systems may be similar if not the same. The mouse vomeronasal receptor does not seem to be specialized for urine odor reception. But AOB neurons have greater responsiveness to urine odors than receptors, when using aqueous stimuli.

The mouse vomeronasal organ always samples odorants by periodic vasomotor movement. An impinging vapor like acetic acid and an aqueous solution like urine lead to strong continued contractions of vessels and to cessation of periodic odor sampling. This is followed by enlargement of the lumen of organ and the retention of considarable odorant within it. Vessel contraction over the olfactory mucosa has also been observed in the tortoises and may possibly be a trigeminal reflex (Tucker, 1962). It has been shown previously that the hamster's vomeronasal organ receives volatile and non-volatile molecules (Clancy et al., 1984; Meredith and O'Connell, 1984) and it would appear that this is similarly the case in the mouse. The present data suggest that the mouse vomeronasal system is also likely to be essential for the reception of costituents of non-volatile urine.

REFERENCES

Clancy, A. N., Macrides, F., Singer, A. G., and Agosta, W. C., 1984, Male hamster copulatory responses to a high molecular weight fraction of vaginal discharge: effects of vomeronasal organ removal, Physiol. Behav., 33:653.

Ichikawa, M., 1989, Recovery olfactory behavior following removal of accessory olfactory bulb in adult rat, Brain Res., 498:45.

Meredith, M., and O'Connell, R. J., 1979, Efferent control of stimulus access to the hamster vomeronasal organ, J. Physiol., 286:301.

Meredith, M., Marques, D. M., and O'Connell, R. J., 1980, Vomeronasal pump: Significance for male hamster sexual behavior, Science, 207:1224.

O'Connell, R. J., and Meredith, M., 1984, Effects of volatile and nonvolatile chemical signals on male sex behaviors mediated by the main and accessory olfactory system, Behav. Neurosci., 98:1083.

Tucker, D., 1962, Olfactory, vomeronasal and trigeminal receptor responses to odorants, in: "Olfaction and Taste," Y. Zotterman, ed., Pergamon Press, Oxford.

LHRH-IMMUNOCYTOCHEMISTRY IN THE NERVUS TERMINALIS OF MAMMALS

Helmut A. Oelschläger and Hynek Burda

Dept.of Anatomy, J.W.Goethe-University
Theodor Stern-Kai 7
6000 Frankfurt am Main 70, FRG

INTRODUCTION

Apart from the chemoreceptor systems (olfactory, vomeronasal nerves), the septum of the vertebrate nose is innervated by the paired nervus terminalis. Fiber bundles run through rostromedial foramina of the cribriform plate and pass between and beneath the olfactory bulbs to enter the CNS in the area of the lamina terminalis and on the basal surface of the prosencephalon. The nervus terminalis in adult mammals is a ganglionated fiber plexus which obviously comprises two or more subpopulations of neurons. While one of these subpopulations is characterized by the occurrence of acetylcholinesterase (AchE), the other fraction of neurons expresses luteinizing hormone-releasing hormone (LHRH) and seems to be involved in mating behavior and reproduction (Jennes and Stumpf, 1980; Schwanzel-Fukuda and Silverman, 1980; Wirsig and Leonard, 1986; Schwanzel-Fukuda and Pfaff, 1989). During prenatal ontogenesis, LHRH is first detected in neurons of the terminalis nerve; this fact has led to the supposition that the nerve even helps to organize the brain-pituitary-gonadal axis (Schwanzel-Fukuda et al., 1985, 1987, 1988; Jennes, 1989). From late fetal and early postnatal stages of rodents onward, the number of LHRH-immunoreactive (-ir) cells is reduced.

THE NERVUS TERMINALIS IN BATS

In the juvenile and adult big brown bat (<u>Eptesicus fuscus</u>), the LHRH component of the nervus terminalis is restricted to the cranial vault. A moderate number of cells and many fibers are dispersed within the leptomeninx along the surface of the olfactory bulb and olfactory tubercle, the periseptal region, diagonal band, lateral olfactory tract, the prepiriform and periamygdaloid regions, the hippocampus, and the basal diencephalon as far caudalward as the interpeduncular fossa. Just rostral to the ventral rhinal fissure, one or more bundles of thick fibers (including a few cell bodies) attach to the brain surface, where they end abruptly but seem to send collaterals caudalward (Fig. 1a). Rostrally, they join a concentration within the fiber plexus along the medioventral surface of the olfactory bulb. Interestingly, a major part of the terminalis ganglia (high up between the olfactory bulbs) remains nearly unlabeled.

Within the CNS, the LHRH-ir cells and fibers are only roughly attributable to specific nuclei or fiber tracts. They are distributed throughout the

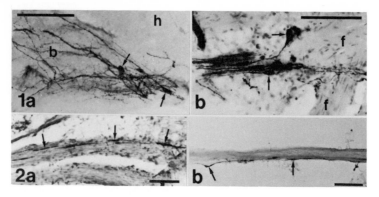

Fig. 1a. Big brown bat (<u>Eptesicus</u> <u>fuscus</u>). Sagittal section. Main terminalis fiber bundle with perikarya (arrows) attaching to olfactory bulb (b) and telencephalic hemisphere (h). Bar = 100 μm.
Fig. 1b. African mole rat (<u>Cryptomys</u> <u>hottentotus</u>). Sagittal section. Termi-nalis fiber bundle colocalized with olfactory fiber bundles (f). Bar = 100 μm.
Fig. 2. <u>Cryptomys</u>. LHRH-ir cells and fibers (arrows) coursing in olfactory fiber bundles. a) sagittal section, b) horizontal section. Bar in a) and b) 100 μm.

olfactory tubercle, diagonal band, preoptic area, the bed nuclei of stria me-dullaris and stria terminalis, the anterior lateral and posterior hypothala-mus, and the tuber cinereum. While the greatest concentration of cells was found within the arcuate nucleus (Fig. 7a), fibers were most abundant in the median eminence, infundibular stalk and the medial habenula. In principle, this distribution of LHRH immunoreactivity corresponds well with findings of King et al. (1984) in <u>Myotis</u> <u>lucifugus</u> and may be characteristic of microchir-optera, i.e., insectivorous bats (Oelschläger and Northcutt, 1991, submitted).

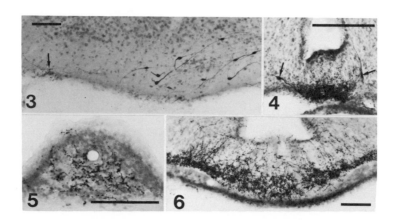

Fig. 3. dto. Horizontal section through rostromedial hemisphere with enter-ing site of terminalis nerve (arrow) and its continuation near the taenia tecta. Bar = 100 μm.
Fig. 4. dto. Transverse section of organum vasculosum laminae terminalis (OVLT). Bar = 100 μm.
Fig. 5. dto. Sagittal section of subfornical organ. Bar = 100 μm.
Fig. 6. dto. Transverse section through median eminence. Bar = 100 μm.

THE NERVUS TERMINALIS IN THE AFRICAN MOLE RAT (<u>Cryptomys</u> <u>hottentotus</u>)

This subterranean rodent so far has been investigated only as to immuno-cytochemistry. The LHRH-ir material of the nervus terminalis is rather scattered but often forms distinct fiber bundles (Fig. 1b) with integrated cell bodies or even smaller ganglia. The perikarya and fibers may also course with fila olfactoria and, perhaps, also with fiber bundles of the vomeronasal nerve (Fig. 2a, b). A few cells were found in fila olfactoria near the septal mucosa. In principle, most of the cells are bipolar, however, there are also tripolar cells. While the shape of the perikarya may vary significantly, the cells within olfactory or vomeronasal fiber bundles are extremely slender and spindle-shaped. Most of the terminalis fibers enter the CNS about the anterior olfactory nucleus and in the area of the precommissural hippocampus (Fig. 3). The fibers proceed to the medial septum, diagonal band, area preoptica and

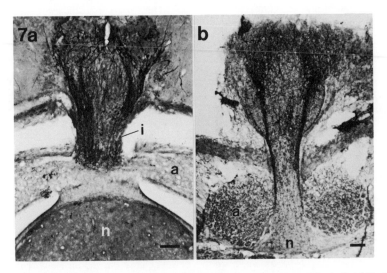

Fig. 7. Horizontal sections through basal hypothalamus with infundibulum (i), adenohypophysis (a) and neurohypophysis (n). a) <u>Cryptomys</u>, b) <u>Eptesicus</u>. Bar in a) and b) = 100 µm.

anterior hypothalamus where additional LHRH-ir cells and fibers are located. Many of these intrinsic neurons project to the organum vasculosum laminae terminalis (OVLT), the median eminence, and the infundibulum (Figs. 4, 6, 7b). Extrahypothalamic LHRH projections, which course in the stria medullaris thalami to the habenula or in the fasciculus retroflexus to the interpeduncular nucleus, are relatively weak; the same is true for LHRH material in the stria terminalis and fornix. In contrast, the subfornical organ (SFO) is markedly labeled (Fig. 5). Dorsal projections ascend to the habenular and epithalamic commissures and course through the optic tectum and the central gray. Other dorsal fibers run in the indusium griseum and to ammon's horn.

COMPARATIVE CONSIDERATIONS

In principle, the quantity of LHRH-ir material (cells and fibers) seems to be distinctly higher in bats (King et al., 1984; Oelschläger and Northcutt, submitted) than in the African mole rat. The two bats so far investigated for LHRH exhibit remarkable similarity in the distribution of cells and fibers, e.g., they both show heavy labeling of perikarya in the arcuate nucleus and strong projections to the medial habenula. In rodents, generally, the LHRH-ir perikarya are concentrated more rostrally; however, there are remarkable differences between members of different groups as to the number of LHRH-ir cells and fibers and their distribution throughout the nervous system (Jennes and Stumpf, 1980; Wray and Hoffman, 1986; Schwanzel-Fukuda et al., 1987). Concerning the outstanding number and density of LHRH-ir cells in the arcuate nucleus, there are surprising similarities between bats and primates. Extrahypothalamic fibers project to the same sites in both bats and primates, as well as in those rodents (hamsters, guinea pigs) that exhibit LHRH perikarya extending into the medial basal hypothalamus (King et al., 1984). At the moment, there is no possibility to correlate the quantity and distribution of LHRH-ir material in the nervus terminalis and brain with the physiology and behavior of the species concerned. The continuum of LHRH neuron populations still reflects the essential ontogenetic impact of the nervus terminalis on the establishment of the brain-pituitary-gonadal axis. Because the LHRH-ir neurons in the adult always exhibit projections to various parts of the limbic system, their significance, e.g., for reproductive behavior and physiology, is all the more obvious.

ACKNOWLEDGMENTS

The authors are indebted to R. Glenn Northcutt (Dept. of Neurosciences, San Diego, California, U.S.A.) and J. Winckler (Dept. of Anatomy, Frankfurt am Main) for generous help and fruitful discussions. Jutta S. Oelschläger is thanked for the accurate preparation of the manuscript and the originals. The senior author gratefully acknowledges the support by the Deutsche Forschungsgemeinschaft (Oe 103/2-1) as well as by the Faculty of Human Medicine, Frankfurt am Main (Dr. Paul und Cilli Weill Foundation, Ebert Foundation).

REFERENCES

Jennes, L., 1989, Prenatal development of the gonadotropin-releasing hormone - containing systems in rat brain, Brain Res. 482:97-108.
Jennes, L., and Stumpf, W.E., 1980, LHRH-systems in the brain of the golden hamster. Cell Tissue Res. 209:239-256.
King, J.C., Anthony, E.L.P., Gustafson, A.W., and Damassa, D.A., 1984, Luteinizing hormone-releasing hormone (LH-RH) cells and their projections in the forebrain of the bat Myotis lucifugus. Brain Res. 298:289-301.
Oelschläger, H.A., and R.G. Northcutt, 1991, Luteinisierendes Hormon - Releasing Hormon (LHRH) in Nervus terminalis und Vorderhirn der Fledermaus Eptesicus fuscus. Eine immuncytochemische Untersuchung. Verh.Anat.Ges. 84 (Anat.Anz.Suppl. 168):499-500.

Oelschläger, H.A., and R.G. Northcutt, 1991, Immunocytochemical localization of luteinizing hormone-releasing hormone (LHRH) in the nervus terminalis and brain of the big brown bat, Eptesicus fuscus, (Journal of Comparative Neurology, in press).

Schwanzel-Fukuda, M., and Silverman, A.J., 1980, The nervus terminalis of the guinea pig: A new LHRH neuronal system, J.Comp.Neurol. 191:213-225.

Schwanzel-Fukuda, M., and Pfaff, D.W., 1989, Origin of luteinizing hormone-releasing hormone neurons, Nature (Lond.) 338:161-164.

Schwanzel-Fukuda, M., Morrell, J.I., and Pfaff, D.W., 1985, Ontogenesis of neurons producing luteinizing hormone-releasing hormone (LHRH) in the nervus terminalis of the rat, J.Comp. Neurol. 238:348-364.

Schwanzel-Fukuda, M., Fadem, B.H., Garcia, M.S., and Pfaff, D.W., 1988, Immunocytochemical localization of luteinizing hormone-releasing hormone (LHRH) in the brain and nervus terminalis of the adult and early neonatal gray short-tailed opossum (Monodelphis domestica). J.Comp.Neurol. 276:44-60.

Schwanzel-Fukuda, M., Garcia, M.S., Morrell, J.I., and Pfaff, D.W., 1987, Immunocytochemical localization of luteinizing hormone-releasing hormone (LHRH) in the nervus terminalis and brain of the mouse, J.Comp.Neurol. 255:231-244.

Wirsig, C.R., and Leonard, C.M., 1986, Acetylcholinesterase and luteinizing hormone-releasing hormone distinguish separate populations of terminal nerve neurons, Neuroscience 19:719-740.

Wray, S., and Hoffman, G., 1986, A developmental study of the quantitative distribution of LHRH neurons within the central nervous system of postnatal male and female rats. J.Comp.Neurol. 252:522-531.

OLFACTION IN RATS WITH TRANSECTION OF THE LATERAL OLFACTORY TRACT

Burton M. Slotnick

Department of Psychology, The American University
Washington, D.C. 20016

INTRODUCTION

The lateral olfactory tract (LOT) is the major, although not the
only, projection pathway from the olfactory bulb to the anterior olfactory nucleus and olfactory cortex. Transection of the LOT produces
deficits in the ability to smell. However, some reports indicate that
rats or hamsters with LOT transection are anosmic or have severe olfactory deficits (Cattarelli, 1982; Devor, 1973; Long and Tapp, 1970;
Marques, O'Connell, Benimoff, and Macrides, 1982; Thompson, 1980) while
others (Gervais and Pager, 1982; Slotnick, 1985; Slotnick, 1990; Slotnick
and Berman, 1980; Slotnick and Risser, 1990; Swann, 1934) indicate much
less severe deficits after similar lesions. Except for a few studies
(Bennett, 1968; Slotnick and Schoonover, 1984), most prior investigations
of neuroanatomical correlates of olfactory function have not used quantitative olfactometric methods or psychophysical tests and it is difficult
to compare or evaluate their outcomes. This problem is addressed in the
present study by assessing odor detection, two-odor discrimination, intensity-difference and absolute detection thresholds of rats in which the
lateral olfactory tract was transected.

METHODS

Adult albino Wistar rats were maintained in a temperature- and
humidity-controlled vivarium in individual plastic cages. All animals
were maintained on a 10 ml/day water depreviation schedule except for the
first 3 postoperative days when ad lib. water was given. The rats were
tested using the olfactometer and procedures described by Slotnick and
Schoonover (1984).

In Experiment 1, rats were first trained on a simple detection task
in which a 0.5% (of vapor saturation) amyl acetate vapor served as S^+ and
clean air served as S^-. All animals acquired this detection task in the
first 200 trials and maintained stable performance of 90-100% correct
responding in the remaining 600 trials of preoperative training. Rats
were then anesthetized with i.p. sodium pentobarbital and operated on.
Each had the right olfactory bulb removed and received a lesion in the
left hemisphere. In 5 rats, the LOT was sectioned using a knife made from
a razor blade (Group LOT). In 5 additional rats, the knife cut transected

Chemical Signals in Vertebrates VI, Edited by R.L. Doty and
D. Müller-Schwarze, Plenum Press, New York, 1992

37

the LOT and extended into the left hemisphere to transect all or most of the piriform cortex and the ventral and lateral aspect of the anterior limb of the anterior commissure (Group LOT/PC). In 6 control rats, a 3-4 mm long knife cut was made in the cortex dorsal and parallel to the rhinal fissure. Fourteen days after surgery the rats were tested for retention of the detection task and trained to discriminate between 0.1% ethyl acetate (S$^+$) and 0.1% isopropyl acetate vapors (S$^-$).

For Experiment 2, rats were first trained on the amyl acetate detection task. They were then operated on as described above except that 5 rats received a discrete lesion of the LOT (Group LOT) and 7 rats served as lesion controls. Fourteen days after surgery the rats were tested for retention of the amyl acetate detection task and then tested for their amyl acetate intensity-difference threshold. For the intensity-difference threshold, the S$^+$ (standard) stimulus was 0.1% amyl acetate and the S$^-$ (comparison) stimuli were (in separate sessions) 0.01% - 0.09% amyl acetate in 0.01% steps. Finally, all rats were tested for their absolute detection threshold. In separate sessions, the concentration of amyl acetate vapor (S$^+$) was decreased in half log unit steps. Air served as the S$^-$ stimulus. The psychophysical procedures followed those described by Slotnick and Schoonover (1984).

For both experiments, rats were given 100 trials/day and criterion performance for all tests was set at 90% correct responding in a block of 20 trials. After completion of behavioral tests, animals were perfused with saline and 10% formalin and their brains sectioned at 50 microns on a freezing microtome. Sections were stained with cresyl violet and the extent of the lesion was reconstructed on drawings made from the Slotnick and Hersch (1980) atlas of the rat olfactory system.

RESULTS

Anatomical Results

Each rat had the left olfactory bulb completely removed. Control rats had a narrow 2-4 mm incision approximately 1-2 mm dorsal and parallel to the rhinal fissure in the lateral orbital cortex in the right hemisphere.

In the LOT group, the lesions in the left hemisphere were similar to those in the lesioned control group. In the right hemisphere, the LOT was completely transected in each rat at the level of the rostral 1/3 of the olfactory tubercle. In most cases the lesions were fairly discrete and transected the tract with little damage to layer 2 of piriform cortex. The lesions were similar in LOT/PC rats except that the knife extended deeper into the hemisphere, transected the LOT, most or all of the piriform cortex, extended into the lateral olfactory tubercle and damaged the lateral and/or ventral aspect of the anterior limb of the anterior commissure.

Behavioral Results

Experiment 1. As shown in Table 1, rats with discrete LOT lesions (Group LOT) made slightly but significantly (p < 0.01) more errors than controls in the detection task but acquired the 2-odor problem as quickly as did controls. However, rats with deeper piriform cortical lesions (Group LOT/PC) made many more errors than the control and LOT groups in both the detection and discrimination tasks (p < 0.002, each comparison). Rats in Group LOT rapidly reacquired the detection task in their first postoperative session, but most rats in Group LOT/PC performed at chance for several hundred trials before meeting criterion.

Table 1.   Errors to criterion.   Values represent means (ranges).

Subject Group	Psychophysical Task	
	Detection	Discrimination
Controls	5.0 (0–14)	42.3 (18–67)
LOT	33.0 (10–57)	36.0 (8–54)
LOT/PC	373.8 (127–672)	225.0 (134–301)

Experiment 2.   The intensity-difference threshold of LOT rats (mean = 0.69) was significantly ($p < 0.002$) higher than that of controls (mean = 0.29; Table 2).   Experimental rats performed nearly as well as controls on the initial 2 intensity-difference tasks but none reached criterion on the 0.1% vs. 0.05% problem, while all controls reached criterion on the 0.1% vs. 0.06% problem.

The absolute detection threshold of LOT rats (0.0028%) was significantly ($p < 0.01$) higher than that of controls (0.000015%).   As in the intensity-difference task, experimental rats performed nearly as well as controls on the initial problems of the series.

Table 2.   Intensity-difference and absolute threshold values. Numbers represent means (ranges).

Subject Group	Threshold Measurement	
	Delta I[1]	Absolute Threshold[2]
Controls	.29 (.20–.36)	.000015 (.0005–.000003)
LOT	.69 (.59–.80)	.0028 (.0001–.0055)

[1]Fractional difference in intensity

[2]Percent vapor saturation

DISCUSSION

Rats with discrete transection of the LOT made few errors in retention of a simple odor detection task and acquired a 2-odor discrimination as well as did controls.   However, significantly greater deficits were produced by the deeper knife cuts in Group LOT/PC.   In these rats, the lesion not only transected the LOT but also the longitudinal association fibers interconnecting the anterior olfactory nucleus with olfactory cortex (Broadwell, 1975; Haberly and Price, 1978; Luskin and Price, 1982, 1983; Powell, Cowan and Raisman, 1965). The relatively severe deficits from LOT transection reported in prior studies may stem from transection of these additional olfactory efferents to olfactory cortex.

The absolute detection threshold and intensity difference threshold obtained for the 7-control rats in this study are similar to those reported by Slotnick and Schoonover (1984) in 20 normal and unilaterally bulbectomized rats. The psychophysical tests demonstrate that discrete transection of the LOT results in a decrease in odor sensitivity of approximately 2.2 orders of magnitude for detection and an approximate 2 fold decrease in ability to discriminate between concentrations of the same odor.

While rats with transection of the LOT and the consequent partial deafferentation of olfactory cortex and diencephalon (Price, 1977; Price and Slotnick, 1983; Price, Slotnick and Revial, 1991) have decreased olfactory sensitivity, they retain significant olfactory function. The lesion certainly spared bulbofugal terminations in the anterior olfactory nucleus and the olfactory tubercle. Thus, residual function is probably mediated by the further projections of one or both of these structures.

REFERENCES

Bennett, M.H., 1968, The role of the anterior limb of the anterior commissure in olfaction. Physiol. Behav., 3:507.
Broadwell, R.D., 1975, Olfactory relationships of the telencephalon and diencephalon in the rabbit. II. An autoradiographic study of the efferent connections of the anterior olfactory nucleus. J. Comp. Neurol., 164:389.
Cattarelli, M., 1982, Transmission and integration of biologically meaningful olfactory information after bilateral transection of the lateral olfactory tract in the rat. Behav. Brain Res., 6:313.
Devor, M., 1973, Components of mating dissociated by lateral olfactory tract transection in male hamsters. Brain Res., 64:437.
Gervais, M. and Pager, J., 1982, Functional changes in waking and sleeping rats after lesions in the olfactory pathways. Physiol. Behav., 29:7.
Haberly, L.B. and Price, J.L., 1978, Association and commissural fiber systems of the olfactory cortex. II. Systems originating in the olfactory peduncle. J. Comp. Neurol., 181:781.
Long, C.J. and Tapp, J.T., 1970, Significance of olfactory tracts in mediating response to odors in the rat. J. Comp. Physiol. Psychol., 72:435.
Luskin, M.B. and Price, J.L., 1982, The distribution of axon collaterals from the olfactory bulb and the nucleus of the horizontal limb of the diagonal band to the olfactory cortex, demonstrated by double retrograde labeling techniques. J. Comp. Neurol., 209:249.
Marques, D.M., O'Connell, R.J., Benimoff, N. and Macrides, F., 1982, Delayed deficits in behavior after transection of the olfactory tracts in hamsters. Physiol. Behav., 28:353.
Powell, T.P.S., Cowan, W.M. and Raisman, G., 1965, The central olfactory connections. J. Anat., 99:791.
Price, J.L., 1977, Structural organization of the olfactory pathways. In: "Proceedings of the Sixth International Symposium on Olfaction and Taste,". Information Retrieval, London.
Price, J.L. and Slotnick, B.M., 1983, Dual olfactory representation in the rat thalamus: An anatomical and electrophysiological study. J. Comp. Neurol., 214:63.

Price, J.L., Slotnick, B.M. and Revial, M.F., 1991, Olfactory projections to the hypothalamus. J. Comp. Neurol., 306:447.

Slotnick, B.M., 1985, Olfactory discrimination in rats with anterior amygdala lesions. Behav. Neurosci., 99:956.

Slotnick, B.M., 1990, Olfactory Perception. In: "Comparative Perception. Volume 1: Basic Mechanisms," W. Stebbins and M. Berkley, eds., Wiley, New York.

Slotnick, B.M. and Berman, E.J., 1980, Transection of the lateral olfactory tract does not produce anosmia. Brain Res. Bull., 5:441.

Slotnick, B.M. and Hersch, S., 1980, A stereotaxic atlas of the rat olfactory system, Brain Res. Bull., 5 (Supplement 5):1.

Slotnick, B.M. and Risser, J.M., 1990, Odor memory and odor learning in rats with lesions of the lateral olfactory tract and mediodorsal thalamic nucleus. Brain Res., 529:23.

Slotnick, B.M. and Schoonover, F.W., 1984, Olfactory thresholds in normal and unilaterally bulbectomized rats. Chem. Senses, 9:325.

Swann, H.G., 1934, The function of the brain in olfaction. II. The results of destruction of olfactory and other nervous structures upon the discrimination of odors, J. Comp. Neurol., 59:175.

Thompson, R., 1980, Some subcortical regions critical for retention of an odor discrimination in albino rats. Physiol. Behav., 24:915.

PHOSPHATASE ACTIVITY OF RAT OLFACTORY AND VOMERONASAL EPITHELIAL TISSUE

E.S. Chukhray[1], M.N. Veselova[1], O.M. Poltorack[1],
V.V. Voznessenskaya[2], E.P. Zinkevich[2], and C.J. Wysocki[3]

[1]Moscow State University, Russia
[2]A.N. Severtzov Institute of Evolutionary Animal
 Morphology and Ecology, Moscow V-71, Russia
[3]Monell Chemical Senses Center, Philadelphia, PA

INTRODUCTION

High levels of alkaline phosphatase activity are characteristic of the olfactory epithelium in numerous vertebrate species (Gladisheva, Kukushkina and Martynova, 1980) and have been demonstrated histochemically in the vomeronasal organ of the mouse (Bannister and Cuchieri, 1972; Cuchieri, 1974), golden hamster (Taniguchi, Taniguchi and Mikami, 1986) and buffalo (Saksena and Chandra, 1980). Phosphorylation, transport and receptor functions of this enzyme have been discussed (Poltorac, Chukhray and Veselova, 1991; Chukhray Poltorac and Veselova, 1991). Furthermore, high levels of phosphatase activity have been described for the opioid μ-receptor (Roy, Lee and Loch, 1986). Based on this and characteristics such as a wide range in substrate specificity and a considerable number of activators and inhibitors (alcohols, thiols, metallic ions, etc.,), we suggest that proteins with high levels of alkaline phosphatase activity may bind with different kinds of odorants and may be involved in receptor functions within the olfactory system.

Two forms of alkaline phosphatase within the rat olfactory epithelium can be distinguished via their kinetic properties (Chukhray et al., 1991). The soluble form is more likely to be involved in transport functions while the immobilized form may be involved in receptor functions. We have made comparative measurements of kinetic and catalytic properties of proteins with high phosphatase activity that are located on the external cell membranes of olfactory, vomeronasal and respiratory epithelia.

MATERIALS AND METHODS

Test subjects were 8 male Wistar rats (Stolbovaya nursery, Russia). Vomeronasal, olfactory and respiratory (from the trachea) epithelia were removed and transferred to 0.9% NaCl solution maintained at 4°C.

Determination of Alkaline Phosphatase Activity

The activity of alkaline phosphatase was determined at pH 7-8 in borax, glycine and tris-HCl buffers. The initial substrate concentration of p-nitrophenylphosphate was 9×10^{-4} M. The increase of p-nitrophenol (the product of p-nitrophenylphosphate hydrolysis) was measured by a double

Chemical Signals in Vertebrates VI, Edited by R.L. Doty and
D. Müller-Schwarze, Plenum Press, New York, 1992

Table 1. Extinction coefficients (C_{ext}) of p-nitrophenol at 400 nm, 25°C, 3 ml quartz cuvette, l = 1 cm.

pH	7.5	8.0	8.2	8.5	8.8	9.0	9.2	9.5
Σ, OU x 1/mM	1.5	1.6	1.6	2.0	2.2	2.4	1.4	2.4

beam spectrophotometer (Hitachi-124) at 400 nm (sensitivity was less than 0.02 - 0.03 optical units [OU]). 1.5 ml of buffer, 1 ml of 0.1 M $MgCl_2$ and 0.5 ml of substrate were added to vomeronasal, olfactory and respiratory epithelium. Enzyme hydrolysis was conducted at 25°C with continuous mixing. Every 2 min aliquots of solutions were obtained and the concentration of the reaction product was measured (Fig. 1). The velocity of the reaction was determined by estimating the angle of the slope (OU/min). The obtained values were divided into the extinction coefficient (C_{ext}) under corresponding pH values to obtain the reaction velocity values in μM/min (Table 1). Entries in Table 2 are based on the functions in fig. 1 and represent initial velocity as a measure of enzyme activity. The error in determining the initial velocity was less than 6% and included errors in measuring optical density and weight (δ = 1%) and other apparatus (δ = 2%).

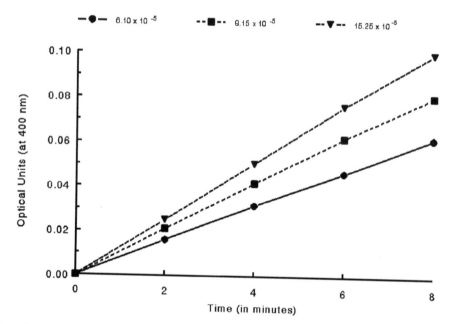

Fig. 1.   Kinetic curve of p-nitrophenylphosphate hydrolysis reaction in borate buffer, pH = 8.7, elicited by rat olfactory membrane alkaline phosphatase after 11 days of storage in the borate buffer at 4°C. Initial substrate concentration is indicated in the legend.

Table 2. Activity of immobilized alkaline phosphatase on the external surface structures of olfactory epithelium ($\approx$ 30 mg) that was stored in borate buffer for 11 days (1 sample).

$[S]_o$ x 10^{-5} M	V x 10^3 (OU/min)	V (μM/min)
6.10	12	5.45
9.15	20	9.10
15.25	50	11.40

Titration of Active Centers of the Immobilized Alkaline Phosphatase

The membrane form of alkaline phosphatase is inhibited by sodium o-vanadate (K = 1 x 10^{-6}M). The inhibition is reversible and alkaline phosphatase activity returns to normal after removing the inhibitor. We used this inhibitor for titration of active centers of the immobilized form of the enzyme that is found in vomeronasal and olfactory epithelia. As can be seen from Fig. 2a, in solution, as the concentration of the inhibitor increases, enzyme activity decreases. The same functions (in coordinates of equation 1) are presented at Fig. 2b. From this work, the following equations apply:

$$[I]_0 \frac{V_\infty - V_0}{V - V_0} = [E]_0 + \frac{1 + K_s^{-1}[S]_0}{K_i^{-1}(1 + K_{is}^{-1}[S]_0)} \times \frac{V_\infty - V_0}{V_\infty - V} \tag{1}$$

where
$[I]_0$ = initial concentration of inhibitor;
$[E]_0$ = initial concentration of enzyme;
$[S]_0$ = initial concentration of substrate;
V_0, V, and V_∞ are initial, current and maximal velocities of the the reaction as $[I]_0 \rightarrow \infty$.

Equation 1 can be represented by:

$$[I]_0 y = [E]_0 + k(x) \tag{2}$$

where

$$x = \frac{V_\infty - V_0}{V_\infty - V}; \quad y = \frac{V_\infty - V_0}{V - V_0}; \quad k = \frac{1 + K_s^{-1}[S]_0}{K_i^{-1}(1 + K_{is}^{-1}[S]_0)}$$

According to equation 2 the values of the Y-axis that intersect the functions are the concentrations of alkaline phosphatase active centers. These values are 6.5 μM and 12.5 μM for functions 1 and 2 respectively.

RESULTS

The mucus covering the surface of the olfactory, vomeronasal and respiratory epithelia contains a soluble form of alkaline phosphatase.

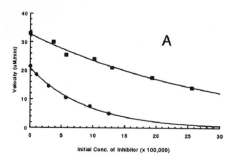
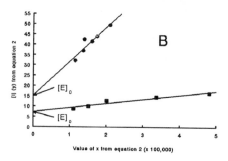

Fig. 2.   Titration of olfactory alkaline phosphatase active centers by
sodium o-vanadate on day 5 of storage of membranes in borate
buffer, pH = 8.7, 4°C.   Initial substrate concentrations were
30.4 x 10^{-5}M (●) and 45 x 10^{-5}M (■).   (A) Relationship between
alkaline phosphatase activity and initial substrate
concentration.   (B) The same expressed in coordinates of
equation 2 in text.

Pulsing changes of alkaline phosphatase activity elicited by p-nitrophenylphosphate were found in the olfactory epithelium but not in vomeronasal or
respiratory tissue.   Immobilized proteins with high phosphatase activity
were observed in olfactory and vomeronasal epithelial tissue, but not in
respiratory epithelium.   The immobilized form of alkaline phosphatase in
olfactory and vomeronasal epithelia shows high conformational lability
typical for receptor proteins.   Flexion points can be seen on "enzyme
activity - substrate concentration" functions for the immobilized form of
the vomeronasal and olfactory alkaline phosphatase.   These correspond to
conformational modifications as a result of enzyme activation.   The same
functions for respiratory alkaline phosphatase do not display such flexion
points, i.e., there were no conformational modifications as a result of
substrate concentration.   Two forms of alkaline phosphatase, defined by
their kinetic properties, were distinguished.   Form 1 has a low Michaelis
constant (similar to the known opioid μ-receptor phosphatase activity) and
form 2 has an effective Michaelis constant one order of magnitude higher.
The maximum velocity of the hydrolysis reaction mediated by form 2 is
approximately 5 times more than that for form 1.   The conformational alterations of alkaline phosphatase for olfactory and vomeronasal epithelia are
likely to be followed by the formation of additional active centers (in
accord with the results of titration with sodium vanadate).

The immobilized alkaline phosphatase of vomeronasal and olfactory
epithelia was used to produce heterogeneous biocatalysts with stable
characteristics.   The preparation of the bioactalyst included a complete
wash-out of the soluble form of alkaline phosphatase from the surface of
the tissue and the conduct of heterogeneous reactions at pH 8.0 - 9.5 in
the presence of Mg^{++}.   A borate buffer (pH 8.2 - 8.8) containing Mg^{++} (0.03
M) at 4°C was the preferred condition to preserve prepared samples of the
biocatalysts.

CONCLUSION

The investigation of kinetic and catalytic properties of alkaline
phosphatase of olfactory, vomeronasal and respiratory epithelia determined

46

the occurrence of an immobilized form of the enzyme with high activity in
receptor tissue only. Some characteristics of this protein (conformational
lability, kinetic properties, etc.) are typical for receptor proteins. It
may be appropriate to use the stable biocatalyst as a model for investigating olfactory and vomeronasal mechanisms of reception.

REFERENCES

Bannister, C. H., and Cuschieri, A., 1972, The fine structure and enzyme
    histochemistry of vomeronasal receptor cells, in: "Olfaction and
    Taste IV," D. Schneider ed. Wissenschaftliche Verlagsgesellschaft MBH,
    Stuttgart.
Chukhray, E. S., Poltorack, O. M., and Veselova, M. N., 1991, in press,
    Multifunctionality of alkaline phosphatase receptor and transport
    function, J. Phys. Chem. (in Russian).
Cuchieri, A., 1974, Enzyme histochemistry of the olfactory mucosa and
    vomeronasal organ in the mouse, J. Anat., 118, 3:477.
Gladysheva, O. S., Kukushkina, D. M., and Martynova, G. I., 1980, Functions
    and properties of olfactory epithelium protein, in: "Sensory Systems.
    Olfaction and taste," Nauka Publishers, Leningrad (in Russian).
Poltorack, O. M., Chukhray, E. S., and Veselova, M. N., 1991, in press,
    Multifunctionality of alkaline phosphatase, J. Phys. Chem. (in
    Russian).
Pryakhin, A. N., and Poltorack, O. M., 1982, Titration of enzyme active
    centers by use of the reversible inhibitors, Vestnik of Moscow
    University. Chem., 23, 3:233 (in Russian).
Roy, S., Lee, N. M., and Loch, H. H., 1986, Mu-opioid receptor is
    associated with phosphatase activity, Biochem. Biophys. Res. Comm.,
    140, 2:660.
Saksena, A. C., and Chandra, G., 1980, Gross histological and certain
    histochemical observations on the vomeronasal organ of Buffalo
    Bubalus-Bubalis, Indian J. Anim. Health, 19, 2:99.
Taniguchi, K., Taniguchi, K., and Mikami, S. I., 1986, Developmental
    studies on enzyme histochemistry of the three olfactory epithelia in
    the golden hamster, in: "Ontogeny of olfaction; principles of
    olfactory maturation in vertebrates," W. Breipohl, ed., Springer-
    Verlag, Berlin-New York.
Veselova, M. N., Chukhray, E. S., Poltorack, O. M., and Tischenko, S. M.,
    1986, Alkaline phosphatase stabilized by butanol competitive
    inhibition by o-sodium vanadate, Vestnik of Moscow State University.
    Chem., 26, 3:256 (in Russian).

PHYSIOLOGY AND PHARMACOLOGY OF THE ACCESSORY OLFACTORY SYSTEM

H. Kaba, C.-S. Li, E. B. Keverne*, H. Saito and K. Seto

Department of Physiology, Kochi Medical School, Nankoku
Kochi 783, Japan and *Sub-Department of Animal Behaviour
University of Cambridge, Madingley, Cambridge CB3 8AA, U.K.

INTRODUCTION

The accessory olfactory system originating in the vomeronasal organ is important in a variety of chemosensory primer effects including the acceleration of puberty, induction of oestrus, and pregnancy block in female mice following exposure to male urinary odours (pheromones) (Keverne, 1983). A considerable body of evidence has accumulated showing that pheromones received via the vomeronasal organ alter activity in tuberoinfundibular (TI) dopaminergic neurons, which in turn regulate pituitary prolactin release, and thereby lead to the reproductive effects described above. Injections of the dopamine agonist bromocriptine reproduced the actions of male pheromones in female mice with identical timing (Keverne, 1983). Conversely, the blocking of dopaminergic transmission by pimozide prevented the pheromonal action of a strange male in newly mated female mice (Marchlewska-Koj and Jemiolo, 1978). Further evidence for dopamine involvement in the context of pregnancy block has been demonstrated by measuring hypothalamic DOPA after pheromone exposure and blockade of dopamine synthesis (Rosser, Remfry and Keverne, 1989).

The purpose of this paper is to summarize recent our studies in female mice using electrophysiological techniques which have revealed the functional pathway of the centrally projecting accessory olfactory system and possible neurotransmitters mediating synaptic transmission in this pathway.

TI dopaminergic neurons as part of the final common pathway of the accessory olfactory system

Subjects for this study were Balb/c female mice which were ovariectomized for at least four weeks prior to being implanted subcutaneously with silastic capsules containing 0.5 µg oestradiol. Unit recordings were made under chloral hydrate anaesthesia (400 mg/kg, i.p.) 5-14 days after the capsule implantation, which has been shown to increase the percentage of TI arcuate neurons responding to stimulation of the accessory olfactory bulb (AOB) (Li, Kaba, Saito and Seto, 1989).

Electrical stimulation of the AOB orthodromically excited about 30% of TI arcuate neurons which were antidromically stimulated from the median eminence. The response was induced in an all-or-none fashion. The mean latency to onset of this effect measured from peri-stimulus-time histogram was 50 ms. No inhibitions followed AOB stimulation among the cells tested. Electrophysiological techniques have served as powerful tools for estimating the activity of mesotelencephalic dopaminergic neurons (Bunny, 1979). However, there have been some difficulties in making similar estimations in TI dopaminergic neurons. One reason for this is that perikarya of TI dopaminergic neurons are not as densely packed as those in the substantia nigra, and they are dispersed among non-dopaminergic cells (Moore and Demarest, 1982). In an attempt to identify TI dopaminergic as opposed to other types of TI arcuate neurons, we examined the effectiveness of 6-hydroxydopamine (6-OHDA) infusions into the median eminence in interrupting impulse flow (Li, Kaba, Saito and Seto, 1990a), since 6-OHDA has been demonstrated to be effective in blocking the axonal conduction of catecholaminergic neurons but not other types of neurons after acute administration into the medial forebrain bundle (Guyenet and Aghajanian, 1978) or the dorsal noradrenergic bundle (Aghajanian, Cedarbaum and Wang, 1977). The 6-OHDA disrupted the antidromic responses of 50% of the TI arcuate neurons with excitatory inputs from the AOB, whereas 5,7-dihydroxytryptamine (5,7-DHT) did not. The specificity of these changes is indicated by the fact that 5,7-DHT but not 6-OHDA has been found to block propagation in serotonergic axons (Wang and Aghajanian, 1977). The findings presented here therefore clearly show that accessory olfactory information is transmitted to TI dopaminergic neurons.

Excitatory amino acid receptors in the amygdala mediating AOB-induced excitation of TI arcuate neurons

It is now established that the output neurons of the AOB project to the ipsilateral corticomedial nuclei of the amygdala. We investigated the effectiveness of the local anaesthetic lignocaine, infused into the amygdala, in blocking the response of TI arcuate neurons to AOB stimulation, thereby confirming that the route of the AOB input to TI arcuate neurons is through the amygdala (Li et al., 1989). The principal output neurons of the AOB, the mitral cells, have been suggested to utilize glutamate or a closely related substance as a neurotransmitter (Fuller and Price, 1988). We therefore examined the ability of excitatory amino acid antagonists, infused into the amygdala, to block the excitatory response of TI arcuate neurons to AOB stimulation (Li et al., 1990a). Four drugs were tested: the broad spectrum antagonist kynurenate, specific N-methyl-D-aspartate (NMDA) antagonist D,L-2-amino-5-phosphonovalerate (AP5), preferential non-NMDA antagonist γ-D-Glutamylaminomethylsulphonate (GAMS) and D,L-2-amino-4-phosphonobutyrate (AP4). The finding that intra-amygdala infusions of the broad spectrum antagonist kynurenate (3 nmol) reversibly abolished the AOB-induced excitation of TI arcuate neurons suggest that a glutamate-related substance is the neurotransmitter of AOB mitral cells which project to the amygdala. The AOB-induced excitation was antagonized, in addition, by intra-amygdala infusions (3 nmol) of the specific NMDA antagonist AP5, the preferential non-NMDA antagonist GAMS, or AP4, elaborating mediation by excitatory amino acid receptors within the amygdala.

To further characterize the excitatory amino acid involved, we examined the ability of excitatory amino acid agonists, infused into the amygdala, to change firing activity of TI arcuate neurons with excitatory inputs from the AOB (Li et al., 1990a). Both NMDA and kainate (3 nmol for both) markedly enhanced the activity of all the neurons tested, whereas quisqualate (3 nmol) was without effect in any of the neurons tested. The ability of NMDA and kainate to enhance firing activity of TI arcuate neurons showing AOB-induced excitation offers further evidence for the involvement of excitatory amino acid receptors in this response. The failure of intra-amygdala infusions of quisqualate to produce changes in firing activity of TI arcuate neurons suggests that amygdaloid quisqualate receptors are not involved in the AOB-induced excitation of these neurons and that the results obtained with the same dose of NMDA and kainate are unlikely to be due to some non-specific effect.

Taken together, these findings lead us to the tentative conclusion that multiple receptor subtypes, namely NMDA, kainate and AP4 receptors within the amygdala may be implicated in the AOB-induced excitation of TI arcuate neurons.

Stria terminalis mediating AOB-induced excitation of TI arcuate neurons

Anatomical studies have provided evidence that fibres from the amygdaloid complex travel through the stria terminalis to the ventromedial hypothalamus, medial preoptic area and bed nucleus of the stria terminalis (De Olmos and Ingram, 1972; Kevetter and Winans, 1981). There is evidence for a component also travelling through the ventral amygdaloid pathway (Lehman and Winans, 1983). This ventral pathway appears to be essential for copulating behaviour which is dependent on the vomeronasal and olfactory chemoreception in the male hamster (Winans, Lehman and Powers, 1982). To determine whether and to what extent the excitatory response of TI arcuate neurons to AOB stimulation is via the stria terminalis, we investigated the effectiveness of lignocaine infusion into the stria terminalis in blocking the AOB-induced excitation, and that of stria terminalis stimulation in evoking a response with a shorter latency (Li et al., 1990a). Lignocaine, infused into the stria terminalis, reversibly blocked the AOB-induced excitation of all the TI arcuate neurons tested, indicating that the excitatory transmission from the AOB to TI arcuate neurons is mediated entirely through the stria terminalis. This conclusion is substantiated by the finding that TI arcuate neurons responding to AOB stimulation also responded with a shorter latency to stria terminalis stimulation (Li et al., 1990a).

CCK-B receptors in the medial preoptic area mediating AOB-induced excitation of TI arcuate neurons

Despite anatomical evidence for projections from the corticomedial amygdala receiving AOB input to the ventromedial hypothalamus, Gunnet and Moore (1988) failed to detect the effect of electrical stimulation of the ventromedial hypothalamus on activity of TI dopaminergic neurons. The medial preoptic area has been recognized as an antisurge key control centre for prolactin secretion (Wiersma and Kastelijn, 1990). We therefore examined if the medial preoptic area mediates AOB-induced excitation of TI arcuate neurons (Li, Kaba, Saito and Seto, in press). Lignocaine infusions into the medial preoptic area reversibly blocked the AOB-induced excitation of all the TI arcuate neurons tested, indicating that the medial preoptic area is an essential relay for AOB-induced excitation of TI arcuate neurons. This conclusion is favoured by the finding that TI arcuate neurons excited by AOB stimulation were also excited with a shorter latency by medial preoptic stimulation (Li et al., in press).

Combined axonal transport/immunohistochemical studies have demonstrated the presence of cholecystokinin (CCK)-containing neurons projecting from the medial amygdala to the medial preoptic area (Simerly, 1990). Therefore, we further examined the involvement of CCK receptors in the medial preoptic area mediating AOB-induced excitation of TI arcuate neurons (Li et al., in press). The CCK-B receptor antagonist L365,260 (0.3, 0.6 and 0.9 pmol), infused into the medial preoptic area, blocked the AOB-induced excitation of TI arcuate neurons in a dose-dependent manner, whereas the CCK-A receptor antagonist L364,718 (0.9 pmol) had no effect. Furthermore, CCK-8 (0.6 pmol), infused into the medial preoptic area, enhanced the spontaneous firing activity of TI arcuate neurons with excitatory inputs from the AOB. These results demonstrate that excitatory inputs from the AOB to TI arcuate neurons are through the activation of CCK-B receptors within the medial preoptic area.

Current source-density analysis of synaptic events in the AOB

Female mice form a long-term olfactory memory to the urinary odours of the male that mates with them (Keverne, 1983; Kaba, Rosser and Keverne, 1988). This olfactory memory is of critical biological importance, since it prevents any subsequent exposure to this male's pheromones from initiating neuroendocrine mechanisms that would terminate pregnancy (Keverne and Rosser, 1986). Our studies have shown that the relatively primitive trilaminar structure of the AOB has the capacity for synaptic changes of importance for the recognition/memory and subsequent gating of biologically significant odours (Brennan, Kaba and Keverne, 1990). As a first step to approaching the electrophysiological correlates of this recognition/memory, we examined synaptic events in the AOB following stimulation of the vomeronasal organ using a one-dimensional current source-density (CSD) analysis in anaesthetized intact female mice (Kaba and Keverne, in press). The one-dimensional CSD analysis revealed two major sinks (inward membrane currents), one in the external plexiform layer and the other in the glomerular layer. The onset latencies for glomerular layer sinks were always shorter than those for external plexiform layer sinks.

Additional evidence distinguishing between two spatially and temporally distinct neuronal events was obtained from pharmacological experiments. Local infusions of the broad spectrum excitatory amino acid antagonist kynurenate (5 nmol) into the AOB completely blocked the generation of the external plexiform layer sink without effect on the glomerular layer sink in any of the animals tested. Based on anatomical and physiological considerations, it is suggested that the inward membrane current in the external plexiform layer is generated by the mitral to granule cell dendrodendritic synapse and mediated by excitatory amino acid receptors and that the inward membrane current in the glomerular layer underlies monosynaptic EPSPs evoked in the apical dendrites of mitral cells by vomeronasal afferents.

GABAergic mechanisms in the AOB regulating activity of TI arcuate neurons

The basic synaptic organization in the AOB is the same as in the main olfactory bulb (Mori, 1987), where the mitral/tufted cell relay neurons form reciprocal dendrodendritic synapses with the intrinsic granule cells; the granule cells release inhibitory GABA onto the mitral/tufted cell secondary dendrites at these synapses, where the mitral/tufted cells are excitatory to the granule cells (Halasz and Shepherd, 1983). In the main olfactory bulb, there is strong evidence that the GABAergic granule cells act to control mitral/tufted cell excitability (Halasz and Shepherd, 1983). We examined the effect of GABA transmitter blockade in the AOB on firing activity of TI arcuate neurons (Li, Kaba, Saito and Seto, 1990b). The effect of local infusions of bicuculline or saline into the AOB was tested on TI arcuate neurons which were also orthodromically stimulated from the AOB. Bicuculline (0.1 nmol) infused into the AOB increased spontaneous firing rates in all the TI arcuate neurons tested. In freely behaving female mice, local infusions of bicuculline into the AOB following mating have been shown to cause a direct block to pregnancy (Kaba and Keverne, 1988). This could be explained by the effect of the bicuculline antagonizing the GABA-mediated feedback inhibition on the mitral cells of the AOB, resulting in an over-excitation of the accessory olfactory system. This over-excitation would activate TI dopaminergic neurons as part of the final common pathway of the accessory olfactory system by way of the amygdala, causing a return to oestrus. The present results provide direct support for this explanation and further suggest that GABAergic inhibitory feedback to the mitral cell of the AOB is crucial to close control of mitral cell output to TI arcuate neurons.

CONCLUDING REMARKS

We have undertaken experiments to determine the neural pathways involved in the actions of primer pheromones. Incorporating the findings presented here with those of others, the following sequence of events is proposed to occur in female mice after exposure to the male's pheromones. The vomeronasal organ contains the receptor neurons for primer pheromones. The receptor neurons send their axons to the AOB mitral cells, the first relay neurons. The projections of the AOB mitral cells activate excitatory amino acid receptors within the amygdala, from where the signal is channelled through the stria terminals. The stria terminalis neurons release CCK acting on CCK-B receptors located in the medial preoptic area, thereby causing excitation of TI dopaminergic arcuate neurons as part of the final common pathway of the accessory olfactory system. The neural transmission from the vomeronasal receptor neurons to TI dopaminergic neurons is critically regulated by GABAergic feedback mechanisms within the AOB. Such a sequence of neural events can account for the reproductive effects so far described as a consequence of vomeronasal chemoreception.

ACKNOWLEDGEMENTS

We thank Merk Sharp and Dohme Research Laboratories (West Point, USA) for the gift of L364,718 and L365,260. This work was supported by Grants-in-Aid for Scientific Research to H.K. from the Ministry of Education, Science and Culture of Japan.

REFERENCES

Aghajanian, G. K., Cedarbaum, J. M., and Wang, R. Y., 1977, Evidence for norepinephrine-mediated collateral inhibition of locus coeruleus neurons, Brain Res., 136: 570.

Brennan, P., Kaba, H., and Keverne, E. B., 1990, Olfactory recognition: a simple memory system, Science, 250: 1223.

Bunney, B. S., 1979, The electrophysiological pharmacology of midbrain dopaminergic systems, in: "The Neurobiology of Dopamine," A. S. Horn, J. Korf, and B. H. C. Westerink, eds., Academic Press, New York.

De Olmos, J. S., and Ingram, W. R., 1972, The projection field of the stria terminalis in the rat brain. An experimental study, J. Comp. Neurol., 146: 303.

Fuller, T. A., and Price, J. L., 1988, Putative glutamatergic and/or aspartatergic cells in the main and accessory olfactory bulbs of the rat, J. Comp. Neurol., 276: 209.

Gunnet, J. W., and Moore, K. E., 1988, Effect of electrical stimulation of the ventromedial nucleus and the dorsomedial nucleus on the activity of tuberoinfundibular dopaminergic neurons, Neuroendocrinology, 47: 20.

Guyenet, P. G., and Aghajanian, G. K., 1978, Antidromic identification of dopaminergic and other output neurons of the rat substantia nigra, Brain Res., 150: 69.

Halasz, N., and Shepherd, G. M., 1983, Neurochemistry of the vertebrate olfactory bulb, Neuroscience, 10: 579.

Kaba, H., and Keverne, E. B., 1988, The effect of microinfusions of drugs into the accessory olfactory bulb on the olfactory block to pregnancy, Neuroscience, 25: 1007.

Kaba, H., and Keverne, E. B., Analysis of synaptic events in the mouse accessory olfactory bulb with current source-density techniques, Neuroscience, in press.

Kaba, H., Rosser, A. E., and Keverne, E. B., 1988, Hormonal enhancement of neurogenesis and its relationship to the duration of olfactory memory, Neuroscience, 24: 93.

Keverne, E. B., 1983, Pheromonal influences on the endocrine regulation of reproduction, Trends Neurosci., 6: 381.

Keverne, E. B., and Rosser, A. E., 1986, The evolutionary significance of the olfactory block to pregnancy, in: "Chemical Signals in Vertebrates 4," D. Duvall, D. Muller-Schwarze, and R. M. Silverstein, eds., Plenum Press, New York.

Kevetter, G. A., and Winans, S. S., 1981, Connections of the corticomedial amygdala in the golden hamster. I. Effects of the "vomeronasal amygdala", J. Comp. Neurol., 197: 81.

Lehman, M. N., and Winans, S. S., 1983, Evidence for a ventral non-strial pathway from the amygdala to the bed nucleus of the stria terminalis in the male golden hamster, Brain Res., 268: 139.

Li, C.-S., Kaba, H., Saito, H., and Seto, K., 1989, Excitatory influence of the accessory olfactory bulb on tuberoinfundibular arcuate neurons of female mice and its modulation by oestrogen, Neuroscience, 29: 201.

Li, C.-S., Kaba, H., Saito, H., and Seto, K., 1990a, Neural mechanisms underlying the action of primer pheromones in mice, Neuroscience, 36: 773.

Li, C.-S., Kaba, H., Saito, H., and Seto, K., 1990b, GABAergic mechanisms are involved in the control of tuberoinfundibular arcuate neurons by the accessory olfactory bulb, Neurosci. Lett., 120: 231.

Li, C.-S., Kaba, H., Saito, H., and Seto, K., Cholecystokinin: critical role in mediating olfactory influences on reproduction, Neuroscience, in press.

Marchlewska-Koj, A., and Jemiolo, B., 1978, Evidence for the involvement of dopaminergic neurons in the pregnancy block effect, Neuroendocrinology, 26: 186.

Moore, K. E., and Demarest, K. T., 1982, Tuberoinfundibular and tuberohypophyseal dopaminergic neurons, in: "Frontiers in Neuroendocrinology," Vol. 7, W. F. Ganong, and L. Martini, eds., Raven Press, New York.

Mori, K., 1987, Membrane and synaptic properties of identified neurons in the olfactory bulb, Prog. Neurobiol., 29: 275.

Rosser, A. E., Remfry, C. J., and Keverne, E. B., 1989, Restricted exposure of mice to primer pheromones coincident with prolactin surges blocks pregnancy by changing hypothalamic dopamine release, J. Reprod. Fertil., 87: 553.

Simerly, R. B., 1990, Hormonal control of neuropeptide gene expression in sexually dimorphic olfactory pathways, Trends Neurosci., 13: 104.

Wang, R. Y., and Aghajanian, G. K., 1977, Inhibition of neurons in the amygdala by dorsal raphe stimulation: mediation through a direct serotonergic pathways, Brain Res., 120: 85.

Wiersma, J., and Kastelijn, J., 1990, Electrophysiological evidence for a key control function of the medial preoptic area in the regulation of prolactin secretion in cycling, pregnant and lactating rats, Neuroendocrinology, 51: 162.

Winans, S. S., Lehman, M. N., and Powers, J. B., 1982, Vomeronasal and olfactory CNS pathways which control male hamster mating behavior, in "Olfaction and Endocrine Regulation," W. Breipohl, ed., IRL Press, London.

ROLE OF LIPIDS OF RECEPTOR MEMBRANES IN ODOR RECEPTION

Shuichi Enomoto, Takayuki Shoji,
Mutsuo Taniguchi and Kenzo Kurihara

Faculty of Pharmaceutical Sciences
Hokkaido University, Sapporo, 060 Japan

INTRODUCTION

It is generally believed that an olfactory response is induced by binding of an odorant to a specific receptor protein in olfactory receptor membranes (Buck and Axel, 1991). On the other hand, it has been shown that nonolfactory systems such as the turtle trigeminal nerve (Tucker, 1963), the Helix ganglion, the fly taste nerve, the frog taste cell, and the neuroblastoma cell respond to various odorants (Kashiwayanagi et al., 1984; 1985). These nonolfactory systems do not seem to provide specific receptor proteins for odorants.

In general, odorants are hydrophobic and interactions of odorants with lipid layers of olfactory receptor membranes play an important role in the generation of olfactory responses. In previous studies, we have shown that azolectin liposomes respond to various odorants (Nomura et al., 1987a,b). In this paper, we describe conditions which increase the sensitivity of liposomes to odorants (Enomoto et al., 1991) and the effects of changes in lipid composition on specificity of liposomes to various odorants. We also show that phosphatidylserine (PS)-containing liposomes exhibit large responses, especially to fatty acids.

In order to examine the role of lipids in the in vivo olfactory system, we have applied PS-containing liposomes to the turtle olfactory epithelium and have found that PS treatment greatly increases the responses to fatty acids. On the basis of these findings, a possible mechanism of odor reception and discrimination is presented.

RESULTS AND DISCUSSION

Liposomes having high sensitivity to odorants

Figure 1 shows the odorant-induced membrane potential changes [which were monitored with a voltage dependent fluorescent dye 3,3'-dipropylthiocarbocyanine iodide (diS-C_3(5)] of liposomes containing phosphatidylcholine (PC) and PS in different ratios. The sensitivity of liposomes to odorants varies with odorant type. The minimum concentration of amyl acetate to induce the response (referred to as threshold) in the PC liposomes is about 10^{-4} M. The addition of 10 or 20% PS

Chemical Signals in Vertebrates VI, Edited by R.L. Doty and
D. Müller-Schwarze, Plenum Press, New York, 1992

lowers the threshold to about 10^{-9} M and increases the magnitude of the response. The olfactory thresholds for amyl acetate were determined to be 10^{-4} M for the frog and 10^{-7} M for the turtle. Hence, the threshold of the PC-PS liposomes (PS/PC=0.2) to amyl acetate is comparable to or lower than the olfactory thresholds in these animals. It was calculated that adsorption of a few molecules of amyl acetate on a single liposome elicits detectable changes in the membrane potential.

As in the case of amyl acetate, the addition of PS to PC greatly lowered the threshold for beta-ionone and increased the magnitude of the response. On the other hand, the effects of the addition of PS on the response to citral were not simple; the order of the magnitude of the response to citral at equimolar was PS/PC=0.1 > PC > PS/PC=0.2 > PS/PC=0.05. Thus, the specificity of liposomes to various odorants greatly depended on lipid composition.

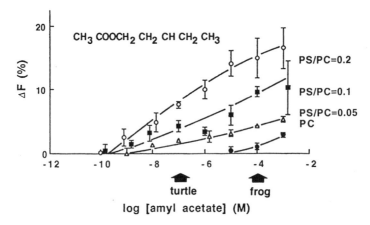

Fig. 1. Changes in the fluorescence intensity of voltage-dependent fluorescent dye, diS-C_3(5) added to suspension of liposomes composed of PC and PS in different ratios as a function of concentration of amyl acetate. Arrows represent threshold concentrations of amyl acetate in the frog and turtle olfactory systems.

Addition of proteins to liposomes also changed the specificity of liposomes to odorants. We examined the effects of the addition of proteins such as concanavalin A (Con A) and membrane proteins of canine erythrocytes to liposomes on odor specificity. Incorporation of Con A to PS/PC=0.2 liposomes greatly increased the response to citral and decreased that to amyl acetate. The magnitudes of responses of PS/PC=0.2 liposomes to nonanol, beta-ionone, menthol and menthone were greatly increased by incorporation of membrane proteins of the canine erythrocytes, while those to amyl acetate and octanol were greatly decreased.

The liposomes discriminate odorants having different functional groups. Figure 2 shows the magnitude of the responses to aliphatic compounds having different functional groups in PS/PC=0.2 liposomes. The compounds of the same concentration (0.1 mM) induce different magnitudes of response. Fatty acids such as butyric acid, valeric acid and isovaleric acid induce large responses in the PS-containing liposomes, whereas they do not induce such large responses in liposomes containing no PS.

Detailed analysis of the adsorption sites for various odorants in lipid membranes was carried out using various fluorescence dyes which monitor the membrane fluidity changes in different regions of the membrane. Different odorants were absorbed on different regions in the membranes having complex lipid compositions, whereas different odorants were absorbed on similar regions in the membranes having a simple lipid composition (Kashiwayanagi et al., 1990).

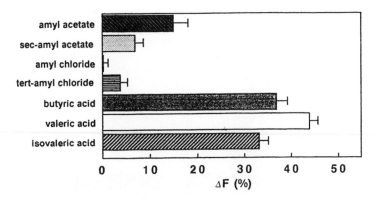

Fig. 2. Changes in the fluorescence intensity of diS-C$_3$(5) added to PS/PC=0.2 liposomes in response to various odorants of 10^{-4} M concentration.

Selective Enhancement of the Turtle Olfactory Responses to Fatty Acids by Treatment of the Olfactory Epithelium with PS-Containing Liposomes

As shown above, the addition of PS to PC greatly enhanced the responses of liposomes to fatty acids. In order to examine whether or not this holds in the olfactory system, PS-containing liposomes were applied to the turtle olfactory epithelium and its effects on the olfactory responses were observed (Taniguchi et al., in preparation). The results are shown in Fig. 3. Here R_o and R represent the magnitude of the response to an odorant before and after PS-treatment, respectively. The PS-treatment enhanced the responses to fatty acids such as valeric acid, isovaleric acid and butyric acid by a factor of 4-5 times. The response to anisol was enhanced about 1.5 times by the treatment. The treatment did not affect the responses to amyl acetate, sec-amyl acetate, and tert-amyl chloride. The PS-treatment lowered the threshold of the response to valeric acid by a factor of about 100, and enhanced the responses across the whole concentration range examined. The enhanced responses to fatty acids by the treatment returned to the original level approximately 10 hours after the PS-treatment.

The findings that the olfactory responses to fatty acids were greatly enhanced by the PS-treatment closely resembles the findings with the liposomes. It seems that PS is incorporated into the olfactory receptor membranes and modifies the receptor sites for the fatty acids. These results suggest that lipids are important for the reception of odorants. In a separate study in our laboratory (Hanada et al., in preparation), it was shown that an increase of temperature of the turtle olfactory epithelium up to 40°C abolishes the ability of the receptor to discriminate among odorants of similar structure. These results suggested that the membrane fluidity changes in the lipid layer of the receptor membrane greatly affects the structure of the receptor sites for odorants, supporting the idea that the lipids are important in odor reception.

The present findings are in accord with the following hypothesis for odor reception and discrimination. The composition of the lipids and proteins of each olfactory cell membrane is different from cell to cell. The magnitude of depolarization in response to an odorant is different from cell to cell. The depolarization is transformed into nerve impulses. The qualities of odors are recognized by firing patterns among various olfactory axons and the quality of odors is recognized in the brain. In this scheme, hydrophobic pockets of the olfactory receptor membranes composed of lipids and proteins are postulated to be absorption sites for odorants. Variation in combinations of lipids and proteins provides many different adsorption sites for odorants.

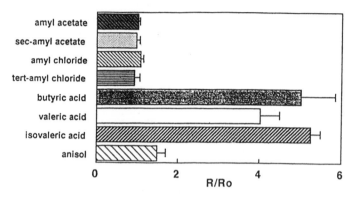

Fig. 3.   Effects of PS-liposomes on the turtle olfactory bulbar responses to various odorants. R_o and R represent the relative magnitude of the response to an odorant before and after the PS-treatment.

REFERENCES

Buck, L., and Axel, R., 1991, A novel multigene family may encode odorant receptors: a molecular basis for odor recognition, Cell, 65:175.
Enomoto, S., Kashiwayanagi, M., and Kurihara, K., 1991, Liposomes having high sensitivity to odorants, Biochem. Biophys. Acta., 1062:7.
Kashiwayanagi, M., and Kurihara, K., 1984, Neuroblastoma cell as model for olfactory cell: Mechanism of depolarization in response to various odorants, Brain Res., 293:251.
Kashiwayanagi, M. and Kurihara, K., 1985, Evidence for non-receptor odor discrimination using neuroblastoma cells as a model for olfactory cells, Brain Res., 359:97.
Kashiwayanagi, M., Suenaga, A., Enomoto, S., and Kurihara, K., 1990, Membrane fluidity changes of liposomes in response to various odorants: Complexity of membrane composition and variety of adsorption sites for odorants, Biophys. J., 58:887.
Nomura, T., and Kurihara, K., 1987a, Liposomes as a model for olfactory cells: Changes in membrane potential in response to various odorants, Biochemistry, 26:6135.
Nomura, T., and Kurihara, K., 1987b, Effects of changed lipid composition on responses of liposomes to various odorants: Possible mechanism of odor discrimination, Biochemistry, 26:6141.
Tucker, D., 1963, Olfactory, vomeronasal and trigeminal receptor responses to odorants, in: "Olfaction and Taste.", Y. Zotterman, ed., Macmillian, New York.

THALAMOCORTICAL MECHANISMS AND SMELL: INSIGHT FROM AMNESIA

Robert G. Mair

Department of Psychology
University of New Hampshire
Durham, NH

INTRODUCTION

In 1963, Powell, Cowan, and Raisman described a pathway from the pyriform lobe to the mediodorsal nucleus (MDn) of the thalamus. Although previous reports had provided evidence of direct olfactory projections to frontal neocortex in dogs and rabbits (cf. Allison, 1953), this report revolutionized thinking about central olfactory processes by emphasizing the importance of thalamocortical projections in olfaction. Powell et al. (1963), pg. 712, states:

"First, it shows that the thalamus is related to the olfactory system in much the same way as it is to other sensory pathways, and from this it follows that part, at least, of the orbito-frontal area of the cortex may be thought of as a sensory projection field for olfaction -- although clearly not in the same way as the primary sensory areas."

Subsequent studies in rodents, monkeys, and man have provided evidence confirming an olfactory function for pathways from MDn to orbito-frontal cortex. In man, olfactory impairments have been described in patients with unilateral temporal lobectomies, affecting pyriform cortex (Eskenazi et al., 1983, 1986; Jones-Gotman & Zatorre, 1988); Korsakoff's disease, a form of amnesia associated with lesions of MDn (Jones et al., 1975a,b, 1978; Mair et al., 1980, 1986; Gregson et al., 1981); as well as in patients with unilateral removals of orbito-frontal cortex (Potter & Butters, 1980; Jones-Gotman & Zatorre, 1988). These conditions are associated with distinct patterns of cognitive impairment and disrupt different portions of pathways interconnecting pyriform cortex, MDn, and orbitofrontal cortex. Nevertheless, they are reported to have comparable effects on olfactory function, disrupting measures of odor discrimination and identification while sparing the capacity to detect weak concentrations of odorants. The observation of deficient discrimination and recognition with sparing of primary sensory capacity is consistent with perceptual impairments produced by lesions of higher-order centers in other sensory modalities (Milner and Teuber, 1968).

In rodents, lesions of MDn and its connections to ventral agranular frontal cortex have been reported to disrupt olfactory discrimination learning while sparing performance of detection tasks measuring absolute

Chemical Signals in Vertebrates VI, Edited by R.L. Doty and
D. Müller-Schwarze, Plenum Press, New York, 1992

sensitivity (Eichenbaum et al, 1980, 1983; Sapolsky and Eichenbaum, 1980; Slotnick and Kaneko, 1981; Staubli et al., 1987). On the surface, this seems similar to findings in humans. In each of these studies, however, lesioned animals were eventually able to perform at criterion and thus demonstrate a preserved capacity for olfactory discrimination sufficient to carry out the tasks in question. In several of these studies, observations were made that are more consistent with a cognitive than a simple discriminative impairment. Slotnick and Kaneko (1981) argued that thalamocortical pathways are involved in complex aspects of olfactory learning based on their observation that rats with MDn lesions were impaired in their ability to reverse, but not to initially learn, a go/no go discrimination between propyl acetate and ethyl acetate. Staubli et al. (1987) reported that the effects of thalamic lesions on an olfactory maze learning task were reduced considerably by prior training on the same task with different pairs of odorants. Based on these data, they argued that the MDn is needed to learn procedures related to the task but not to represent information about particular odor cues. The rodent literature does not lend itself to a simple interpretation; nevertheless, it presents a fairly compelling case that thalamic lesions can impair processes related to olfactory learning while leaving a more basic capacity for sensory discrimination intact. A similar analysis could be applied to findings in humans. For instance, normal sensitivity has been demonstrated in patients with Korsakoff's disease by a detection task which used a single odorant and a blank and incorporated procedures designed to eliminate differences due to differential rates of learning (Mair et al., 1980). By contrast, impairments in recognition and identification have used larger set of odorants and procedures that place considerable demands on working memory (Jones et al., 1975a,b, 1978; Mair et al., 1980, 1986; Gregson et al., 1981).

LEARNING, MEMORY & MEDIAL THALAMUS

There is abundant evidence that lesions of medial thalamus that include MDn can disrupt learning and memory in other sensory modalities. Thus, the implication of thalamocortical systems in processes related to olfactory learning and memory, rather than in sensory discrimination, calls into question the view that the MDn plays a role specifically related to smell. In humans, lesions of medial thalamus have been correlated to disorders of learning and memory associated with Korsakoff's disease (Victor et al., 1989), tumors (McEntee et al., 1976), trauma (Squire and Moore, 1979), and infarct (Cramon et al., 1985). Analyses of experimental lesions in animals have also provided evidence that lesions of medial thalamus can disrupt memory, although they have not led to consensus on the critical substrate of these disorders (cf. Markowitsch, 1982).

Over the past six years, we have adopted a dual approach to this problem by (a) developing an animal model of Korsakoff's disease using rats recovered from a subacute bout of pyrithiamine-induced thiamine deficiency (PTD) and (b) comparing deficits produced by PTD treatment with more circumscribed radiofrequency (RF) and excitotoxic lesions. The results of these studies have led to the view that lesions of medial thalamus disrupt working memory for tasks in which stimulus information must be held "on-line" for a limited period of time or in which animals must discriminate the temporal recency of stimuli. Rats recovered from PTD treatment exhibit lesions in two consistent locations: in the mammillary bodies centered on the medial nucleus; and in medial thalamus centered on the internal medullary lamina (IML) in a bilaterally symmetric pattern that encroaches on adjacent portions of MDn (Mair et al., 1985, 1988, 1991a,b; Knoth and Mair, 1991). The lesions produced by this treatment apparently result from an excitotoxic reaction to excess glutamate and can be blocked by the glutamate antagonist MK-801 (Langlais and Mair, 1990; Robinson and Mair, 1991).

Rats with IML lesions produced by PTD treatment require more trials than controls to learn many tasks. However, when given sufficient training, they can _eventually_ perform at criterion or as well as controls in discrimination learning based on light/dark, place, olfactory, or auditory cues (Mair et al., 1988; 1991a,b) and in locating a platform hidden in a consistent location in the Morris water maze task (Langlais et al., 1991). The effects of PTD treatment on discrimination learning can be illustrated by the place serial reversal learning task (Figure 1, data from Mair et al., 1991b). In initial learning (IL), rats were trained to select the non-preferred of two ports in an operant chamber until a criterion of 8 consecutive correct responses was reached. On the session after criterion was reached, contingencies were switched (for the first reversal problem) so that rats were reinforced for selecting the opposite port as in IL. Whenever criterion was reached in this and subsequent problems, training was ended for the day and rats were moved to the next reversal problem (numbered 1 to 7 in Figure 1) until training was completed for IL and seven reversals. The PTD group made significantly more errors in reaching criterion for each of the problems and thus showed evidence of a learning impairment. Nevertheless, they showed a preserved capacity in reaching criterion for each of the problems and in the extent of positive transfer between problems. This transfer effect is important as it indicates that the animals were able to acquire a strategy (associating place cues with reinforcement) that was not tied to a fixed stimulus-response relationship. Based on these and other data, we have argued that the PTD model exhibits a spared capacity for reference memory or the ability to respond based on a consistent rule or a psychological structure _once_ _acquired_ (Mair et al., 1991a,b).

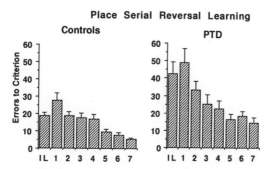

Fig. 1.   Errors to reach criterion for initial learning (IL) and seven subsequent reversals of a discrimination learning task in which reinforcement was associated with one (of two) spatially distinct choice ports. The side reinforced was switched between each successive learning problem. Data replotted from Mair et al., 1991b.

PTD rats have shown impairments in delayed conditional discrimination learning based on cues that are consistent with a deficient capacity for working memory, or the ability to remember information that must be updated on a trial by trial basis. When trained after treatment, PTD animals fail, even with extended training, to learn delayed non-matching to sample (DNMTS) as well as matching to sample tasks at levels readily attained by controls (Mair et al., 1985, 1988, 1991a). When trained prior to treatment, PTD treated animals have been found to be deficient in both spatial delayed alternation (Mair et al., 1985) and DNMTS (Knoth and Mair, 1991; Robinson and Mair, 1991). Following extensive pre-treatment training, rats given PTD treatment perform significantly worse than controls (and substantially below pre-treatment performance) when retrained at retention intervals varying from 0.1 to 15 seconds (Knoth and Mair, 1991; Robinson and Mair, 1991). Animals given PTD treatment and protected from lesions by MK-801 performed comparably to controls on all measures of DNMTS performance (Robinson and Mair, 1991).

An alternate approach to measuring DNMTS performance is to vary the retention interval to determine the critical delay at which a given level of accuracy is obtained (cf. Mair et al., 1990). Robinson and Mair (1991) used a modified staircase to do this, increasing the retention interval when animals performed at better than 75% accuracy and decreasing it when performance fell below this level. When tested after extensive pre-treatment and post-treatment training, rats recovered from PTD treatment were found to be significantly impaired, falling below 75% accuracy at shorter retention intervals, relative to either controls or rats given PTD treatment but protected from lesions by MK-801 (Figure 2). Analyses of individual performances showed two important features that were spared by this treatment. First, all animals were able to perform the task with at least 88% accuracy when retention intervals were relatively short. This shows a preserved capacity to perform procedural demands of the task. Second, the performance of every animal showed a consistent dependence on retention interval, improving when the interval was shortened and worsening when it was lengthened. This finding shows that DNMTS performance was critically dependent on retention interval and suggests that the impairment of the PTD group reflects an abnormal decay in working memory. Since the staircase method was applied to only one level of performance, these data do not distinguish between the possibilities that this decay results from poor performance at minimal delays (deficient encoding) or increased rate of temporal decay (rapid forgetting).

To determine the critical neural substrate causing the DNMTS deficits in the PTD model, Mair et al. (1990) pretrained rats on this task and then studied the effects of RF lesions in areas commonly associated with pathology in the PTD model. Lesions were directed at the following targets: medial portions of the IML thalamic site (within 0.6 mm of midline); lateral portions of the IML thalamic site (0.4 to 1.6 mm off midline); mammillary bodies; mammillary bodies combined with the medial IML site; and the fornix (at the level of posterior septum). Following recovery from surgery, animals with lateral IML lesions performed significantly worse than controls and all other lesion groups at retention intervals varying from 0.1 to 15 sec. In subsequent training with the staircase procedure, they fell below 75% accuracy at significantly shorter intervals than the controls and all of the other lesion groups (Figure 2). Like the PTD animals, rats with lateral IML lesions were able to perform DNMTS with high accuracy at short intervals and they showed a consistent relationship between accuracy and retention interval. Taken together, the data indicate that the DNMTS impairments of the PTD model can be fully replicated by RF lesions of the lateral IML site, and can be prevented by using MK-801 to protect animals in PTD from the thalmic lesions associated with this treatment.

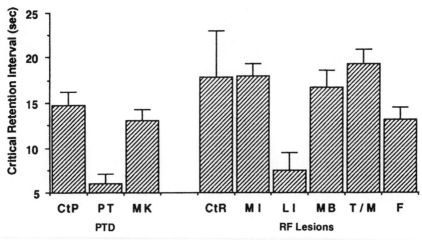

Fig. 2. Mean critical retention interval for 75% accuracy in DNMTS
as determined by the staircase procedure. Error bars
indicate standard error of the mean. Data are shown for:
a thiamine deficiency study including PTD (PT), control
(CtP), and PTD + MK-801 (MK) treatments (data from Robinson and Mair, 1991); and for a radiofrequency lesion study
including: sham operated controls (CtR), the medial IML
site (MI), the lateral (IML) site (LI), the mammillary
bodies (MB), medial IML combined with mammillary bodies
(T/M), and fornix (F) lesions (data from Mair, et al.,
1990).

THE MDn: WORKING MEMORY OR OLFACTION?

The lateral IML site includes central and lateral portions of MDn,
the full extent of centrolateral and paracentral nuclei, and portions of
adjacent nuclei. When this area is lesioned by either PTD or RF methods,
rats exhibit impairments of working memory, as measured by the place
DNMTS task. Although PTD treatment disrupts DNMTS performance and slows
the rate at which many tasks are learned, the extensive thalamic lesions
produced by this treatment apparently spare reference memory, leaving
intact the capacity to perform discriminations, once acquired. This
pattern of impairment is consistent with many of the effects of thalamic
lesions on the sense of smell that have been presumed to result from the
disruption of central olfactory mechanisms. Rats with MDn lesions are
reported to be slow in learning a variety of odor mediated tasks, although they are apparently unimpaired in their capacity to perform
olfactory discriminations when given sufficient training. Similarly,
humans with lesions affecting olfactory thalamocortical pathways are
impaired in their capacity to perform odor recognition memory and identification tasks, yet retain an ability to discriminate between weak
odorants and blanks, when given extended training in detection tasks.

The apparent parallel between deficits observed for olfactory and
non-olfactory tasks begs the question of whether the MDn should be
considered an olfactory structure. Does MDn serve as a specific conduit
for olfactory information to reach cortex, or does it process olfactory
information along with inputs from other sensory modalities in some
function related to working memory? The possibility even remains that it
might do both, with projections to agranular insular cortex serving a
primary function related to olfaction and projections to anterior
cingulate and area 2 of frontal cortex serving a function primarily

related to working memory (cf. Eichenbaum et al., 1983). These distinctions will only be resolved with more systematic studies of frontal cortical lesions and the development of olfactory tasks that can distinguish between deficits in working and reference memory.

REFERENCES

Allison, A.C., 1953, The morphology of the olfactory system in the vertebrates, Biol. Rev., 28:195-244.

Cramon, D.Y. von, Hebel, N., and Schuri, U., 1985, A contribution to the anatomical basis of thalamic amnesia, Brain, 108:993.

Eichenbaum, H., Clegg, R.A., and Feeley, A., 1983, Reexamination of functional subdivisions of rodent prefrontal cortex, Exp. Neurol., 79:434.

Eichenbaum, H., Shedlack, K.J., and Eckmann, K.W., 1980, Thalamocortical mechanisms in odor-guided behavior. I. Effects of lesions of the mediodorsal thalamic nucleus and frontal cortex on olfactory discrimination in the rat, Brain Behav. Evol., 17:255.

Eskenazi, B., Cain, W.S., Novelly, R.A., and Friend, K.B., 1983, Olfactory functioning in temporal lobectomy patients, Neuropsychologia, 3:365.

Eskenazi, B., Cain, W.S., Novelly, R.A., and Mattson, R., 1986, Odor perception in temporal lobe epilepsy patients with and without temporal lobectomy, Neuropsychologia, 24:553.

Gregson, R.A.M., Free, M.I., and Abbott, M.W., 1981, Olfaction in Korsakoffs, alcoholics and normals, Brit. J. Clinical Psych., 20:3.

Jones, B.P., Moskowitz, H., and Butters, N., 1975a, Olfaction discrimination in alcoholic Korsakoff patients, Neuropsychologia, 13:173.

Jones, B.P., Moskowitz, H., and Butters, N., and Glosser, G., 1975b, Psychophysical scaling of olfactory, visual, and auditory stimuli by alcoholic Korsakoff patients, Neuropsycholgia, 13:387.

Jones, B.P., Butters, N., Moskowitz, H., and Montgomery, K., 1978, Olfactory and gustatory capacities of alcoholic Korsakoff patients, Neuropsychologia, 16:332.

Jones-Gotman, M. and Zatorre, R.J., 1988, Olfactory identification deficits in patients with focal cerebral excision, Neuropsychologia, 26:387.

Langlais, P.J. and Mair, R.G., 1990, Protective effect of the glutamate antagonist MK-801 on pyrithiamine induced lesions and amino acid changes in rat brain, J. Neurosci., 10:1664.

Langlais, P.J., Mandel, R.J., and Mair, R.G., 1991, Thalamic lesions, learning impairments, intact retrograde memory in a rat model of Korsakoff's disease, unpublished manuscript.

Mair, R.G., Anderson, C.D., Langlais, P.J., and McEntee, W.J., 1985, Thiamine deficiency depletes cortical norephinephrine and impairs learning processes in the rat, Brain Res., 360:273.

Mair, R.G., Anderson, C.D., Langlais, P.J. and McEntee, W.J., 1988, Behavioral impairments, brain lesions, and monoaminergic activity in the rat following recovery from a bout of thiamine deficiency. Behav. Brain Res., 27:223.

Mair, R.G., Capra, C., McEntee, W.J., and Engen, T., 1980, Odor discrimination and memory in Korsakoff's psychosis, J. Exp. Psy.: Perc. Perf., 6:445.

Mair, R.G., Doty, R.L., Kelly, K., Wilson, C., Langlais, P.J., McEntee, W.J., and Vollmecke, T., 1986, Multimodal sensory discrimination deficits in Korsakoff's psychosis, Neuropsychologia, 24:831.

Mair, R.G., Knoth, R., Rabchenuk, S., and Langlais, P.J., 1991b, Impairments of olfactory, auditory, and spatial serial reversal learning in rats recovered from pyrithiamine induced thiamine deficiency (PTD) treatment, Behav. Neurosci., 105:360.

Mair, R.G., Lacourse, D.M., Koger, S.M., and Fox, G.D., 1990, In the rat, RF lesions of thalamus and fornix produce different patterns of impairment on a delayed non-matching to sample task, Neurosci. Abst., 16:608.

Mair, R.G., Otto, T., Knoth, R., Rabchenuk, S., and Langlais, P.J., 1991a, An analysis of aversively conditioned learning and memory in rats recovered from pyrithiamine induced thiamine deficiency (PTD) treatment, Behav. Neurosci., 105:351.

Markowitsch, H.J., 1982, Thalamic mediodorsal nucleus and memory: A critical evaluation of studies in animals and man, Neurosci. Biobehav. Rev., 6:351.

McEntee, W.J., Biber, M.P., Perl, D.P., and Benson, D.F., 1976, Diencephalic amnesia: A reappraisal, J. Neurol. Neurosurg. Psychiat., 39:436.

Milner, B., and Teuber, H-L., 1968, Alteration of perception and memory in man: Reflections in methods, in "Analysis of Behavioral Changes," L. Weiskrantz, ed., Harper and Row, New York.

Potter, H. and Butters, N., 1980, An assessment of olfactory deficits in patients with damage to prefrontal cortex, Neuropsychologia, 18:621.

Powell, T.P.S., Cowan, W.M., and Raisman, G., 1963, Olfactory relationships of the diencephalon, Nature, 199:710.

Robinson, J.K. and Mair, R.G., 1991, MK-801 protects rats from brain lesions and behavioral impairments following pyrithiamine-induced thiamine deficiency (PTD), Neurosci. Abst., 17:781.

Sapolsky, R.M. and Eichenbaum, H., 1980, Thalmo-cortical mechanisms in odor-guided behavior II. Effects of lesions of the mediodorsal thalamic nucleus and frontal cortex on odor preferences and sexual behavior in the hamster, Brain Behav. Evol., 17:276.

Slotnick, B.M. and Kaneko, N., 1981, Role of mediodorsal thalamic nucleus in olfactory discrimination learning in rats, Science, 214:91.

Squire, L.R. and Moore, R.Y., 1979, Dorsal thalamic lesion in a noted case of human memory dysfunction, Ann. Neurol., 6:506.

Staubli, U., Schottler, F., and Nejat-Bina, D., 1987, Role of dorsomedial thalamic nucleus and piriform cortex in processing olfactory information, Behav. Brain Res., 25:117.

Victor, M., Adams, R.D., and Collins, G.H., 1989, "The Wernicke-Korsakoff Syndrome," F.A. Davis, Philadelphia.

THE HIPPOCAMPUS AND THE SENSE OF SMELL

Howard Eichenbaum and Tim Otto

Department of Psychology
University of North Carolina at Chapel Hill
Chapel Hill, N.C.

Brodal, in a 1947 review from which we boldly borrow the title of the present paper, outlined some of the critical evidence that ultimately led to the demise of the notion that the hippocampus was a part of the olfactory brain, or "rhinencephalon" as it was called according to the prevailing view of the time. Since then it has become abundantly clear that the hippocampus processes information from many input sources (cf. Deacon et al., 1983). Nevertheless, converging data from neuroanatomical, physiological, and behavioral studies indicate that the olfactory system projects heavily onto and has especially immediate access to the hippocampal system, suggesting that the olfactory-hippocampal pathway may be particularly useful for explorations of sensory-limbic interactions leading to the higher order coding of perceptual information. As will be described below, the intimate anatomical associations between the olfactory and hippocampal systems are paralleled by 1) the critical role played by the hippocampal system in odor-guided learning and memory, 2) the strong influence of olfactory processing over the physiological activity in the hippocampus both at the level of rhythmic EEG activity and at the level of neuronal firing patterns, and 3) the role these physiological processes may play in the induction of synaptic plasticity supporting memory formation. Thus, in the spirit of a "renaissance of the rhinencephalon" (Macrides, 1977), we will argue that olfaction is a particularly advantageous model system for studies of "sensory" processing by the hippocampus across behavioral, neuronal, and synaptic levels of analysis. Our data on studies at each of these levels of analysis will be discussed in turn (see also Otto and Eichenbaum, 1992b).

THE NEUROPSYCHOLOGY OF HIPPOCAMPUS IN ODOR-GUIDED LEARNING

The projections of olfactory bulb and piriform cortex include three main targets in the limbic system: the amygdala, the dorsomedial thalamic nucleus and associated orbital prefrontal cortex, and the entorhinal cortex and hippocampus (for review see Otto and Eichenbaum, 1992a). Damage to, or olfactory deafferentation of, no one of these areas produces anosmia. However, each of these divergent central pathways supports critical aspects of odor perception and learning (Bermudez-Rattoni et al., 1982, 1983; Devor, 1973; Macrides et al., 1976; Eichenbaum et al., 1986; Slotnick, 1985; Eichenbaum et al., 1980; Staubli et al., 1984). Our efforts in recent years have been directed towards understanding the particular role

Chemical Signals in Vertebrates VI, Edited by R.L. Doty and
D. Müller-Schwarze, Plenum Press, New York, 1992

of the hippocampus in odor-guided learning. The results of these efforts indicate that damage to the hippocampal system produces a pattern of behavioral impairment that mirrors the global amnesia observed in humans following damage to the hippocampal system. Numerous recent studies have shown that amnesia in humans is a selective disorder of memory - that certain aspects of memory representation are severely impaired after damage to the hippocampal system while other aspects of learned performance are completely spared (Squire, 1987). The results of these studies indicate that amnesia is a deficit in declarative memory, the kind of memory representation that allows one to compare and contrast items in memory and to generate inferences from prior experience to novel situations. The kind of learning spared in amnesia is procedural memory, characterized as the acquisition of skills and adaptations that are expressible only through reactivation of the particular processes engaged during learning. We developed the hypothesis, consistent with a wide range of data on the consequences of hippocampal system damage in rats, monkeys and humans, that the hippocampal system participates in the construction of representations based on relations among items stored in memory (see Eichenbaum et al., 1992a,b for detailed accounts). Our view was that such a relational representation would support the comparison of items in memory, including those not previously experienced together, thus supporting a capacity for the use of memories in novel situations. An initial series of experiments was aimed at testing this hypothesis by comparing the performance of intact rats and rats with damage to the hippocampal system on odor-guided learning tasks in which the demands for comparison of cues and novel use of memories was varied.

To assess the importance of comparing memory cues we developed a series of tests in which different groups of intact rats and rats with disconnection of the hippocampal system produced by fimbria-fornix transection (FX) were presented with an identical set of odor discrimination problems, but the memory processing demands were varied so as to encourage or hinder rats in comparing and contrasting odor cues and hence to encourage or hinder the encoding of relations between them in memory (Eichenbaum et al., 1988). Thus, in one version of the task, odor cues were presented simultaneously and the response required comparison of alternative go-left or go-right choices (simultaneous odor discrimination; see Figure 1A). Under these conditions, FX rats were severely and persistently impaired in discrimination learning over the series of discrimination problems. Similar results are obtained when the hippocampaus is disconnected from olfactory input by ablation of the entorhinal cortex (Staubli et al., 1984). In the other version of the task, odor cues were presented separately across trials, hindering odor comparisons, and the instrumental requirement was simply to complete or discontinue a single behavioral response, eliminating the response choice (successive odor discrimination, see Figure 1B). Under these conditions, FX rats were superior to normal animals in learning the same discriminations they had failed to learn under different processing demands (see also Eichenbaum et al., 1986); again, similar results are obtained following lesions of entorhinal cortex (Otto et al., 1991b).

To further our understanding of the nature of differences in memory representation in intact and FX rats, we pursued a follow-up experiment using the simultaneous odor discrimination task (Eichenbaum et al., 1989). Our investigation exploited the finding that, although FX rats usually performed very poorly on this task, they occasionally succeeded in learning individual discrimination problems at least as rapidly as normal animals. We trained pairs of normal rats and FX rats on a series of discrimination problems until the FX rat of each pair learned two problems at the normal rate. We then challenged rats to use the learned odor representations in a novel situation by asking them to identify familiar odors in combinations not previously experienced. To do this we intermixed within a series of

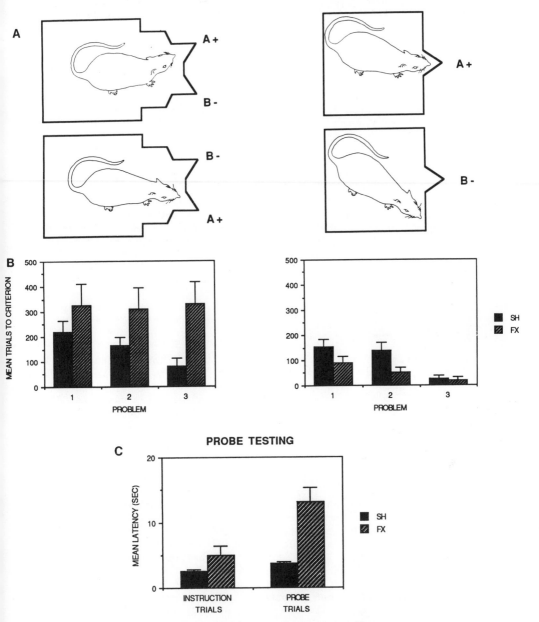

SIMULTANEOUS DISCRIMINATION

A

A +

B -

B -

A +

SUCCESSIVE DISCRIMINATION

A +

B -

B

MEAN TRIALS TO CRITERION

■ SH
▨ FX

PROBLEM

PROBLEM

PROBE TESTING

C

MEAN LATENCY (SEC)

INSTRUCTION
TRIALS

PROBE
TRIALS

■ SH
▨ FX

Figure 1. A. A schematic illustration of the two versions of odor discrimination training. In each case the rat must discriminate one S+ (in this case, odor "A") from one S- (odor "B"). B. Mean number of trials required to reach a criterion of 18 correct responses in 20 consecutive trials for a group of normal rats (SH) and a group of rats with lesions of the fornix (FX) across three problems. C. Performance of SH and FX rats on odor discrimination trials from the instruction problems or the probe trials composed of mispaired odors S+ and S- from the instruction trials. (From Eichenbaum et al., 1989).

concurrent trials on the two instruction problems occasional probe trials composed of an S+ odor from one problem "mispaired" with the S- odor from the other. Both normal and FX rats continued to perform well on the trials composed of the odor pairings used on instruction trials (Figure 1C), and normal rats performed accurately on the probe trials, even when they were first presented. In contrast, FX rats performed at chance levels on the probe trials when they were introduced, as if presented with novel stimuli. Thus, even when successful in learning, FX rats could demonstrate their memory for odors only in repetition of the learning event.

Confirming evidence for our account and additional insight into the nature of memory representation in FX rats was derived from the observation that FX rats had an abnormal pattern of stimulus sampling during simultaneous discrimination, as revealed by their response latencies. Each normal rat had a bimodal distribution of response latencies, and each mode was associated with one of the positions where the S+ was presented and the response was executed. Our interpretation of this pattern of response latencies is that normal rats consistently approached and sampled one odor port first, then either performed a nose-poke there, or approached and sampled the other odor port. In contrast, each FX rat had a unimodal distribution of response latencies, and the pattern was the same regardless of S+ position. This pattern of response latencies suggests that FX rats sample each odor pair presented as a stimulus compound. Note that this strategy would require that the rat be able to differentiate odor stimulus compounds made up of different configurations of the same odor pairs for each discrimination problem. For just those stimulus compounds that could be differentiated, FX rats succeeded in learning individual associations and employed them sequentially across trials, as they had done effectively in the successive odor discrimination task.

More recently we have employed another strategy for investigating the role of the hippocampus in encoding the relations among items in memory by developing odor-guided tasks that require (rather than merely encourage or hinder) explicit comparisons among multiple discriminative cues. Accordingly, we have developed an odor-guided continuous delayed non-match to sample (cDNM) task that is similar in memory demands to the visually-guided delayed non-match to sample task commonly used to investigate memory in primates. On each trial of odor guided cDNM, rats were presented with one odor chosen randomly from a relatively large set. Successful performance required that they remember across a variable memory delay the odor presented on the immediately preceding trial, and respond for water reinforcement only if the odor presented on the current trial was different (i.e., a non-match). Because correct performance in this task requires the comparison of current information to a stored representation of information presented previously, and so we predicted that rats with hippocampal system damage would be impaired. Indeed, the pattern of performance on cDNM by rats after aspiration of entorhinal and perirhinal cortex closely paralleled delay-dependent memory deficits characteristic of hippocampal-system damage in humans and non-human primates (Squire, 1987). Thus, at very short memory delays (imposing very little memory demand) rats with entorhinal/ perirhinal cortex lesions performed as well as normal subjects, but when longer memory delays were introduced, their performance fell dramatically (Otto & Eichenbaum, 1991).

THE BEHAVIORAL PHYSIOLOGY OF THE HIPPOCAMPUS IN ODOR-
GUIDED LEARNING

The above-described studies provide strong evidence that the hippocampus participates critically in the encoding of the relations among multiple cues. To reveal the way in which these features of odor memories are encoded by cellular activity within the hippocampus, we have employed a second line of investigation seeking to characterize hippocampal involvement in performance of odor-guided tasks. These studies involve two approaches to exploring the behavioral physiology of the hippocampus.

One approach has focused on the behavioral correlates of the hippocampal EEG during olfactory discrimination. Initial analyses revealed that cycles of hippocampal information processing and of stimulus sampling become entrained during odor discrimination. Macrides, Forbes, and Eichenbaum (1982) found that the theta rhythm, a 4-7 Hz sinusoidal EEG pattern reflecting the excitability cycles in hippocampal neurons (Rudell et al., 1980), synchronized with the sniffing cycle as the rat sampled the stimulus during odor discrimination performance. Further analyses indicated that the degree of entrainment was maximal just prior to achieving criterion performance or just after a reversal of stimulus valence, suggesting that the synchronized cycles of odor sampling and central excitability represent epochs of olfactory information processing within hippocampal circuitry.

Our second approach entails characterizations of the activity of single hippocampal neurons in relation to critical events in odor processing during learning in each of the three odor-guided tasks described above. In a series of studies we have identified several classes of cells that fire selectively during isolated behavioral events in each task. Of greatest interest was a class of CA1 pyramidal cells that were maximally active during the odor sampling period in each of these tasks (Eichenbaum et al., 1987; Wiener et al., 1989; Otto et al., 1991a). Some of these cells were active throughout the period of odor sampling and response generation. Other cells demonstrated quite striking specificities in the contingencies associated with increased firing. In simultaneous odor discrimination, these cells responded selectively during sampling of unique configurations of particular odor cues (Figure 2A); these same cells were relatively quiescent during the presentation of other configurations of the same cues and to configurations of other odors even though the rat occupied the identical spatial location on all such trials (Wiener et al., 1989). In successive odor discrimination the activity of some of this same class of cells was dependent upon specific odor sequences (Figure 2B); for example, some cells responded during the odor sampling period only on S+ trials that were preceded by S- trials (Eichenbaum et al., 1987). Thus in each of the two variates of olfactory discrimination, the functional correlates of hippocampal unit firing reflected those critical relations between odor stimuli that were appropriate for each task.

Further data supporting the hypothesis that the hippocampus participates in the processing of relationships among cues comes from a recent study (Otto et al., 1990) examining CA1 pyramidal cell activity during performance of the odor-guided cDNM task described above. These analyses revealed that during the odor sampling period some cells fired differentially on match trials or, alternatively, on non-match trials. Furthermore, the differential activity of most of these cells was independent of the particular odors involved. Thus the activity of these neurons reflected the outcome of comparisons (i.e. "same" or "different") between multiple stimuli rather than the specific combinations of past and present odor cues that were the subject of particular comparisons.

OLFACTORY LEARNING AND HIPPOCAMPAL SYNAPTIC PLASTICITY

Within the same studies in which we examined how memory representations are encoded by individual hippocampal neurons we have also tried to address directly the mechanisms of cellular plasticity that underlie the consolidation of these memories. The search for specific memory traces of

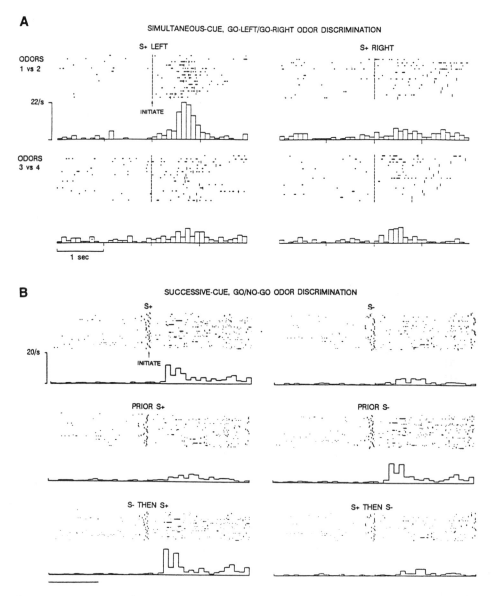

Figure 2. Raster diagrams and cumulative histograms of the firing of single hippocampal pyramidal cells during the odor sampling period in simultaneous (A) and successive (B) discrimination. Each raster line shows the time of occurrence of unit spikes as dots and the trial initiation and nose-poke as tic marks. The summary histogram is normalized to the maximal firing rate (spikes/sec) in a 100 msec bin. Note that in A this cell fired preferentially to a particular combination of an odor and its location (i.e. odor pair 1 vs 2, S+ on left). The cell depicted in B fired preferentially to a specific sequence of odors (i.e. and S+ preceded by an S-).

discrete learning events has daunted investigators of memory for decades. Any particular memory trace is quite likely a proverbial "needle in the haystack" of memories, in that it may be reasonably assumed that the fraction of cells participating critically in the storage of a single memory is infinitesimal, thereby making it technically infeasible to detect their associated lasting neurochemical and neurophysiological changes. Accordingly, the search for the physiological substrate of memory storage may best proceed within a system with a dense source of afferent fibers projecting to a brain area known to play a critical role in memory consolidation. This is indeed the case for the rodent olfactory-hippocampal pathway. Primary olfactory information reaches the hippocampus in as few as two synapses via a dense projection from entorhinal cortex (Otto and Eichenbaum, 1992a; Lynch, 1986). Correspondingly, as described above, hippocampal physiology is strongly engaged during odor learning, entrained to the cycle of stimulus sampling, and involved in encoding virtually every aspect of the learning event. Moreover, because we can identify particular cells activated during each of these events, we may be able to reveal, at the single cell level, mnemonic (storage) mechanisms associated with identified integrative (encoding) processes.

In evaluating the plausibility of any proposed physiological mnemonic storage device one must first consider whether its operating characteristics match those of behaviorally-defined memory. Thus, like memory itself, a physiological memory mechanism must be long-lasting, rapidly induced, strengthened by repetition, associative, and selective. Further, this mechanism must be found within brain regions critical to memory formation, and be blocked by manipulations which also impair memory. Accumulating evidence suggests that long-term potentiation (LTP), a selective strengthening of synaptic efficacy, meets each of these criteria (for review see Morris et al., 1990; Lynch 1986).

The discovery of LTP initiated two important lines of research attempting to relate LTP directly to memory. One approach has focused on evaluating whether the conditions optimal for LTP induction are characteristic of natural patterns of hippocampal activity. A second approach has sought to determine whether naturally-occurring alterations of hippocampal excitability occur as a result of learning. Our work has focused on the first of these, taking advantage of recent reports that hippocampal LTP is preferentially induced by a specific combination of three stimulation parameters: high-frequency bursts, activity 100-200 ms prior to a burst (Larson et al., 1986; Rose and Dunwiddie, 1986), and burst delivery near the positive peak of the ongoing dentate theta rhythm (Pavlides et al., 1988).

To determine whether any or all of these patterns of activity occur naturally during odor learning, we performed a detailed analysis of the temporal firing patterns of hippocampal neurons in rats engaged in simultaneous odor discrimination (Otto et al., 1991a). These analyses revealed that all of these three stimulation parameters are reflected simultaneously in the endogenous firing patterns of CA1 pyramidal cells, and that these patterns emerge preferentially during episodes of likely mnemonic processing (Figures 3A-D). Specifically, many CA1 pyramidal cells discharged in high-frequency bursts during the period of stimulus sampling and analysis. Further, for the overwhelming majority of these cells, the bursts occurred near the positive peak of the ongoing dentate theta rhythm. Finally, these bursts were typically preceded by neural activity at latencies of 100-200ms. These three patterns of activity, collectively referred to as "theta-bursting" (Lynch et al., 1988), were observed to occur far more often during odor sampling and analysis than during any other behavioral event. Thus, at least in the hippocampus, the conditions appropriate for inducing LTP commonly occur during olfactory learning, time-locked to behavioral events associated with mnemonic processing.

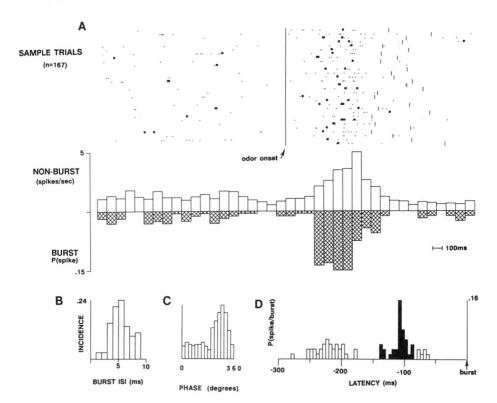

Figure 3. Analyses of a CA1 neuron that theta-burst during odor sampling.
A. Raster diagram of single spike (small dots) and burst (large dots) firing across trials. Summary histograms indicate average firing rates for
non-burst (spikes/sec) and burst (probability of a spike/bin) activity.
B. Distribution of interspike intervals (ISI) within identified bursts.
C. Distribution of spikes within the theta cycle (0=dentate theta peak).
Bursts tend to occur just before the dentate theta peak. D. Distribution
of firing preceding a burst. The darkened portion represents latencies
corresponding to frequencies between 7 and 12.5 Hz. A chi-square analysis
revealed a significant tendency for firing to precede bursts at latencies
within this interval ($X2 = 139.0$, $p < .001$) (Taken from Otto et al.,
1991a).

While we consider it likely that theta-bursting could induce alterations in synaptic strength, it is at present unclear whether naturally-
occurring synaptic enhancement actually accompanies these patterns, and if
so, where the anatomical locus of this change occurs. With respect to the
locus of synaptic enhancement associated with theta-bursting, two possibilities come immediately to mind. First, it is possible that theta-bursting
in CA1 cells reflects the result of afferent (CA3) activity, and by hypothesis an enhancement of synaptic efficacy in the CA1 region itself. This
account is consistent with views suggesting that synaptic enhancement
within the hippocampus serves as a memory "buffer", temporarily storing
memory prior to (or perhaps in addition to) to its permanent storage elsewhere (Rawlins, 1985), or as an "index" to the location of long-term memories stored in neocortex (Teyler and DiScenna, 1986). Alternatively,
theta-bursting might reflect processes leading to potentiation in CA1
targets, for example, the anterior olfactory nucleus (Van Groen and Wyss,

1990) or neocortex (Jay et al., 1989). Consistent with this hypothesis is the observation of LTP in rat prefrontal cortex induced by tetanic stimulation of the CA1/subicular region of temporal hippocampus (Laroche et al., 1990).

CONCLUSIONS

The combination of characteristics of hippocampal-dependent memory emphasized here, relational representation and representational flexibility, suggests that the hippocampal system supports an organizational scheme for experiences in odor-guided learning that may be imagined as an "odor space", a map of significant relations among odor-items (and other items) in memory that permits the inferential use of memories in novel situations. The functional correlates of hippocampal slow wave and neuronal activity are consistent with this account and suggest that the hippocampal representation of odor space is encoded in terms of conjunctions of specific multiple odor-events that occur across time or space. Finally, our findings also provide compelling evidence that conditions suitable for the natural enhancement of synaptic strength are found in the hippocampus during olfactory learning, and that these patterns of activity emerge preferentially during periods of mnemonic processing. It is at present unclear, however, whether these patterns of neural activity are associated with increases in hippocampal synaptic efficacy, and indeed where the locus of odor memories lies. While these and many other important questions remain unanswered, it is clear that the olfactory-hippocampal system provides a fruitful domain for their investigation.

REFERENCES

Bermudez-Rattoni, F., Keifer, S.W., Grijalva, C.V., and Garcia, J., 1982, Basal and central amygdala involvement in the acquisition of taste and odor aversions, Soc. Neurosci. Abst., 4, 501.
Bermudez-Rattoni, F., Rusiniak, K.W., and Garcia, J., 1983, Flavor-illness aversions: Potentiation of odor by taste is disrupted by application of novocaine to the amygdala, Behav. Neurol. Biol., 37:61-75.
Brodal, A., 1947, The hippocampus and the sense of smell. A review, Brain, 70:179-222.
Devor, M., 1973, Components of mating behavior dissociated by lateral olfactory tract transection in male hamsters, Brain Res, 64:437-441.
Eichenbaum, H., Cohen, N.J., Otto, T., and Wible, C.G., 1992b, Memory representation in the hippocampus: Functional domain and functional organization. In Memory: Organization and Locus of Change, L.R. Squire, G. Lynch, N.M. Weinberger, and J.L. McGaugh, Eds., Oxford University Press.
Eichenbaum, H., Fagan, A., and Cohen, N.J., 1986, Normal olfactory discrimination learning set and facilitation of reversal learning after medial-temporal damage in rats: Implications for an account of preserved learning abilities in amnesia, J. Neurosci., 6: 1876-1884.
Eichenbaum, H., Fagan, A., Mathews, P., and Cohen, N.J., 1988, Hippocampal system dysfunction and odor discrimination learning in rats: Impairment or facilitation depending on representational demands, Behav. Neurosci., 102:331-339.
Eichenbaum, H., Kuperstein, M., Fagan, A., and Nagode, J., 1987, Cue-sampling and goal-approach correlates of hippocampal unit activity in rats performing an odor-discrimination task, J. Neurosci., 7:716-732.
Eichenbaum, H., Mathews, P., and Cohen, N.J., 1989, Further studies of hippocampal representation during odor discrimination learning. Behav. Neurosci., 103:1207-1216.
Eichenbaum, H., Otto, T., and Cohen, N.J., 1992, The hippocampus: What does it do? Ann. Rev. Neurosci., in press.

Eichenbaum, H., Shedlack, K., and Eckmann, K., 1980, Thalamocortical mechanisms in olfaction, I. Effects of lesion of the mediodorsal thalamic nucleus and frontal cortex on olfactory discrimination in the rat, Brain Behav. Evol. 17:255-275.

Jay, T.M., Glowinski, J., and Thierry, A.M., 1989, Selectivity of the hippocampal projection to the prelimbic area of the prefrontal cortex in the rat, Brain Res. 202:337-340.

Laroche, S., Jay, T.M., and Thierry, A., 1990, Long-term potentiation in the prefrontal cortex following stimulation of the hippocampal CA1/subicular region, Neurosci. Letters, 114:184-190.

Larson, J., and Lynch, G., 1986, Induction of synaptic potentiation in hippocampus by patterned stimulation involves two events, Science, 232:985-988.

Larson, J., Wong, D., and Lynch, G., 1986, Patterned stimulation at the theta frequency is optimal for the induction of hippocampal long-term potentiation, Brain Res. 441:111-118.

Lynch, G., 1986, Synapses, Circuits, and the Beginnings of Memory. Cambridge MA:MIT Press.

Lynch, G., Muller, D., Seubert, P., and Larson, J., 1988, Long-term potentiation: Persisting problems and recent results, Brain Res. Bull. 21:363-372.

Macrides, F., 1977, Dynamic aspects of central olfactory processing. In Chemical Signals in Vertebrates, D. Muller-Schwarze and M.M. Mozell (Eds.). New York: Plenum, pp. 449-514.

Macrides, F., Eichenbaum, H.B., and Forbes, W.B., 1982, Temporal relationship between sniffing and the limbic theta rhythm during odor discrimination reversal learning, J Neurosci. 2:1705-1717.

Macrides, F., Firl, A.C., Schneider, S.P., Bartke, A., and Stein, D.G., 1976, Effects of one stage or serial transections of the lateral olfactory tracts on behavior and plasma testosterone levels in male hamsters, Brain Res., 109:97-109.

Morris, R.G.M., Davis, S., Butcher, S.P., 1990, Hippocampal synaptic plasticity and NMDA receptors: A role in information storage? Philosophic Trans. Roy. Soc., London, 329, 187-204.

Otto, T., and Eichenbaum, H., 1991, Dissociable roles of orbitofrontal cortex and the hippocampal system in an odor-guided delayed non-matching to sample task, Soc. Neurosci. Abst., 17.

Otto, T., and Eichenbaum, H., 1992a, Olfactory learning and memory in the rat: A "model system" for studies on the neurobiology of memory. In The Science of Olfaction, M. Serby, K. Chobor (Eds). New York: Springer-Verlag, in press.

Otto, T., and Eichenbaum, H., 1992b, Toward a comprehensive account of hippocampal function: Studies of olfactory learning permit an integration of data across multiple levels of neurobiological analysis. In Neuropsychology of Memory, N. Butters & L.R. Squire (Eds.), In Press.

Otto, T., Eichenbaum, H., Wiener, S.I., and Wible, C.G., 1991a, Learning-related patterns of CA1 spike trains parallel stimulation parameters optimal for inducing hippocampal long term potentiation, Hippocampus, 1, 181-192.

Otto, T., Eichenbaum, H., and Wible, C.G., 1990, Behavioral correlates of hippocampal unit activity in an odor-guided delayed non-matching to sample task, Soc. Neurosci. Abst., 16, 263.

Otto, T., Schottler, F., Staubli, U., Eichenbaum, H., and Lynch, G., 1991b, The hippocampus and olfactory discrimination learning: Effects of entorhinal cortex lesions on learning-set acquisition and on odor memory in a successive-cue, go/no-go task, Behav. Neurosci., 105, 111-119.

Pavlides, C., Greenstein, Y.J., Grudman, M., and Winson, J., 1988, Long-term potentiation in the dentate gyrus is induced preferentially on the positive phase of theta rhythm, Brain Res., 439, 383-387.

Rawlins, J.N.P., 1985, Associations across time: The hippocampus as a temporary memory store, Brain Behav. Sci., 8, 479-496.

Rose, G.M., and Dunwiddie, T.V., 1986, Induction of hippocampal long-term potentiation using physiologically-patterned stimulation, Neurosci. Lett., 69, 244-248.

Rudell, A., Fox, S., and Ranck, J.B., Jr., 1980, Hippocampal excitability phase-locked to the theta rhythm in walking rats, Exp. Neurol., 68, 87-96.

Slotnick, B.M., 1985, Olfactory discrimination in rats with anterior amygdala lesions, Behav. Neurosci., 99, 956-963.

Squire, L.R., 1987, Memory and Brain. New York, NY: Oxford.

Staubli, U., Ivy, G., and Lynch, G., 1984, Hippocampal denervation causes rapid forgetting of olfactory information in rats, Proc. Nat. Acad. Sci. USA, 81, 5885-5887.

Teyler, T.J., and DiScenna, P., 1986, The hippocampal memory indexing theory, Behav. Neurosci, 100, 147-154.

VanGroen, T., and Wyss, J.M., 1990, Extrinsic projections from area CA1 of the rat hippocampus: Olfactory, cortical, subcortical, and bilateral hippocampal formation projections, J. Comp. Neurol., 303, 1-14.

Wiener, S.I., Paul, C.A., and Eichenbaum, H., 1989, Spatial and behavioral correlates of hippocampal neuronal activity, J. Neurosci., 9, 2737-2763.

THE STRUCTURE OF ENVIRONMENTAL ODOR SIGNALS: FROM TURBULENT DISPERSION TO MOVEMENT THROUGH BOUNDARY LAYERS AND MUCUS

Paul A. Moore[1], Jelle Atema[2], and Greg A. Gerhardt[1]

[1]Departments of Psychiatry and Pharmacology, University of Colorado Health Sciences Center, 4200 East Ninth Ave., Denver, CO 80262, [2]Boston University Marine Program, Marine Biological Laboratory, Woods Hole, MA 02543

INTRODUCTION

Chemical signals play a major role in the lives of many animals, such as orientation to odor sources, the identification of food, and selection of mates. During the journey from the odor source (e.g., food source) to the "receiver" (either receptor appendage, organ, or cell), the chemical signal is influenced by many different dispersal processes including molecular diffusion, bulk advection, boundary layer flows, and impulsive flows created from sampling structures. The purpose of this report is to outline the present state of knowledge of the structure of chemical signals under the processes mentioned above and how these function together to form the environmental odor signals to which animals respond.

The information and data presented in the first two sections are gathered from aquatic situations. However, these results obtained in an aquatic medium are applicable to terrestrial environments. The major difference between the physical dynamics of turbulent dispersal in water and air are time and space scale differences which can be bridged with a scaling factor (Vogel, 1981). The aquatic environment provides an easier medium in which to measure and model odor distributions due to smaller spatial scales and better measuring techniques. Additionally, diffusion calculations and measurements are easier in water because there is no partitioning of stimulus molecules at air-water/mucus interfaces.

THE STRUCTURE OF TURBULENT ODOR PLUMES

For years, many studies assumed that odor dispersion could be effectively described by time-averaged or Gaussian distribution models (Sutton, 1953; Bossert and Wilson, 1963). These models work for those organisms that operate either at small spatial scales, e.g., bacteria (Berg and Purcell, 1977), or sample over long time intervals, e.g., tsetse flies (Bursell, 1984). For animals that operate at other time and space scales, such models are a poor predictor of animal behavior (Elkinton et al., 1984). Murlis and Jones (1981) and Moore and Atema (1991) have shown that odor distributions are quite heterogeneous as compared to the time-averaged models.

The structure of odor plumes is mainly influenced by turbulence produced by the mechanical forces within a moving fluid. These mechanical forces create large scale eddies (compared to the initial size of the odor plume) that transfer their energy to successively smaller eddies until the energy is dissipated. This cascade of eddie sizes is called the Kolmogoroff scale and has a lower size limit which is determined by friction, viscous forces, and fluid velocity (Pedlosky, 1987). Below this limit, molecular diffusion is the dominant dispersal process.

The spatial range of eddies in the environment is important because it is the interaction between the turbulent eddies and the odor plume that creates the size and length of concentration fluctuations within the odor plume (Miksad and Kittredge, 1979). As the plume travels down-current, it expands relative to the size of eddies within the fluid medium. Initially, when the plume diameter is smaller than the smallest eddies, they cause the plume to meander. As the plume size expands to match the scale of eddies present, the plume is broken into separate patches of odor. This results in a fluctuating odor signal (Murlis and Jones, 1981; Moore and Atema, 1991). The final stage of plume growth occurs when the plume expands to sizes larger than the largest eddies. At this point, eddies begin to redistribute the odor within single patches and begin to homogenize the odor between patches. This results in signals that fluctuate less and have fewer periods of no concentration (Murlis and Jones, 1981; Murlis, 1986; Moore and Atema, 1991).

Although turbulent odor plumes produce odor signals that seem chaotic, there are biological and computational methods of analysis that can track changes occurring within the structure of odor plumes (Moore and Atema, 1988). Which structural features are detectable and used by the animal for information about it's chemical environment? To answer this question and form hypotheses, we must measure different turbulent odor signals at biologically relevant time and space scales and then analyze those results with biologically relevant spatial and temporal filters. Only then can those hypotheses be tested with specific behavioral studies.

EFFECTS OF BOUNDARY LAYERS ON ODOR SIGNALS

In order to sense odor patterns, an organism must have a chemo-receptive organ within the flow field delivering the odor signal. Any receptor structure placed in a moving fluid (e.g., vertebrate nose, reptilian tongue, catfish barbel) has a boundary layer surrounding it due to the "no-slip" condition along the surface of the organ (Vogel, 1981; Tritton, 1977). The "no-slip" condition states that fluid velocity at any solid surface is, for all practical purpose, zero. Thus, a gradient of decreasing fluid velocities is formed between the free flow velocity a great distance away from the surface and zero velocity at the surface. Within this boundary layer, turbulence dominated dispersal slowly gives way to diffusion dominated processes. The changing dispersal processes result in a change in the structure of the chemical signal (Moore et al., 1991).

The degree of change (or filtering) will depend upon the boundary layer thickness which in turn depends upon the physical conditions of tne flow within and around the receptor structures. The flow conditions will depend upon the morphology of the receptor structure and fluid velocities. Therefore, knowledge of boundary layer conditions of morphologically different receptor structures is needed to understand the structure of chemical signals arriving at receptor cell surfaces.

Recent studies have shown that the morphology of each receptor structure results in a distinct boundary layer (Moore et al., 1991). The boundary layer surrounding a receptor structure will act as a (low pass) smoothing filter for incoming odor signals; many aspects of the odor signal are changed by the boundary layer (Moore et al., 1991: Figures 3 and 6). Since the thickness and structure of the boundary layer will depend upon the morphology of the chemoreceptor appendage and the fluid velocity, identical odor pulses in the environment will be different in the microscale environment of different chemosensory appendages.

Animals can control the filtering produced by the boundary layer by either altering the morphology of receptor structures or controlling the fluid motion surrounding receptor structures by actively sampling the environment. There are many examples of animals controlling fluid motions by precise or stereotyped sampling behaviors. Crustaceans can control the boundary layer thickness by flicking their antennules (Snow, 1973). Catfish will periodically flick their barbels rapidly and vertebrates (aquatic and terrestrial) will sniff, thereby increasing the fluid velocities in the nasal cavity.

Periodic sampling can create a digitized record of the environment. This is different from the olfactory organ ventilation in some fish (isosmates; see Døving et al., 1977), which have continual movement of water over the appendage or receptor cell surfaces. We hypothesize two possible functions of chemosensory sampling. One, digitized sampling can enhance information extraction and receptor filtering from a turbulent odor signal. Two, discrete sampling is a necessary consequence to deliver fluid flow to the low flow environment of dense receptor packing. Preliminary computer simulation of discrete and continuous sampling on the extraction of information from odor signals shows that there is no difference in the type of information available under these two sampling methods (Moore and Atema, 1988). Although this is a preliminary analysis, this would indicate that discrete sampling is primarily used as a method of delivering stimulus to a large number of receptor cells, and not to enhance information extraction. Further experimentation is needed to test this hypothesis.

MOVEMENT AND DIFFUSION THROUGH MUCUS TO RECEPTOR SITES

After the signal chemical passes through the boundary layer and before it can bind to a receptor site, it must move through the medium surrounding the receptor dendrites. In most vertebrates, this medium is mucus. (In most invertebrates, this is receptor lymph). The mixed fluid/gelatinous layers that make up the mucus and overlay olfactory epithelium are where many perireceptor processes occur (Getchell et al., 1984). These processes, which include diffusion, potential binding of signal chemicals to transporter proteins, and degradation of chemicals by enzymes, can greatly effect the structure of the chemical signal before it binds to receptor sites. Movement through the mucus layer is mainly by diffusion, but until recently the diffusion coefficients of chemicals within the mucus layer have been unknown. The accurate diffusion coefficients of chemicals are important for determining chemical dynamics within the mucus layer, including boundary layer thickness and the location of receptor binding sites.

Preliminary measurements of the diffusion coefficients of a tracer (dopamine) and an odorant (vanillin) in the mucus layers of spotted salamanders are 0.29×10^{-5} and 0.21×10^{-5} cm^2/s, respectively (Friedemann et al., 1991). These values are an order of magnitude slower than those previously used for models of odorant diffusion (Getchell et al., 1984;

DeSimone et al., 1981). In fact, the solution diffusion coefficients used in previous models were at least a factor of 3 faster than values for small organic molecules (Gerhardt and Adams, 1982). This has several important implications for the structure of odor signals entering the microenvironment around receptor sites. With these diffusion coefficients, odor signals within the mucus layer will be slow-rising and long-lasting odor pluses. Because most receptor cells only respond during the initial part of an odor presentation and then adapt, response thresholds of receptor cells (as number of molecules) may be lower than what is now believed. In addition, with latencies of 25 to 2000 ms in salamander olfactory cells (Getchell et al., 1980), odor molecules will only travel approximately 1-11 μm into the mucus in this amount of time. Thus, receptor sites have to be located within these depths. Finally, if receptor sites are located this close to the air-mucus interface, odorant binding proteins may play a more important role in the removal of odorants from receptor sites than delivery of odorants to them. Thus, these measurements have shown that the chemical dynamics of odorants in mucus layers is different than previous models have protrayed. Further work is needed in this area for a more thorough understanding of chemical dynamics.

SUMMARY

The structure of odor signals as they arrive at receptor sites is due to many processes, e.g. turbulence, boundary layers, diffusion. Each of these processes plays a dominant role during some point in the transport of chemicals from the source to the receptor site. At the macroscopic level, turbulence results in a patchy odor signal in both space and time. As the chemical stimulus approaches the receptor organ or appendage, turbulence gives way to first laminar flow and then diffusion processes. This transition occurs in the boundary layers surrounding the receptor organ or appendage. In this boundary layer, the patchiness in the odor signal caused by turbulence is modified and reduced by the slower flow regime in the boundary layer. (Essentially, the fluctuating odor signal is "smoothed" as it passes through the boundary layer.) Once through the boundary layer, the odor must diffuse through the medium surrounding the actual receptor sites. This process further slows down the movement of chemicals to receptor sites. Thus, the odor signal that was seemingly chaotic in the free flow of the environment is extensively low-pass filtered before it reaches the receptor sites. Detailed knowledge of how each of these processes affects the structure of odor signals is critical to understanding the true structure of environmental odor signals and the biologically relevant stimulus energies for chemoreceptors.

ACKNOWLEDGEMENTS

This work was supported by a BUMP Alumni research award and ADMH Drug Abuse Training Fellowship (AA07464) to P.M., an NSF grant (BNS-8812952) to J.A., and USPHS (AG00441 and AG06434) to G.G.

REFERENCES

Atema, J., 1985, Chemoreception in the sea: adaptation of chemoreceptors and behavior to aquatic stimulus conditions, Soc. Exp. Biol. Symp., 39:387-423.
Berg, H.C., and Purcell, E.M., 1977, Physics of chemoreception, Biophys. J., 20:193-219.

Bossert, W.H., and Wilson, E.O., 1963, The analysis of olfactory communication among animals, J. Theor. Biol., 5:443-469.

Bursell, E., 1984, Observations on the orientation of tsetse flies (Glossina pallidipes) to wind-borne odours, Physiol. Entomol., 9:133-137.

DeSimone, J.A., Heck, G.L., and Price, S., 1981, Physiochemical aspects of transduction by chemoreceptor cells, in: Perception of Behavioral Chemicals, C.M., Norris, ed., Elsevier Biomedical Press, NY, NY.

Døving, K.B., Dubois-Dauphin, M., Holley A., and Jourdan, F., 1977, Functional anatomy of the olfactory organ of fish and the ciliary mechanism of water transport, Acta. Zool. (Stockh), 58:245-255.

Elkinton, J.S., Cardé, R.T., and Mason, C.J., 1984, Evaluation of time-average dispersion models for estimating pheromone concentration in a deciduous forest, J. Chem. Ecol., 10:1081-1108.

Friedemann, M.N., Moore, P.A., Finger, T.E., Silver, W.L., and Gerhardt, G.A., 1991, Perireceptor events: direct determination of diffusion coefficients in the olfactory mucus layers of salamanders, Chem. Senses (Achems Abst.), (In press).

Gerhardt, G.A., and Adams, R.N., 1982, Determination of diffusion coefficients by flow injection analysis, Anal. Chem., 54:2618-2620.

Getchell, T.V., Margolis, F.L., and Getchell, M.L., 1984, Perireceptor and receptor events in vertebrate olfaction, Prog. Neurobiol., 23:317-345.

Getchell, T.V., Heck, G.L., DeSimone, J.A., Price, S., 1980, The location of olfactory receptor sites: inferences from latency measurements, Biophys. J. 29:397-412.

Miksad, R.W., and Kittredge, J., 1979, Pheromone aerial dispersion: a filament model, 14th Conf. Agric. For. Met., Am. Met. Soc., 1:238-243.

Moore, P.A., and Atema, J., 1988, A model of a temporal filter in chemoreception to extract directional information from a turbulent odor plume, Biol. Bull., 174:355-363.

Moore, P.A., and Atema, J., 1991, Spatial information in the three-dimensional fine structure of an aquatic odor plume, Biol. Bull., (Submitted).

Moore, P.A., Atema, J., and Gerhardt, G.A., 1991, Fluid dynamics and microscale chemical movement in the chemosensory appendages of the lobster, Homarus americanus, Chem. Senses, (Submitted).

Moore, P.A., Gerhardt, G.A., and Atema, J., 1989, High resolution spatio-temporal analysis of aquatic chemical signals using microelectro-chemical electrodes, Chem. Senses, 14:829-840.

Murlis, J., and Jones, C.D., 1981, Fine-scale structure of odour plumes in relation to insect orientation to distant pheromone and other attractant sources, Phys. Ent., 6:71-86.

Murlis, J., 1986, The structure of odour plume, in: Mechanisms in Insect Olfaction, T.L. Payne, M.C. Birch, and C.E.J., Kennedy, eds., Claredon Press.

Pedlosky, J. 1987, Geophysical Fluid Dynamics, Springer-Verlag.

Snow, P.J., 1973, The antennular activities of the hermit crab, Pagurus alaskiensis (Benedict), J. Exp. Biol., 58:745-766.

Sutton, O.G., 1953, "Micrometeorology," McGraw-Hill, N.Y., N.Y.

Tritton, D.J., 1977, "Physical fluid dynamics," Van Nostrand Reinhold (UK) Co, Ltd., Wokingham, England.

Vogel, S., 1981, "Life in moving fluids: the physical biology of flow," Princeton University Press, Princeton, N.J.

THE TRIGEMINAL NERVE SYSTEM AND ITS INTERACTION WITH OLFACTORY AND TASTE SYSTEMS IN FISHES

Galina V. Devitsina and Lila S. Chervova

Department of Ichthyology, Biological Faculty, Moscow State University, Moscow, Russia

INTRODUCTION

The trigeminal nerve of the Vertebrata is the most sensitive nerve in the head region. As in the case of higher vertebrates, this nerve is well developed in fishes and is represented by three branches - the nervus maxillaris, the nervus mandibularis, and the nervus ophthalmicus superficialis. In contrast to terrestrial Vertebrata, little study has been made of the structure and function of the trigeminal nerve in fishes. Its wide-spread development in fishes suggests it has substantial sensory capacities and important biological functions.

MATERIALS AND METHODS

Two species were studied - the freshwater carp, Cyprinus carpio L., and the White Sea cod, Gadus morhua marisalbi Derjugin. The local projections of the trigeminal nerve branches to the Gasserian ganglion were studied by marking the neurons with a cobalt labelling technique (Chervova and Devitsina, 1981). Functional properties of the trigeminal nerve were investigated electrophysiologically by recording the impulse activity of single nerve fibers and whole nerves. Chemical solutions (0.5 ml) were infused into the olfactory sac via a canula through which a constant flow of water was maintained (Chervova et al., 1985). Mechanical stimulation was applied by touching the olfactory sac and skin surface with a fine rod or by falling water drops. Thermostimulation was done by changing the temperature of the water fed into thermostatically controlled canula.

RESULTS AND DISCUSSION

Spontaneous impulse activity was recorded in single trigeminal nerve fibers of both species. Such activity was either regular or irregular and differed in frequency from 0.1 to 40 impulses/sec. A large number of "silent" fibers were also found. Up to 90% of the fibers were extremely sensitive to tactile stimulation, and the frequency of the responses was related to the intensity of the mechanostimulation. In response to the stimulation of the olfactory epithelium and the surrounding skin surface, the silent and spontaneously active fibers produced intense trains of

Chemical Signals in Vertebrates VI, Edited by R.L. Doty and
D. Müller-Schwarze, Plenum Press, New York, 1992

impulses which generally did not exceed the time of stimulation. In the experiments with parallel recording of the electrical activity of the olfactory and trigeminal fibers, water drops falling at a frequency of 60/min were used for tactile stimulation. Interestingly, both systems responded to each drop with a series of impulses. When the frequency of drops was doubled, the olfactory system retained its rhythm while the trigeminal system produced a new one. With further increases of stimulation frequency, the trigeminal response rhythm also correspondingly increased, whereas in the olfactory fibers a prolonged inhibition occurred.

Among the trigeminal nerve fibers were ones which reacted to the change in the water temperature around the head. A temperature increase 3-5°C above that to which a fish was adapted caused an impulse increase, whereas a temperature decrease caused inhibition. The greater the difference in temperature, the shorter the latency period and the more intense the reaction. Among the thermosensitive fibers we studied, there was a large group with high differential thresholds. They were, as a rule, both thermo- and mechanosensitive. Ones that were only thermosensitive occurred more rarely (5%) and their differential threshold was not high (1°C). Some of these responded only to a temperature increase or to a temperature decrease.

To study the chemosensitivity of the trigeminal nerve, we selected the smallest branches which innervated the olfactory organ and adjacent tissue. While registering the activity of these branches, we discovered that all the chemical stimuli we applied produced trigeminal responses. The typical reaction to a chemostimulus was a prolonged change in the impulse activity (20-25 sec). Different substances caused reactions differing in length and intensity, and a selective sensitivity to some irritants was noticed. For example, clear responses to agents excreted by both stressed and unstressed conspecifics, extracts and infusions of common food stuffs, and some amino acids were observed. There were also substantial reactions to changes in the salinity of water. The effectiveness of the applied stimuli was the same for the two branches of the trigeminal nerve we investigated (Fig. 1).

In recording the activity of single fibers, a large number were found to be both chemo- and mechanosensitive. Others were specialized chemosensory fibers. Fibers differed in both sensitivity and response spectra. As a rule, most responses were those of increased impulse frequency, although the increase was followed by depression and vice versa. Different stimulants (for instance, two amino acids) could cause different frequency and duration responses in the same fiber. At the same time, the same stimulant could produce both depression and stimulation in different fibers, and some fibers were capable of distinguishing not only between separate amino acids, but between their stereoisomers.

The ability of the trigeminal system to perceive and differentiate among the amino acids and other naturally-occuring stimuli illustrates its participation in the analysis of biological relevant signals of the environment, such as those found in exometabolites of all the hydrobionts and also those contained in food and in intra- and interspecific signals of fishes. The water medium facilitates the simultaneous influence of chemical signals on all chemosensory systems. This is one reason for our interest in interactions among such systems and their ability to regulate and change the reactions of one another.

We also found that the activity of the trigeminal system effects the function of the olfactory and taste systems. Electrostimulation of the n. maxillaris and n. mandibularis branches of the trigeminal nerve

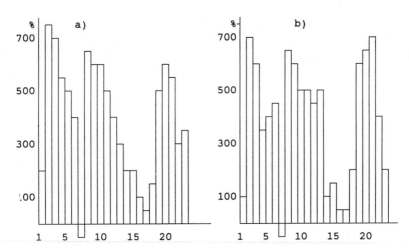

Fig. 1. Relative response intensity in the trigeminal nerve
branches under chemostimulation. (a) R. nasalis, (b) r.
buccalis V.1 - sea water, 2 - fresh water, 3 - camphor
0.001 M, 4 - sucrose 0.01 M, 5 - acetic acid 0.001 M, 6 -
sodium chloride 0.01 M, 7 - quinine 0.0001 M, 8 - L-glutamic acid 0.001 M, 9 - L-asparagine 0.001 M, 10 - L-
alanine 0.001 M, 11 - L-histidine 0.001 M, 12 - L-serine
0.001 M, 13 - L-arginine 0.001 M, 14 - β-phenylalanine
0.001 M, 15 - L-tryptophane 0.001 M, 16 - D-arginine 0.001
M, 17 - D-leucine 0.001 M, 18 - vanillin 0.001 M, 19 -
water containing conspecific exometabolites, 20 - water
containing conspecific exometabolites of stressed fish, 21
- extract of lugworm, 22 - extract of mussels, 23 - extract of pine bark.

modified the background and the elicited activity of the branches of the
facial nerve responsible for taste sensitivity. Transection of the
rootlets of the trigeminal nerve eliminates these effects, suggesting an
absence of interaction between the trigeminal and facial nerves at the
level of the Gasserian ganglion. When both nerves are intact, the
application of GABA to the trigeminal nerve nuclei area in the medulla
eliminated the effect of trigeminal stimulation. When the brain region
was washed clean, the effects were restored. These data indicate that
the interaction between the trigeminal and facial nerves occurs centrally. In the experiments on transecting the trigeminal branches innervating the olfactory organ and on recording the electrical responses in
the olfactory tracts, it was observed that the olfactory system is under
a constant inhibitory influence of the trigeminal system. The olfactory
system, in turn, also affected the activity of the trigeminal system.
Thus, afferent activity in the central endings of the trigeminal nerve is
sharply reduced after cutting the ipsilateral olfactory tracts.

While studying the innervation of the olfactory organ in the two
species we discovered that there exists a double innervation of the organ
through the trigeminal system. From the n. ophthalmicus superficialis a
thin branch r. nasalis is separated which terminates in the olfactory
sac. R. buccalis - the most medial of the three branches comprising the
medial stem of the n. maxillaris - sends a thin and short nerve terminat-
ing in the inlet of the olfactory organ. Several receptive loci of the

nerves were found by applying a spot tactile stimulation of the olfactory organ and the adjacent part of the skin (Chervova, 1985).

No direct contacts of the olfactory receptors with the trigeminal terminals were detected in experiments by alternately applying electrophysiological stimulation to the peripheral endings of the olfactory and trigeminal nerves. In the conditions of parallel recording of the trigeminal and olfactory fiber activity under corresponding stimulation, it was found that the trigeminal system responds with a shorter latency, even though the olfactory system is 6 times more sensitive (Belousova et al., 1983). Thus, we can assume that the trigeminal system is in close interrelation with other chemosensory systems; this interaction takes place centrally. It is well known that an important characteristic of all sensory systems is the organization of its central projections.

To explore details of this organization and to estimate the proportion of the trigeminal afferents that are related to the olfactory organs, we investigated the projections of these branches to the sensory trigeminal ganglion using the cobalt labelling technique. In each of the ganglion regions, the percentage of labelled and unlabelled neurons was calculated. The projections of all the branches were largely local; i.e., the ganglion is organized according to the principle of topical projections (Devitsina and Chervova, 1983) which is typical of the structures responsible for spatial neural analysis.

The present data indicate that, in fishes, the trigeminal nerve is a polymodal sensory system. An important function of this system is the perception and analysis of environmental chemical signals. It was shown in the present research that the trigeminal nerve receptors are capable of reacting not only to typical trigeminal stimuli but to ecologically relevant chemical signals - exometabolites, natural extracts, amino acids, and a variety of tastants and odorants. These data suggest that the trigeminal system is more than an afferent component of the common chemical sense, being a well-developed specialized chemosensory system which participates in the detection and perception of biologically relevant stimuli and in the formation of a wide range of specialized behavioral reactions.

REFERENCES

Belousova, T.A., Devitsina, G.V., and Malukina, G.A., 1983, Functional peculiarities of fish trigeminal system, Chem. Senses, 8:121.
Chervova, L.S., 1985, Electrophysiological study of the trigeminal nerve innervating the olfactory organ of the White Sea cod, Gadus morhua marisalbi Der. (Gadidae), Voprosy Ichthyol., 25:694 (in Russian).
Chervova, L.S., and Devitsina, G.V., 1981, The use of cobalt ionophoresis technique for investigation of the structure of the trigeminal ganglion in fishes, J. Evol. Biochem. Physiol., 17:316 (in Russian).
Chervova, L.S., and Devitsina, G.V., and Malukina, G.A., Amino acids as stimuli of the trigeminal nerve receptors of the White Sea cod, Gadus morhua marisalbi Der. (Gadidae), Voprosy Ichthyol., 25:320 (in Russian).
Devitsina, G.V., and Chervova, L.S., 1983, The structural peculiarities of the cod trigeminal ganglion, (1983), J. Evol. Biochem. Physiol., 19:293 (in Russian).

TRANSDUCTION MECHANISM IN VERTEBRATE OLFACTORY SYSTEMS

Kenzo Kurihara, Takayuki Shoji, Shuichi Enomoto, Mutsuo Taniguchi, Takuya Hanada, Makoto Kashiwayanagi

Faculty of Pharmaceutical Sciences, Hokkaido University Sapporo 060, Japan

It is generally considered that odorants bind to specific receptor proteins (Buck and Axel, 1991) and activate cAMP-sensitive channels located in olfactory cilia (Pace et al., 1985; Nakamura and Gold, 1987). The present paper discusses whether or not the above mechanism holds in the in vivo olfactory system.

1. Role of cilia in reception of odorants

In a previous study (Kashiwayanagi et al., 1988), we eliminated the carp olfactory cilia by "ethanol-calcium shock" and examined its effects on olfactory bulbar responses to amino acids. The results indicated that deciliation did not meaningfully affect the responses to amino acids, suggesting that the olfactory cilia may not be necessary for receptor neuron function in the carp.

The vomeronasal organ is a chemoreceptor which is distinguished from the main olfactory organ. Vomeronasal cells lack cilia and possess microvilli, while olfactory cells have long cilia. We have compared the sensitivities of the turtle vomeronasal organ to various odorants with those of the olfactory organ (Shoji and Kurihara, 1991). The threshold concentrations for various odorants in the vomeronasal organ were similar to those of olfactory organs, despite the fact the vomeronasal organ receptor cells lack cilia.

2. Effects of changed ionic environment on responses to odorants

In order to test the possibility that ion transport across the cation channels localized at olfactory cilia is involved in in vivo olfactory transduction mechanism, we have examined the effects of changed ionic environment on the olfactory responses.

Carp olfactory response

Application of amino acids dissolved in deionized water to the EDTA-treated olfactory epithelium did not induce any response (Yoshii and Kurihara, 1983). The addition of various species of salts to the stimulating solution reversibly restored the response. All the inorganic cations examined (Li^+, NH_4^+, K^+, Ca^{2+}, Mg^{2+}, Co^{2+}, Mn^{2+}, Cd^{2+}) and organic cations ($Tris^+$, $choline^+$, bis-Tris $propane^{2+}$), which are impermeable to membranes, were effective in returning the

Chemical Signals in Vertebrates VI, Edited by R.L. Doty and
D. Müller-Schwarze, Plenum Press, New York, 1992

response to amino acids. The divalent cations were effective at much lower concentrations than the monovalent cations. It is noted that 2 mM KCl restored a full olfactory responses to amino acids, suggesting ionic permeability at the apical membranes of the olfactory cells, including cilia, does not contribute to the depolarization of the cells. Ion dependence of the carp olfactory responses was explained in terms of binding of the cations to receptor membranes (Nomura and Kurihara, 1989).

Turtle olfactory response

In order to examine effects of salts on the turtle olfactory responses, the turtle olfactory epithelium was perfused with a salt free solution (Shoji et al., 1991). To examine whether or not salts in mucus covering the olfactory epithelium are sufficiently eliminated by perfusing the epithelium with the salt free solution, the olfactory bulbar responses to salts themselves were recorded after the epithelium was perfused with the salt free solution. The minimum concentrations of the responses (thresholds) were between 10^{-7} and 10^{-6} M for $CaCl_2$, between 10^{-6} and 10^{-5} M for NaCl and 10^{-5} and 10^{-4} M for KCl, and the responses were increased with an increase in salt concentrations. These results indicate that salts on the surface of the olfactory epithelium were sufficiently eliminated by perfusion of the epithelium with the salt free solution.

We have compared dose-dependence curves for the responses to n-amyl acetate, citral, β-ionone, and cineole in the salt free solution with those in the 100 mM NaCl solution. The results obtained indicated that there was no essential difference between the responses in the salt free solution and those in 100 mM NaCl solution in the whole concentration range examined. We also examined the effects of elimination of salts on the responses to 13 odorants. The responses to all odorants tested were practically unchanged by removal of salts.

The effects of salt concentration on the magnitude of the responses to various odorants were examined. The results indicated that the responses were independent of NaCl or $CaCl_2$ concentration. The replacement of NaCl with choline chloride or Tris chloride did not affect the responses to all odorants tested.

The present results suggested that activation of cation channels located at apical membranes of turtle olfactory cells including ciliary membranes does not contribute to in vivo olfactory transduction.

Frog olfactory responses

The dependence of the frog olfactory bulbar responses on salt concentration was also examined (Kashiwayanagi et al., 1991). The dependence greatly varied from odorant to odorant. The responses to odorants such as 1-carvone and isoamylacetate were essentially unchanged by removal of NaCl, while those to odorants such as citral and β-ionone were greatly decreased by removal of NaCl. The responses to the latter odorants did not show the NaCl requirement at pH 10 or 37 °C. The results obtained suggested that changes in ion permeability at the apical membrane of the frog olfactory cells is not involved in generation of the in vivo olfactory responses to at least certain odorants.

Turtle <u>vomeronasal</u> <u>responses</u>

The effects of salts on the odor responses in the vomeronasal system were examined (Shoji and Kurihara, 1991). In all 7 odorants tested, there was no essential difference between the responses to the odorants in the salt free solution and those in the 100 mM NaCl solution cross the entire concentration range examined.

The effects of salt concentration on the magnitude of the responses were examined. The results indicated that the responses to various odorants were independent of NaCl or $CaCl_2$ concentration. The replacement of NaCl with choline chloride, N-acetyl-D-glucosamine chloride, Bis-Tris propane dichloride and Na_2SO_4 did not affect the responses to odorants. The effects of a Na channel blocker, amiloride, and Ca channel blockers, diltiazem and verapamil, on the magnitudes of vomeronasal responses were tested. The magnitudes of responses were unchanged by the presence of 0.5 mM amiloride, 0.5 mM diltiazem and 0.5 mM verapamil. Thus the ion dependence of the turtle vomeronasal system on the responses to odorants was similar to that of the turtle olfactory system.

The present results suggested that in the carp, frog and turtle olfactory system and the turtle vomeronasal system, ion transport across the apical membranes of the cells does not contribute to <u>in vivo</u> transduction. There are several possible transduction mechanisms in there systems. One is that activation of adenylate cyclase or phosphoinositide turnover in response to odorants occurs at the membranes of olfactory knobs or the apical membranes of vomeronasal cells and the second messengers produced activate ionic channels located at cell membranes below the tight junction. Another possible mechanism is a physiochemical one (Kurihara et al., 1986). That is, adsorption of odorants on the receptor membranes leads to changes in the phase boundary potential at the receptor membranes. The depolarization is electrotonically propagated to the cell body membranes, which activates ionic membrane channels. It is unknown at present which mechanism is involved in transduction.

3. <u>Role of lipids of olfactory receptor membranes in odor reception</u>

<u>Liposomes</u> <u>having</u> <u>high</u> <u>sensitivity</u> <u>to</u> <u>odorants</u>

It is generally believed that an olfactory response is induced by the binding of an odorant to a specific receptor protein in olfactory receptor membranes. On the other hand, it has been known that nonolfactory systems such as the turtle trigeminal nerve, the Helix ganglion, the fly taste nerve, the frog taste cell and the neuroblastoma cells (Kashiwayanagi and Kurihara, 1984) respond to various odorants. These nonolfactory systems do not seem to provide specific receptor proteins for odorants. In general, odorants are hydrophobic and the interaction of odorants with lipid layers of olfactory receptor membranes seems to play an important role in generation of olfactory responses. We found that liposomes sensitively respond to various odorants (Nomura and Kurihara, 1987ab).

The sensitivity of liposomes to odorants varies with the species of odorante (Enomoto et al., 1991). The minimum concentration of amyl acetate to induce the response (referred to as threshold) in the phosphatidylcholine (PC) liposomes was about 10^{-4} M. Addition of 10%

or 20% phosphatidylserine (PS) lowered the threshold to about 10^{-9} M and increased the magnitude of the response. The olfactory thresholds for amyl acetate were determined to be 10^{-4} M for the frog and 10^{-7} M for the turtle. Hence, the threshold of the PC-PS liposomes (PS/PC=0.2) to amyl acetate was comparable to or lower than the olfactory thresholds in these animals. It was calculated that adsorption of less than a few molecules of amyl acetate on a single liposome elicits detectable changes in the membrane potential.

The specificity of liposomes to odorants was greatly dependent on the lipid composition. The response to fatty acids such as valeric acid, isovaleric acid and butyric acid became remarkably large when PS was added to PC. The specificity of liposomes to odorants was also affected by addition of proteins. For example, incorporation of concanavalin A to PS/PC=0.2 liposomes greatly increased the response to citral and decreased that to amyl acetate.

Detail analysis of the adsorption sites for various odorants in lipid membranes was carried out using various fluorescent dyes which monitor the membrane fluidity changes in different regions of the membrane. It was shown that different odorants are adsorbed on different regions in the membranes having complex lipid composition, whereas different odorants are adsorbed on similar region in membranes having simple lipid composition (Kashiwayanagi et al., 1990).

Enhancement of turtle olfactory responses to certain odorants by treatment of the epithelium with liposomes

As described above, addition of PS-containing liposomes to the epithelial surface resulted in enhanced responses to fatty acids. In order to examine whether or not this occurs in the olfactory system, PS-containing liposomes were applied to the turtle olfactory epithelium and its effects on the olfactory responses were observed (Taniguchi et al., in preparation). The PS-treatment enhanced the responses to fatty acids such as valeric acid, isovaleric acid and butyric acid by a factor of 4-5 times. The response to anisol was enhanced about 1.5 times by the treatment. The treatment did not affect the responses to other odorants examined.

The enhanced olfactory responses to fatty acids by PS-treatment closely resemble the enhanced responses observed with the liposomes. It seems that PS is incorporated into the olfactory receptor membranes and modifies the receptor site for the fatty acids. These results suggest that lipids are important for the reception of odorants.

Effects of membrane fluidity changes on the ability of turtle olfactory receptors to discriminate odorants

If the lipid layer in the olfactory receptor membrane plays an important role in odor reception, changes in temperature, which will cause membrane fluidity changes, may affect the structure of the receptor sites. We have examined the effects of temperature changes on the ability of the turtle olfactory receptors to discriminate odors using a cross adaptation method (Hanada et al., in preparation). Temperature was changed by perfusing the olfactory epithelium with Ringer solution of different temperatures. The cross adaptation experiment was carried out as follows. For example, first 1 mM geraniol (trans-isomer) was applied to the olfactory epithelium and after the response to geraniol was adapted to a spontaneous level, 1 mM nerol (cis isomer) was applied. These experiments were carried out at 5, 18 and 40 ℃. At 5 ℃, the response to nerol applied secondarily was not suppressed by previous application of geraniol, suggesting that the receptor site for geraniol is different from that for nerol. The response to geraniol was partly suppressed at 18 ℃ and greatly suppressed at 40 ℃. It is noted that both responses to geraniol and nerol when applied alone increased with an increase in temperature.

Similar results to those obtained with geraniol and nerol were obtained with other pairs of odorants having similar structures (tans-3-hexenol and cis-3-hexenol, l-carvone and d-carvone, camphor and cineole). On the other hand, suppression of the response to odorants applied secondarily at 40 ℃ was not appreciable when pairs of odorants having different structures (l-limonene and cineol, anisol and cineol) were used.

The results obtained with geraniol and nerol, for example, are explained as follows. In general, phase transition of the lipid layer of biological membranes occurs near body temperature. Hence it is likely that the phase transition of the lipid layer is related to the elimination of the odor-discriminating ability at 40 ℃. Geraniol and nerol are adsorbed on lipid layer of the olfactory receptor membranes. At 5 or 18 ℃, the lipid structure is rather rigid and geraniol and nerol are adsorbed on different sites. At 40 ℃, the fluidity of the lipid layer is increased, the receptor sites for both odorants become flexible and the receptor sites for respective odorants accept both odorants. On the other hand, the structures of receptor sites, those for anisol and cineol, for example, are greatly different from each other. The receptor site for anisol does not accept cineol and vice versa even when the membrane fluidity is increased. There is a possibility that conformational changes of receptor proteins are induced by a temperature increase and, therefore, the specificity of the receptor site of the protein is changed. In general, the specificity of proteins is not, however, changed by such a small change in temperature; hence, it is unlikely that the specificity of proteins is changed by the alteration in temperature.

It is possible that the desensitization of the olfactory response is induced by a decrease of cAMP level or by the inhibition of cAMP-sensitive channels by Ca^{2+} (Kurahashi, 1990). Both cAMP and Ca^{2+} easily diffuse in the cytosol of olfactory cells and, hence, the channels located over the whole area of the cells are desensitized. In other words, desensitization occurs at the cellular level. According to this hypothesis, the fact that the response to nerol appeared after the response to geraniol was desensitized at 5 ℃ suggests one cell has the receptor site for geraniol and another cell has that for nerol. This hypothesis cannot, however, explain the fact that the response to nerol did not appear after that of geraniol at 40 ℃. The present results are more reasonably explained by assuming that one cell has receptor sites for both geraniol and nerol, and that desensitization occurs at each receptor site.

The selective enhancement of the turtle olfactory responses to the fatty acids by the PS-treatment and the abolishment of the odor-discriminating ability of the turtle olfactory receptors by a temperature increase suggest that odor is not recognized by binding of a ligand to a specific protein, as seen in an interaction between enzyme and substrate or between a transmitter and its receptor. The present results are in accord with the following mechanism for odor reception and discrimination. Composition of the lipids and proteins of each olfactory cell membrane is assumed to be different from cell to cell. The magnitude of depolarization in response to an odorant is different from cell to cell. The depolarization is transformed into nerve impulses. The qualities of odors are recognized by firing pattern among various olfactory axons and the quality of odors is recognized in the brain. In this scheme, hydrophobic pockets of the olfactory receptor membranes composed of lipids and proteins are postulated to be adsorption sites for odorants. Variation in combinations of lipids and proteins provides many different adsorption sites for odorants. A diverse family of proteins found in the olfactory epithelium (Buck and Axel, 1991) will contribute to the variation in the combinations of lipids and proteins.

REFERENCES

Buck, L., and Axel, R., 1991, A novel multigene family may encode odorant receptors: a molecular basis for odor recognition, Cell, 65:175.

Enomoto, S., Kashiwayanagi, M., and Kurihara, K., 1991, Liposomes having high sensitivity to odorants, Biochim. Biophys. Acta, 1062:7.

Kashiwayanagi, M., Horiuchi, M., and Kurihara, K., 1991, Differential ion dependence of frog olfactory responses to various odorants, Comp. Biochem. Physiol., 100A:287.

Kashiwayanagi, M., and Kurihara, K., 1984, Neuroblastoma cell as model for olfactory cell: Mechanism of depolarization in response to various odorants, Brain Research, 293:251.

Kashiwayanagi, M., Shoji, T., and Kurihara, K., 1988, Large olfactory response of the carp after complete removal of olfactory cilia, Biochem. Biophys. Res. Commun., 154:437.

Kashiwayanagi, M., Suenaga, A., Enomoto, S., and Kurihara, K., 1990, Membrane fluidity changes of liposomes in response to various odorants: Complexity of membrane composition and variety of adsorption sites for odorants, Biophys. J., 58:887.

Kurahashi, T., 1990, The response induced by intracellular cyclic AMP in isolated olfactory receptor cells of the newt, J. Physiol., 430:355.

Kurihara, K., Yoshii, K., and Kashiwayanagi, M., 1986, Transduction mechanisms in chemoreception, Comp. Biochem. Physiol., 85A:1.

Nakamura T., and Gold, H., 1987, A cyclic nucleotide-gated conductance in olfactory receptor cilia, Nature, 325:442.

Nomura, T., and Kurihara, K., 1987a, Liposomes as a model for olfactory cells: Changes in membrane potential in response to various odorants, Biochemistry, 26:6135.

Nomura, T., and Kurihara, K., 1987b, Effects of changed lipid composition on responses of liposomes to various odorants: Possible mechanism of odor discrimination, Biochemistry, 26:6141.

Nomura, T., and Kurihara, K., 1989, Similarity of ion dependence of odorant responses between lipid bilayer and olfactory system, Biochim. Biophys. Acta, 1005:260.

Pace, U., Hanski, E., Salomon, Y., and Lancet, D., 1985, Odorant-sensitive adenylate cyclase may mediate olfactory reception, Nature, 316:255.

Shoji, T., Kashiwayanagi, M., and Kurihara, K., 1991, Turtle olfactory responses are unchanged by perfusing olfactory epithelium with salt free solution, Comp. Biochem. Physiol., 99A: 351-356.

Shoji, T., and Kurihara, K., 1991, Sensitivity and transduction mechanisms of responses to general odorants in turtle vomeronasal system, J. Gen. Physiol., 98:909.

Yoshii, K., and Kurihara, K., 1983, Role of cations in olfactory reception to amino acids, Brain Research, 274:239.

SECTION TWO

BODY FLUIDS AND SCENT GLAND CHEMISTRY AND HISTOLOGY

ANALYSIS OF THE SECRETIONS FROM THE FLANK GLANDS OF 3 SHREW SPECIES

AND THEIR POSSIBLE FUNCTIONS IN A SOCIAL CONTEXT

Debora Cantoni[1] and Laurent Rivier[2]

[1]Institut de Zoologie et d'écologie animale, Université de
Lausanne, 1015 Lausanne
[2]Institut de Médecine légale, Université de Lausanne, rue
du Bugnon 21, 1005 Lausanne

INTRODUCTION

Numerous studies have demonstrated that rodent social odours play an
important role in many aspects of rodent social behavior, affecting mate
selection, reproduction, parent-offspring interactions, and general
social interactions (Brown, 1985). In contrast, social odours and the
social organization of insectivores, especially shrews, have been poorly
studied, probably because of the secret habits of these animals and the
difficulty of maintaining them in captivity. Studies of both captive
shrews (Crowcroft, 1955; Vogel, 1969; Baxter and Meester, 1980; 1982) and
wild shrews (Croin Michielsen, 1966; Platt, 1976; Hawes, 1977; Genoud,
1978; 1981; Ricci and Vogel, 1984) suggest that species belonging to the
genus Crocidura can be aggressive or tolerant towards any conspecific,
whereas species belonging to the genus Sorex are generally only very
aggressive. These same authors noticed the existence of some degree of
home range overlap for species belonging to the genus Crocidura and
strict territoriality for species belonging to the genus Sorex.

Concerning social odours, the odoriferous lateral flank glands of
shrews were first noticed by Geoffroy-Saint-Hillaire (1815). These
lateral flank glands have since been described for all investigated
species and are known to be of various degrees of importance in both
sexes. Histological studies (Johnsen, 1914; Eadie, 1938; Pearson, 1946;
Murariu, 1973) and physiological studies (Dryden and Conaway, 1967;
Rissman, 1987; 1989) have shown that the activity of these glands is
greater in males than in females. Furthermore, there appears to be a
connection between flank gland activity and the reproductive condition of
both sexes. Despite the world-wide distribution of shrews and their
apparently different social structures, very little is known about the
functions of the lateral flank glands, or any other skin glands, in
Soricid species. Different hypotheses have been suggested but most are
highly speculative. One generally accepted hypothesis is that odours of
scent glands are used in mate location.

In order to examine the role of flank gland secretions on pair
formation, the social organization and mating system of free ranging
shrews belonging to subfamily Crocidurinae and Soricinae (Crocidura
russula, Sorex coronatus, Neomys fodiens) were studied during the breed-

ing season using the radio-isotope tracking technique (Ricci and Vogel, 1984; Cantoni and Vogel, 1989). Comparative gas chromatographic analysis of the flank gland secretions of males and females of the 3 shrew species were undertaken in parallel.

MATERIALS AND METHODS

The study was carried out in the neighbourhood of Lausanne (Switzerland) at Préverenges, a residential area for C. russula, and along the river "la Morges" at Clarmont for S. coronatus and N. fodiens. At monthly intervals shrews were trapped using Longworth traps, identified, weighed, sexed, marked by toe-clipping and released again at the point of capture. The present study took place during the March to July-August breeding season. During this period the trapped shrews consisted predominantly of adults, with the exception of a few juveniles from June onwards.

Gas chromatography analysis

The flank glands of trapped adult shrews were each rubbed with 2 pieces of filter pater (Whatman 40, 3X3 mm). The 4 filters were extracted in 100 μl dichloromethane inside a hermetically sealed vial. When the flank glands were not visible, their position was located according to Murariu (1973) and rubbed in the manner described above. Samples (3 μl) were injected (250° C) on to a gas chromatograph (Hewlett Packard 5790) capillary column (25 m x 0,2 mm) containing a methyl silicone phase (OV-101) and eluted from 130° C to 280° C during 26 minutes (12 PS, pressure at the head of the column). These eluting compounds were detected by a flame ionisation detector maintained at 300° C.

The same elution program was used for all analyses, thereby permitting comparison of the different chromatograms, taking into account the number of peaks and peak retention time of the detected substances. These analyses were repeated for samples from each shrew species.

Radioactive tracking technique

During the breeding season, the behavior of several shrews from each species was monitored simultaneously using the tracking technique described by Ricci and Vogel (1984) and modified by Cantoni (1990). The following procedure was used. (1) Shrews were trapped and identified. (2) At any time, either 2 or 3 neighbouring individuals were marked with an ear tag (0.1g) bearing a filament of radioactive tantalum (^{182}Ta). Filaments of 3 different activity levels (100-200, 250-350, 400-450 μCi) were used. Marked animals could be reliably distinguished using a portable scintillation counter. Thus, resting behaviour with other marked shrews could be determined and was monitored continuously. (3) After release, the shrews were monitored continuously for several days with 20 Geiger-Müller sensors set at different sites regularly visited by the shrews. Preliminary detection was undertaken manually with a portable scintillation counter (Cantoni, 1990). This system permitted the simultaneous monitoring of the movement of different animals and included both resting and activity patterns, as well as social interactions, among 2 or 3 radioactively tagged shrews.

The following behavioural features were measured on a daily basis: (1) Total individual activity, total individual rest and home range size, and (2) social interactions. Three parameters of social interaction were considered: (a) coincident activity, defined as the time spent by two radioactively marked shrew simultaneously frequenting the same point; (b)

coincident rest, defined as the time spent by two radioactively marked
shrews simultaneously frequenting the same nest; and (c) home range
overlap, defined as the total number of square meters frequented during
the same day by two radioactively marked shrews.

RESULTS

The lateral flank glands of C. russula and S. coronatus males became
active during late February or early March and remained active throughout
the breeding season. All N. fodiens males trapped at this time of the
year presented no active lateral scent glands and no odoriferous secretions could be detected. Female shrews of these three species exhibited
no glandular activity.

Chromatograms obtained for samples extracted from the flank glands
of males of the three shrew species (Fig. 1) showed important differences
in the number and type of detected substances. The chromatograms obtained for C. russula individuals differed significantly (Table 1), with
only a few peaks with the same retention time in all sample extracts.
For S. coronatus males, an homology of about 50% was observed in the
number of detected substances between different S. coronatus individuals
(Table 1). Extracts from N. fodiens flank glands contained very few
substances when compared with the other species (Table 1), but exhibited
good homology between different individuals of N. fodiens. In the females of all 3 species either none or very few substances were detected.

Table 1.   Number of substances detected in the secretions of the later-
            al flank glands of shrews during the breeding season. Values
            represent the number of substances detected in the secretions
            of 1) males, 3) females, 2) the number of substances common
            to males.

Substances	C. russula	S. coronatus	N. fodiens
1) Males	58-75 N=18	66-104 N=7	4-43 N=3
2) =	7	53	4(+)
3) Females	0 N=13	3-4 N=2	8 N=1

The study of social organization during the breeding season of the
three shrew species revealed distinct mating systems (Fig. 2). For C.
russula, a monogamous mating system was established. Members of a pair
shared a common nest and defended a common territory (Fig. 2) until one
of the 2 partners disappeared (Cantoni & Vogel 1989; Cantoni 1990).
Coincident rest occurred frequently between the 2 partners while coincident activity consisted, predominantly, of pursuit of the female by the
male.

S. coronatus females were strictly territorial whereas a large home
range overlap was observed between males and females (Fig. 2) and between
males. During the breeding season males kept intensive watch over one
female and occasionally visited a second or third, but no coincident rest
was observed and only limited coincident activity occurred (Fig. 2).
This behavior illustrates the establishment of a monogamous-type mating
system with a tendency towards a polygamous mating system (Cantoni,

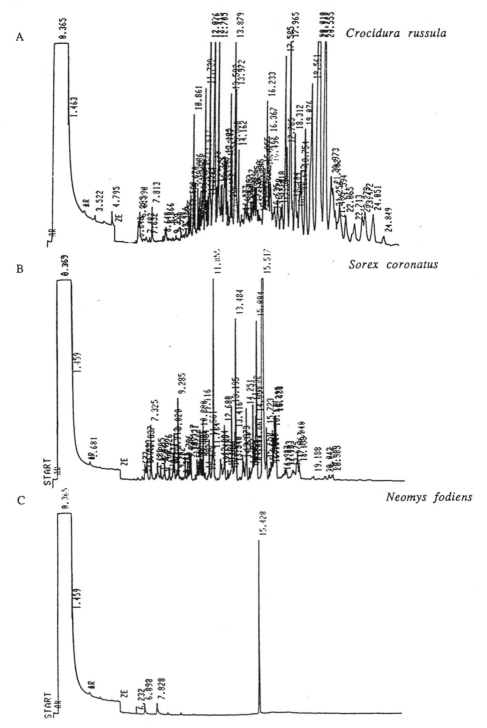

Fig. 1. Typical chromatograms obtained for the flank gland secretions of 3 shrew species. The abcissa represents the
elution time (minutes) and the ordinate indicates the
intensity of the detected substances. The first revealed
peak of each trace is the solvent front.

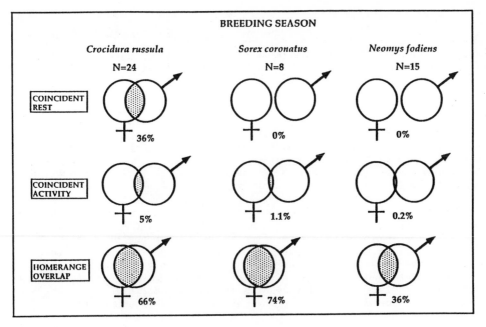

Fig. 2. Parameters describing shrew social interactions during the
breeding season for the three species. Coincident rest,
coincident activity, and home range overlap are expressed
as a percentage of the total. Overlap between the symbols
corresponds to the indicated percentages.

1990). N. fodiens females were strictly territorial, whereas the males
wandered during this period of the year, indicating the establishment of
a "promiscuous overlap" mating system (Cantoni, 1990). No coincident
rest was observed and only limited coincident activity was seen in N.
fodiens (Fig. 2).

DISCUSSION

These data demonstrate the different social strategies of the three
shrew species (C. russula, S. coronatus and N. fodiens) during the
breeding season, differing mainly in the role of the male during pair
formation. Indeed, males of C. russula share a common nest and a common
territory with a specific female, also spending some coincident rest and
activity with juveniles (Cantoni, 1990), demonstrating the formation of a
solid bond between paired males and females and suggesting parental
investment on the part of the male. S. coronatus males, however, spend
no coincident rest with any female, but regularly visit the resting
places of a particular female. No common territory was defended and
males also frequented the territory of other females, but less than those
of their own particular female. This behavior illustrates a more flexible bond between paired males and females, where the male plays no role
in the care of the offspring or in territorial defense. An extreme case
is shown by N. fodiens males, which establish no bond with any female and
play no role in caring for the offspring or defending a territory.

Several hypotheses can be proposed to explain the evolution of such different mating systems in these three shrew species. Each species has different ecological requirements and constraints. Switzerland lies at the northern limit of the distribution of C. russula (paleotropical origin), where it survives usually only in the neighbourhood of human habitations. This could explain the improved reproductive success of a female by the presence of the male at the nest, especially during the first days following parturation. Therefore, in C. russula, the formation of a solid pair bond ensures the female of the relatively permanent presence of the male and ensures the male of paternity. S. coronatus lives in a relatively predictable habitat (forests), where resources are homogeneously distributed, resulting in a generally monogamous mating system (Wittenberger, 1981). Nevertheless, in such habitats predation is quite heavy (by owls), with males being more vulnerable, probably because of their larger home range and the greater activity relative to the females (Cantoni, 1990). This could explain the tendency of this species to shift towards a polygamous mating system, where the reproductive success of a female will not be affected by the absence of a male. N. fodiens lives in relative unpredictable habitats where resources (macrobenthic invertebrates) are generally heterogeneously distributed and where changes in habitat due to climatic conditions are quite frequent. This could explain the evolution of a promiscuous mating system, where females are successful in offspring care without the help of the male (Vehrencamp and Bradbury, 1984).

One feature of the spatial organization common to the three species during the breeding season is the strict territoriality between females. This behavior is well known for rodents (Brown, 1962) and probably ensures sufficient food to cover the requirement of their litters (Barnard, 1983; Poole, 1985).

Concerning social odours, the general hypothesis is that odoriferous secretions play a role in mate location (Holst, 1985). Concerning the present results, this hypothesis should be partially rejected or at least modified. Indeed, females of the three species presented no or very few odoriferous secretions from their lateral flank glands, indicating that the location of females by males is probably based on other factors, such as markers found in urine or fecal pellets. In considering the males, three different results have been obtained. The flank glands secretions of C. russula males contain many substances, the majority of which are specific to each male. These data suggest that each male produces a particular odour which can be specifically identified by his female. This hypothesis has been confirmed in experiments conducted in captivity and in the field in which each C. russula female was confronted with a choice (in the form of a Y tunnel) between flank gland secretions from her male and those from a neighbouring male. In about 80% of the cases, females chose their partner's odour (Cantoni et al., in prep.). Analysis of the odoriferous secretions of S. coronatus males revealed numerous substances, of which more than 50% were common to all males of the species. This result suggests that the odours of individual S. coronatus males show a far greater degree of homology in their composition than the odours produced by C. russula males. S. coronatus females would therefore be confronted by a more general male odour, which could explain the more flexible bond between partners. Further investigation is required, however, to determine whether females are, in fact, more tolerant to other males, as this hypothesis would seem to indicate. Apparently N. fodiens males produce only limited odoriferous secretions which exhibit a high degree of composition homology between males. Therefore, females of the species would not be able to recognize individual males by their odour. This conclusion is in accordance with the observed social organization during the breeding season. A further strategy for species with a comparable mating system would be the production of a universal male odour. Further investigations in other Soricid species on this point will be interesting.

In conclusion, results obtained for the 3 shrew species studied here suggest that the odoriferous flank gland secretions of the male play a role in pair formation and are correlated with the stability of the bonds between partners according to the specificity of the odours produced by the males.

REFERENCES

Barnard, C.J., 1983, "Animal Behaviour: Ecology and Evolution", Croom Helm, London.
Baxter, R.M. and Meester, J., 1980, Notes on the captive behaviour of five species southern African shrews, Säugetierk. Mitt., 28:55.
Baxter, R.M. and Meester, J., 1982, The captive behaviour of the red musk shrew, Crocidura flavescens (Geoffroy, I. 1827) (Soricidae: Crocidurinae), Mammalia, 46:10.
Brown, L.E., 1962, Home range in small mammal communities, in: "Survey of Biological Processes 4", B. Glass, ed., Academic Press, New York.
Brown, R.E., 1985, The rodents I: Effects of odours on reproductive physiology (primer effects) in "Social Odours in Mammals", Vol. I, R.E. Brown and D.W. Macdonald, ed., Clarendon Press, Oxford.
Cantoni, D. and Vogel, P., 1989, Social organization and mating system of free-ranging greater white-toothed shrews, Crocidura russula, Anim. Behav., 38:205.
Cantoni, D., 1990, Etude en milieu naturel de l'organisation sociale de trois espèces de musaraignes, Crocidura russula, Sorex coronatus et Neomys fodiens, (Mammalia, Insectivora, Soricidae), "Ph. D. Thesis", Lausanne.
Croin Michielsen, N., 1966. Intraspecific and interspecific competition in the shrews Sorex araneus L. and Sorex minutus L., Arch. néerl. Zool., 17:73.
Crowcroft, P., 1955, Notes on the behaviour of shrews, Behavior, 8:63.
Dryden, G.L. & Conaway, C.M., 1967, The origin and hormonal control of scent production in Suncus murinus, J. Mammal., 48:420.
Eadie, W.R., 1983, The dermal glands of shrews, J. Mammal., 19:171.
Genoud, M., 1978, Etude d'une population urbaine de musaraignes musettes (Crocidura russula Hermann, 1970), Bull. Soc. Vaud. Sc. Nat., 74:25.
Genoud, M., 1981, Contribution à l'étude de la stratégie énergétique et de la distribution écologique de Crocidura russula (Soricidae, Insectivora) en zone tempérée, Ph D. thesis, Lausanne.
Geoffroy-Saint-Hillaire, M., 1815, Mémoire sur les glandes odoriférantes des musaraignes, Mém. Mus. Hist. Nat., 6:299.
Hawes, M.L., 1977, Home range, territoriality, and ecological separation in sympatric shrews, Sorex vagrans and Sorex obscurus, J. Mammal., 58:351.
Holst, D.V., 1985, The primitive eutherians I: Orders Insectivora, Macroscelidea, and Scandentia, in "Social Odours in Mammals", Vol I, R.E. Brown and D.W. Macdonald, ed., Oxford University Press, Oxford.
Johnsen, S., 1914, Über die Seitendrüsen der Soriciden, Anat. Anz., 46:139.
Murariu, D., 1973, Données macro- et microscopiques sur les organes glandulaires latéraux chez Sorex araneus L., Neomys fodiens Shreb. et Crocidura leucodon Herm. de Roumanie, Trav. Mus. Hist. Nat. Gregoire Antipa, 13:445.
Platt, W.J., 1976, The social organization and territoriality of short-tailed shrew (Blarina brevicauda) populations in old-field habitats, Anim. Behav., 24:305.

Pearson, O.P., 1946, Scent glands of the short-tailed shrew, Anat. Rec., 94: 615.

Poole, T., 1985, "Social Behaviour in Mammals", Chapman & Hall, Blackie, New York.

Ricci, J.C., and Vogel, P., 1984, Nouvelle méthode d'étude en nature des relations spatiales et sociales chez Crocidura russula (Mammalia, Soricidae), Mammalia, 48:281.

Rissman, E.F., 1987, Gonadal influences on sexual behavior in the male Musk shrew (Suncus murinus), Horm. and Behav., 21:132.

Rissman, E.F., 1989, Male related chemical cues promote sexual receptivity in female Musk shrew, Behav. and Neural Biol., 51:114.

Vehrencamp, S.L. and Bradbury, J.W., 1984, Mating systems and ecology, in: "In Behavioural Ecology. An Evolutionary Approach," J.R. Krebs and N.B. Davies, ed., second edition, Blackwell Scientific Publication, Oxford.

Vogel, P., 1969, Beobachtung zum intraspezifischen Verhalten der Hausspitzmaus (Crocidura russula Hermann, 1870), Rev. Suisse Zool., 76:1079.

Wittenberger, J.F., 1981, "Animal social behavior", Duxbury Press, Boston.

CHEMICAL ANALYSIS OF PREY-DERIVED VOMERONASAL STIMULANTS

Dalton Wang[1], Ping Chen[1] and Mimi Halpern[2]

Departments of [1]Biochemistry and [2]Anatomy and Cell Biology
State University of New York Health Science Center at
Brooklyn, 450 Clarkson Avenue, Brooklyn, NY 11203

INTRODUCTION

Earthworms (Lumbricus terrestris) are one of the major prey of garter snakes (Thamnophis sp.). Accurate recognition of this prey involves detection of chemical substances specific to earthworms. Garter snakes respond to earthworm preparations by tongue flicking and attack (Wilde, 1938; Burghardt, 1966; Halpern Kubie, 1980), responses is mediated by the vomeronasal system (Burghardt and Pruitt, 1975; Halpern and Frumin, 1979; Kubie and Halpern, 1979). In an attempt to understand the mole-molecular details of the attractive properties of compounds and their function/structure relationships and signal transduction in vomeronasal system of garter snakes, our laboratory has undertaken an extensive effort to isolate, purify, and characterize the chemo-attractive agents. We have isolated several proteins from earthworm wash (Wang, Chen, Jiang and Halpern, 1988) and electric shock-induced earthworm secretion (Jiang, Inouchi, Wang and Halpern, 1990) that elicit attach by snakes. A 20 kDa protein from electric shock-induced secretion was characterized and found to bind in a saturable and reversible manner to membranes obtained from the sensory epithelium of the vomeronasal organ (Jiang et al., 1990). When this protein is applied to the vomeronasal epithelium of the garter snake, an increase in the neural firing rate is detected in the accessory olfactory bulb (Jiang et al., 1990). In this paper we present chemical, physical, and biological properties of another snake-attractive protein, a low molecular weight protein (LMW), obtained from earthworm wash (EWW).

EXPERIMENTAL PROCEDURES

Materials

The following chemicals were obtained commercially: Molecular weight protein standards and pronase from Boeringer and Mannheim (Indianapolis); Isogel kit and protein standards for pI from FMC Bio-products (Rockland); and methylamine hydrochloride, urea, mercapto-ethanol from Sigma (St. Louis). All reagents were of highest grade commercially available. Earthworms were purchased from Connecticut

Valley Biological Supply, South Hampton, MA; and garter snakes from various animal suppliers.

Isolation and Purification of the LMW Snake-attractive Protein

The aqueous earthworm wash (EMW) was prepared according to the method described earlier (Wang et al., 1988; Jiang et al., 1990). Such a preparation contains a large number of protein, as revealed by poly-acrylamide gel electrophoretic analysis (Fig. 1). The protein labeled as B-2 in the Figure was isolated and purified by means of semiprepa-rative polyacrylamide gel electrophoresis with both Tris-glycine and Tris-borate systems according to the method of Laemmli (1970).

Figure 1.   Electrophoretogram of proteins of aqueous wash of earthworms.

The separation was made on 10% acrylamide separating gel with a 3% acrylamide stacking gel. The running buffer was 0.025 M Tris and 0.5 M glycine, pH 8.6. Proteins were stained with Coomassie Brilliant Blue.

Analytic Procedures

Determination of isoionic point (pI) -- An IsoGel (pH range 3-7) was used to determine the pI of the purified chemoattractant. The anolyte was 0.5 M acetic acid, pH 2.6; the catholyte was 0.1 M histidine (free base). The electrofocusing was performed 500 V (current 20 mA) for 90 min at 4° C. Molecular weight (MW) determination -- This was determined by SDS polyacrylamide gel electrophoresis using 10% acrylamide. Amino acid composition determination -- The amino acid composition was analyzed on Waters Amino Acid Analyzer. N-Terminal amino acid residue determination -- The NH_2-terminal amino acid of the protein was analyzed by the dansylation procedure of Gray (1972). Protein concentration determination -- Protein concentration was estimated by the method of Bradford (1976). Carbohydrates were assayed by phenol-sulfuric acid method of Dubois, Gilles, Hamilton, Rebers and Smith (1956). Pronase digestion -- The purified protein was dissolved in 1.0 ml of 0.15 M ammonium bicarbonate (pH 7.8) containing 1.0 mM calcium chloride. This solution was divided into two equal portions, one of which was delivered into a test tube containing 0.2 mg of Pronase. Both solutions were left at room temperature for 1 hr and used immediately in the snake bioassay or stored in a freezer for subsequent use. A Pronase control solution was similarly treated.

Snake Bioassay

The snake bioassay, an all-or-none test, was performed essentially as described earlier (Wang et al., 1988). Artificial earthworm bits (sections of Creme Lure plastic worms) were glued to circular glass dishes, covered with 100 μl of the test sample or the control solution, and placed in a snake's cage. The snake's behavior was monitored visually. A positive response was recorded when the snake approached and attacked the worm bit. If the snake approached the dish twice without attacking the worm bit, the response was considered negative. A Chi square test was used to determine if responses to a sample differed significantly from those to the control solution.

RESULTS

Electorphoretic Analysis

Figure 2 shows the results of polyacrylamide gel electrophoresis analysis of isolated snake-attractive protein. This protein, which gave a single protein component under both non-denatured and denatured conditions (Fig. 2, lane 2 and lanes 3-6, respectively), has been puri-fied to homo-geneity. Furthermore, the purified protein appears to contain a single polypeptide chain since it showed a similar mobility on gels under both SDS non-reducing or SDS reducing conditions (Fig. 2, lanes 5 and 7, respectively).

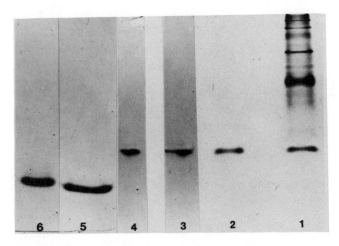

Figure 2.   Electrophoretogram of the purified B-2 snake-attractive protein.

From right to left, lane 1, proteins in the earthworm wash; lanes 2-6 purified B-2 protein. Lanes 1 and 2, under non-denaturing conditions with Tris-Cl⁻ buffer system; lanes 3, under same conditions but in the presence of 5 M urea; lanes 4 and 5, under SDS-denaturing non-reducing conditions in the presence and absence of urea, respectively; and lane 6, under SDS reducing conditions.

The Determination of pI

This protein is a rather acidic protein with an isoionic point of 3.3.

Molecular Weight Determinaion

The relative molecular mass of the snake-attractive protein was determined by SDS polyacrylamide gel electrophoresis. This protein is rather small have an estimated relative molecular mass of 3,000 daltons (Fig. 3). It gave a negative phenol-sulfuric reaction for carbohydrate.

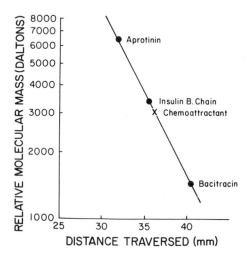

Figure 3.   The estimation of relative molecular mass of the purified chemoattractant by SDS polyacrylamide gel electrophoresis.

The molecular weight protein standards used were cytochrome C (12,500), aprotinin (6,500), insulin Kette B (3,400), and bacitracin (1,450).

Amino Terminal Residue

The amino terminal residue of this purified protein could not be determined by either the method of dansylation (Gray, 1972) or Edman degradation (Niall, 1973), suggesting it has a blocked amino terminal residue.

Amino Acid Composition

This protein, as expected, is rich in acidic amino acids and low in basic amino acids. It is also rich in glycine but has no methionine or lysine (Table 1). The minimum molecular weight calculated on the basis of amino acid composition is 3649. This value is in relatively good agreement with the value of 3000 estimated by the method of SDS poly-acrylamide gel electrophoresis.

Table 1. Amino acid composition of B-2 snake-attractive protein.

Amino acid	Residues/histidine*
Asp	4.07 (4)
Glu	3.43 (3)
Ser	2.48 (2)
Gly	4.13 (4)
His	0.05 (1)
Arg	1.41 (1)
Thr	2.74 (3)
Ala	4.32 (4)
Pro	1.79 (2)
Tyr	0.83 (1)
Val	2.22 (2)
Met	0.03 (0)
Cys	0.68 (1)
Ile	1.86 (2)
Leu	2.60 (3)
Phe	2.03 (2)
Lys	0.17 (0)

* Numerals in parentheses are amino acid residues expressed as integers.

Table 2. The attractivity of the purified low molecular weight (LMW) protein to garter snakes.

Compound	Protein concentration	Snake Bioassay No. of positive responses/total trials tested
	(μg/ml)	
Water control		0/21*
Purified LMW protein	10.8	7/7**
	5.4	7/7**
	2.7	6/7**
	1.4	6/7**
	0.7	1/7

*   The figure represents the total number of control trials.
**   chi square test, $p < 0.001$.

Chemoattractive Activity

This protein is a rather potent chemoattractant to garter snakes. The minimum concentration needed to yield a positive bioassay was 1.4 ug/ml (Table 2).

Proteolytic Digestion of the Purified Chemoattractive Protein

Figure 4 shows the effect of pronase digestion on the chemo-attractivity of the purified protein. This purified protein was very sensitive to the proteolytic action of pronase and lost practically all its attractivity. This loss of attractivity from proteolytic digestion suggests that the chemoattractive property resides in the peptide chain and is not due to some nonproteinaceous component absorbed or attached to the isolated protein.

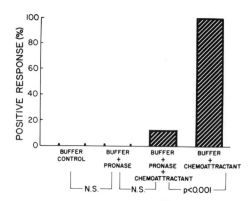

Figure 4.   The effect of pronase digestion on the chemoattractivity of purified LMW protein to garter snakes.

The concentration of the purified chemoattractive LMW protein was 0.3 mg/ml. The number of snakes for each test solution (or test control) was eight. p value was estimated by Chi square analysis; N.S., no significance.

Figure 5.   The effect of methylamine on purified LMW chemoattractive protein.

The sample was electrophoretically separated on 10% acrylamide mini-slab gel with a 3% acrylamide stacking gel. Running buffer was 0.025 M Tris and 0.5 M glycine, pH 8.6. Proteins were stained with Coomassie Brilliant Blue.

Effect of Methylamine on Oligomeric Formations

When the purified chemoattractive protein was treated with methylamine or subjected to cycles of freezing and thawing, it gradually formed oligomers (Fig. 5), but it remained as monomer at room temperature; i.e., it showed a single band with the same mobility as the original protein on polyacrylamide gel (data not shown). These oligomers could not be reversed back to the monomer under SDS-denaturing conditions in the presence or absence of a reducing agent. These results suggest that the oligomers were formed by some kind of cross-linkage other than disulfide. The isoelectric point of a major oligomer was found to have a pI of 4.1. Although the nature of this cross-linking reaction is not known, the change in the value of the pI suggests that the reaction involves either or both acidic and basic amino acid residues.

CONCLUSIONS AND FUTURE RESEARCH

Earthworms contain substances that attract garter snakes, their natural predator. The response to such substances is mediated by the snake's vomeronasal system (Schulman, Erichsen and Halpern, 1987). Previously, a sulfhydryl-containing chemoattractant from earthworm wash (Wang et al., 1988) and a 20 kDa chemoattractant from electric shock-induced earthworm secretion (Jiang et al., 1990) had been isolated. We have now isolated another potent chemoattractive protein from aqueous earthworm wash. This chemoattractant has been purified to homogeneity, as evidenced by its appearance as a single protein band on non-denaturing, SDS-denaturing, urea-SDS-denaturing, or isoelectrofocusing gels. It has a blocked amino terminal, and shows a calculated minimum molecular mass of 3649 daltons. The fact that this protein was no longer active after pronase digestion suggests that the chemoattractive property is not due to some nonproteinaceous component attached or absorbed to the isolated protein. This attractant is immunologically unrelated to either the sulfhydryl-containing chemoattractive protein isolated from earthworm wash or the 20 kDa chemoattractive protein from electric shock-induced secretion (Wang, Chen and Halpern, 1990, 1991). Biologically, on a molar basis, this LMW protein is an order of magnitude more potent than the 20 kDa chemoattractive protein isolated from electric shock-induced earthworm secretion (Jiang et al., 1990). The minimum concentration needed to yield a positive assay is 3.9×10^{-7} M for the 20 kDa chemoattractant from electric shock-induced earth-worm secretion and 3.8×10^{-8} M for the LMW chemoattractant from earthworm wash. The LMW chemoattractive protein has a unique property, in that it forms oligomers apparently through a covalent cross-linking reaction. The covalent linkage is not a disulfide, but the nature of the cross-linkage is unkown and remains to be determined. We are continuing our effort to isolate and characterize additional snake-attractive proteins from earthworms.

ACKNOWLEDGEMENT

The research was supported by National Institutes of Health Grant NS 11713.

REFERENCES

Bradford, M.A., 1976, A rapid and sensitive method for the quantitation of microgram quantities of protein utilizing the principle of protein dye binding. Anal. Biochem. 72: 248-254.

Burghardt, G.M., 1966, Stimulus control of the prey attack response in naive garter snakes. Psychol. Sci. 4: 37-38.

Burghardt, G.M. and Pruitt, C.H., 1975, Role of the tongue and senses in feeding of naive and experienced garter snakes. Physiol. Behav. 14: 185-194.

Dubois, M., Gilles, K.A., Hamilton, J.K., Rebers, P.A., and Smith, F., 1956, Colorimetric method for determination of sugars and related substances. Anal. Chem. 28: 355-358.

Gray, W.R., 1972, End-group analysis using dansyl chloride. in: Methods in Enzymology 25: 121-138.

Halpern, M. and Frumin, N., 1979, Roles of the vomeronasal and olfactory systems in prey attack and feeding in adult garter snakes. Physiol. Behav. 22: 1183-1189.

Halpern, M. and Kubie, J.L., 1980, Chemical access to the vomeronasal organs of garter snakes. Physiol. Behav. 24: 367-371.

Jiang, X.C., Inouchi, J., Wang, D. and Halpern, M., 1990, Purification and characterization of a chemoattractant from electric shock-induced earthworm secretion, its receptor binding, and signal transduction through the vomeronasal system of garter snakes. Jour. Biol. Chem. 265: 8736-8744.

Kubie, J.L. and Halpern, M., 1979, The chemical senses involved in garter snake prey trailing. J. Comp. Physiol. Psychol. 93: 648-667.

Laemmli, U.K., 1970, Cleavage of structural proteins during the assembly of the head of bacteriophage T4. Nature 227: 680-685.

Niall, H.D., 1973, Automated Edman degradation: The protein sequenator. In: Methods in Enzymology. Hirs, C.H.W. and Timasheff, S.N., eds. 27: 942-1010, Academic Press, New York.

Schulmen, N., Erichsen, E. and Halpern, M., 1987, Garter snake response to the chemoattractant in earthworm alarm pheromone is mediated by the vomeronasal system. Ann. N.Y. Acad. Sci. 510: 330-331.

Wang, D., Chen, P., Jiang, X.C, and Halpern, M., 1988, Isolation from earthworms of a proteinaceous chemoattractant to garter snakes. Arch. Biochem. Biophys. 267: 459-466.

Wang, D., Chen, P. and Halpern, M., 1990, Immunological analysis of earthworm derived chemoattractive proteins. Chemical Senses (abstract).

Wang, D., Chen, P. and Halpern, M., 1991, Immunological analysis of earthworm derived chemoattractive proteins. Comparative Biochem. Physiol. (in press).

Wilde, W.S., 1938, The role of Jacobson's organ in the feeding reation of the common garter snake, Thamnophis sirtalis sirtalis (Linn) J. Exp. Zool. 77: 445-465.

GAS CHROMATOGRAPHIC ANALYSIS AND ESTROUS DIAGNOSTIC POTENTIAL

OF HEADSPACE SAMPLING ABOVE BOVINE BODY FLUIDS

G.F. Rivard and W.R. Klemm

Department of Veterinary Anatomy and Public Health
Texas A&M University
College Station, TX 77843-4458

INTRODUCTION

The complete expression of bull sexual behavior depends on cues from both volatile and non-volatile portions of biological fluid samples from a cow in estrus (Rivard and Klemm, 1990). By displaying those behaviors, bulls demonstrate their ability to perceive specific volatile signals (Klemm et al., 1987). Our investigations in cycling cattle have shown that the female releases a pheromone two to three days prior to estrus. As determined by bull behavioral assay, blood contains this pheromone. Glands releasing this pheromone are located at the mucocutaneous junction of the vulva, from which a clear secretion can be expelled by gentle squeezing of the glands (Rivard and Klemm, 1989). We now postulate that there are estrous-specific volatile compounds above bovine body fluids such as blood, vulval skin gland secretion (VSG), and follicular fluid (FF) that correlate with progesterone/estrogen blood levels just prior to estrus.

We monitored variations among blood volatiles of three cows for four estrous cycles by static headspace sampling (HS) gas chromatography (GC). We compared these chromatographic peak profiles to ovarian FF and VSG secretion collected from cycling cows. These experiments support our prediction that estrus is associated with a progressive change in volatile components that regulate the attractiveness of the cow and the bull's sexual behaviors. Moreover, the HSGC peak profile analysis approach may have potential value as a pro-estrous diagnostic.

MATERIALS AND METHODS

Animal care and detection of oestrus

Mature, cycling cows and an androgenized-ovariectomized cow (ovex) of beef breed were managed as a group and housed in a concrete pen with access to coastal hay and fresh water. Four estrous cycles were synchronized by superovulation (Sirard et al., 1985) repeated each 21 days. Rectal palpation findings were used to determine follicular size and the presence of corpus luteum. The ovex cow was used to facilitate visual detection of estrus (Kiser et al., 1977). Also, the animals were observed for 30-min periods early in the morning and late in the afternoon for signs of estrus.

Chemical Signals in Vertebrates VI, Edited by R.L. Doty and
D. Müller-Schwarze, Plenum Press, New York, 1992

Collection of samples

Sampling started for seven consecutive days including day 0 (expected day of estrus), i.e., d-3, d-2, d-1, d-0, d+1, d+2, d+3, and continued every third day until 18 days following d+1, at which time daily collections were repeated. Blood samples (10 ml + 10% Na citrate) were collected by venipuncture and poured in a 30 ml serum vial sealed with a hand-operated crimper (Supelco). FF samples were collected by direct aspiration and transferred in serum vials. VSG secretions were aspirated in a glass capillary tube with a microdispenser (Drummond). All bottles and capillaries were kept frozen at 70°C.

Static HS sampling and gas chromatography

One ml HS from warmed and agitated samples was aspirated and quickly injected into a splitless injection port of an HP 5890 gas chromatograph. VSG samples were injected directly. Temperature of the injector was 150°C and of the FID was 250°C. A polar HP-20M Carbowax column (30 m x 0.53 mm I.D., film thickness 1.3 μm) was used for cow samples labelled White (n=40) and Pink (n=40) and for FF samples (n=10). Column temperature was held at 40°C for 8 min, then increased to 110°C at 4°C/min for 2.5 min. Cow samples labelled Red (n=40), and the VSG secretion samples (n=6), were performed on two in-line non-polar columns, HP-5 (20 m x 0.53 mm I.D., film thickness 2.65 μm) and BP-1 (SGE) (25 m x 0.53 mm I.D., film thickness 5.0 μm). Temperature profile was 30°C held for 2 min, increased to 130°C at 4°C/min, then raised to 195°C at 10°C/min for 1.5 min.

Radioimmunoassay

Progesterone (P) and oestradiol-17B (E_2) were measured by radioimmunoassays (RIA) using procedures of Hansen et al. (1988).

RESULTS AND DISCUSSION

"Day 0", defined as estrus day at which standing behavior was observed, was always associated with the lowest P level in a cycle at $\pm$ 1 day interval. Proestrus was characterized by a steep decline in P secretion, starting 3-5 days before day 0. Also, a rapid increase in secretion of E_2 occurred 3-5 days before day of estrus (Fig. 1), as expected (Clapper et al. 1990). Despite hot weather during the study, the mean estrous cycle (n=4) length was 20 $\pm$ 2 days; cows (n=3) were synchronized within $\pm$ 2 days. The day of estrus was always included in the middle ($\pm$ 2 day) of the consecutive daily sampling, except for the last cycle of cow 'Pink' (Fig. 1C). Volatile organics in vapor phase above blood revealed a profile of peaks consisting of peak #1, #2, #3, and #4 without variation in retention time among cows, but peak weights changed among estrous cycles. Peak #3 appeared the day $\pm$ 1 before estrogen peaking, and its peak weight increased until day 0 $\pm$ 1, then dropped to low or no value.

The HSGC analysis of follicular fluid revealed 18-24 peaks (Fig. 2A). The first part of the chromatogram showed the same peaks #1, #2, #3, and #4 as noticed with blood-volatile profiles. For four out of five FF samples from slaughter, we retrieved peak #3 in amounts comparable to the pro-estrous samples. Also, FF samples (n=5) from super-ovulated cows eluted similar amount of peak #3, except for FF Jul 21 that revealed a high amount. FF Jul 21 was the only sample that consistently elicited the three-set linked sexual behaviors in bull behavioral assay. By considering the P and E_2 values (Fig. 2C & D), we suppose that timing in follicular development or oocyte maturation stage might be important in peak #3 abundance.

Direct GC injection of VSG secretion disclosed complex profiles of 32 to 44 peaks. The ratio of peaks A, B, and E to reference peak M increased two days before estrus and dropped at day 0 (Fig. 3). This finding indicates metabolic changes in vulval gland secretions that might be responsible for cows' attractiveness and bull sexual behaviors.

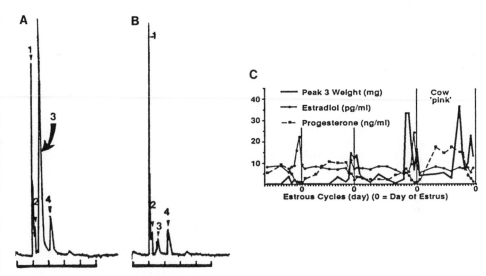

Fig. 1. Chromatogram of HS sampling above blood from cow 'Pink' four days (A) and eight days (B) before the day of estrus showing the estrous-specific peak #3. Plot of peak #3 amount in relation to E_2 and P values for 4 estrous cycles of cow 'pink' (C).

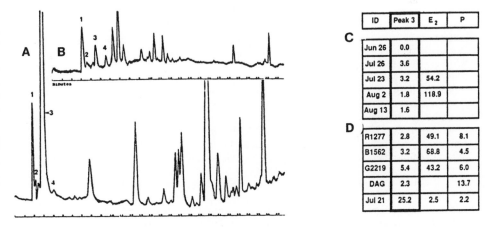

ID	Peak 3	E_2	P
Jun 26	0.0		
Jul 26	3.6		
Jul 23	3.2	54.2	
Aug 2	1.8	118.9	
Aug 13	1.6		
R1277	2.8	49.1	8.1
B1562	3.2	68.8	4.5
G2219	5.4	43.2	6.0
DAG	2.3		13.7
Jul 21	25.2	2.5	2.2

Fig. 2. Chromatogram of HS sampling above FF Jul 21 from superovulated cow (A) and FF Jul 23 from slaughter (B). Peak #3 values (mg) from HSGC of pooled FF obtained at slaughter from antral follicles (C) and from large follicles of superovulated cows (D), with corresponding E_2 and P values.

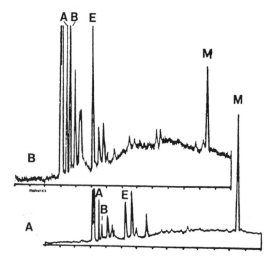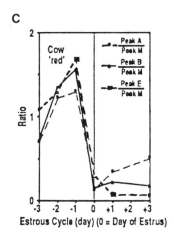

Fig. 3. GC chromatogram of VSG at day of estrus (A) and at day before estrus (B). Plot of ratio for peaks A, B, E to reference peak M (C).

Blood and FF volatile spectral profiles are simple to analyze and change as a cow enters and goes through estrus. Peak #3, as an estrous-specific volatile associated with increased E_2 and low P values, is a good candidate as a marker of proestrus and oocyte maturation phases. It may be reliable enough to be a diagnostic tool for further pheromonal analysis or to detect estrus in cattle. Also blood, FF, and VSG secretion appear to represent a ready source of estrous-specific volatiles.

ACKNOWLEDGMENTS

This research was supported by grants from the U.S.D.A.

REFERENCES

Clapper, J.A., Ottobre, J.S., Ottobre, A.C., and Zartman, D.L., 1990, Estrual rise in body temperature in the bovine. Temporal relationships with serum patterns of reproductive hormones, Anim. Reprod. Sci., 23:89-98.

Hansen, T.R., Randel, R.D., and Welsh, T.H., 1988, Granulosa cell steroidogenesis and follicular fluid steroid concentrations after the onset of oestrus in cows, J. Reprod. Fert., 84:409-16.

Kiser, T.E., Britt, J.H., and Ritchie, H.D., 1977, Testosterone treatment of cows for use in detection of estrus, J. Anim. Sci., 44:1030-5.

Klemm, W.R., Hawkins, G.N., and De Los Santos, E., 1987, Identification of compounds in bovine cervico-vaginal mucus extracts that evoke male sexual behavior, Chem. Senses, 12:77-87.

Rivard, G. and Klemm, W.R., 1989, Two body fluids containing bovine estrous pheromone(s), Chem. Senses, 14:273-4.

Rivard, G. and Klemm, W.R., 1990, Sample contact required for complete bull response to oestrous pheromone in cattle, in: "Chemicals Signals in Vertebrates V," D.W. MacDonald, D. Muller-Schwarze, and S.E. Natynczuk, eds., Oxford University Press, Oxford.

Sirard, M.A., Lambert, R.D., Beland, R., and Bernard, C., 1985, The effects of repeated laparoscopic surgery used for ovarian examination and follicular aspiration in cows, Anim. Reprod. Sci., 9:25-30.

LIPOCALYCINS ASSOCIATED WITH MAMMALIAN PHEROMONES

Alan G. Singer[1] and Foteos Macrides[2]

[1]Monell Chemical Senses Center
3500 Market Street, Philadelphia, PA 19104
[2]Worcester Foundation for Experimental Biology
222 Maple Avenue, Shrewsbury, MA 01545

APHRODISIN

The female golden hamster produces a substance that is emitted in vaginal discharge around the time of estrus and stimulates sexual behavior in male hamsters. This pheromone may be demonstrated in a bioassay using an anesthetized male as a surrogate female, which is placed in the cage of a normal male (Macrides et al., 1984a). If estrous vaginal discharge containing the pheromone is applied to the hindquarters of the surrogate female, a normal male will typically make several intromission attempts consisting of distinct bouts of pelvic thrusting directed at the surrogate female. The number of these bouts can be used as a measure of the activity of the pheromone in vaginal discharge, as demonstrated by a dose-response relation (Macrides et al., 1984b).

The major protein aphrodisin, when isolated from vaginal discharge, elicits copulatory behavior from males in levels comparable to those elicited by the unfractionated vaginal discharge (Singer et al., 1986). The behavioral response to aphrodisin in high molecular weight fractions occurs only if the male can make contact with the stimulus source, and the response is mediated by the vomeronasal organ (Clancy et al., 1984; Singer et al., 1984). The proteinaceous nature of the vomeronasal stimulus aphrodisin has been confirmed by the loss of behavioral activity after the protein was degraded with proteolytic enzymes or heat (Singer et al., 1986). Other proteins in the vaginal discharge have no activity in the behavioral assay (Singer et al., 1989).

The amino acid sequence of aphrodisin is 40% identical with the sequence of rat odorant binding protein (Pevsner et al., 1988), and 31% identical with the sequence of rat probasin (Spence et al., 1989). These 20 kDa extracellular proteins are members of the lipocalycin family of proteins, which was named for the ability of some of its members to shelter relatively small lipophilic ligands in a interior fold between two polypeptide sheets composed of antiparallel beta strands (Sawyer and Richardson, 1991), as demonstrated by the X-ray crystal structures of serum retinol binding protein (Cowan et al., 1990), and insect bilin binding protein from two species (Holden et al., 1987; Huber et al., 1987). Some of the other lipocalycins, for example the pyrazine and odorant binding proteins (Bignetti et al., 1988; Pevsner et al., 1990), beta-lactoglobulin

(Futterman and Heller, 1972; Hemley et al., 1979), alpha-1-acid glyco-protein (Kremer et al., 1988), and apolipoprotein D (Peitsch and Boguski, 1990), have been shown to bind a variety of relatively small lipophilic molecules, such as odorants, retinoids, steroids, and bile pigments in vitro, although ligands have not been found in the isolated proteins. Another lipocalycin, the enzyme prostaglandin D_2 synthase, catalyzes the specific isomerization of prostaglandin H_2 to prostaglandin D_2 and therefore presumably binds these eicosanoids (Nagata et al., 1991). In general these lipocalycins (or alpha-2u-globulins, as they are also known) appear to bind with low specificity a variety of lipophilic molecules having about ten to twenty carbon atoms. In spite of the name, however, only four of the twenty odd proteins in this family actually have lipophilic ligands that have been detected in the purified protein. It is therefore by no means certain that chemically purified aphrodisin should have a low molecular weight ligand solely by virtue of its membership in the lipocalycins.

The early evidence from the ultraviolet and fluorescence emission spectra (unpublished), as well as the retention of activity with the protein on dialysis or gel filtration in aqueous media, suggested that the purified, behaviorally active vaginal discharge protein does not include a ligand (Singer et al., 1986). More recently however, gel filtration experiments in aqueous acetonitrile buffers have indicated that a mixture of ligands may be present in the purified protein (unpublished). After gel filtration, eluting with 20% acetonitrile in water, activity is absent in the high molecular weight fraction, which appears to be chemically unal-tered, native aphrodisin; but we have not yet been able to restore its activity by recombining it with the low molecular weight fractions. If aphrodisin does have a ligand essential for activity, then this ligand must be tightly bound to survive repeated dialysis and gel permeation chromatog-raphy in aqueous buffers containing no organic solvents, such as aceto-nitrile. The ligand itself apparently has no activity or it may decompose when it is separated from the protein. In either case it is clear that the female hamster pheromone requires the intact polypeptide chain, and it apparently also requires a ligand for biological activity.

THE MAJOR URINARY PROTEINS

In the house mouse a similar female pheromone that stimulates a rapid increase in circulating levels of LH in males is not absolutely dependent on protein. Like the female hamster pheromone, the mouse pheromone is probably transmitted by contact with the stimulus source and the response is mediated by the vomeronasal organ (Maruniak and Bronson, 1976; Coquelin et al., 1984). Female mice excrete a high concentration (5 mg/ml) of major urinary proteins in their urine (Thung, 1956). These proteins are lipo-calycins with about 25% identity to aphrodisin (Held et al., 1987; Shahan et al., 1987). When we began our investigation of the chemical nature of the pheromone in female mouse urine, it seemed probable that the pheromonal activity would prove to be associated with these proteins. Dialysis experiments indicated that the activity indeed was associated with protein, but in contrast to the hamster results, the activity was partly dissociated from the protein by dialysis, and it could be completely dissociated from the protein by gel filtration in aqueous solvents. Most significantly, the activity was not destroyed by enzymatic degradation of the protein (Singer et al., 1988).

Apparently the protein is not necessary for the activity of the female mouse pheromone, but it is possible that it enhances the biological ac-tivity. This remains to be experimentally demonstrated by dose-response measurements comparing the responses to various doses of ligand and pro-tein-ligand complex. Recalling that the activity of aphrodisin was destroyed by proteolytic enzymes, and in as much as sequence similarity suggests homologous function in the two proteins, we can speculate that the ligand of aphrodisin is more tightly bound to the protein and is more unstable than the mouse ligand when it is separated from the protein. Some earlier chemical work on the pheromones responsible for the primer effects of male mouse urine suggested that the major urinary protein had pheromonal activity (Vandenbergh et al., 1975; Marchlewska-Koj, 1977), but as we observed for the female mouse pheromone, bioassay of the products of further chemical fractionations indicated that relatively lower molecular weight stimuli also are active (Vandenbergh et al., 1976; Novotny et al., 1980; Marchlewska-Koj, 1981), consistent with the idea that the protein is serving to transport a ligand that acts as the pheromone.

The common occurrence of lipocalycins and other uncharacterized extracellular 20 kDa proteins in urine, saliva, and scent gland secretions associated with pheromonal effects does fuel speculation that these proteins are pheromones or pheromone binding proteins in mammals (Held et al., 1987; Mancini et al., 1989). The major urinary proteins have been extensively investigated as favorable subjects in which to elucidate the mechanisms of hormonal regulation of protein synthesis, and in the course of this work several results have emerged incidentally that are concordant with a role for these proteins in pheromonal communication.

The major urinary proteins in the male rat and the mouse consist of several closely related variants encoded by a family of 20 to 30 genes (Hastie et al., 1979; Kurtz, 1981; Bishop et al., 1982). The various sequences are differentially regulated by multiple hormones; and they occur in developmentally and sexually specific combinations in a number of skin glands producing external secretions, as well as in urine (Shaw et al., 1983; Gubits et al., 1984; MacInnes et al., 1986). Histological studies have found close association between major urinary protein and lipids in preputial, meibomian and perianal glands, and have suggested that these proteins are combined with lipid in peroxisomes before they are secreted (Mancini et al., 1989). Their propensity for combination with hydrophobic molecules apparently is involved in a nephropathy that is specific to male rats and results from exposure to a component of unleaded gasoline and other chemicals, such as limonene, that are bound to the protein (Borghoff et al., 1990). And finally, the physiological potency of this protein is indicated by its ability to stimulate the pituitary-testicular axis in male rats (Ghosh et al., 1991), but in spite of all this and much more work on the major urinary proteins, their function is not yet established.

OTHER PROTEINS

Aphrodisin and the major urinary proteins are the only known lipocalycins yet demonstrated to be associated with pheromones in mammals, but there are some examples of proteins that occur in skin gland secretions or urine with known or potential pheromonal function. These uncharacterized proteins have some of the salient properties of the lipocalycins, that is they are extracellular, abundant proteins with a molecular mass of about 20 kDa and a relatively high negative charge. An abundant, androgen dependent protein with the same molecular weight as the major urinary protein occurs in bank vole urine (Kruczek and Marchlewska-Koj, 1985). Other minimally characterized 20 kDa proteins are found in the secretions of perianal glands of a South American primate (Belcher et al., 1990), and in human armpit secretions (G. Preti and A. Spielman, personal communication).

A partially characterized, abundant, negatively charged (pI: 4.78, 5.35), extracellular protein, pheromaxein, binds the known pheromone androstenol and related steroids in boar submaxillary gland saliva (Melrose et al., 1971; Booth and White, 1988). This is the only pheromone binding protein in mammals for which the ligand is known, and this ligand is a stable, readily available compound. It therefore would be of considerable interest to know whether this protein is a lipocalycin and what is its effect on the pheromonal action of androstenol on female pigs. If the effect of the protein on pheromonal activity could be measured in a reliable bioassay, and if the protein were characterized, it would then be possible to design experiments to determine the importance of the various potential interactions between transport protein, ligand, and nasal chemoreceptors.

This is what we hope to do with aphrodisin in the hamster when we have characterized the putative ligand. There are many questions that remain unanswered about the occurrence and function of these proteins and their ligands in mammalian chemical signals.

ACKNOWLEDGMENTS

Research in our laboratories was supported by NIH grants HD19764, BRSG 507 RR05825, and NS12344.

REFERENCES

Belcher, A. M., Epple, G., Greenfield, K. L., Richards, L. E., Kuderling, I., and Smith, A. B., 1990, Proteins: biologically relevant components of the scent marks of a primate (Saguinus fuscicollis), Chem. Senses, 15:431.

Bignetti, E., Cattaneo, P., Cavaggioni, A., Damiani, G., and Tirindelli, R., 1988, The pyrazine-binding protein and olfaction, Comp. Biochem. Physiol., 90B:1.

Bishop, J. O., Clark, A. J., Clissold, P. M., Hainey, S., and Francke, U., 1982, Two main groups of mouse major urinary protein genes, both largely located on chromosome 4, EMBO J., 1:615.

Booth, W. D., and White, C. A., 1988, The isolation, purification and some properties of pheromaxein, the pheromonal steroid-binding protein, in porcine submaxillary glands and saliva, J. Endocrinol., 118:47.

Borghoff, S. J., Short, B. G., and Swenberg, J. A., 1990, Biochemical mechanisms and pathobiology of alpha-2u-globulin nephropathy, Annu. Rev. Pharmacol. Toxicol., 30:349.

Clancy, A. N., Macrides, F., Singer, A. G., and Agosta, W. C., 1984, Male hamster copulatory responses to a high molecular weight fraction of vaginal discharge: effects of vomeronasal organ removal, Physiol. Behav., 33:653.

Coquelin, A., Clancy, A. N., Macrides, F., Noble, E. P., and Gorski, R. A., 1984, Pheromonally induced release of luteinizing hormone in male mice: involvement of the vomeronasal system, J. Neurosci., 4:2230.

Cowan, S. W., Newcomer, M. E., and Jones, T. A., 1990, Crystallographic refinement of human serum retinol binding protein at 2A resolution, Proteins, 8:44.

Futterman, S., and Heller, J., 1972, The enhancement of fluorescence and the decreased susceptibility to enzymatic oxidation of retinol complexed with bovine serum albumin, beta-lactoglobulin, and the retinol-binding protein of human plasma, J. Biol. Chem., 247:5168.

Ghosh, P. K., Steger, R. W., and Bartke, A., 1991, Possible involvement of hypothalamic monoamines in mediating the action of alpha-2u-globulin on the pituitary-testicular axis in rats, Neuroendocrinology, 53:7.

Gubits, R. M., Lynch, K. R., Kulkarni, A. B., Dolan, K. P., Gresik, E. W., Hollander, P., and Feigelson, P., 1984, Differential regulation of alpha-2u-globulin gene expression in liver, lachrymal gland, and salivary gland, J. Biol. Chem., 259:12803.

Hastie, N. D., Held, W. A., and Toole, J. J., 1979, Multiple genes coding for the androgen-regulated major urinary proteins of the mouse, Cell, 17:449.

Held, W. A., Gallagher, J. F., Hohman, C. M., Kuhn, N. J., Sampsell, B. M., and Hughes, R. G., 1987, Identification and characterization of functional genes encoding the mouse major urinary proteins, Mol. Cell. Biol., 7:3705.

Hemley, R., Kohler, B. E., and Siviski, P., 1979, Absorption spectra for the complexes formed from vitamin-A and beta-lactoglobulin, Biophys. J., 28:447.

Holden, H. M., Rypniewski, W. R., Law, J. H., and Rayment, I., 1987, The molecular structure of insecticyanin from the tobacco hornworm Manduca sexta L. at 2.6 A resolution, EMBO J., 6:1565.

Huber, R., Schneider, M., Mayr, I., Muller, R., Deutzmann, R., Suter, F., Zuber, H., Falk, H., and Kayser, H., 1987, Molecular structure of the bilin binding protein (BBP) from Pieris brassicae after refinement at 2.0 A resolution, J. Mol. Biol., 198:499.

Kremer, J. M. H., Wilting, J., and Janssen, L. H. M., 1988, Drug binding to human alpha-1-acid glycoprotein in health and disease, Pharm. Rev., 40:1.

Kruczek, M., and Marchlewska-Koj, A., 1985, Androgen-dependent proteins in the urine of bank voles (Clethrionomys glareolus), J. Reprod. Fert., 75:189.

Kurtz, D. T., 1981, Rat alpha-2u-globulin is encoded by a multigene family, J. Mol. Appl. Genet., 1:29.

MacInnes, J. I., Nozik, E. S., and Kurtz, D. T., 1986, Tissue-specific expression of the rat alpha-2u-globulin gene family, Mol. Cell. Biol., 6:3563.

Macrides, F., Clancy, A. N., Singer, A. G., and Agosta, W. C., 1984a, Male hamster investigatory and copulatory responses to vaginal discharge: an attempt to impart sexual significance to an arbitrary chemosensory stimulus, Physiol. Behav., 33:627.

Macrides, F., Singer, A. G., Clancy, A. N., Goldman, B. D., and Agosta, W. C., 1984b, Male hamster investigatory and copulatory responses to vaginal discharge: relationship to the endocrine status of females, Physiol. Behav., 33:633.

Mancini, M. A., Majumdar, D., Chatterjee, B., and Roy, A. K., 1989, Alpha--2u-globulin in modified sebaceous glands with pheromonal functions: localization of the protein and its mRNA in preputial, meibomian, and perianal glands, J. Histochem. Cytochem., 37:149.

Marchlewska-Koj, A., 1977, Pregnancy block elicited by urinary proteins of male mice, Biol. Reprod., 17:729.

Marchlewska-Koj, A., 1981, Pregnancy block elicited by male urinary peptides in mice, J. Reprod. Fert., 61:221.

Maruniak, J. A., and Bronson, F. H., 1976, Gonadotropic responses of male mice to female urine, Endocrinology, 99:963.

Melrose, D. R., Reed, H. C. B., and Patterson, R. L. S., 1971, Androgen steroids associated with boar odour as an aid to the detection of oestrus in pig artificial insemination, Br. Vet. J., 127:497.

Nagata, A., Suzuki, Y., Igarashi, M., Eguchi, N., Toh, H., Urade, Y., and Hayaishi, O., 1991, Human brain prostaglandin D synthase has been evolutionarily differentiated from lipophilic-ligand carrier proteins, Proc. Natl. Acad. Sci. USA, 88:4020.

Novotny, M., Jorgenson, J. W., Carmack, M., Wilson, S. R., Boyse, E. A., Yamazaki, K., Wilson, M., Beamer, W., and Whitten, W. K., 1980, Chemical studies of the primer mouse pheromones, in: "Chemical Signals," D. Muller-Schwarze, and R. M. Silverstein, eds., Plenum Press, New York.

Peitsch, M. C., and Boguski, M. S., 1990, Is apolipoprotein D a mammalian bilin-binding protein?, New Biologist, 2:197.

Pevsner, J., Reed, R. R., Feinstein, P. G., and Snyder, S. H., 1988, Molecular cloning of odorant-binding protein: member of a ligand carrier family, Science, 241:336.

Pevsner, J., Hou, V., Snowman, A. M., and Snyder, S. H., 1990, Odorant-binding protein: characterization of ligand binding, J. Biol. Chem., 265:6118.

Sawyer, L., and Richardson, J. S., 1991, Using appropriate nomenclature, Trends Biochem. Sci., 16:11.

Shahan, K., Gilmartin, M., and Derman, E., 1987, Nucleotide sequences of liver, lachrymal, and submaxillary gland mouse major urinary protein mRNAs: mosaic structure and construction of panels of gene-specific synthetic oligonucleotide probes, Mol. Cell. Biol., 7:1938.

Shaw, P. H., Held, W. A., and Hastie, N. D., 1983, The gene family for major urinary proteins: expression in several secretory tissues of the mouse, Cell, 32:755.

Singer, A. G., Clancy, A. N., Macrides, F., and Agosta, W. C., 1984, Chemical studies of hamster vaginal discharge: male behavioral responses to a high molecular weight fraction require physical contact, Physiol. Behav., 33:645.

Singer, A. G., Macrides, F., Clancy, A. N., and Agosta, W. C., 1986, Purification and analysis of a proteinaceous aphrodisiac pheromone from hamster vaginal discharge, J. Biol. Chem., 261:13323.

Singer, A. G., Clancy, A. N., Macrides, F., Agosta, W. C., and Bronson, F. H., 1988, Chemical properties of a female mouse pheromone that stimulates gonadotropin secretion in males, Biol. Reprod., 38:193.

Singer, A. G., Clancy, A. N., and Macrides, F., 1989, Conspecific and heterospecific proteins related to aphrodisin lack aphrodisiac activity in male hamsters, Chem. Senses, 14:565.

Spence, A. M., Sheppard, P. C., Davie, J. R., Matuo, Y., Nishi, N., McKeehan, W. L., Dodd, J. G., and Matusik, R. J., 1989, Regulation of a bifunctional mRNA results in synthesis of secreted and nuclear probasin, Proc. Natl. Acad. Sci. USA, 86:7843.

Thung, P. J., 1956, Proteinuria in mice and its relevance to comparative gerontology, Experientia Suppl., 4:195.

Vandenbergh, J. G., Whitsett, J. M., and Lombardi, J. R., 1975, Partial isolation of a pheromone accelerating puberty in female mice, J. Reprod. Fert., 43:515.

Vandenbergh, J. G., Finlayson, J. S., Dobrogosz, W. J., Dills, S. S., and Kost, T. A., 1976, Chromatographic separation of puberty accelerating pheromone from male mouse urine, Biol. Reprod., 15:260.

MORPHOGENIC AND HISTOLOGIC PATTERNS AMONG THE POSTEROLATERAL GLANDS OF

MICROTINE RODENTS

Frederick J. Jannett, Jr.

Department of Biology
Science Museum of Minnesota
30 East 10th Street
St. Paul, Minnesota 55101

INTRODUCTION

The posterolateral scent glands of voles and lemmings (Rodentia: Muridae: Microtinae) were recognized early as species characters (Miller, 1896) and were thought to occur in discrete body regions (Quay, 1968). But a pharmacological dose of testosterone induced posterolateral glands on two species not usually possessing them (Jannett, 1975), and posterolateral glands occur among species, as a group, in an essentially continuous zone of glandular responsiveness between flanks and rump (Jannett, 1990). Three populations of Microtus pennsylvanicus, a species not characteristically possessing posterolateral glands, have recently been found (Tamarin, 1981; Boonstra and Youson, 1982; Jannett, unpubl.). Scent producing characters can be used to combine the evolutionary and population approaches to the study of a chemical communication system (Schluter, 1989).

This report is of the effects of a pharmacological dose of testosterone propionate (TP) on posterolateral glands of two species of special interest, M. brandti and Synaptomys cooperi. Microtus (Lasiopodomys) brandti, formerly reported to have hip glands (Quay, 1968), represents a subgenus of vole with flank glands not previously studied, and Synaptomys spp. are the only lemmings with flank glands. It reviews the results of similar experiments on other species and recent histologic observations of glandular structure. The posterolateral glands of microtine rodents are all composed of sebaceous glands (Quay, 1968). Considerable differences exist, however, in morphogenic and histologic patterns.

MATERIALS AND METHODS

The Microtus brandti were adult males in the breeding colony of the Institute of Evolutionary Animal Morphology and Ecology, Moscow. They were supplied grain, produce, and water ad libitum. Synaptomys cooperi were trapped in Cook Co., Minnesota. When secured, each was an adult or a subadult subsequently housed in the laboratory for six months before experimental treatment. They were provided rabbit chow, Sphagnum moss, and water ad libitum.

Chemical Signals in Vertebrates VI, Edited by R.L. Doty and
D. Müller-Schwarze, Plenum Press, New York, 1992

The procedure of Jannett (1975) was followed. A 75 mg pellet of testosterone propionate (TP) ("Oreton", Schering Corporation, Bloomfield, NJ, USA) was implanted subcutaneously at the nape in an intact subject. Subjects were housed individually (16L:8D) for the following 21 days and then sacrificed, and carcasses were fixed in 10% formalin. Skin was subsequently cut and peeled back, and one flank gland from each specimen was measured (l, w) to the nearest 0.1 mm with calipers. Length is along the anteroposterior axis of the body and width is perpendicular to it. The glands were all elliptic and area was calculated using the formula for an ellipse.

Five experimental male $\underline{M}$. brandti received TP, and five control males received no such implant. Six male and four female $\underline{S}$. cooperi (two nulliparous and two probably parous) each received an implant. Individuals of the latter species are hard to secure and these were obtained over four years. Results are compared to a literature report of the natural flank gland in this species.

RESULTS

The mean (SEM) size of the Microtus brandti flank glands in the group receiving TP was 9.6 (0.3) mm in length, 7.7 (0.5) mm in width, and 58.4 (3.6) mm^2 in calculated area. Two control males did not have posterolateral glands, and the three flank glands in the control group averaged 9.1 (0.3) mm in length, 5.7 (0.4) mm in width, and 43.0 (3.8) mm^2 in calculated area. There was no significant difference in length in a comparison of treatment groups when the five experimental males were compared with the three control males with glands [$t(6)$ = -1.2, p = 0.27]. However, the width was significantly different between treatments [$t(6)$ = -2.6, p < 0.04], as was area [$t(6)$ = -2.8, p < 0.04]. When all five control males were compared to the experimental males, the same pattern was shown; i.e., there were significant differences in width and area but not length. All flank glands remained discrete patches as viewed macroscopically.

One parous female $\underline{S}$. cooperi failed to develop a macroscopically obvious posterolateral gland. The glands of the other females averaged 8.7 (0.7) mm in length, 5.0 (0.6) mm in width, and 37.2 (6.7) mm^2 in area. The glands of the six males averaged 8.2 (0.5) mm in length, 4.3 (0.3) mm in width, and 31.2 (3.4) mm^2 in area. All glands were discrete ellipses.

DISCUSSION

Morphogenesis

Posterolateral glands are generally better developed on males than on females (Quay, 1968). They show a dose response to TP in $\underline{M}$. montanus (Jannett, 1978). In $\underline{M}$. (Aulacomys) richardsoni and Lagurus curtatus, they regress after castration and are replaced by exogenous TP (Jannett and Jannett, ms.; Jannett, ms a, respectively). In $\underline{L}$. curtatus, intact males given the TP implant have flank glands comparable to those on castrated males given the TP implant, so there would seem to be no need to castrate males to observe the size and general form of glands presumably maximally stimulated by the 75 mg dose.

The flank glands of all four species [$\underline{S}$. cooperi, $\underline{L}$. curtatus, $\underline{M}$. (A.) richardsoni and $\underline{M}$. (L.) brandti] retain their general elliptic shape when stimulated by the same pharmacological dose. They remain about the same size as naturally occurring glands. Stimulated flank glands of $\underline{S}$. cooperi reported herein had lengths 7 to 10 mm; Connor (1959) reported lengths 3 to

8 mm on six adult field-trapped males. The flank glands of M. brandti reported here grew slightly in width but not in length. Those of male L. curtatus did not grow significantly in either area or weight (Jannett, ms a). There was no consistently significant response across replicates in the area or weight of flank glands of M. richardsoni challenged by this dose (Jannett and Jannett, ms). In contrast, the hip glands on very old individuals or those given large doses of TP in M. agrestis and M. montanus (Clarke and Frearson, 1972; Jannett, 1978) grew much larger and changed in shape so that they extended onto the thighs and reached from posterior flank to posterior hip.

These results suggest that there is a morphogenic organizer which constrains the shape of the posterolateral glands in the flank region and that there is no such mechanism in the hip region.

Histologic Differences

Material recently examined for histologic patterns (Jannett, ms b) included posterolateral glands previously described of four species (A. terrestris, M. richardsoni, M. gregalis) (Vrtis, 1930; Quay, 1968; Sokolov, 1982), induced glands (M. longicaudus, M. pennsylvanicus, M. gud and M. pinetorum), naturally occurring glands pharmacologically stimulated by TP (S. cooperi, L. curtatus, A. terrestris, M. richardsoni, M. brandti, M. fortis, and M. maximowiczii), and previously undescribed glands of Myopus schisticolor, L. curtatus, Alticola spp., Microtus brandti, and M. xanthognathus.

As previously reported, posterolateral glands of M. richardsoni have a small set of sebaceous glands in addition to hypertrophied acini within the organ. However, the small units occur only at the margins of the organ, not throughout. The specialized flank gland appears to subtend the surrounding skin. The dimorphism was also found in M. xanthognathus but in no other species.

The discreteness of the posterolateral gland was histologically most prominent in M. richardsoni and M. xanthognathus, less so in Arvicola terrestris and M. brandti. The most gradual change in size of acini from posterolateral gland to surrounding skin was in the four species on which glands were induced.

A gradation in size from larger acini anteriorly in the gland to smaller posteriorly was most obvious in hip glands, e.g., on M. montanus.

There were species with small posterolateral glands (hip or flank) with small component acini. There were also large posterolateral glands with large acini in both regions. There were no very small posterolateral glands composed of highly complex or very large acini. Nor were there any naturally occurring posterolateral glands very large in area but with only slightly enlarged and simple acini.

Increased size of individual sebaceous glands was accompanied by branching, but in one species, M. brandti, even a pharmacological dose of TP did not completely obscure the discreteness of the individual gland associated with a follicle.

The largest and most complex glands had other modifications of the epidermis, namely, loss or near loss of hair and hypertrophy of the stratum malpighii and stratum corneum. Dermal elements were also reduced.

Induced glands had enlarged acini but no other obvious epidermal response. The normally small acini in the flank gland of S. cooperi remained relatively small after implantation of the pharmacological dose of TP, a pattern also seen in Clethrionomys glareolus after implantation of this dose (Sokolov et al., 1989).

ACKNOWLEDGEMENTS

This work was supported by the Ford Foundation Ecology of Pest Management Program at Cornell University, the New York Zoological Society, the New York Cooperative Wildlife Research Unit, the Society of the Sigma Xi, the Minnesota Department of Natural Resources, the Science Museum of Minnesota, the Severtsov Institute, and the Academies of Science of the USA and Russia.

I thank R.S. Courley, A. Magoun, and M.N. Meyer for specimens, and N. Abramson, O. Rossolimo, P.P. Strelkov, and I. Ya. Pavlinov for access and assistance in collections. I especially thank B. Blake and M. Serbenyuk for implanting testosterone in voles. I thank J. Fleming for sectioning, and S. Nowland for typing the manuscript.

REFERENCES

Boonstra, R., and Youson, J.H., 1982, Hip glands in a field population of Microtus pennsylvanicus, Can. J. Zool., 60:2955.

Clarke, J.R., and Frearson, S., 1972, Sebaceous glands on the hindquarters of the vole, Microtus agrestis, J. Repro. Fert., 31:477.

Connor, P.F., 1959, The bog lemming Synaptomys cooperi in southern New Jersey, Publ. Mus.. Michigan State Univ., Biol. Ser., 1:161.

Jannett, F.J., Jr., 1975, "Hip glands" of Microtus pennsylvanicus and M. longicaudus (Rodentia; Muridae), voles "without" hip glands, Syst. Zool., 24:171.

Jannett, F.J., Jr., 1978, Dosage response of the vesicular, preputial, anal, and hip glands of the male vole, Microtus montanus (Rodentia: Muridae), to testosterone propionate, J. Mamm., 59:772.

Jannett, F.J., Jr,, 1990, Posterolateral gland positions among microtine rodents, in: "Chemical Signals in Vertebrates 5", D.W. Macdonald, D. Muller-Schwarze, and S.E. Natynczuk, eds., Oxford University Press.

Jannett, F.J., Jr., ms a, The histology of naturally occurring and experimentally induced posterolateral glands of microtine rodents.

Jannett, F.J., Jr., ms b, The testosterone dependence of the major scent glands of the sagebrush vole, Lagurus curtatus.

Jannett, J.A., and Jannett, F.J. Jr., ms, Effects of gonadectomy and testosterone on scent glands and marking behaviors of the Richardson water vole (Microtus richardsoni).

Miller, G.S., Jr., 1896, Genera and subgenera of voles and lemmings, N. Amer. Fauna, 12:1.

Quay, W.B., 1968, The specialized posterolateral sebaceous glandular regions in microtine rodents, J. Mamm., 49:427.

Schluter, D., 1989, Bridging population and phylogenetic approaches to the evolution of complex traits, in: "Complex Organismal Functions: Integration and Evolution in Vertebrates," D.B. Wake, and G. Roth, eds., John Wiley and Sons, New York.

Sokolov, V.E., 1982, "Mammal skin," University of California Press, Berkeley, California.

Sokolov, V.E., Skurat, L.N., Serbenyuk, M.A. and Jannett, F.J. Jr., 1989, Influence of testosterone on skin glands of the European bank vole, Clethrionomys glareolus, Zool. Zh., 68:116.

Tamarin, R.H., 1981, Hip glands in wild-caught Microtus pennsylvanicus, J. Mamm., 62:421.

Vrtis, V., 1930, Glandular organ on the flanks of the water rat, their development and changes during breeding season, Biol. Spisy, Brno, 9(4):1.

SOURCES OF OESTROUS ODOURS IN CATTLE

G.C. Perry and Susan E. Long

Department of Animal Husbandry
University of Bristol
Langford, Bristol, UK

INTRODUCTION

Oestrus detection in the absence of a bull is relatively poor and averages 40% in Britain (Milk Marketing Board, 1986), but 95-100% of oestrous cows running with bulls in small herds are recognised and mounted (Blockley, 1976). Some bulls have also been shown to identify pro-oestrous cows (Kerruish, 1955; Reinhardt, 1983; Stevens, 1983; MAFF, 1984). The breeding efficiency of cattle in commercial situations will therefore increase if improved oestrus detection methods are developed.

It has been shown that bulls routinely use olfactory and gustatory cues when investigating cows (Williamson et al., 1972; Blockley, 1976; Reinhardt, 1983), and can distinguish between oestrous and non-oestrous mucus and urine (Hart et al., 1946). Additional support for an olfactory role in oestrus identification comes from studies in which urine (Ladewig and Hart, 1981), vaginal swabs (Kiddy et al., 1978; Kiddy and Mitchell, 1981) and milk (Kiddy et al., 1980) from oestrous and non-oestrous cows was differentiated by dogs and cats. None of the cited work, however, provides clear evidence of the precise source of the odours since the fluids could contain secretions from several sources.

Field data

A detailed analysis of a field study was undertaken to obtain clues regarding the precise source of the signalling material (French et al., 1989). The results revealed that bulls interacted more with pro-oestrous cows than other cows and interest began to increase 4 days before oestrus. Most of the olfactory and gustatory behaviours were directed towards the perineal region of the cows. This suggested the reproductive tract, perineal skin or urine as possible source(s) of the attractant(s). Faeces were discounted as an odour source since very few incidences of faecal investigation were recorded.

Histology and histochemistry of the reproductive tract

Intact reproductive tracts from multiparous cows at various stages of the oestrous cycle were examined post mortem (Blazquez et al., 1987a). The stage of cycle at slaughter was estimated from the appearance of the ovaries, reproductive tracts and histological examination of the corpus

Chemical Signals in Vertebrates VI, Edited by R.L. Doty and
D. Müller-Schwarze, Plenum Press, New York, 1992

luteum. The tracts were excised from the vestibule to the cervix. The epithelial lining, glands and secretory ducts were examined in detail. Although histochemical staining indicated the presence of mucus in the major and minor vestibules and the Gartner's ducts, no cyclical changes were observed and it was therefore concluded that the vestibule and vagina were unlikely to be the source of the semiochemical.

The earlier behavioural observations indicated licking of the perineal area by the bull, therefore attention focused on the skin around the perineum. The perineal sweat and sebaceous glands are well developed in adult cattle (Ortmann, 1960) and there are precedents for odour production by morphologically specialised skin glands (Mykytowycz, 1970).

Perineal skin

Skin samples were taken post mortem from sexually mature cows, steers of approximately the same weight, and sexually immature (6 and 8 weeks old) female calves. Samples were removed from the perineal region and two other muco-cutaneous sites, the lower lip and the eyelids. A fourth sample was taken from the neck. Quantitative morphological examinations were carried out to determine the volumes of sweat and sebaceous glands per unit area of skin surface (Blazquez et al., 1987b). In adult cows the volumes of sweat and sebaceous glands in the vulva were significantly greater than those of the neck, lips or eyelids. Adult cows also had much greater volumes of perineal skin glands than the 6 and 8 week old female calves or steers. There was no significant difference in neck skin gland dimensions between the animal categories.

The difference in perineal skin gland volumes between adult and young females suggested possible association with puberty. Skin biopsies were removed from female cattle aged 6, 9, 15 and 18 months and sebaceous gland volumes estimated. The 15 and 18 month old animals had significantly greater sebaceous gland volumes than the 6 and 9 month counterparts and were of adult proportions. Concurrent estimates of plasma progesterone levels in blood samples taken on three occasions around the time of skin sampling showed adult levels (in excess of 4 ug/ml) in the 15 and 18 month old females in contrast to low levels (less than 1 ug/ml) in the two younger groups. The conclusion from this study was that the skin glands in the perineal region of cows are specialised and at puberty they differentiate from similar glands in males and young calves.

A preliminary study showed that the perineal skin glands enlarged in pre-pubescent female calves following oestrogen administration. This was unexpected since oestrogen reduces sebaceous gland size (Strauss et al., 1983). A trial was conducted in which a small group of 16-week old calves was given oestrogen implants and killed 14 days later. A second group given blank implants served as controls (Blazquez et al., 1987c). Plasma concentrations of oestradial-17ß were measured in all animals on the day of treatment and days 3, 8 and 14. On days 3, 8 and 14 the concentrations in the experimental group were within the physiological range for adult cows in oestrus. After implantation the experimental group had consistently greater plasma concentrations of oestradial-17ß than the control group. Histological examination showed that sebaceous gland volumes in the perineal region was greater in the experimental than control group whereas similar glands in the neck had significantly lower volumes. Sweat gland volumes in the perineal region were also greater in the experimental animals than the control group but there was no difference between the groups in sweat gland volumes in the neck. The response of the perineal skin glands, therefore, suggested a similar role to that seen in other morphologically specialised sebaceous glands such as those in the gerbil abdominal gland (Glenn and Gray, 1965).

Two further trials were conducted to strengthen the association between the perineal glands and oestrus signalling.

Adrenaline has been shown to cause a localised discharge of sweat glands in cattle (Findlay and Jenkinson, 1964). Two cows with synchronised oestrous cycles were housed with a bull. During the mid-cycle period one cow received an intradermal injection of adrenaline in the perineal region. This resulted in sweat gland discharge (Blazquez et al., 1988). Bull-cow interactions were recorded on the day before, day of and day following treatment. On the day of treatment the bull directed significantly more attention to the treated cow than to the water injected control animal. This was repeated four times with similar results. It is suggested that the adrenaline caused secretion of residual semiochemical from a previous oestrous cycle which induced courtship attention from the bull.

In the second trial, the response of the perineal sweat glands to high ambient temperatures was measured (Blazquez et al., 1991). The earlier field data did not suggest an increased bull interest in cyclic, non-oestrous cows in hot weather conditions; therefore, it was important to show that when temperatures are high the glands do not discharge secretion containing residual amounts of semiochemical.

Cutaneous moisture evaporation rates (CME) were measured from sweat glands in the lumbo-dorsal and perineal regions of dioestrous cows and bulls. The measurements were carried out in a climate chamber at dry bulb temperatures of $10^{\circ}C$ and $36^{\circ}C$. The results showed no differences in CME between sites and sexes at $10^{\circ}C$. At $36^{\circ}C$ the sweating rate in the lumbo-dorsal site significantly increased for both sexes but in the cows the perineal region sweating rate did not increase, suggesting a thermal insensitivity.

In a subsequent trial, CME rates of the perineal and lumbo-dorsal regions were measured through the oestrous cycle of two cows. The measurements were taken before and after exposure of the cow to a vasectomised bull who was allowed free access for a limited period each day. Perineal CME reached a maximum on the day of oestrus (identified by bull service) in both cows. There was no effect of stage of cycle on lumbo-dorsal CME rate.

SUMMARY

1. The perineal skin glands in the cow are highly specialised because of their (i) very large size,
    (ii) sexual dimorphism,
(iii) development at puberty,
  (iv) response to oestrogen,
    (v) low thermal sensitivity.

2. Increased discharge generates components of bull courtship behaviour.

3. Increased secretion coincides with oestrus.

REFERENCES

Blazquez, N. B., Batten, E. H., Long, S. E. and Perry, G. C., 1987a, Histology and histochemistry of the bovine reproductive tract caudal to the cervix. Part 1. The vestibule and associated glands. Br. Vet. J. 143:337.

Blazquez, N. B., Batten, E. H., Long, S. E. and Perry, G. C., 1987b, A quantitative morphological examination of bovine vulval skin glands. J. Anat. 155:153.

Blazquez, N. B., Batten, E. H., Long, S. E. and Perry, G. C., 1987c, Effect of oestradiol-17ß on perineal and neck skin glands in heifer calves. J. Endocrinol. 115:43.

Blazquez, N. B., French, J. M., Long, S. E. and Perry, G. C., 1988, A pheromonal function for the perineal skin glands in the cow. Vet. Rec. 123:49.

Blazquez, N. B., Long, S. E., Mayhew, T.M., Perry, G. C., Prescott, N. J. and Wathes, C. M. (1991) Perineal skin glands in the cow: sweat gland discharge and morphometry (submitted).

Blockley, M. A., 1976, Sexual behaviour of bulls at pasture, a review. Theriogenol. 6:387.

Findlay, J. D. and Jenkinson, D. McEwan, 1964 Sweat gland function in the Ayrshire calf. Res. Vet. Sci. 5:109.

French, J. M., Moore, G. F., Perry, G. C. and Long, S. E., 1989, Behavioural predictors of oestrus in domestic cattle, Bos taurus. Anim. Behav. 38:913.

Glenn, E. M. and Gray, J., 1965, Effect of various hormones on the growth and histology of the gerbil (Meriones inguiculatus) abdominal sebaceous gland pad. Endocrinol. 76:1115.

Hart, G. H., Mead, S. W. and Regan, W. M., 1946, Stimulating the sex drive of bovine males in artificial insemination. Endocrinol. 39:221.

Kerruish, B. M., 1955 The effect of sexual stimulation prior to service on the behaviour and conception rate of bulls. Br. J. Anim. Behav. 3:125.

Kiddy, C. A. and Mitchell, D. S., 1981. Estrus related odors in cows, time of occurrence. J. Dairy Sci. 64:267.

Kiddy, C. A., Canley, H. H. and Hawk, H. W., 1980, Identification by trained dogs of milk samples from estrous cows. J. Dairy Sci., Suppl. 63:91.

Kiddy, C. A., Mitchell, D. S., Bolt, D. J. and Hawk, H. W., 1978, Detection of estrus-related odors in cows by trained dogs. Biol. Reprod. 19:385.

Ladewig, J. and Hart, B. L., 1981, Demonstration of estrus-related odors in cows' urine by operant conditioning of rats. Biol. Reprod. 24:1165.

MAFF, 1984, The cattle industry and welfare in Great Britain. ADAS/MAFF Booklet 2482:43.

Milk Marketing Board, 1986, Checkmate 1985. Farm Information Unit, Report 47, A. H. Poole and S. J. Mabey, eds., Milk Marketing Board Farm Management Services Information Unit, Reading, UK.

Mykytowycz, R., 1970, The role of skin glands in mammalian communication, in: "Advances in Chemoreception", J. W. Johnson, Jr., D. G. Moulton and A. Turk, eds., Appleton-Century-Crofts, New York.

Ortmann, R., 1960, Die Analregion der Säugetiere. Handbuch der Zoologie 8:63.

Reinhardt, V., 1983, Flehmen, mounting and copulation among members of a semi-wild cattle herd. Anim. Behav. 31:641.

Stevens, K., 1983, The role of olfaction during oestrus detection in sheep and cattle. Ph.D. thesis, University of Bristol.

Strauss, J. S., Downing, D. T. and Ebling, F. J., 1983, Sebaceous glands, in: "Biochemistry and Physiology of the Skin", L. A. Goldsmith, ed., Oxford United Press, Oxford.

Williamson, N. B., Morris, R. S., Blood, D. C., Cannon, C. M. and Wright, P. J., 1972, A study of oestrous behaviour and oestrous detection methods in a large commercial dairy herd. 2. Oestrous signs and behaviour patterns. Vet. Rec. 91:58.

SECTION THREE

DEVELOPMENT OF CHEMOSENSORY SYSTEM STRUCTURE AND FUNCTION

DEVELOPMENT OF OLFACTORY AND TASTE RESPONSES TO CHEMICAL SIGNALS

IN ACIPENSERID FISHES

Alexander O. Kasumyan

Department of Ichthyology
Biological Faculty
Moscow State University
Moscow, Russia

INTRODUCTION

The functional development of chemoreceptor systems in ontogeny is one of the scantily explored aspects of chemoreception in fishes (Appelbaum, 1980). Many authors investigating the development of olfaction in fishes have discovered in young larvae an ability to react to biologically relevant odorants. Detailed studies of the time when this ability appears and the time when juveniles evidence sensitivity to such stimuli have been undertaken on cyprinids using food odours (Kasumyan and Ponomarev, 1990) and alarm pheromones (Smith, 1977; Waldman, 1982; Pachschenko and Kasumyan, 1983, 1986).

This paper describes a comparative study of the development, in acipenserids, of the ability to smell and taste (intraorally and extra-orally) natural and artificial food substances. Behavioral tests were used to estimate the level of development of the chemoreceptory systems by determining the threshold concentrations of the chemicals and the fishes' reaction to them (e.g., whether the fish were repelled, or attracted, deterred or stimulated).

MATERIALS AND METHODS

The test fishes were larvae and juveniles of 5 species of acipenserids: Russian sturgeon (Acipenser gueldenstaedti), Siberian sturgeon (A. baeri), stellate sturgeon (A. stellatus), green sturgeon (A. medirostris) and hausen (Huso huso). Acipenserids were chosen because of their good olfactory sensitivity and because they have taste buds not only in their oral cavity but also on their barbels. This provided an opportunity to study their intra- and extraoral taste systems separately.

Determination of olfactory responses

Special aquaria with closed water circulation were used (Kasumyan and Ponomarev, 1986). Water from an aquarium went to a biofilter, the last section of which contained granulated activated carbon for additional cleaning the water of the test agents. From the biofilter, the water went through pipes into parallel compartments of the aquarium. In the course of the experiments, the supply of clean water in one of the compartments was replaced by the test solution for 2-3 minutes.

Solutions of amino acids (1-isomers, Fluka AG) and also water extracts of live daphnia (Daphnia pulex) or chironomid larvae (Chironomidae) were used as olfactory stimuli. The strength of the behavioral reaction was estimated according to a specially-designed 5-point scale which took into account the degree of disturbance of the initial organization of the fish in the aquarium, their ability to localize the compartment with the test solution, and their display of special behavioural movements. Fish were made anosmic, in some trials, by cauterization of the olfactory rosettes. The completeness of the destruction of the rosettes was evaluated using a microscope.

Determination of taste responses

Some fish (10-20) were housed in an aquarium 10-15 min before the experiment. Special food pellets (50-100) were then placed in the middle of the aquarium. The pellets were made either of starch or of agar-agar. Commercial starch was preliminarily purified. For preparing the gel, 12 g of starch were mixed with 100 ml of clean water (for blank pellets) or the test solution (for test pellets). In the experiments using the starch pellets, the test substances were sodium chloride (10, 1 and 0.1%), calcium chloride (10, 1 and 0.1%), citric acid (5, 1, 0.1 and 0.01%) and sucrose (25, 10 and 1%). The agar-agar pellets were made of 2% gel. After dissolving, agar-agar solutions of amino acids and the dye (chromium oxide) were added to the hot gel. In the experiments with agar-agar pellets, the test substances were alanine, arginine, asparagine, valine, histidine, glycine, glutamine, lysine, methionine, norvaline, proline, serine, threonine, phenylalanine, cysteine (0.1 M), aspartic acid, glutamic acid, isoleucine, leucine, tryptophane (0.01 M) and tyrosine (0.001 M). The gel used for the preparation of blank pellets contained only chromium oxide.

In the course of the experiment (which lasted from 5 to 10 minutes) the number of bites was registered while the fishes were taking the pellets from the bottom of the aquarium. After the experiment, the remaining pellets were counted. The size of the aquarium, the number of fish, the number of placed pellets, and the duration of the experiment were the same in all experiments of the same series.

Biting of the pellets occurred only after preliminary touching with one of the four barbels which are covered with taste buds. By counting the number of bites it was possible to make conclusions about the effectivenes of the test agent for the extraoral taste sensitivity. In order to estimate the effectiveness of the test agent for intraoral taste sensitivity, the percentage of consumed pellets was counted according to the number of bites.

RESULTS

The development of olfactory responses

Experiments with larvae of Russian, Siberian and stellate sturgeons were carried out at the time when the change over to exogenous feeding occured ("start feed" age). At that time the larvae were about 17-19 mm long and 8-10 days old, counted from the moment of hatching. The introduction of the extract of chironomid and daphnium into the aquarium did not change any of the behaviors of these larvae. Young fish of the second age group were about 22mm long and 12-15 days old. At that time, they were free of the yolk sack and had changed over fully to exogenous feeding. The introduction of the extract (1 g/l) into the aquarium caused a pronounced behavioral reaction: within 60-90 sec., 50-60% of the acipenserid larvae moved closer to the bottom and nearer to the compartment where the extract was supplied. They evidenced loop-like movements of different lengths. The frequency of bites increased. The 0.1 and 0.01 g/l extract concentrations caused weaker reactions, and in some experiments there was no reaction at all. In juveniles 25-27 days old and 28.5-31.0 mm long, the concentration 0.1 g/l caused a stronger reaction, the initial signs of which could be noticed 5-15 sec after the beginning of the experiment. After about 60 sec (when the reaction reached maximal intensity), 70-80% of the juveniles congregated at the bottom near the entrance to or inside the compartment with the extract. All the juveniles actively and repeatedly moved along circular and S--shaped trajectories, sharply intensifying the biting frequency. The reaction faded in 3-4 min after cutting the supply of the extract to the aquarium. Responses of juveniles to the 0.01 and 0.001 g/l concentrations were also observed, though the reaction was much weaker. The reaction of juveniles 32-35 days old and 35-45 mm long to high concentrations of extract was much more strong than that of the previous age group. Weak responses were observed to other extract concentrations (0.0001 g/l). The degree of responsiveness to the extract in young fishes of 60 days old and 70-80 mm long was no different from that of the 32-35 day olds.

In young Russian, Siberian and stellate sturgeons of 2-3 days of age (80-120 mm long), a strong food searching reaction was produced by glycine and alanine (0.1 mM). The threshold concentration for the amino acids was approximately 0.001 mM. The amino acids serine, proline, cysteine, leucine, isoleucine, asparagine, aspartic acid, lysine, glutamine, glutamic acide, tyrosine, histidine-HCl, arginine-HCl, norvaline, valine, threonine, phenylalanine, methionine and betaine were ineffective at the 0.1 mM concentration. Glycine at 0.1 mM also caused a behavioral reaction in the young green sturgeons (250-300 mm) and hausens (150-200 mm). Alanine at 0.1 mM was ineffective for the hausen. Young Russian sturgeons were capable of reacting to 0.01 mM glycine solutions after the change over to full exogenous feeding. By the age of 35 days they reacted to the 0.001 mM glycine.

After cauterizing the olfactory rosettes, the juvenile acipencerids reacted neither to the extracts of chironomid larvae nor the solutions of amino acids at normally effective concentrations.

The development of taste sensitivity

The development of taste sensitivity to sour, bitter, sweet and salty stimuli was investigated in the Siberian sturgeon using the starch pellets. It was discovered that the 8-10 day-old, 17-19 mm-long larvae bit the pellets containing 10% and 1% calcium chloride less often (p < 0.05) than the blank pellets. Pellets containing various concentration of sucrose, citric acid, and sodium chloride were bit by the fishes at the same frequency as the blank ones. The relative consumability of the pellets (a measure of the intraoral taste sensitivity) was lower for the pellets containing 0.1% and 10% calcium chloride (p < 0.01) and 10% sodium chloride (p < 0.05). In the remaining cases, no differences from the blank pellets were found. Larvae that had fully changed to exogenous feeding and were 15-17 days old (22-24 mm long) demonstrated the same reaction to the test agents as the previous age group. Juveniles 36-38 days old and 49-52 mm in length gave the same reactions to the pellets containing calcium chloride, sodium chloride and sucrose. However, the behavior of the fish to the sour stimuli sharply changed: the number of bites directed to the pellets containing 5%, 1% and 0.1% citric acid increased by 3.5, 3.7 and 1.9 times, respectively. Biting of the 0.01% citric acid pellets did not differ much from that of the blank ones. The relative consumability of the pellets with 5% and 1% citric acid was much lower than that of the blank ones (p < 0.01). The 0.1% and 0.01% concentration pellets received no more or less attention than the blank ones.

The determination of the taste sensitivity to amino acids was carried out for young Russian sturgeons the using agar-agar pellets. It was established that the increase in the number of bites in fishes 12-15 days old and 21-25 mm long was caused by 11 amino acids in the following manner: cysteine > histidine > glutamine > asparagine > glutamic acid > lysine > glycine > aspartic acid > methionine > serine > threonine. Phenylalanine, isoleucine, alanine, norvaline, arginine, proline, tryptophane, valine, tyrosine and leucine did not cause a statistically significant increase in biting frequency. The increase in the relative pellet consumability was caused only by one amino acid - methionine (p<0.05). In older juveniles 35-40 days old and 60-80 mm in length, the increase in biting frequency was caused by 16 amino acids in this order: aspartic acid > glutamic acid > cysteine > histidine > threonine > lysine > arginine > asparagine > glutamine > serine > tryptophane > phenilalanine > alanine > methionine > leucine > isoleucine. Glycine, tyrosine, valine, proline and norvaline were ineffective. The order of the effectiveness of the amino acids for increasing consumption was: cysteine > aspartic acid > tryptophane > glutamic acid > arginine. Other amino acids proved ineffective in altering the intraoral taste sensitivity of the young Russian sturgeons.

DISCUSSION

The experiments described in this paper show that responsiveness to olfactory and gustatory stimuli appears very early in the ontogeny of acipenserid fishes. Evidently the extraoral taste function in these fishes starts developing first, because the larvae of "start feed" age exhibit the same wide range of responsiveness to amino acids and to sweet, bitter and salty stimuli as older juveniles. Young fish of these two age groups differ much more in the range of the effective stimuli for intraoral taste responsiveness, which suggests that the functional development of this sensory system is slower. Olfactory development in the ontogeny of acipenserid fishes starts later than taste development and lasts longer. The time period of the development of the sensitivity to food odours in acipenserids is quite different from that of cyprinids (Kasumyan and Ponomarev, 1990), which points out the variation in developmental parameters in different fish species and the peculiarities of their ecology. Acipenserids differ from other species in their responsiveness to sour, bitter, sweet and salty stimuli (Appelbaum, 1980) and also to amino acids (Caprio, 1982; Marui and Kiyohara, 1987; Jones, 1989). This suggests that each type of fish has its own specific set of effective taste stimuli, whether attractants or repellents.

REFERENCES

Appelbaum, S., 1980, Versuche zur Geschmacksperzeption einiger Susswassserfische im larvalen und adulten Stadium, Arch. Fischereiwiss., 31, 2:105-114.
Caprio, J., 1982, High sensitivity and specificity of olfactory and gustatory receptors of catfish to amino acids, in: "Chemoreception in Fishes", T.J. Hara, ed., Elsevier, Amsterdam.

Jones, K.A., 1989, The palatability of amino acids and related compounds to rainbow trout, _Salmo gairdneri_ Richardson, _J. Fish Biol._, 34, 1:149-160.

Kasumyan, A.O., Ponomarev, V.Yu., 1986, Study of the behaviour of zebrafish, _Brachidannio rerio_, in response to natural chemical food stimuli, _J. Ichthyol._, 26, 5:96-105.

Kasumyan, A.O., Ponomarev, V.Yu., 1990, The development of food search behaviour to natural chemical signals in the ontogeny of cyprinid fishes, _Voprosy Ichthyologyi_, 30, 3:447-456.

Marui, T., Kiyohara, S., 1987, Structure-activity relationships and response features for amino acids in fish taste, _Chem. Senses_, 12, 2:265-275.

Pashschenko, N.I., Kasumyan, A.O., 1983, Some morpho-functional peculiarities of the olfactory organ in ontogeny of the minnow, _Phoxinus phoxinus_ (Cypriniformes, Cyprinidae), _Zoological J._, 62, 3:367377.

Pashschenko, N.I., Kasumyan, A.O., 1986, Morpho-functional features of the olfactory organ of cyprinid fishes. I. Morphology and function of the olfactory organ during ontogeny of the white amur, _Ctenopharyngodon idella_, _J. Ichthyol._, 26, 5:96-105.

Smith, R.J.F., 1977, Chemical communication as adaptation: alarm substance of fish, _in_: "Chemical signals in vertebrates", D. Müller-Schwarze and M.M. Mozell, eds., Plenum Press, New York.

Waldman, B., 1982, Quantitative and developmental analyses of the alarm reaction in the zebra danio, _Brachidanio rerio_, _Copeia_, 1:1-9.

DEVELOPMENT OF THE OLFACTORY AND TERMINALIS SYSTEMS IN WHALES AND DOLPHINS

Helmut A. Oelschläger

Dept.of Anatomy, J.W.Goethe-University
Theodor Stern-Kai 7
6000 Frankfurt am Main 70, FRG

INTRODUCTION

In the early ontogenesis of mammals, the chemoreceptor systems of the nose (olfactory and accessory olfactory complex) originate from the olfactory placode together with the terminalis system (Oelschläger and Buhl, 1985a, b; Buhl and Oelschläger, 1986; Oelschläger, 1988, 1989). The embryonal chemoreceptor epithelia send axons and possibly Schwann cells to the rostrobasal part of the telencephalon. Here they induce the formation of an olfactory bulb, which is apparent in an evagination of the telencephalic wall, and of an accessory olfactory bulb, which may be reduced later together with the vomeronasal nerve and Jacobson's organ. The terminalis system, however, the functional implications of which are only imperfectly known, is represented by neuroblasts which migrate from the olfactory placode to the rostromedial extremity of the basal telencephalic hemisphere, adjacent to the olfactory bulb. While the terminalis neuroblasts send out central processes to the brain wall, they obviously retain connections with the dorsomedial part of the olfactory placode. In the adult stage of mammals the terminalis system, which consists of a ganglionated fiber plexus, connects the olfactory and respiratory mucosa of the nasal septum with rostral and basal parts of the prosencephalon. The neurons belong to at least two subpopulations of cells which differ in the size of their perikarya and nuclei. Whether these two types of neurons are identical with cell populations characterized with histochemical and immunocytochemical methods (Bojsen-Møller, 1975; Wirsig and Leonard, 1986; Silverman and Krey, 1978; Oelschläger and Burda, this volume; Oelschläger and Northcutt, 1991, submitted), however, is not clear. Taken as a whole, the terminalis system may innervate arteries of the basal forebrain and of glands and blood vessels in the mucous membrane lining the septal wall of the nasal cavity (Johnston, 1914; Larsell, 1918, 1950). A fraction of mostly bipolar terminalis neurons immunoreactive for luteinizing hormone-releasing hormone (LHRH) obviously are involved in the establishment and maturation of the brain-pituitary-gonadal circuit (Jennes and Stumpf, 1980; Schwanzel-Fukuda et al., 1981, 1985; Oelschläger and Northcutt, 1991, submitted). In the ontogenesis of rats, LHRH immunoreactivity first appears in peripheral (nasal) and central (meningeal) ganglion cells of the nervus terminalis (Jennes and Stumpf, 1980; Schwanzel-Fukuda and Silverman, 1980; Schwanzel-Fukuda et al., 1985, 1988; Jennes, 1989). Correspondingly, after the establishment of the hypothalamo-hypophyseal-gonadal axis in rat and mouse, the number of LHRH neurons seems to decrease from late fetal stages onward (Schwanzel-Fukuda et al., 1987; Schwanzel-Fukuda and Pfaff, 1989).

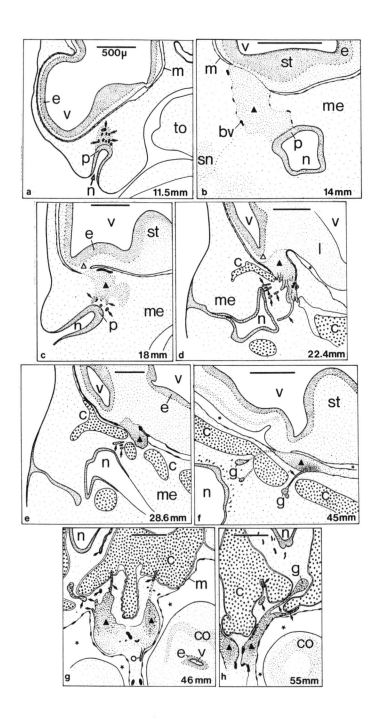

TOOTHED WHALES (ODONTOCETI)

The early development of the chemoreceptor and terminalis systems in odontocetes corresponds fairly well to the situation found in other mammals (Figs. 1a–d, 2a–c). However, the nasal pit does not develop into a chambered nasal cavity and Jacobson's organ, the vomeronasal nerve, and the accessory olfactory bulb have not been found in toothed whales (Oelschläger and Buhl, 1985a, b; Buhl and Oelschläger, 1986; Oelschläger et al., 1987). A centrifugal migration of neuroblasts (Pearson, 1941; Sinclair, 1966; Bossy, 1980; Schwanzel–Fukuda et al., 1985, 1988) from the telencephalon to the future ganglion terminale was not observed in our investigations on odontocetes. Figures 1 and 2 show the embryonal and early fetal development of the harbor porpoise nose. As the chemoreceptor and terminalis axons arrive at the primitive meninx (Figs. 1b, c, 2b), the latter seems to be dissolved in some places and the axons enter the telencephalic wall in order to induce its evagination (primordial olfactory bulb) and thus help to form the definitive olfactory bulb of mammals (Figs. 1d, 2c). While human embryos of 23 mm crown-rump length (CRL) exhibit the separation of terminalis ganglion and olfactory bulb, no clear separation was found in toothed whales (Figs. 1d, e, 2d). Instead, the whole olfactory/terminalis complex changes in shape and staining characteristics. Obviously the chemoreceptor axons and Schwann cells are reduced while the terminalis neuroblasts show advanced maturity (enlargement and enhanced acidophilia of perikaryon). At about 45 mm CRL (Figs. 1f, 2e), the whole placodal material gives the impression of a ganglion which is gradually uncoupled from the telencephalon through the reappearance of the thin layer of primitive meninx (Figs. 1d–f, 2d). In toothed whales, the main terminalis ganglion divides into smaller clusters of neurons which persist until the adult stage, embedded in the dura mater of the ethmoid region (Ridgway et al. 1987). While the olfactory complex (Fig. 4b–e) is totally reduced (adult toothed whales are anosmic), the number of terminalis neurons in dolphins attains a maximum within mammalia (Sinclair, 1951; Buhl and Oelschläger, 1986; Oelschläger et al., 1987; Ridgway et al., 1987; Demski et al., 1990). In a narwhal of 137 mm total length (TL), each of the two main terminalis ganglia comprised about 10,000 neurons which obviously belonged to two subpopulations. In early fetal stages (Fig. 1f–h), the central (meningeal) terminalis ganglia are still connected to peripheral (submucosal) ganglia via terminalis fiber bundles running through persisting foramina of the cribriform plate. It is presumed that in adult dolphins these topographical relations are retained.

BALEEN WHALES (MYSTICETI)

More than in odontocetes, the development of the nose in mysticetes corresponds to the situation found in mammals, in general (Oelschläger 1989).

Fig. 1. Survey on the embryonal and early fetal development of the nose in the harbor porpoise. a) and b) Stage of axonal sprouting and migration of terminalis neuroblasts; axons not shown. a) sagittal, b) transverse section. c) Fusion of placodal material and telencephalon (definitive olfactory bulb), sagittal section. d) Chondrification of cribriform plate. Stadium optimum of olfactory bulb; sagittal section. e) and f) Transformation and uncoupling of placodal material, reduction of olfactory component; sagittal sections. g) and h) Persistence of meningeal terminalis ganglia; horizontal sections. bv, blood vessel; c, cribriform plate; co, cortical plate; e, ependyma; g, peripheral terminalis ganglion; l, lamina terminalis; m, primitive meninx; me, mesenchyme; n, nasal cavity; p, olfactory placode; sn, nasal septum; st, primitive striatum; to, tongue; v, ventricular system; arrows, fila olfactoria/ terminalia; light triangle, telencephalic component (primordium) of olfactory bulb; black triangle, placodal component of olfactory bulb and/or main terminalis ganglion; open circle, rootlet of terminalis nerve; asterisk, artifact.

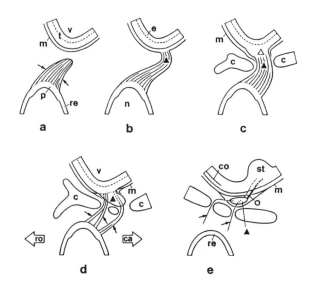

Fig. 2. Embryonal and early fetal development of the olfacto-terminalis complex in toothed whales. For explanation see text. ca, caudal; re, respiratory mucosa; ro, rostral; t, telencephalic wall. For other abbreviations and symbols see Fig. 1.

While bundles of axons from chemoreceptor cells (fin whale of 20 mm CRL) grow into the direction of the rostrobasal telencephalic hemisphere, dense clouds of cells leave the medial part of the olfactory placode. The cells stain dark as the placode itself, and therefore can easily be distinguished from the light olfactory bulb, although the forming main ganglion attaches to the bulb medially (Fig. 3). The ganglion cells (thousands of neurons) again are premature in comparison, e.g., with neuroblasts in the telencephalic wall and show relatively large perikarya and nuclei, and one to three nucleoli. In the fin whale of 60 mm CRL, the peripheral processes of the terminalis ganglion run between the septal mucosa and the nasal septum and the central processes enter the brain in the septal region. In contrast to toothed whales, fin whales of 60 and 105 mm CRL show relatively complicated nasal cavities with two nasal folds and a broad distribution of olfactory fiber bundles. In the further course of mysticete ontogeny, the nose is far less modified than in odontocetes (Fig. 4f). Due to allometric phenomena, the olfactory bulb becomes a finger-like evagination of the telencephalon that shows a marked olfactory ventricle and the beginnings of laminar organization (blue whale of 170 mm TL). Intermediate and more caudal olfactory structures (tractus olfactorius, regio retrobulbaris, tuberculum olfactorium) are also obvious. A juvenile minke whale of 5.80 m (Weber, 1928) exhibited complicated nasal cavities with a total of five turbinals in each. Adult baleen whales retain a reduced but certainly functional main olfactory system with three rows of olfactory foramina in each side of the cribriform plate and two or three ethmoturbinals which presumably are covered with olfactory epithelium. They have inconspicuous olfactory bulbs and very slender olfactory tracts which resembles the situation in pinnipeds and humans. The vomeronasal nerve and accessory olfactory bulb are lacking from the beginning (Oelschläger, 1989).

EVOLUTIONARY ASPECTS

Comparison of nasal organization in toothed and baleen whales reveals some profound differences (Fig. 4). In odontocetes, the modification of the mammalian nose is far more advanced than in baleen whales: The nostrils have

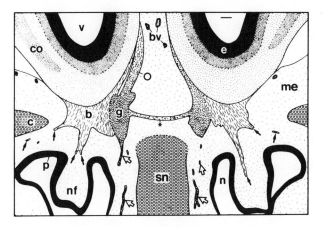

Fig. 3. Olfacto–terminalis complex in a 60 mm CRL fetus of the fin whale. Transverse section. b, olfactory bulb; g, main terminalis ganglion; nf, nasal fold; asterisk, transverse fiber bundle; black arrows, fila olfactoria; open arrows, fila terminalia. For other abbreviations and symbols see Fig. 1.

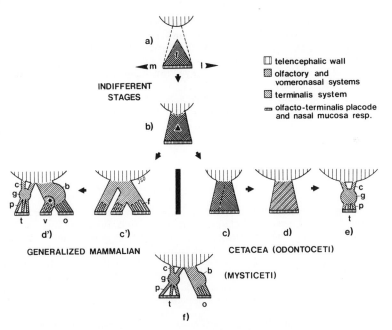

Fig. 4. Development of the olfacto–terminalis complex in generalized mammalia (a–d'), in toothed whales (a–e), and in baleen whales (f). b, definitive olfactory bulb; c, central terminalis rootlet; f, fila olfactoria; g, definitive terminalis ganglion; l, lateral; m, medial; o, olfactory system; p, peripheral terminalis fiber; t, terminalis system; v, vomeronasal system; black triangle, olfactory bulb anlage; black dot, accessory olfactory bulb. Courtesy: Verlag S. Karger, Basel (Switzerland)

been fused to a single blowhole, the nasal cavity is transformed to a "respiratory tube" and surrounded by a large melon and accessory air sacs that are obviously involved in the generation of sonar signals (Norris, 1964). Baleen whales have only a small spermaceti organ (Heyning and Mead, 1990) and do not emit ultrasound. Moreover, they lack specific odontocete features involved in the protection of the dolphin's ear against sound emission and allowing the perception of echoes of their own signals (echo-location) via the hypertrophied fat body in the lower jaw (Oelschläger, 1990). The fact that, while in adult odontocetes the rostral olfactory system is totally reduced, the terminalis system reaches a maximum in these animals, clearly indicates that the nervus terminalis is only distantly related to the olfactory system functionally. Ridgway et al. (1987) and Demski et al. (1990) found, in adult dolphins, numerous terminalis fiber strands and ganglia with up to 1,000 large round neurons enclosed by capsules of satellite cells. It has been discussed whether this hypertrophied terminalis system could represent a fast-conducting afferent pathway for the control of the emission of sonar signals in the accessory air sacs of the nasal tract, which could be triggered by the efferent (motor) part of the well-developed facial nerve and the blowhole musculature (Oelschläger, 1989). Baleen whales have not yet been studied as to their mature terminalis system. Because they lack most of the specific odontocete features mentioned above it seems plausible that they will not show a hypertrophy of the terminalis system as seen in toothed whales.

ACKNOWLEDGMENTS

The author is indebted to Milan Klima (Department of Anatomy, University of Frankfurt am Main, FRG), William B. Perrin (Southwest Fisheries Center, La Jolla, San Diego, U.S.A.), J.F. Willgohs (Zoologisk Museum, University of Bergen, Norway), P. Deimer (Hamburg, FRG), and to E.C. Boterenbrood (Hubrecht Laboratory, University of Utrecht, The Netherlands) for generous supply with prenatal cetacean specimens. I am very grateful to Kirsten K. Osen (Department of Anatomy, University of Oslo, Norway) for kind support and hospitality during my stay in her laboratory.
My sincere thanks go to Jutta S. Oelschläger for her skillful help in the preparation of the manuscript and the original figures.

REFERENCES[*]

Buhl, E.H., and Oelschläger, H.A., 1986, Ontogenetic development of the nervus terminalis in toothed whales. Evidence for its non-olfactory nature, Anat.Embryol. 173: 285-294.

Demski, L.S., Ridgway, S.H., and Schwanzel-Fukuda, M., 1990, The terminal nerve of dolphins: Gross structure, histology and luteinizing hormone – releasing hormone immunocytochemistry, Brain Behav.Evol. 36: 249-261.

Heyning, J.E., and Mead, J.G., 1990, Evolution of the nasal anatomy of cetaceans, in: "Sensory Abilities of Cetaceans", J. Thomas and R. Kastelein, eds. (NATO ASI Series A, Vol. 196) Plenum Press, New York, pp. 67-79.

Oelschläger, H.A., 1988, Persistence of the nervus terminalis in adult bats: A morphological and phylogenetical approach, Brain Behav.Evol. 32: 330-339.

Oelschläger, H.A., 1989, Early development of the olfactory and terminalis systems in baleen whales, Brain Behav.Evol. 34: 171-183.

Oelschläger, H.A., 1990, Evolutionary morphology and acoustics in the dolphin skull, in: "Sensory Abilities of Cetaceans, J.Thomas and R.Kastelein, eds. (NATO ASI Series A, Vol. 196) Plenum Press, New York, pp. 137-162.

Oelschläger, H.A., and Buhl, E.H., 1985b, Development and rudimentation of the peripheral olfactory system in the harbor porpoise Phocoena phocoena (Mammalia: Cetacea), J.Morphol. 184: 351-360.

Oelschläger, H.A., and Burda, H. (this volume), LHRH—immunocytochemistry in the nervus terminalis of mammals.

Oelschläger, H.A., und Northcutt, R.G., 1991, Luteinisierendes Hormon – Releasing Hormon (LHRH) in Nervus terminalis und Vorderhirn der Fledermaus Eptesicus fuscus. Eine immuncytochemische Untersuchung. Verh.Anat.Ges. 84 (Anat.Anz.Suppl. 168): 499–500.

Oelschläger, H.A., and Northcutt, R.G., 1991, Immunocytochemical localization of luteinizing hormone-releasing hormone (LHRH) in the nervus terminalis and brain of the big brown bat, Eptesicus fuscus (Journal of Comparative Neurology, in press).

Oelschläger, H.A., Buhl, E.H., and Dann, J.F., 1987, Development of the nervus terminalis in mammals including toothed whales and humans, in: "The Terminal Nerve (Nervus Terminalis). Structure, Function and Evolution", L.S. Demski and M. Schwanzel-Fukuda, eds., Ann.NY Acad.Sci. 519: 447–464.

Ridgway, S.H., Demski, L.S., Bullock, T.H., and Schwanzel-Fukuda, M., 1987, The terminal nerve in odontocete cetaceans, in: "The Terminal Nerve (Nervus Terminalis). Structure, Function and Evolution", L.S. Demski and M. Schwanzel-Fukuda, eds., Ann.NY Acad.Sci. 519: 201–212.

Sinclair, J.G., 1951, The terminal olfactory complex in the porpoise. Tex. J.Sci. 3: 251.

Sinclair, J.G., 1966, The olfactory complex of dolphin embryos, Tex.Rep.Biol. Med. 24: 426–431.

---

* For other references see Oelschläger et al., 1987; Oelschläger, 1989, 1990; Oelschläger and Burda, this volume; Oelschläger and Northcutt, 1991.

INTERACTION BETWEEN TYROSINE HYDROXYLASE EXPRESSION AND OLFACTORY AFFERENT INNERVATION IN EMBRYONIC RAT OLFACTORY BULB IN VIVO AND IN VITRO

Harriet Baker[1] and Albert I. Farbman[2]

[1]Cornell University Medical College at the Burke Medical Research Institute, White Plains, NY
[2]Northwestern University, Evanston, IL

INTRODUCTION

A regulatory role for the olfactory bulb in the development and maintenance of the olfactory epithelium has been well documented. Removal of the olfactory bulb or sectioning of the olfactory nerves results in the death of mature receptor cells. Sensory cells are reconstituted from stem cells, found in the basal layer of the epithelium, which divide, differentiate and send out axons to reinnervate the olfactory bulb (Graziadei and Monti Graziadei, 1978). In vitro studies (Chuah and Farbman, 1983) have demonstrated that the presence of the olfactory bulb is necessary for the normal maturation of olfactory receptor neurons. Similarly, there is evidence for a role of the olfactory receptor innervation in the development of the olfactory bulb. Either deafferentation or odor deprivation in neonates results in abnormal development of the olfactory bulb, including reduced size and fewer neurons (for review, see Brunjes and Frazier, 1986). In addition, there is evidence for an inductive role of the olfactory afferent innervation. Supernumerary olfactory bulbs can be produced by transplantation of additional olfactory bulbs, and glomerular-like structures can be induced by regenerating afferent fibers which innervate the forebrain of bulbectomized mice (Graziadei et al., 1980; reviewed in Brunjes and Frazier, 1986).

Afferent innervation also appears to play a role in the maintenance of neurotransmitter phenotype in the adult olfactory bulb. Either deafferentation or neonatal and adult odor deprivation produce decrements in the expression of the dopamine phenotype in the intrinsic juxtaglomerular dopamine neurons (Nadi et al., 1981; Baker et al, 1983; Brunjes, et al, 1985; Kosaka et al., 1987; Baker, 1990b; Ehrlich, 1990, Stone et al. 1990; Stone et al., 1991). Synthesis of the peptide, substance P, may be similarly altered (Kream et al., 1984). A number of lines of evidence indicate that the changes in dopamine biosynthetic capacity occur without neuronal loss; that is, the neurons remain, but no longer express the phenotype (Kosaka et al, 1987; Baker, 1990b, Stone et al., 1990; Stone et al., 1991).

No clear correlation exists between afferent innervation and the differentiation of neurotransmitter phenotype during olfactory embryogenesis. The role of afferent innervation in the regulation of the

Chemical Signals in Vertebrates VI, Edited by R.L. Doty and
D. Müller-Schwarze, Plenum Press, New York, 1992

dopamine phenotype in the adult olfactory bulb suggested that the initial expression of phenotype might be similarly regulated during development. Previous correlations, with respect to the spatial and temporal aspects of afferent innervation and dopamine expression, were the synthesis of data obtained from different studies (Allen and Akeson, 1985; Farbman and Squinto, 1985; Monti Graziadei et al., 1980; McClean and Shipley, 1988; Matsutani, 1988; Specht et al., 1981). However, the findings suggested that afferent innervation of the olfactory bulb, as indicated by the presence of immunoreactivity for olfactory marker protein (OMP), preceded dopamine expression. The latter was demonstrated utilizing antisera to tyrosine hydroxylase, the first enzyme in the dopamine biosynthetic pathway. The studies described here investigated, in the same animals, the spatial-temporal relationship between afferent innervation and expression of the dopamine phenotype. To further examine these questions and to investigate the possible trophic role of afferent innervation, explant cultures of embryonic olfactory epithelium and olfactory bulb were utilized. Lastly, the relationships were addressed between afferent innervation, the expression of tyrosine hydroxylase, and the formation of glomeruli.

METHODS AND RESULTS

Development in vivo

Embryos were obtained from timed-pregnant Sprague-Dawley rats (the animals were sperm positive on embryonic day 1, E1). Embryos (E14 and E15) were either immersed in fixative (4% formaldehyde generated from paraformaldehyde, contained in 0.1M phosphate buffer) or were perfused with the same fixative (E16 and older). Sections were prepared in a cryostat and stained according to previously published procedures (Stone et al., 1991).

At E14 and E15 lightly labelled OMP-containing cells were observed in the olfactory epithelium. By E15, the anlage of the olfactory bulb was clearly outlined by the incoming receptor afferent fibers. Although TH-immunoreactivity was extensive in other brain regions and in peripheral ganglia, immunoreactivity was not observed in the anlage of olfactory bulb. At E16, afferent innervation was more extensive, and what appeared to be migrating OMP-immunoreactive cells were observed. TH-immunoreactivity still was not present. By E18, the afferent innervation completely surrounded the olfactory bulb and, for the first time, a few TH-immunoreactive cells were observed in the presumptive glomerular region. At E19, a large increase was observed in the number of TH-immunoreactive cells. At E20, the first glomeruli were discernable as clusters of OMP-immunoreactive terminals. Their number and definition increased significantly by P1. At E22, heavily labelled TH-immunostained cells were observed in the glomerular region and more lightly-labelled cells occurred in the mitral and external plexiform layers. OMP-immunolabelled fibers also were found in these ectopic sites suggesting that the afferent fibers might induce TH expression.

Development in vitro

The region of the presumptive olfactory bulb (POB) and the olfactory epithelium were explanted from E15 embryos and co-cultured for seven days (approximately E22) according to the methods of Chuah and Farbman (1983). The cultures were fixed, sectioned on a cryostat and immunostained as described above. Cultures also were prepared of either POB alone or POB and epithelium, which were separated and subsequently reapposed. The en bloc POB and epithelium co-cultures exhibited significant development of

OMP-containing olfactory receptor cells and processes. Many TH-immunoreactive cells were observed in the olfactory bulb tissue but without a
periglomerular distribution. Cultures of POB alone and those prepared by
separating and reapposing POB and epithelium contained many fewer TH-
immunostained cells. Table I (below) compares the number of TH-containing cells in the three types of cultures and the number of TH-labelled
cells observed during development in vivo.

TABLE I. Number of TH-Containing Cells

AGE/TYPE OF CULTURE	NUMBER OF CELLS
E19	1,196 ± 70[*]
E22	2,725 ± 270
P1	5,840 ± 616
POB plus Epithelium	2,803 ± 1,001
POB alone	589 ± 262
POB/Epithelium recombined	606 ± 210

[*]Data expressed as mean ± SEM

Note that the number of cells in the E22 embryos is similar to that
observed in the POB and epithelium co-cultures.

DISCUSSION

These data demonstrate a direct temporal and spatial correlation
between the presence and distribution of receptor afferent innervation
and the expression of TH-immunoreactivity in the olfactory bulb. That
is, TH-expression occurs only when afferent innervation is present.
However, TH-immunoreactive cells appear about 3-4 days after afferent
innervation (E18 versus E14-15). This temporal disparity could arise
through a number of mechanisms. First, neurogenesis of the TH-cells may
not occur before E18 since the majority of external granule cells are
born from P0 to P7 (Bayer, 1983). Second, the amount of afferent innervation may not be sufficient to support TH-expression. The ectopically
located TH-cells found in the external plexiform and mitral cell layers
display much lighter labelling and receive less afferent innervation than
the darkly-labelled TH-cells in the glomerular region. The in vitro
experiments support this hypothesis. The en bloc co-cultures, which
contain extensive OMP-innervation, also exhibit many TH-cells, especially
when compared to either the cultures of bulb alone or those where the POB
and epithelium were separated and recombined before culturing. Development of the receptor epithelium and the number of afferent fibers in the
bulbar tissue was decreased in the latter cultures when compared to the
en bloc cultures of bulb and epithelium.

Significantly, while afferent innervation, here defined as the
presence of afferent fibers containing OMP, is present by E15, synaptic
specializations reportedly do not occur in the rat until E18 (Farbman,
1986), the stage at which TH first appears. Thus, these and other
observations suggest that factors, in addition to the presence of afferent innervation, are necessary to the expression of the dopamine phenotype both during development and in the adult. For example, in odor
deprived animals, TH expression is reduced but synaptic specializations
remain within the glomeruli (Benson et al., 1984; Spencer and Baker,

unpublished observations). A previous report suggested that the release of a molecule such as calcitonin gene-related peptide might act trophically and induce the dopamine phenotype (Denis-Donini, 1989). Subsequent analysis suggested that, in vivo, CGRP did not exhibit the appropriate tissue distribution to play a role in TH-expression during development or in adults (Baker, 1990a; Biffo et al., 1990). However, in conjunction with the experiments reported here, the data do suggest that the induction of TH expression may require a trophic molecule. Furthermore, the release of this molecule may be dependent on functional receptor cell innervation of the olfactory bulb. That is, activity-dependent processes may be involved.

Lastly, our findings suggest that glomerular-like structures can be distinguished prenatally, that is, by E19, the embryonic age at which large numbers of TH neurons are first observed. TH-containing neurons appear to be distributed around these immature glomeruli. Thus, while glomeruli form early in development, large numbers of them are distinguishable only by P1 as previously indicated by Pomeroy et al. (1990). The coincident formation of glomeruli and TH expression suggests that some of the same developmental cues may be operating in these phenomena.

In summary, these data suggest that TH-expression is correlated with the presence of afferent innervation. However, the mere presence of the afferent is not sufficient to induce TH-synthesis. Activity dependent processes, alone or in conjunction with the release of trophic factors, may be required for phenotypic expression.

ACKNOWLEDGEMENTS

Supported by grants AG09686 and DC00080.

REFERENCES

Allen, W.K., and Akeson, R., 1985, Identification of an olfactory receptor neuron subclass: cellular and molecular analysis during development, Dev. Biol., 109:393-401.
Baker, H., 1990a, Calcitonin gene-related peptide in the developing mouse olfactory system, Dev. Brain Res., 54:295-298.
Baker, H., 1990b, Unilateral, neonatal olfactory deprivation alters tyrosine hydroxylase expression but not aromatic amino acid decarboxylase or GABA immunoreactivity, Neuroscience, 36:761-771.
Baker, H., Kawano T., Margolis, F.L., and Joh, T.H., 1983, Transneuronal regulation of tyrosine hydroxylase expression in olfactory bulb of mouse and rat, J. Neurosci., 3:69-78.
Bayer, S.A., 1983, [^{3}H] thymidine-radiographic studies of neurogenesis in the rat olfactory bulb, Exp. Brain Res., 50:329-340.
Benson, T.E., Ryugo, D.K., and Hinds, J.W., 1984, Effects of sensory deprivation on the developing mouse olfactory system: a light and electron microscopic, morphometric analysis, J. Neurosci., 4:638-653.
Biffo, S., Delucia, R., Mulatero, B., Margolis, F.L., and Fasolo, A., 1990, Carnosine-, calcitonin gene-related peptide- and tyrosine hydroxylase-immunoreactivity in the mouse olfactory bulb following peripheral denervation, Brain Res., 528:353-357.
Brunjes, P.C., and Frazier, L.I., 1986, Maturation and plasticity in the olfactory system of vertebrates, Brain Res. Rev., 11:1-45.
Brunjes, P.C., Smith-Crafts, L.K., and McCarty, R., 1985, Unilateral odor deprivation: effects on the development of olfactory bulb catecholamines and behavior, Dev. Brain Res., 22:1-6.

Chuah, M.I., and Farbman, A.I., 1983, Olfactory bulb increases marker protein in olfactory receptor cells, J. Neurosci., 3:2197-2205.

Denis-Donini, S., 1989, Expression of dopaminergic pheno types in the mouse olfactory bulb induced by the calcionin gene-related peptide, Nature, 339:701-703.

Ehrlich, M.E., Grillo, M., Joh, T.H., Margolis, F.L., and Baker, H., 1990, Transneuronal regulation of neuronal specific gene expression in the mouse olfactory bulb, Mol. Brain Res., 7:115-122.

Farbman, A.I., 1986, Prenatal development of mammalian olfactory receptor neurons, Chemical Senses, 11:3-18.

Farbman, A.I., and Squinto, L.M., 1985, Early development of olfactory receptor cell axons, Dev. Brain Res., 19:205-213.

Graziadei, P.P.C., and Monti Graziadei, G.A., 1978, The olfactory system: a model system for the study of neurogenesis and axon regeneration in mammals, in: "Neuronal plasticity," C.W. Cotman, ed., pp 131-153, Raven Press, New York.

Graziadei, P.P.C., Levine, R.R., and Monti Graziadei, G.A., 1980, Regeneration of olfactory axons and synapse formation in the forebrain after bulbectomy in neonatal mice, Proc. Nat. Acad. Sci. USA, 75:5230-5234.

Kosaka, T., Hama, K., Wu, J.-Y., Nagatsu, I., 1987, Differential effect of functional olfactory deprivation on the GABAergic and catecholaminergic traits in the rat main olfactory bulb, Brain Res., 413:197-203.

Kream, R.M., Davis, B.J., Kawano, T., Margolis, F.L., and Macrides, F., 1984, Substance P and catecholaminergic expression in neurons of the hamster main olfactory bulb, J. Comp. Neurol., 222:140-154.

Matsutani, S., Senba, E., and Tohyama, M., 1988, Neuropeptide- and neurotransmitter-related immunoreactivities in the developing rat olfactory bulb, J. Comp. Neurol., 72:331-342.

McLean, J.H., and Shipley, M.T., 1988, Postmitotic, postmigrational expression of tyrosine hydroxylase in olfactory bulb dopaminergic neurons, J. Neurosci., 8:3658-3669.

Monti Graziadei, G.A., Stanley, R.S., and Graziadei, P.P.C., 1980, The olfactory marker protein in the olfactory system of the mouse during development, Neurosci., 5:1239-1252.

Nadi, N.S., Head, R., Grillo, M., Hempstead, J., Granno-Reisfeld, N., and Margolis, F.L., 1981, Chemical deafferentation of the olfactory bulb; plasticity of the levels of tyrosine hydroxylase, dopamine and noepinephrine, Brain Res., 213:365-377.

Pomeroy, S.L., LaMantia, A.-S., and Purves, D., 1990, Postnatal construction of neural circuitry in the mouse olfactory bulb, J. Neurosci., 10:1952-1966.

Specht, L.A., Pickel, V.M., Joh, T.H., and Reis, D.J.,1981, Light-microscopic immunocytochemical localization of tyrosine hydroxylase in prenatal rat brain. II. Late Ontogeny, J. Comp. Neurol., 199:255-276.

Stone, D.M., Wessel, T., Joh, T.H., and Baker, H., 1990, Decrease in tyrosine hydroxylase, but not aromatic L-amino acid decarboxylase, messenger RNA in rat olfactory bulb following neonatal, unilateral odor deprivation, Mol. Brain Res., 8:291-300.

Stone, D.M., Grillo, M., Margolis, F.L., Joh, T.H., and Baker, H., 1991, Differential effect of functional olfactory bulb deafferentation on tyrosine hydroxylase and glutamic acid decarboxylase messenger RNA levels in rodent juxtaglomerular neurons, J. Comp. Neurol., (In Press).

ONTOGENETIC OLFACTORY EXPERIENCE AND ADULT SEARCHING BEHAVIOR IN THE CARNIVOROUS FERRET

Raimund Apfelbach

University of Tübingen, Dept. of Zoology
Auf der Morgenstelle 28, D-7400 Tübingen, FRG

INTRODUCTION

Several studies on prey catching behavior of mustelids have shown that optical and acoustical stimuli are important for eliciting hunting reactions in the European polecat (Mustela putorius) and its domesticated form the ferret (Mustela putorius f. furo L.) (Goethe, 1940; Räber, 1944; Eibl-Eibesfeldt, 1956, 1963; Wüstehube, 1960; Gossow, 1970; Apfelbach and Wester, 1977). Yet, in this species both sensory modalities are surpassed in importance by the olfactory system. Behavioral studies indicate that adult ferrets respond reliably with food searching behavior only when the odor of known prey is offered (Apfelbach, 1973). This suggests that odor serves as an acquired sign stimulus for prey identification and selection in ferrets. In subsequent experiments it was demonstrated that the preference for a specific prey odor is not due to the length of prior feeding experience but due to imprinting during a postnatal sensitive phase occuring between the second and fourth month of life (Apfelbach, 1978). During this phase animals react strongly to novel odors (prey odors and pure chemical odors) with distinct searching behavior. Animals that were introduced to a novel food after their fourth month paid decreasing attention to it. At one year of age, the animals showed no response to the odors of unknown prey objects or any other odor, although they continued to respond readily to that of known prey with searching behavior.

The present work was, therefore, undertaken to study the influence of early olfactory experience (e.g. olfactory enrichment and olfactory deprivation) during the sensitive phase for olfactory food imprinting on adult searching behavior and odor preference.

MATERIAL AND METHODS

In a series of experiments, aimed at elucidating the role of olfactory cues for prey selection in ferrets, a total of 24 newly weaned young were divided into three groups and all were fed on a diet of dead chicks up to day 60, after which each group was fed differently until day 90. The control group (group C) was fed with dead chicks and mice; the enriched group (group E) received chicks, mice and in addition rats, rabbits, beef and different types of dog chow; the deprived group (group D) was fed only with

Chemical Signals in Vertebrates VI, Edited by R.L. Doty and
D. Müller-Schwarze, Plenum Press, New York, 1992

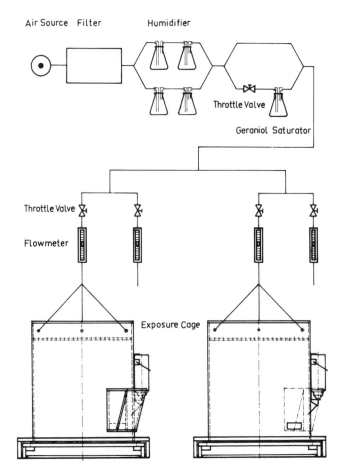

Fig. 1. Housing conditions to induce odor deprivation.

chicks. After day 90, all groups received dead chicks only. To enhance olfactory deprivation, animals of group D were kept in an artifical olfactory environment saturated with geraniol odor (exposure cage, Fig. 1). The continuous overexposure to a single odor masks the ability of the animal to experience other odors in the environment, bringing about a relative state of olfactory deprivation (Doving and Pinching, 1973; Apfelbach and Weiler, 1985).

For the behavioral studies animals which were at least two years old were tested for their preference toward known and unknown prey odors and toward different pure chemical odors (geraniol, linalool) using a Y-maze, which was made of plexiglass (Fig. 2). Air/odor flow to the Y-maze was regulated with the use of flowmeters (F), which were located before and after the odor saturators (Odors). Each end of the Y-maze was provided with Teflon valves (V) to control the direction flow of the odors in each respective end. Odors were randomly introduced into the right or left end before the animal was allowed to leave the start box. The Y-maze was provided with infrared sensors at different areas (A – G) to monitor the behavior of the animal. In addition, the behavior was observed via a TV-camera positioned above the Y-maze. With this arrangement initial latency time, decision time, locomotory speed and odor preference were quantified.

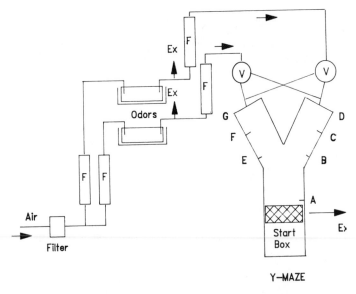

Fig. 2. Experimental set-up to detect odor preferences.

RESULTS

The results of the preference tests are given in Fig. 3. In the choice situation of known prey odor versus unknown prey odor, control animals (C) showed a preference of 33% to the unknown prey odor (67% to the known prey odor) where as enriched animals (E) 61%, and deprived animals (D) only 24%. The differences between C and E, and C and D are statistically significant with $p < 0.01$ and $p < 0.001$ respectively (Mann-Whitney U-test). Unknown chemical odors were less attractive than unknown prey odors.

Figure 4 shows the time needed before a decision was made to investigate an odor. In the choice situation of known prey odor versus unknown prey odor, controls needed on the average 4.1 s, enriched animals 2.2 s (difference C to E: $p < 0.01$), deprived animals 14.8 s (difference C to D: $p < 0.001$). When a pure chemical odor was offered instead of unknown prey odor, decision times remained basically the same.

DISCUSSION AND SUMMARY

The data show that ferrets depend mainly on olfactory cues when searching for prey. However, knowledge of prey odors is not innate in this species but has to be acquired via an imprinting process during a sensitive phase lasting throughout the third month of postnatal life. If the ferret is exposed to only a small number of prey objects during the sensitive phase, it will develop an olfactory search image which focuses food search behavior for specific prey. As shown, under laboratory conditions the search image is not modifiable in later ages. Therefore, the ferret will respond little to novel prey odors and even less to pure chemical odors which may convey even less olfactory information. The animal can be regarded as a food specialist. On the other hand, if the ferret is exposed to a wide range

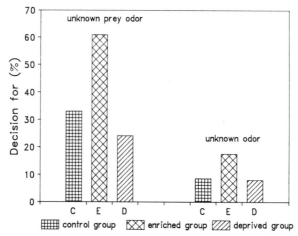

Fig. 3. Graphs showing that enriched animals (E) show higher
tendency to explore novel odor than control (C) and
deprived animals (D). (Decision for = % of cases where
animals explored the unknown odor.

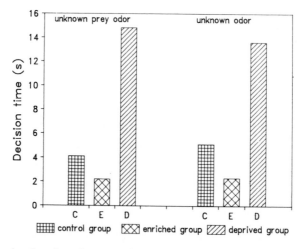

Fig. 4. Graphs showing that enriched animals (E) decided
faster to explore novel odor than control (C) and
deprived animals (D).

of different prey objects during the sensitive phase, it will develop an olfactory search image which is thought to be modifiable by prevailing prey abundance, so as to optimize the predator's feeding efficiency. The ferret is still capable of exploring novel odors even if it is imprinted to specific prey objects (which will be preferred in a choice situation). In this case, the ferret can be considered to be more like a food generalist.

The North American black-footed ferret (Mustela nigripes) is closely related to the European polecat (ferret). In nature it is a food specialist feeding mainly on prairi dogs. A close relationship between the endangered black-footed ferret and the decreasing number of prairie dogs has been proposed (Hillman and Clark, 1980). From this, it is tempting to suggest that, in nature, the black-footed ferret became a food specialist because it is practically imprinted only to prairie dogs. Noteworthy also are data indicating that depending on the area, the European polecat behaves as a food specialist or as a food generalist. In some Swiss areas it feeds mainly on amphibians (Weber, 1987) while in some eastern parts of Germany it feeds on a comparatively wide range of invertebrates and vertebrates (Ansorge, 1990).

REFERENCES

Ansorge, H., 1990, Ernährungsökologischer Vergleich von fünf Carnivoren-Arten der Oberlausitz, 64. Hauptvers. Dtsch. Ges. Säugetierkde., Osnabrück.
Apfelbach, R., 1973, Olfactory sign stimulus for prey selection in polecats (Putorius putorius L.), Z. Tierpsychol., 33:273.
Apfelbach, R., 1978, A sensitive phase for the development of olfactory preference in ferrets (Mustela putorius f. furo L.), Z. Säugetierkde., 43:289.
Apfelbach, R. and Weiler, E., 1985, Olfactory deprivation enhances normal spine loss in the olfactory bulb of developing ferrets. Neurosci. Letters, 62:169.
Apfelbach, R. and Wester, U., 1977, The quantitative effect of visual and tactile stimuli on the prey-catching behaviour of ferrets (Putorius furo L.), Behav. proc., 2:187.
Doving, K.B. and Pinching, A.J., 1973, Selective degeneration of neurons in the olfactory bulb following prolonged odour exposure, Brain Res., 52:115.
Eibl-Eibesfeldt, I., 1956, Angeborenes und Erworbenes in der Techni des Beutetötens (Versuche am Iltis, Putorius putorius L.), Z. Säugetierkde., 21:135.
Eibl-Eibesfeldt, I., 1963, Angeborenes und Erworbenes im Verhalten einiger Säuger, Z. Tierpsychol., 20:705.
Goethe, F., 1940, Beiträge zur Biologie des Iltis, Z. Säugetierkde., 15:180.
Gossow, H., 1970, Vergleichende Verhaltensstudien an Marderartigen. I. Über Lautäußerungen und zum Beuteverhalten, Z. Tierpsychol., 27:405.
Hillman, C.N. and Clark, T.W., 1980, Mustela nigripes, Mamm. Species, 126:1.
Räber, H., 1944, Versuche zur Ermittlung des Beuteschemas an einem Hausmarder (Martes foina) und einem Iltis (Putorius putorius), Rev. Suisse Zool., 51:293.
Weber, D., 1987, Zur Biologie des Iltisses (Mustela putorius L.) und den Ursachen seines Rückganges in der Schweiz, Ph.D. Thesis, University of Basel/Switzerland.
Wüstehube, C., 1960, Beiträge zur Kenntnis besonders des Spiel- und Beuteverhaltens einheimischer Musteliden, Z. Tierpsychol., 17:579.

OLFACTORY RECEPTORS IN ACIPENSERIDAE FISH LARVAE

Galina A. Pyatkina

I.M. Sechenov Institute
St. Petersburg 194223 Russia

INTRODUCTION

Basic aspects of the development of the olfactory system of Acipenseridae occurs during the period between the embryo's hatching and its
period of independent feeding. It was found that actively feeding
Acipenseridae larvae have the same types of olfactory receptor cells as
observed in adults (Pyatkina, 1976). However, structural peculiarities
of developing olfactory receptor cells have not been studied during the
postnatal period. Therefore, we carried out an electron microscopic and
cytochemical study of the developing olfactory receptor cells of the
Acipenseridae. The ratio of ciliated biopolar receptor cells to microvillar cells was calculated at different stages of larval development.

MATERIALS AND METHODS

One to twelve-day-old Acipenseridae larvae (beluga, Huso huso;
servruga, Acipenser stellatus; sturgeon, Acipenser guldenstaedti; sterlet, Acipenser ruthenus) were studied at the periods of hatching, transition to mixed feeding and active feeding.

For structural studies, each olfactory mucosa was fixed for 2-3 h
in 6% glutaraldehyde solution, washed in a buffer, fixed in OsO_4 solution, dehydrated in a graded series of ethanol concentrations and embedded in EPON-812. Ultrathin sections were examined using a JEM-100B
electron microscope.

For the cytochemical studies, each olfactory muscosa was fixed in
2.5% glutaraldehyde solution for 30 minutes, washed in 0.1M cacodylate
buffer (pH 7.4), and placed into an incubation medium for 30 minutes for
the study of adenylate cyclase (AC) distribution using adenylyl-imido-
diphosphate as a substrate. The reaction product was revealed as a
granular electron-dense deposit. We applied the method using lanthanum
and ruthenium red for cytochemical identication of glycosaminoglycans in
the olfactory epithelia.

All cells whose apical processes extended into the olfactory cavity
were counted to determine the ratio of the ciliated receptor cells to
the microvillar cells. Longitudinal sections were taken. The cell

Chemical Signals in Vertebrates VI, Edited by R.L. Doty and
D. Müller-Schwarze, Plenum Press, New York, 1992

counts were made in groups of 5-8 animals using the 1st, 7th and 10th days as the beginning points of each stage. Only one section was taken from each animal.

RESULTS AND DISCUSSION

Olfactory receptor cells develop in a similar way and have the same stage of development in various forms of Acipenseridae larvae. Receptor cells were observed early during embryogenesis (Pyatkina, 1987). In the olfactory epithelium of 4- to 5-day-old sturgeon and sevruga embryos, two types of differentiating olfactory cells are observed among the supporting cells -- the ciliary and microvillar cells. At this age, neither had connections with higher olfactory centers. However, Larvae at the time of hatching have approximately equal number of ciliary and microvillar cells (Table I).

Table 1. The ratio of different types of receptor cells in the olfactory epithelium of Acipenseridae larvae (Mean $\pm$ SD).

Species	Age after Hatching (days)	Ratio of Number of Ciliar to Microvillar Receptor Cells in the Epithelium	a	Number of Animals n
Acipenser guldenstaedti	1	1.08 $\pm$ 0.10		5
	7	1.53 $\pm$ 0.22	0.01	6
	10	2.27 $\pm$ 0.70	0.01	6
Huso huso	1	1.12 $\pm$ 0.13		8
	7	1.85 $\pm$ 0.48	0.01	6
	10	2.36 $\pm$ 0.60	0.01	5
Acipenser stellatus	1	1.06 $\pm$ 0.13		6
	7	1.41 $\pm$ 0.30	0.05	6
	10	1.45 $\pm$ 0.34	0.01	8
Acipenser ruthenus	1	1.00 $\pm$ 0.35		5
	7	0.82 $\pm$ 0.19	0.01	6
	10	0.64 $\pm$ 0.15	0.05	7

a -- significance level of comparison of 7- and 10-day-old groups to the 1-day-old group.

In beluga larvae, the increase in the ciliary to microvillar cell ratio occurs sooner than in the other larvae studied. Beluga differs from the other Acipenseridae in having faster rate of development of all vital functions (Gershanovich, 1983). Beluga's olfactory organ already has, by the 9th day of age, an adult-like olfactory epithelium. Thus, species specific variations in the rate of differentiation of the olfactory system are observed.

In anadromous species (beluga, sturgeon, sevruga), ciliary cells predominate over microvillar cells by the mixed feeding stage, unlike the case in the fresh water forms (e.g., sterlet). This difference becomes much more clear by the time of active feeding (Table 1). The different relative number of these cells in the anadromous and freshwater larvae probably results from their different ecological adaptations.

During the yolk stage (1-5 days), the olfactory cells are not connected with the olfactory bulb. The central processes of olfactory cells actively grow during the mixed feeding stage (7-9 days). Nevertheless, at this stage of development the olfactory cells are still separated from the bulb. The connection between the olfactory cells and the bulb appears only during the active feeding stage (10-12 days). Our preliminary data indicates that the first synapses with the olfactory bulb of beluga larvae occur early in this state (8-10 days). Glycocalyx appears on the surface of the olfactory epithelium (Fig. 1) with its thickness increasing from 3 to 60 nm across the early stages of the development. Glycocalyx seems to be a diffusional barrier which determines the penetration rate of odorous molecules. The important role of this layer is supported by cytochemical data showing the AC reaction product in glycocalyx, axonemes and knobs of the olfactory cells (Fig. 2).

Fig. 1.   Glycocalyx on the surface of the olfactory cilia.

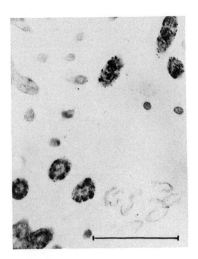

Fig. 2.   Cytochemical detection of AC in the olfactory cilia of Acipenseridae at the active feeding stage.

Our morphological results correlate with biochemical experiments, which show that the supramembrane layer and cytoplasmic proteins may play a role receptors. It is known that AC is activated by odorants only in the presence of guanine nucleotides with the help of G-proteins (Parfenova and Etingof, 1988). Immunocytochemical results show that G-proteins can take part in the olfactory transduction even in early stages of development in rats (Mania-Farnell and Farbman, 1990). Therefore, we assume that Acipenseridae larvae are capable of odour discriminating only during the active feeding stage; i.e., the period when the olfactory cell knobs with cilia or microvilli are completely developed and when the synapses are formed.

REFERENCES

Gershanovich, A.D., 1983, Temperature influencing energy metabolism, growth and food needs of young beluga Huso huso and thorn Acipenser nudiventris Lov (Acipenseridae), Voprosi Ichtiol., 23:238 (in Russian).

Mania-Farnell, B., and Farbman, A.I., 1990, Immunohistochemical localization of guanine nucleotide-binding proteins in rat olfactory epithelium during development, Brain Res., 51:103.

Parfenova, E.V., and Etingof, R.N., 1988, On GTP-binding proteins participation in the olfactory reception in vertebrates, Biochem., 53:498 (in Russian).

Pyatkina, G.A., 1976, Receptor cells of different types and their quantative ratio in the olfactory organ of Acipenserida larvae and sexually matured ones, Cytologia, 18:1444 (in Russian).

Pyatkina, G.A., 1987, Development of the receptor cells of the olfactory organ in vertebrates and humans, in: "Sense organ systems", G.V. Gershuny, ed., Sci., L. (in Russian).

SECTION FOUR

SEMIOCHEMICALS AND THE MAJOR HISTOCOMPATIBILITY COMPLEX

EVOLUTIONARY AND IMMUNOLOGICAL IMPLICATIONS

OF THE ROLE OF THE MHC IN OLFACTORY SIGNALLING

Rachael Pearse-Pratt[1], Heather Schellinck[2],
Richard Brown[2] and Bruce Roser[1]

[1]Quadrant Research Foundation, Cambridge CB3 0DJ, UK
[2]Psychology Dept. Dalhousie, Halifax, B3H 4J1, Canada

INTRODUCTION

Amongst its many odour signals, urine emits an odour, unique to each individual, which is directly related to the MHC type (Yamazaki et al., 1976, Singh et al., 1987, Brown et al., 1989). Extreme polymorphism of the class I MHC loci provides sufficient variation within a species to confer uniqueness on individuals at this molecular level (Klein, 1982). However, the way in which this protein polymorphism is expressed as odour polymorphism in the urine is not known. Since the MHC class I molecules are themselves excreted in the urine, they could be thought to provide individually unique signals except that they are non-volatile proteins and therefore not detectable by smell. Because the excreted class I molecules are extensively degraded in the urine (Singh et al., 1987), it is possible that aromatic amino acids, characteristic of the variable sequences in the heavy chain contribute the aromatic signals. However, since most amino acids have no smell, this explanation seems unlikely.

An attempt has been made to explain the linkage between class I phenotype and urine odour by postulating that the immuno-regulatory function of class I defines variations in the commensal bacterial flora (Crozier, 1987). This idea requires that the bacterial flora of each individual releases unique volatile aromatics and also that the commensal phenotype is stable with time. Although this very indirect association between class I phenotype and urine odour may seem unconvincing, recent studies have shown that germ-free and specific pathogen-free rats do not excrete individual odours in their urine (Singh et al., 1990, Roser et al., 1991), a surprising finding but one which is predicted by this hypothesis. We have previously described a complementary hypothesis in which the commensal flora acts as a source of multiple volatile aromatics while the class I molecule acts as an individual-specific carrier which selects and binds a unique cocktail of bacterial volatiles, transports them to the urine and releases them (Roser et al., 1991). We now present molecular and behavioral data which support this explanation.

The details of the assembly of class I during biosynthesis and of their disassembly in the serum and urine provide a plausible chronological sequence for a dual role for class I -- first as a binding and presenting platform for self-peptides in the immune system which maintains tolerance of self antigens and second as a binding and transporting vector for bacterial volatiles which maintains self identity-signalling to the environment.

Finally, we formally demonstrate that class I molecules play a direct role in determining urine odour by behavioral studies showing that the familiar odour of an individual's urine could be temporarily changed to a foreign odour by the excretion of a foreign MHC class I molecule.

RESULTS AND DISCUSSION

Characteristics of class I molecules in blood and urine

The affinity chromatography method of Parham (1979) was used to isolate the RTI.A^a molecule from the serum of DA rats. This showed typical heavy and light chain bands. The heavy chain band (MW 39kD) was smaller than the heavy chain of membrane-associated class I (MW 44kD), while the light chain at 12.5kD was typical of β2 microglobulin. When isolated from rat urine by affinity chromatography, the class I molecules again showed 39kD and 12.5kD bands but also a dense diffuse band with a MW of about 27kD. Since this latter band was specifically bound and eluted from the S-site specific monoclonal antibody MAC-30, it was identified as a cleavage product of the intact molecule bearing the S-site antigenic determinant of RTI.A^a (Singh et al., 1988).

The finding of heavy chain cleavage products raised the question of the fate of excreted class I molecules in urine. This was approached by infusing known amounts of labelled class I molecules into rats and measuring the amounts recoverable in urine. Less than 6.5% of the infused molecules lost from the circulation could be accounted for in urine pooled over 48hr (Singh et al., 1988). Since the size of the heavy chain in serum (39kD) is precisely the size of heavy chains released from cell membranes by papain digestion, this suggests extensive, probably enzymatic, degradation of the excreted molecules in the urine. These data are consistent with the notion that class I molecules enter the circulation by proteolytic cleavage from membrane molecules and undergo further fragmentation by unknown proteases during excretion in the urine. The presence of a major band at 27kD in the urine suggests that the predominant proteolytic event is highly specific and restricted to a single cleavage site at the junction of the α2 and α3 domains. The amino acid sequence of rat class I indeed shows that there is a specific cleavage site in this area.

Papain cleavage of rat class I molecules

One kilogram of liver tissue from DA rats was processed to yield a cell membrane preparation which was digested with papain and the released class I molecule was purified by chromatography on a MAC-30 affinity column. This procedure yielded about 10mg of highly purified A^a class I molecules. SDS PAGE analysis of this molecule gave a picture reminiscent of that seen when class I molecules were purified from urine. In addition to the expected 39kD heavy chain band and the 12.5kD β2M band, additional bands were present with MW's of 23kD and 27kD.

Amino-terminal sequence analysis of the first 17 amino acids of the 39kD band showed it to be authentic class I heavy chain. The first 10 amino acids of the 27kD band were also the amino terminus of class I. It thus constituted the entire α1/α2 domain of the heavy chain while the 23kD band had lost the first 26 amino acids of the α1 domain.

Amino Acid Sequences of Bands

```
Amino acid No.  1       5          10          15
CLASS I         G S H S L R Y F Y T A V S R P G L
39kD            G S - S L - Y F Y T A V S - P G L
27kD            G S - S L - Y F Y T


Amino acid No.  27    30          35          40
CLASS I         Y V D D T E F V R F D S D A
23kD            Y V D D T - F V - F D S D A
```

Molecular mechanism of odorant binding

In order for class I molecules to act as carrier proteins for allele-specific cocktails of volatile aromatics, they need binding sites for hydrophobic residues. X-ray crystallography has revealed just such a site on the distal end of the class I molecules (Garrett et al., 1989). Normally this cleft is occupied by strongly bound "self" peptides (Bjork-man et al., 1987). Recent work has shown that the self peptides are bound in the cleft very early during synthesis of the class I heavy chain and that this binding is required before the α3 domain of the heavy chain associates with the light chain β2M (Townsend et al., 1989, Kvist & Hamann 1990). At this stage the fully assembled class I molecule is transported to and inserted in the cell membrane.

Our own work on the characteristics of the soluble class I molecules in rat serum and urine has shown that there is a degredative sequence of enzymatic cleavages which accompany excretion of class I which appear to be almost a mirror image of the synthesis sequence outlined above. Of

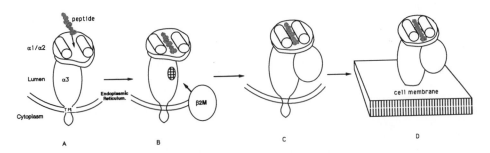

Fig. 1.   Assembly and expression of class I molecule. A. Newly
synthesised heavy chain consists of a platform of β pleat-
ed sheet carrying two α helices defining a peptide binding
pocket which is occupied by a self peptide of about 9
amino acids which stabilises the structure. The α1/α2
domains are supported by a globular α3 domain, a hydrophobic trans-membrane domain and a short intracytoplasmic
domain. B. After peptide binding the heavy chain associates with the β2 microglobulin light chain via a patch of
binding residues on a α domain. C. Binding of β2M induces
a conformational change which enhances the stability of
binding of the peptides and readies the molecule for
transport to the membrane. D. It is anchored in position
in the cell membrane by hydrophobic interaction between
the TM domain and the lipid bilayer. A papain cleavage
site is adjacent to the TM region.

great potential importance is the effect of such cleavages on the allele-specific binding pocket of the heavy chain. This cleavage sequence, which results in isolated α1/α2 domains in the urine, could convert a peptide-presenting structure into a volatile-transporting molecule. In both functions the particular molecules bound in the pocket will vary from individual to individual because the amino acid variations which define the many alleles at class I loci are in fact all located in and around the binding cleft (Garrett et al., 1989).

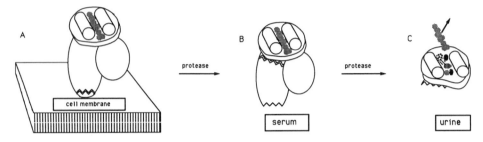

Fig. 2. Disassembly of class I. A. Cleavage of the molecule at the juxta-membranous site removes the hydrophobic TM region so that the molecule does not aggregate and circulates as a soluble monomer (B). C. In the molecule recovered from the urine, further enzymatic cleavage has removed β2M, allowed relaxation of the binding cleft and loss of bound peptide. Such a molecule is capable of binding an allele-specific cocktail of small hydrophobic molecules such as volatile odorants.

Injected class I molecules change the odour of urine

The only definitive test of this carrier hypothesis is to change the odour of the urine of a rat using highly purified class I molecules alone. Additionally such an experiment should show that any new odour given to the urine is specific for the particular allele of the added class I molecule. Because we have no idea at which stage any putative odorant-binding event could occur during excretion of class I, we added these purified molecules to rat urine by injecting them into the circulation of normal rats and allowed them to transit the excretion pathway into the urine which was collected using metabolic cages. This urine was then tested against the urine of non-injected syngeneic rats to determine whether a new odour had appeared in the urine. It was also tested against the urine of rats which normally expressed the same class I molecule A[a] to see whether the new odour was identified by the detector animal as the authentic urine odour of this allele.

As with all our previous work on individual urine odours, these experiments used rats selected to express no genetic variation except for class I. Thus all the rats used were from the PVG congenic series in which all non-MHC genes are identical. We used only virgin males to avoid sexual and oestrous signals and all animals were similarly treated, fed and housed. To be sure we were measuring a signal which had behavioral effects, we used the habituation/dishabituation paradigm previously described (Brown et al., 1987). The results show that the congenic pair PVG.RT1[u] and PVG.r8 which differ only at the class I locus are readily discriminated by PVG detector animals while separate individual animals of either strain cannot be discriminated. The odorant signals causing dishabituation to a familiar (habituated) odour thus maps to the class I locus.

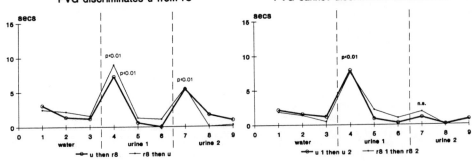

Fig. 3. Discrimination of r8 and PVG.RT1^u strains by odour. A.
Time (Y axis) which PVG detector rats spend in rearing up
to sniff filter papers impregnated with water is short at
trial 1 and rapidly becomes shorter still on trials 2 and
3 (X axis). When urine from a foreign rat is substituted
in trial 4, the time spent sniffing increases, showing
dishabituation. This again habituates in trials 5 and 6
only to dishabituate again in trial 7 when urine from
another strain is substituted. Thus, PVG rats have no
difficulty in recognising the difference between the urine
odour of the congenic pair PVG.RTI^u and PVG.r8. B. When
the second urine sample is from another member of the same
strain, no dishabituation occurs in trial 7. Thus the
differences detected in experiment A maps to the class I
region which is the sole genetic difference between the
strains r8 and u.

When the urine of PVG.RT1^u animals injected with the A^a molecule was
tested against normal PVG.RT1^u urine, the detector animals had no difficulty in detecting that a different odour had appeared in the urine.
This was so whether they smelled injected or non-injected urine first.
When the urine of PVG.RT1^u animals which had been injected with A^a molecules was compared with the urine of normal PVG.r8 rats, a striking
difference emerged. If normal PVG.R8 urine was smelled first, the rats
readily identified a new odour in the urine of the injected rats. This
odour was probably due to the A^u molecule normally expressed by this
strain because testing in the reverse order showed continued habituation
when normal r8 urine followed exposure to injected urine. The most
satisfactory way these results can be explained is that the injected A^a
molecule is detected by PVG rats as the odour <u>characteristic</u> of this
allele. When habituated to this odour they fail to respond to r8 urine
which only carries the odour of A^a.

These data confirm that injected, highly purified class I molecules
excreted in the urine do indeed change the odour profile of the urine and
the odour they confer is characteristic of the allelic form of the class
I molecule. In this context it is important to note that the purified
class I molecule itself offered to rats in a behavioral test does not
elicit any response and appears to be odourless (Roser et al., 1991).
Also, the serum of both rats (Brown et al., 1987) and mice (G.K. Beauchamp, personal communication) emits copious volatile odorant molecules
but they seem to be essentially the same mixture of odorants in all
strains since the sera of different strains cannot be discriminated.
Taken together with the structural modifications undergone by class I

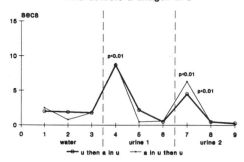
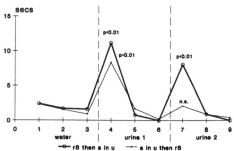

Fig. 4. Injected A^a molecules confer A^a urine odour. A. PVG rats readily detect the different smell in the urine of u rats excreting in their urine A^a molecules injected several hours previously. This urine is discriminated from that of injected u strain rats whichever urine is offered first. B. Rats habituated to injected A^a molecules in the urine of u strain animals as a first exposure do not respond when offered r8 urine on trial 7 (compare with Figure 5). However, habituation to the A^a molecule in r8 urine offered first does not mask the detection of the endogenous odour of the A^u molecule in injected u strain rats. This shows that the injected molecule has transferred an authentic A^a odour to the PVG.RT1^u urine.

molecules in their journey from cell membranes into the urine, a strong case emerges for the role of class I molecules in picking up, via their allelic binding pocket, unique mixture of the odorants in serum and carrying them into the urine to be used in individuality signalling. The origin of the volatile odorants is not known but the failure of urine from germ-free rats to emit such signals strongly implicates commensal bacterial flora as their source.

REFERENCES

Bjorkman P.J., Saper M.A., Samraouchi B., Bennet W.S., Strominger J.L. and Wiley D.C., 1987, Structure of the human class I histocompatability antigen, HLA-A2, <u>Nature</u>, 329:506.
Brown R.E., Singh P. and Roser B. 1987, The major histocompatibility complex and the chemosensory recognition of individuality in rats. <u>Physiol</u>. <u>Behav</u>., 40:65.
Crozier R.H. Genetic Aspects of Kin Recognition: Concepts, Models and Synthesis, pp. 55-73 <u>in</u>: "Kin Recognition in Animals", eds. D.J.C. Fletcher and C.D. Michener, John Wiley & Sons, N.Y.
Garrett T., Sapers M.A., Bjorkman P.J., Strominger J.L. and Wiley D.C., 1989, Specificity pockets for the side chains of peptide antigens in HLA-Aw68, <u>Nature</u>, 342:692.
Klein J., 1982, Histocompatibility Antigens: Structure and Function pp 221-239 <u>in</u>: "Receptors and Recognition series B" eds P. Parham and J. Strominger. Chapman & Hall, London.
Kvist S. and Hamann U., 1990, A nucleoprotein peptide of influenza A virus Stimulates assembly of HLA-B27 class I heavy chanis and β2
Parham P., 1979, Purification of immunologically active HLA-A and B antigens by a series of monoclonal antibody columns, <u>J</u>. <u>Biol</u>. <u>Chem</u>. 254:8709.

Pearse-Pratt R., 1990, Immunoregulatory and olfactory properties of soluble classical Class I MHCX molecules, PhD Thesis, Cambridge University.

Roser B., Brown R.E. and Singh P., 1991, Excretion of transplantation antigens as signals of genetic individuality, pp. 187-209 in: "Chemical Senses, Vol. 3, Genetics of Perception and Communications", eds., C.J. Wysocki and M.R. Kare. Marcel Dekker, N.Y.

Singh P. Brown R.E., and Roser B., 1987, MHC antigens in urine as olfactory recognition cues, Nature, 327:161.

Singh P., Brown R.E. and Roser B., 1988, Class I transplantation antigens in solution in body fluids and in the urine. Individuality signals to the environment, J. Exp Med, 168:195.

Singh P., Herbert J., Roser B., Arnott L., Tucker D.K. and Brown R.E., 1990, Rearing rats in a germ-free environment eliminates their odours of individuality, J. Chem. Ecol., 16:1667.

Townsend A., Ohlen C., Bastin J., Ljunggren H-G., Foster L. and Karre K., 1989, Association of class I major histocompatibility heavy and light chains induced by viral peptides, Nature, 340:443.

INTERACTIONS AMONG THE MHC, DIET AND BACTERIA IN THE PRODUCTION OF

SOCIAL ODORS IN RODENTS

Richard E. Brown and Heather M. Schellinck

Department of Psychology
Dalhousie University
Halifax, N.S. Canada   B3H 4J1

INTRODUCTION

Many mammals can be identified by their individual odors or "olfactory fingerprints" (Brown, 1979) and these individual odors have been
linked to genetic differences at the major histocompatibility complex
(MHC) in both laboratory mice (Yamazak et al., 1991) and rats (Brown et
al., 1987). The MHC regions of the mouse and rat have a similar organization (Brown et al., 1990), each having three different regions, each
of which contributes to the individual's body odor (Brown et al., 1989;
Yamazaki et al., 1990a). The MHC, however, is not the only genetic
region which contributes to individual odors. The sex chromosomes and
the background genes (i.e., non-MHC regions) also contribute to the
individual odors of mice (Yamazaki et al., 1986; Beauchamp et al.,
1990). Since the class I MHC antigens are found in the urine of rats
(Singh et al., 1987, 1988; Roser et al., 1991), we hypothesized that the
class I antigens themselves might be the source of the odor of individuality, but this is not the case as removal of the class I antigen from
the urine of rats did not remove the individuality signal (Brown et al.,
1987). While the intact MHC class I antigens do not appear to be the
source of the odor of individuality, it is possible that volatile frag-
ments of these antigens or other volatiles attached to these antigens
may provide individually unique odors in the urine.

Bacteria play important role in the production of the odors of the
skin and skin glands in many mammalian species (Brown, 1979; Albone,
1984). Bacterial flora may also provide urinary volatiles which are
unique to each individual (Howard, 1977). To test the bacterial hypothesis, we reared MHC congenic rats in germ-free conditions and presented
their urine to subjects in a habituation-dishabituation test. The
results of this experiment indicated that germ-free rearing suppresses
the production of the odor of individuality in rats (Singh et al., 1990;
Roser et al., 1991). We repeated this experiment using an olfactory
discrimination learning paradigm and found that when odors were presented in an olfactometer, it was significantly more difficult to subjects
to discriminate between the urine odors of germfree than conventionally
housed congenic rats. Even when they could discriminate between the
odors of the germfree rats, the discriminations were relearned each day,
indicating that the subjects were not using constant odor cues (Schel-
lick, Brown and Slotnick, 1991). Thus, it would appear that in rats,

Chemical Signals in Vertebrates VI, Edited by R.L. Doty and
D. Müller-Schwarze, Plenum Press, New York, 1992

bacteria interact with genetic differences at the MHC to produce individually distinct urinary odors. The urine of germfree rats is known to differ from that of conventionally housed rats in a number of volatile components (Holland et al., 1983) and some of these volatiles may provide the basis of the individually unique odors. In a gas chromatograph analysis, using the dynamic solvent effect, Natynczuk and Albone (this volume) have found differences between the urines from our germfree and conventionally housed rats.

The bacterial hypothesis, however, does not appear to account for the MCH-related individual odors in mice. The urine odors of MHC congenic strains of mice reared under germfree conditions are just as readily discriminated as those of conventionally housed mice, both in a Y maze with mice as subjects (Yamazaki et al., 1990b) and in an olfactometer with rats as subjects (Schellinck and Brown, this volume). Why the individual urine odors of rats are eliminated by germ-free rearing, but those of mice are not, is unknown. The failure to find an effect of germ-free rearing on the production of individual odors in mice is particularly puzzling as this is the first time that the mouse and rat MCH studies conducted by Yamazaki and ourselves have not given comparable results (see Brown et al., 1990). Nevertheless, our results with germfree rats led us to investigate the possible interactions between MHC type, bacterial flora and diet in the production of individual odors.

BACTERIA AND DIET INFLUENCE URINE ODORS

To account for the finding that conventionally housed, but not germfree, MHC congenic rats are discriminated by their urine odors, we hypothesized that the gastrointestinal (GI) bacteria metabolize dietary products to produce a pool of volatile molecules and that the class I antigens select a cocktail of these volatile metabolites and deliver them to the urine (Singh et al., 1990). This selection may result in a unique urinary odor because each individual has a unique bacterial population or because all individuals have the same bacterial population and dietary or other differences provide the bacteria with different proteins to metabolize.

The GI flora of humans is stable over long periods of time, shows significant individual differences, and is altered by major dietary changes. Furthermore, genetic differences between hosts may influence the bacterial flora of the GI tract. For example, the bacterial flora of monozygotic twins is more similar than that of dizygotic twins; unrelated adults, even if they live together, have significantly different GI flora (Holderman et al., 1976; Van de Merwe et al., 1983). The microflora of the gut is established before weaning and the nature of the neonatal gut microflora depends on the infant's genotype, its mother's antibody production and her commensal bacteria, the proper development of the infant's immune system, and the type of food in its diet. The MHC type of an individual may influence the initial microflora which colonizes the GI tract and this resident microflora then becomes tolerated by the immune system (i.e., treated as self) and does not cause antibody production. Thus, siblings may have similar gut microflora for two reasons: they have genetically related immune systems and they acquire their initial microflora from the same mother. The development of the commensal microflora in the gut is also influenced by the infant's diet (Smith, 1965; Van der Waaij, 1984).

The unique urinary odors of MHC congenic rat strains may, therefore, be the result of an interaction among the MHC type, the GI bacteria and the diet of the animal. If the ability to discriminate between the urinary odors of conventionally housed rats is due to the presence of GI bacteria, then a change in diet should result in a change in urinary odor, even in genetically identical individuals. We tested the hypothesis that dietary changes alter the urine odors of genetically identical mice by collecting urine from conventionally housed adult male C57BL/6J mice fed on two different diets: Hagen hamster food and Purina rat chow, which differ in their proportions of proteins, fats and carbohydrates. Urine was collected by abdominal palpatation while the mice were held over a funnel, so that no feces contaminated the urine, which was frozen at $-20^{\circ}C$ until used as a test stimulus. When adult male Long-Evans rats were presented with urine from these mice in a habituation-dishabituation test (Fig. 1), they could readily discriminate between the urine odors of two mice on different diets, but not between the odors of two mice on the same diet (Schellinck, West and Brown, 1991).

These results demonstrate tha dietary changes alter the urine odors of rodents of the same MHC type and this dietary influence on urine odors may be produced by bacterial action. GI bacteria have the ability to degrade amino acids, and their metabolites are then secreted in the urine (Borud, Midtvedt and Gjessing, 1973). For example, caffeic acid, a constituent of vegetable matter in food, is metabolized to form dihydrocaffeic acid and hydroxypropionic acid by the bacteria of the GI tract and these metabolites are excreted in the urine of conventionally housed, but not germfree, rats (Peppercorn and Goldman, 1972). Thus, individual differences in the urine odors of rats and mice may be due to differences in the urinary volatiles produced by the GI micorflora acting on dietary amino acids.

The volatile constitutents of the urine of conventionally housed rats may also be derived from the sex pheromones of the GI bacteria. The peptide sex pheromones of E. faecalis, for example, bind to receptors on rat neutrophils (Sannomiya et al., 1990) and may also attach to the MHC antigens. It is possible that these bacterial sex pheromones or their metabolites form some of the constituents of the volatile urinary odor of conventionally housed rats and are absent from the urine of bacteria free rats. Whether or not the sex pheromones of commensal bacteria could provide the basis of some mammalian social odors merits further investigation.

DIET, BUT NOT MHC TYPE, INFLUENCES MATERNAL ODORS

Gastrointestinal bacteria are essential for the production of the maternal odor in the feces of lactating female rats, mice and gerbils. Antibiotics, such as tetracycline or neomycine, which inhibit GI bacteria, prevent the production of the maternal odor. Changes in diet also alter the maternal odor as different foods provide different amino acids for these bacteria to metabolize (Leon, 1974, 1975; Skeen and Thiessen, 1977). Moltz and Lee (1981) suggested that the breakdown of cholic acid into a number of metabolites by intestinal micro-organisms provides the basis of the maternal odor. Because germfree rats do not have the bacteria necessary to metabolize dietary amino acids, they secrete higher quantities of these acids in their feces than conventionally housed rats and fewer of their metabolites (Peppercorn and Goldman, 1972; Gustafsson, Norman and Sjovall, 1960). Thus, one would not expect germfree rats to produce maternal odors, but this hypothesis has not been tested.

We examined the interaction of the MHC and diet in the production

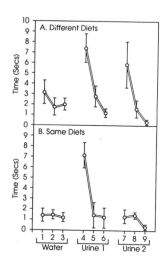

Fig. 1. Mean (+SEM) time spent by rats (n=8) investigating the urine odors of two C57BL/6J mice on (A) different diets or (B) the same diet. Water was presented on the first three 2-minute trials followed by three trials with the first urine sample and three trials with the second urine sample. Urine samples were presented in counter balanced order.

of maternal odors in MHC congenic strains of mice. We maintained two MHC congenic strains of mice (C57BL/6J and B6.AKR-H2^k), which differ at all three MHC regions, on two different diets (Purina and Hagen) and tested their pups for attraction to feces from each of the four different groups of lactating females. Both strains of pups showed a significant preference for the feces of mothers on the same diet as their own mother (Fig. 2). Pups were not attracted more to the maternal odors of lactating females of the same strain as their own mothers than to mothers of the other strain on the maternal diet (Brown and Wisker, 1991). Thus, the effect of diet is more important than genetic differences at the MHC in determining the attractiveness of mouse pups to maternal feces odors, as has been found in studies on spiny mice (Porter and Doane, 1977).

INDIVIDUAL ODORS AS SIGNALS OF HEALTH STATUS

Changes in bacterial flora associated with infectious diseases may result in changes in body odor. Indeed, there is anecdotal evidence that doctors with well-trained noses can detect some diseases by smelling their patients and specific odors have been associated with typhus fever, smallpox, nephritis and the plague (Bedichek, 1960). If the urine odors of rats depend on their commensal gastrointestinal flora, we should be able to manipulate this flora and alter their urine odors. This could be done in two ways: by depleting the bacterial flora through the use of antibiotics, or by altering the bacterial flora. Anti-bacterial agents alter the maternal odors of rats (Leon, 1974) and the body odors of adult male rats (Barnett and Sandford, 1982), and we are now conducting an experiment to investigate whether the depletion of GI bacteria by antibodies eliminates the individual odors of rats. Follow-

ing the depletion of commensal gut flora by antibiotics, it is possible
for the GI tract to be repopulated by different bacteria, leading to a
long-lasting change in the population of GI micro-organisms (Rosebury,
1969). Thus, it should be possible to alter the urinary odors of individuality following antibiotic treatment by innoculating MHC congenic
animals with different micro-organisms. We hope to test this hypothesis
in the near future.

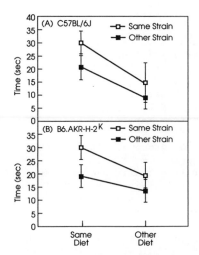

Fig. 2. Mean ($\pm$SEM) attraction scores (time spent near feces odor
minus time spent near no odor) for (A) C57BL/6J (H-2^b) and
(B) B6.AKR-HO2^k pups in tests with feces from lactating
females of the same strain as their mothers or the other
strain which were on the same diet as their mother or the
other diet.

Hamilton and Zuk (1982) hypothesized that secondary characteristics
and elaborate courtship displays advertise the health status of the
displaying animal to potential mates. Support for this hypothesis has
been provided by showing that parasites dull the nuptial coloration and
reduce the effectiveness of courtship displays in red jungle fowl (Zuk
et al., 1990) and stickleback fish (Milinski and Bakker, 1990), and that
injecting male sage grouse with antibiotics to reduce parasitic infection results in increased reproductive success (Boyce, 1990). Since
mammals use urine and scent gland secretions as sex attractants, it is
possible that the odors of these secretions provide information on the
health of the animal. Hamilton and Zuk (1982) suggest that animals
attempting to choose disease free mates should examine the urine and
fecal samples of potential mates. The urine and other odorous secretions of mammals may thus act as carriers of information concerning the
physical condition of the animal, as indicated by the volatile metabolites excreted.

The MHC related odor of the urine may, therefore, be a reflection
of the health status of the animal. If this is so, then the urine odor
of disease animals should be less attractive to conspecifics than the
urine odor of healthy individuals. If diseases influence urine odors
and thus reduce an animal's ability to attract mates, the role of the
MHC may be to maintain a "healthy" urine odor. Thus, the genetic diversity of the MHC may be the result of evolution in response to both
disease and to reproductive mechanisms such as mate selection (see Potts
and Wakeland, 1990).

ACKNOWLEDGEMENTS

This research was supported by NSERC of Canada grant A7441.

REFERENCES

Albone, E.S., 1984, "Mammalian Semiochemistry", Wiley, New York.

Barnett, S.A. and Sandford, M.H.R., 1982, Decrements in "social stress" among wild Rattus rattus treated with antibiotic. Physiol. Behav., 28:483.

Beauchamp, G.K., Yamazaki, K., Duncan, H., Bard, J., and Boyse, E.A., 1990, Genetic determination of individual mouse odour, in: "Chemical Signals in Vertebrates 5," D.W. MacDonald, et al., eds., Oxford University Press, Oxford.

Bedichek, R., 1960, "The Sense of Smell," Michael Joseph, London.

Borud, O., Midtvedt, T., and Gjessing, L.R., 1973, Phenolic metabolites in urine and feces from rats given radioactive ^{14}C-L-DOPA, Acta Pharm. Toxicol., 33:308.

Boyce, M.S., 1990, The red queen visits sage grouse leks. Amer. Zool., 30:263.

Brown, R.E., 1979, Mammalian social odours, Adv. Study Behav., 10:103.

Brown, R.E., Roser, B., and Singh, P.B., 1989, Class I and class II regions of the major histocompatibility complex both contribute to individual odors in congenic inbred strains of rats, Behav. Genet., 19:659.

Brown, R.E., Roser, B., and Singh, P.B., 1990, The MHC and individual odours in rats, in: "Chemical Signals in Vertebrates 5," D.W. Mac Donald et al., eds., Oxford University Press, Oxford.

Brown, R.E., Singh, P.B., and Roser, B., 1987, The major histocompatibility complex and the chemosensory recognition of individually in rats, Physiol. Behav., 40:65.

Brown, R.E. and Wisker, L., 1991, Effects of maternal diet and MHC differences on the attractiveness of the maternal odor in H-2 congenic mice (unpublished manuscript).

Gustafsson, B.E., Norman, A., and Sjovall, J., 1960, Influence of E. coli infection on turnover and metabolism of cholic acid in germ-free rats, Arch. Biochem. Biophysics, 91:93.

Hamilton, W.D. and Zuk, M., 1982, Heritable true fitness and bright birds: A role for parasites? Science, 218:384.

Holdenman, L.V., Good, I.J., and Moore, W.E.C., 1976, Human fecal flora: Variation in bacterial composition within individuals and a possible effect of emotional stress, Appl. Env. Microbiol., 31:359

Holland, M., Rhodes, G., DalleAve, M., Wiesler, D., and Novotny, M., 1983, Urinary profiles of volatile and acid metabolites in germ-free and conventional rats, Life Sciences, 32:787.

Howard, J.C., 1977, H-2 and mating preferences, Nature, 266:406.

Leon, M., 1974, Maternal pheromone, Physiol. Behav., 13:441.

Leon, M., 1975, Dietary control of maternal pheromone in the lactating rat, Physiol. Behav., 14:311.

Milinski, M. and Bakker, T.C.M., 1990, Female sticklebacks use male coloration in mate choice and hence avoid parasitized males, Nature, 344:330.

Moltz, H. and Lee, T.M., 1981, The maternal pheromone of the rat: Identity and functional significance, Physiol. Behav., 26:301.

Peppercorn, M.A. and Goldman, P., 1972, Caffeic acid metabolism by gnotobiotic rats and their intestinal bacteria, Proc. Nat. Acad. Sci. USA, 69:1413.

Porter, R.H. and Doane, H.M., 1977, Dietary-dependent cross-species similarities in maternal chemical cues, Physiol. Behav., 19:129.

Potts, W.K. and Wakeland, E.K., 1990, Evolution of diversity at the Major Histocompatability Complex, Tr. Ecol. Evol., 5:181.

Rosebury, T., 1969, "Life on Man, " Martin Secker and Warburg, London.

Roser, B., Brown, R.E., and Singh, P.B., 1991, Excretion of transplantation antigens as signals of genetic individuality, in: "Chemical Senses. Volume 3," C.J. Wysocki and M.R. Kare, eds., Marcel Dekker, New York.

Sannomiya, P., Craig, R.A., Clewell, D.B., Suzuki, A., Fujino, M., Till, G.O., and Marasco, W.A., 1990, Characterization of a class of non formylated Enterococcus faecalis-derived neutrophil chemotactic peptides: The sex pheromones. Proc. Nat. Acad. Sci. USA, 87:66.

Schellinck, H.M., Brown, R.E., and Slotnick, B.M., 1991, Training rats to discriminate between the odors of individual conspecifics, Anim. Learn. Behav., in press.

Schellinck, H.M., West, A.M., and Brown, R.E., 1991, The ability of rats to discriminate between the urine odors of genetically identical mice on different diets (unpublished manuscript).

Singh, P.B., Brown, R.E., and Roser, B., 1987, MHC antigens in urine as olfactory recognition cues, Nature, 327:161.

Singh, P.B., Brown, R.E., and Roser, B., 1988, Class I transplantation antigens in solution in body fluids and in the urine: Individuality signals to the environment, J. Exp. Med., 168:195.

Singh, P.B., Herbert, J., Roser, B., Arnott, L., Tucker, D.K., and Brown, R.E., 1990, Rearing rats in a germ-free environment eliminates their odors of individuality, J. Chem. Ecol., 16:1667.

Skeen, J.T. and Thiessen, D.D., 1977, Scent of gerbil cuisine, Physiol. Behav., 19:11.

Smith, H.W., 1965, The development of the flora of the alimentary tract in young animals, J. Path. Bact., 90:495.

Van de Merwe, J.P., Stegeman, J.H., and Hazenberg, M.P., 1983, The resident faecal flora is determined by genetic characteristics of the host. Implications for Crohn's disease? Antonie van Leeuwenhock, 49:119.

Van der Waaij, D., 1984, The immunoregulation of the intestinal flora: Experimental investigations on the development and the composition of the microflora in normal and thymusless mice, Microecol. Therapy, 14:63.

Yamazaki, K., Beauchamp, G.K., Bard, J. and Boyse, E.A., 1990a, Single MHC gene mutations alter urine odour constitution in mice, in: "Chemical Signals in Vertebrates 5," D.W. MacDonald, D. Muller-Schwarze, and S.E. Natynczuk, eds., Oxford University Press, Oxford.

Yamazaki, K., Beauchamp, G.K., Bard, J., Boyse, E.A., and Thomas, L., 1991, Chemosensory identity and immune function in mice, in: "Chemical Senses. Volume 3," C.J. Wysocki and M.R. Kare, eds., Marcel Dekker, New York.

Yamazaki, K., Beauchamp, G.K., Imai, Y., Bard, J., Phelan, S.P., Thomas, L., and Boyse, E.A., 1990b, Odortypes determined by the major histocompatibility complex in germfree mice, Proc. Natl. Acad. Sci. USA, 87:8413.

Yamazaki, K., Beauchamp, G.K., Matsuzaki, O., Bard, J., Thomas, L., and Boyse, E.A., 1986a, Participation of the murine X and Y chromosomes in genetically determined chemosensory identity, Proc. Natl Acad. Sci. USA, 83:4438.

Zuk, M., Johnson, K., Thornhill, R., and Ligon, J.D., 1990, Parasites and male ornaments in free-ranging and captive red jungle fowl, Behaviour, 114:332.

MHC-BASED MATING PREFERENCES IN MUS OPERATE THROUGH BOTH SETTLEMENT PATTERNS AND FEMALE CONTROLLED EXTRA-TERRITORIAL MATINGS

Wayne K. Potts[1], C. Jo Manning[1,2] and Edward K. Wakeland[1]

[1]Laboratory of Molecular Genetics, Dept. of Pathology, University of Florida, Gainesville, FL; [2]Dept. of Psychology, University of Washington, Seattle, WA

INTRODUCTION

We have recently demonstrated strong MHC-based mating preferences in semi-natural populations of Mus (Potts et al., 1991). These preferences are primarily responsible for a 27% deficiency of homozygotes (relative to random mating expectations) observed in nine independent experimental populations. The strength of these preferences is sufficient to explain the majority of MHC genetic diversity found in natural populations of Mus. Here we expand our analysis of the nature of these mating preferences. Although the majority of the MHC-related non-random mating appears to be controlled by females traveling to nearby territories to mate with the resident male, a significant proportion (approximately one quarter) is explained by settlement patterns. MHC-related non-random settlement patterns were primarily due to an excess of settlement pairs that share no MHC haplotypes. Because males may exercise control over which females settle on their territories, this component of MHC non-random mating provides an explanation for the presence of male controlled MHC-based mating preferences found in laboratory studies (Yamazaki et al. 1976; Yamazaki et al. 1978; Yamazaki et al. 1988). The prevalence of studies showing male MHC mating preferences (Beauchamp et al. 1988) has been theoretically troubling due to the expectation that females should be the choosier sex (Partridge, 1988; Nei and Hughes, 1991). A companion paper in this volume (Manning et al.) provides the theory and empirical support for why female MHC mate choice may be diminished or lost during inbreeding.

EXPERIMENTAL POPULATIONS

A brief description of the experimental populations follows (described in detail in Potts et al. (1991)). This study was conducted on nine independent populations of half wild individuals with known MHC genotypes. These mice came from generations three and six of original crosses between locally caught wild mice and four different inbred strains (C57BL/10, BALB/c, B10.BR and DBA/1, carrying MHC haplotypes b, d, k and q, respectively). In the F_2 generation, only mice homozygous for the inbred-derived MHC were used to continue this outbred colony. Consequently, half of the genome was wild-derived and the other half was derived from one or more

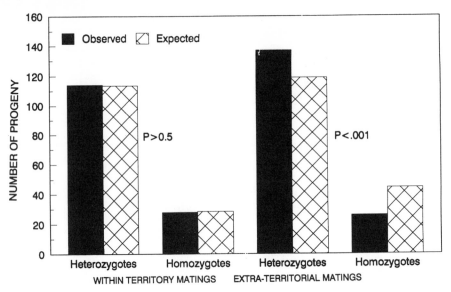

Fig.1   A comparison of observed and expected frequencies of MHC
         heterozygotes and homozygotes for offspring from litters that
         were and were not involved in extra-territorial matings. Data
         from Potts et al., 1991.

inbred strains, but the MHC region was always derived from the inbred
strains.

    Generally 8 males and 16 females, with approximately equal numbers of
MHC homozygotes and heterozygotes, founded each of the nine populations.
When offspring neared weaning age (14-21 days), they were ear punched and a
2 cm tail biopsy taken, from which DNA was extracted and used for MHC geno-
typing by RFLP analysis. The populations were terminated after approximately 150 pups were born. At that time embryos were collected from pregnant females and MHC genotyped. Each animal was given a unique ear punch
which allowed the identification (with the use of binoculars) of mating
pairs and winners and losers of aggressive encounters.

FEMALES CONTROL EXTRA-TERRITORIAL MATINGS

    The mating system of <u>Mus</u> is often cited as the textbook case of
territorial harem polygyny. Gene flow is thought to be limited by a rigid
territorial structure and some authors have thought it appropriate to consider residents of a territory to comprise a deme (Bronson, 1979; Selander,
1970). Our genetic analysis of nine semi-natural populations reveals a
very different story. Over 52% of litters born within a male's territory
contained offspring sired by a neighboring male (Potts et al., 1991). Our
behavioral observations of mating pairs show that these extra-territorial
matings are controlled by females, as all observed extra-territorial
matings (N=13) occurred when a female traveled to a nearby territory and
mated with the resident male. Surprisingly, males showed no mate guarding
behavior of females in estrus.

    Figure 1 compares the observed and expected MHC genotypic frequencies
(heterozygotes and homozygotes) for offspring from litters that did or did
not involve extra-territorial matings. When a female mated with her territorial male the offspring genotype conformed with Mendelian expectations.

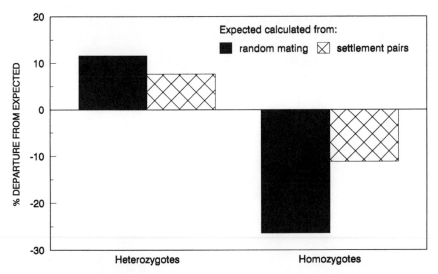

Fig.2   Comparison of observed mean per cent departure from expected of both MHC heterozygotes and homozygotes based on either random mating or settlement pairs.

When females indulged in extra-territorial matings, there was a 41% deficiency of MHC homozygous offspring relative to that expected if she had stayed at home and mated with her territorial male (p<0.001). These results indicate that when females seek extra-territorial matings they do so with males that are more MHC disparate than their own. These female controlled MHC-based mating preferences appear to explain the majority of MHC homozygote deficiencies observed in the semi-natural population experiments.

We believe the original view that <u>Mus</u> territories are equivalent to demes was due to an overinterpretation of Selander's data (Selander, 1970). These data showed a lack of gene flow between buildings and when this was coupled with the well-known aggressive territorial defense, the unjustified conclusion of limited gene flow between territories seeped into the literature (Bronson, 1979).

MHC-RELATED SETTLEMENT PATTERNS

The 27% deficiency of MHC homozygotes observed in the semi-natural populations was based on random mating expectations derived from all females associating freely with all territorial males. If the mean deficiency of MHC homozygous progeny is recalculated based on females mating with their own territorial males, it is reduced from 27% to 19%. This suggests that approximately one quarter of the deficiency of homozygotes (relative to random mating expectations) may be explained by settlement patterns. Figure 2 compares the mean departure from expected MHC heterozygote and homozygote frequencies based on calculations from random mating and settlement pairs. If one uses the number of genotyped offspring from the original nine populations (N=1139) to perform a Chi square evaluation, the difference between expected numbers based on random mating and settlement pairs is significant at the p<0.005 level.

To determine if a particular type of MHC pairing was preferred, the settlement pairs (N=144) were subdivided into four categories on the basis of expected proportions of MHC heterozygotes in the progeny (0, 1/2, 3/4, and 1). For these categories, Figure 3 provides the departure from random settlement expectations. The largest departure from expected was an excess of pairs that shared no MHC haplotypes and consequently were expected to produce no MHC homozygous offspring. This trend was not statistically significant, but it is suggestive and deserves further investigation.

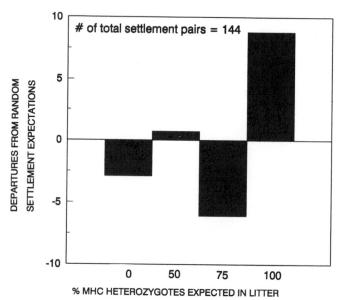

Fig.3   Observed departures from random settlement expectations.

CONCLUSIONS

These results indicate that MHC-based mating preferences in natural populations of <u>Mus</u> are primarily exercised by females. Although these females primarily associate with a single male for the purposes of feeding and nesting, for the purposes of mating their tastes diversify, as over half of all litters were partially sired by neighboring males. We have provided evidence indicating that approximately one quarter of MHC-based mating preferences occur via non-random settlement patterns. This provides an explanation for the existence of male preferences (although there are other theoretical possibilities, e.g. sperm limitations (Dewsbury, 1982)), but it does not explain why female preferences have been more difficult to document in laboratory studies than male preferences. The opposite trend would be expected. One likely answer to this paradox is that strong selection against disassortative mating preferences eliminates (or reduces) female preferences during the inbreeding process (Manning et al., this volume).

REFERENCES

Beauchamp, Gary K., Yamazaki, K., Bard, J. and Boyse, E.A., 1988, Preweaning experience in the control of mating preferences by genes in
the major histocompatibility complex of the mouse, Behav. Genet.,
18(4):537.

Bronson, F.H.,1979, The reproductive ecology of the house mouse, Q. Rev.
Biol., 54:265.

Dewsbury, D.A., 1982, Ejaculate cost and male choice, Am. Nat., 119:601.

Nei, M. and Hughes, A.L., 1991, Polymorphism and evolution of the major
histocompatibility complex loci in mammals, in: "Evolution at the
molecular level:, R.K. Selander, A.G. Clark and T.S. Whittam, eds.,
Sinauer, Sunderland, MA.

Partridge, L., 1988, The rare-male effect: What is its evolutionary
significance?, Philos. Trans. R. Soc. Lond. B Biol. Sci., 319:525.

Potts, W.K., Manning, C.J. and Wakeland, E.K., 1991, Mating patterns in
seminatural populations of Mus infuluenced by MHC genotype, Nature,
352:619.

Selander, R.K., 1970, Behavior and genetic variation in natural populations, Am. Zool., 10:53.

Yamazaki, K., Boyse, E.A., Mike, V., Thaler, H.T., Mathieson, B.J., Abbott,
J., Boyse, J., Zayas, Z.A. and Thomas, L., 1976, Control of mating
preferences in mice by genes in the major histocompatibility complex,
J. Exp. Med., 144:1324.

Yamazaki, K., Yamaguchi, M., Andrews, P.W., Peake, B. and Boyse, E.A.,
1978, Mating preferences of F 2 segregants of crosses between MHC-congenic mouse strains, Immunogenetics, 6:253.

Yamazaki, K., Beauchamp, G.K., Kupniewski, D., Bard, J., Thomas, L. and
Boyse, E.A., 1988, Familial imprinting determines H-2 selective
mating preferences, Science, 240:1331.

MHC CONTROL OF ODORTYPES IN THE MOUSE

Kunio Yamazaki[1], Gary K. Beauchamp[1], Yoshihisa Imai[1],
Judith Bard[2], Lewis Thomas[3] and Edward A. Boyse[2]

[1]Monell Chemical Senses Center, Philadelphia, PA 19104
[2]Department of Microbiology and Immunology, University of
Arizona Health Sciences Center, Tucson, AZ 85724; [3]New York
Hospital/Cornell Medical Center, New York, NY 10021

INTRODUCTION

Odortypes, defined as genetically-determined body odors that enable
individuals of a species to distinguish one another by odor, are specified
in part by polymorphic genes of the Major Histocompatibility Complex or
MHC. The participation of the MHC in individual odor constitution may shed
light on evolutionary and developmental functions of the MHC (Beauchamp et
al., 1986; Boyse et al., 1987, 1991; Yamazaki et al., 1991). These functions have long been suspected to account for many non-immunological correlates of MHC polymorphism which have no obvious connection with the precisely defined immunological functions of the MHC (review in Boyse et al.,
1983), but which may signify more rudimentary functions that preceded the
specific adaptive immunity which is fully manifested only in vertebrates.

THE MHC OF THE MOUSE

The mouse has 20 pairs of chromosomes, with the MHC occupying a segment of chromosome 17. The importance of this group of linked genes can be
gauged from the fact that a similar set of genes probably exists in all
vertebrates (see Klein, 1986). The MHC of mouse, called H-2, comprises
about fifty linked genes and is divided into regions, H-2K (K) H-2I (I),
H-2D (D), Qa, and Tla. The mouse's "MHC type" or "H-2 type" is the total
set of variable alleles of all genes in the MHC region. A given constella-
tion of alleles of Class I, Class II and other less polymorphic genes
within the MHC defines a "haplotype", denoted H-2^b, H 2^k, etc. in the
mouse. The number of such potential haplotypes is huge and makes exact
haplotype matching of unrelated human subjects for HLA almost impossible.

The MHC is best known from studies on tissue transplantation because
incompatibility of MHC types causes rapid rejection of grafts. The fate of
organ transplants depends mainly on MHC compatibility. Throughout the MHC
region there are also genes that determine the degree of response to particular antigens, and other genes expressed selectively in lymphocytes.
Thus, the MHC is concerned in many aspects of how immune cells, lymphocytes
equipped with specific receptors for antigen, handle chemical information
from the environment. We now know that specific MHC genes also determine,
in part, olfactory individuality.

Chemical Signals in Vertebrates VI, Edited by R.L. Doty and
D. Müller-Schwarze, Plenum Press, New York, 1992

CONGENIC MICE

In general, mice of an inbred strain are genetically identical to one another. Mice of an inbred H-2 congenic strain are likewise identical with one another and differ from a selected standard inbred strain only in the vicinity of the MHC (for a fuller description see Beauchamp et al., 1985a). Any difference that distinguishes a pair of H-2 congenic strains, provided that this is shown to be genetic by appropriate segregation tests, must be due to genes in the H-2 region, because this is the only genetic difference between the inbred strain and its congenic partner.

MATING PREFERENCE

The original observation, suggesting MHC odors were involved in chemical communication, made in the congenic breeding rooms at Memorial Sloan-Kettering Cancer Center, was that the male and female of different MHC type paid greater attention to one another and tended to nest together to the relative exclusion of the female whose MHC type was the same as the male's. These chance observations appeared to agree with the suggestion by Thomas (1975) that histocompatibility genes might impart to each individual a characteristic scent. In testing for mating preference, three panels were assembled, patterned on the circumstances of the original observation: a panel of inbred males, a panel of syngeneic females and a panel of MHC-congenic females. For each test, a female in estrus was selected from each female panel and the two females were caged with one selected member of the male panel. The trio was watched until successful copulation, verified by a vaginal plug, had occurred with one of the females, at which time the receptivity of the second female was verified by mating with another male (not belonging to the male test panel).

An MHC-associated mating bias was demonstrated, commonly favoring the female whose MHC-type differed from that of the male (Yamazaki et al., 1976). This was evidently the first example of vertebrate reproduction behavior and selection that has been traced to variation of a particular gene or gene complex. Such preference could act to promote valuable heterozygosity and avoid inbreeding.

In a subsequent study, the origin of mating preference was examined. To determine whether this natural preference for the non-self MHC-haplotype is acquired during early life, the mating preferences of B6 males reared by B6-H-2^k foster parents and of B6-H-2^k males reared by B6 foster parents were studied (Yamazaki et al., 1988). Control males were fostered onto H-2 syngeneic parents. This experimental design seemed most likely to reveal any influence that imprinting on parental MHC-types may have with respect to subsequent choice of a mate.

Within 16 hours of birth, entire litters were removed from their natural parents and transferred to foster parents whose own litters, born at approximately the same time, were simultaneously removed. At 21 days of age, the fostered mice were weaned and the males maintained in stock cages, containing only males of the same genotype and fostering history, until sexual maturity, when tests of mating preference began. The method of testing mating preference was the same as described earlier.

In previous mating tests, male mice showed a tendency to mate with females of an MHC-type different from their own. These preferences for the non-self MHC-haplotype were not significantly altered by fostering on syngeneic parents, which reproduces the genetic relations that obtain in the usual propagation of inbred strains. In contrast, when mice were fostered to parents of a different MHC-type (Fig. 1), the usual mating preferences were reversed: B6 males fostered by B6-H-2^k parents mated preferentially with B6 females, and similarly, B6-H-2^k males fostered by B6 parents mated preferentially with B6-H-2^k females. Thus, MHC selective mating preference is acquired by imprinting on familial MHC types. Whichever MHC type is experienced during the rearing period of 3 weeks becomes the less favored MHC type.

Other investigators have now also reported mating bias based on MHC type. In particular, females also seem to be biased to mate with males of different H-2 types (Egid and Brown, 1990). Importantly, Potts et al. (1991) have recently reported a negative assortative mating bias according to H-2 type in penned mice held under semi-natural conditions, lending support to the hypothesis that such a bias serves to maintain heterozygosity at this locus and to avoid the deleterious effects of inbreeding.

OLFACTORY DISCRIMINATION

It was likely that mating preferences were determined by chemical cues, but it was necessary to directly test for an H-2 associated chemical communication system. To test this, a Y-maze, and later, an automatic olfactometer, were used.

In the Y-maze, air is drawn through two odor boxes, containing urine of H-2 congenic mice. The air is then conducted to the left and right arms of the maze, which are thereby scented differentially by urines of mice whose only genetic difference is H-2. Some mice are trained to run for water reward toward the odor of one H-2 congenic type, whereas others are trained to run to the other, the test subject mice having been deprived of water for 23 hours beforehand. In a second approach, rats are trained to discriminate odors of congenic mice in an automated olfactometer based on the prototype described by Duncan et al. (1992).

Mice could indeed be trained to distinguish between odor of urine of mice differing only at the MHC, which is proof that the MHC is involved in individual odor production (Yamaguchi et al., 1981). The criterion employed in validating a distinction between two alternative odor sources includes a highly significant concordance score, generally 80% or more (chance would be 50%), observed not only in rewarded trials with the familiar odor sources but also in unrewarded blind trials of newly encountered odor sources that duplicate the genetic constitution of the familiar odor sources used in training. Neither the sex nor the genotype of mice chosen for training in the Y-maze, nor which of the alternative urine sources is chosen for reward, significantly influenced proficiency attained in training.

Thus, mice are able to distinguish individuals on the basis of genetic differences at the major histocompatability complex of genes, and urine is a prime odor source (Yamaguchi et al., 1981). Rats are also capable of making these distinctions (Beauchamp et al., 1985b), as are some humans (Gilbert et al., 1986).

The extended MHC of the mouse occupies 2 cM of a total haploid genome of 1600 cM and can accommodate perhaps 50 genes. MHC antibodies identify different genes within the complex, permitting the identification of MHC recombinant mice from which recombinant congenic strains can be derived, differing genetically in only a part of the MHC. Several such recombinant congenic strains, representing three main divisions of the MHC, were all shown to be discriminated in the Y-maze and mating preference, signifying that at least three loci within the MHC autonomously specify an odortype (review in Boyse et al., 1991). And also we have shown that mice have the ability to distinguish the odor of a number of different mutant mice from the odors of non-mutant mice (Yamazaki et al., 1990b), providing direct proof that specific genes within the MHC specify odortype.

CONTRIBUTION OF MICROORGANISMS MHC ODORTYPES

How the MHC genotype influences MHC-determined odortype is unknown. One possible explanation is that odorants are generated by populations of commensal microorganisms, the composition of which is somehow geared to MHC diversity. Alternatively, microorganisms could provide a pool of volatiles for selection by MHC binding molecules, as has been suggested for rats (Singh et al., 1988). Work with rats by Singh et al. (1990) supported a role for microorganisms in MHC-determined odortypes.

This hypothesis was tested for mice in the Y-maze system and the olfactometer system. Germ-free B6 and B6-H-2^k mice were derived from breeders obtained from Sloan-Kettering Institute, according to standard procedures. Following delivery by caesarian section, all germ free mice were kept in total isolation. All standard procedures to monitor and insure germ-free conditions were maintained. To confirm that urine of the germ-free mice had not become contaminated, samples of freshly defrosted urine, and urine that had been used in 6 hours of Y-maze testing, were cultured for bacteria. All samples were negative.

The purpose of the first study was to determine whether mice could be trained to distinguish the urine of germ-free (GF) B6 males from the urine of germ-free (GF) B6-H-2^k males. Clearly such a distinction was possible in both the Y-maze (Fig. 2) and the olfactometer (Fig. 3). Thus, odortypes distinguishing H-2^b and H-2^k genotypes are expressed in the urine of gem-free mice (Yamazaki et al., 1990a).

The purpose of the second study was to investigate the similarity between H-2 determined odortypes expressed by GF mice and H-2 determined odortypes expressed by conventionally-maintained (CV) mice. Mice trained to discriminate between CV donors generalized this to urines of GF animals (Fig. 4). These data suggest that the GF state entails no distinctive alteration in the H-2-determined odortypes expressed by CV mice, as juddged by proficiency of chemosensory distinction in the Y-maze (Yamazaki et al., 1990a).

If bacteria are unlikely to be involved in mouse odortype production, the source of the odor remains a mystery. It is known that soluble MHC molecules, presumably binding antigens, are found in serum, milk, urine and probably other fluids (e.g., Dawson et al., 1974). These molecules may break down to produce odorous metabolites, they may bind odorants or they may act on other tissues to produce the odorants. Their excretion in urine does not, by itself, prove that they are either the odorants themselves, as has been postulated (Singh et al., 1987), or carriers, although the latter is an attractive hypothesis.

DEVELOPMENT OF MHC ODORTYPES

So far, all of our studies of MHC-determined odortypes have involved odors of adult mice. It is known that urine acts as a chemical cue in interactions between parent and infant rodents (Londei et al., 1989). Mother mice are able to discriminate their own pups from alien pups (e.g., Chantrey and Jenkins, 1982); perhaps MHC differences underlie this ability. Cell-surface expression of the mouse MHC Class I antigens becomes detect-

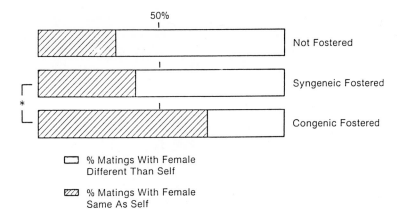

Fig. 1.   Mating preference of males (B6 and B6-H-2^k combined) from
          inbred non-fostered stock, males fostered to parents of the
          same (syngeneic) H-2 types as self, and males fostered to
          parents of a different (congenic) H-2 type than self.   The
          first two groups do not differ, whereas the two fostered
          groups differ (*) significantly, at p < 0.01.

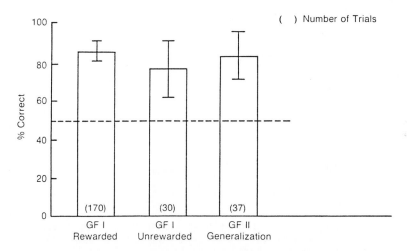

Fig. 2.   Odortypes distinctive of H-2^b and H-2^k genotypes are
          expressed in the urine of GF mice.   Percent correct for four
          trained mice was attained in rewarded trials with urine
          samples from donors of germ-free mouse panels (GF I Reward-
          ed), and was attained in interspersed unrewarded trials
          (which accustom the trained mice to periodic withholding of
          reward for correct choice) (GF I Unrewarded), and, finally,
          was attained in interspersed uniformly unrewarded blind
          trials of coded urine samples from duplicate panels of B6
          and B6-H-2^k GF donors (GF II Generalization) not previously
          encountered by the trained mice.   Data presented here are
          mean percentage correct trials (± 95% confidence interval).

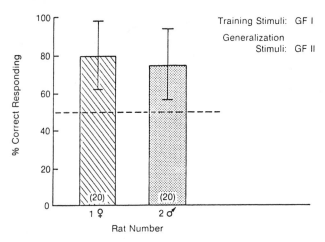

Fig. 3. **Responses of trained rats during generalization trials. Two rats (#1 and #2) were trained to discriminate between urine samples from donors of germ-free mouse panels (GF I) of B6 and B6-H-2[k] and generalized this to urine samples from duplicate panels (GF II) not previously encountered by the trained rats.**

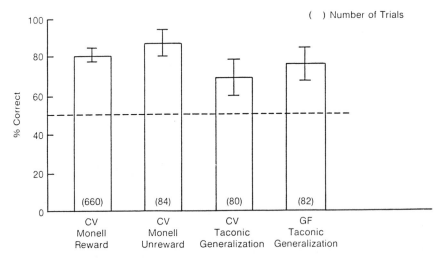

Fig. 4. **Odor types distinctive of H-2[b] and H-2[k] genotypes are similar in CV and GF mice. Percent correct for eleven trained mice was attained in training with respect to Monell CV urine donors in rewarded trials (far left), and was attained in interspersed unrewarded trials (second from left). This also occurred for two generalization tests. Trained mice discriminated appropriately when they were tested with interspersed, uniformly unrewarded blind trials of coded urine samples from SKI specific pathogen-free donors not previously encountered (third from left). Finally, in tests with interspersed uniformly unrewarded, blind trials of coded urine samples from GF donors, not previously encountered (far right). Significant differential responses were observed. Data presented are mean percentage correct trials (± 95% confidence interval).**

able at least as early as the midsomite stage on gestation day 10 (Ozato et al., 1985). However, since the mechanisms by which MHC differences affect odortypes is unknown, it is difficult to predict the time of onset of such odors. The age of onset of MHC-determined odortype was thus examined. We tested the ability of trained mice to discriminate urine odors of pups differing only at the MHC. B6 and B6-H-2^k pups served as urine donors.

The preliminary study showed that odortypes distinctive of H-2 H-2^k genotypes are seen in the urine of pups one day or older (manuscript submitted). Urinary chemical signals influence interactions between parent and infant mice; MHC-determined odortypes may play a prominent role in the mothers' ability to discriminate her own pups from alien pups. Mothers are certainly exposed to the signal since they expend considerable time investigating and licking the anogenital region of their pups and consume a large portion of the urine produced by young pups (e.g., Gubernick and Alberts, 1985). Furthermore, these data support the results of the mouse germ-free studies in that the normal intestinal bacteria flora are not yet present in very young pups.

A final point concerning odortype needs emphasis. Any general theory must take into account the fact that the MHC is not the sole source of olfactory individuality. Our studies have demonstrated that gene(s) on the X and Y chromosomes, as well as other genes scattered throughout the autosomal genome, also contribute to an animals' odortype (Boyse et al., 1991). There may be many different pathways between genes and odorants excreted into the urine.

ACKNOWLEDGEMENTS

This work was supported in part by NIH grant GM 32096 and the Richard Lounsbery Foundation. We thank Ms. Maryanne Curran and Ms. Susanne Corisdeo for excellent technical assistance.

REFERENCES

Beauchamp, G.K., Yamazaki, K. and Boyse, E.A., 1985a, The chemosensory recognition of genetic individuality, Sci. Amer, 253:86.

Beauchamp, G.K., Yainazaki, K., Wysocki, C.J., Slotnick, B.M., L. and Boyse, E.A., 1985b, Sensory recognition of mouse major histocompatibility types by another species, Proc. Natl. Acad. Sci. USA, 82: 4186.

Beauchamp, G.K., Gilbert, A.N., Yamazaki, K. and Boyse, E.A., 1986, Genetic basis for individual discriminations: The major histocompatability complex of the mouse. In Duvall, D., Muller-Schwarze, D. and Silverstein, R.M. (eds.), Chemical Sigmals in Vertebrates IV, Plenum Press, New York, p. 413.

Boyse, E.A., Beauchamp, G.K. and Yamazaki, K., 1983, Polymorphism of the major histocompatibility complex and other genes in relation to sensory perception of genotypes: Some physiological and phylogenetic implications, Human Immunology, 6:177.

Boyse, E.A., Beauchamp, G.K. and Yamazaki, K., 1987, The genetics of body scent, Trends in Genetics, 3:97.

Boyse, E.A., Beauchamp, G.K., Bard, J. and Yamazaki, K., 1991, Behavior and the Major Histocompatibility Complex (MHC), H-2, of the mouse, In Ader, R., Felter, D.L. and Cohen, N. (eds.), Psychoneuroimmunology II, Academic Press, p. 831.

Chantrey, D.F. and Jenkins, B.A.B., 1982, Sensory processes in the discrimination of pups by female mice (Mus musculus). Anim. Behav., 30: 881.

Dawson, J.R., Shasby, S., Amos, D.B., 1974, The serologic detection of HLA antigens in human milk, Tissue Antigens, 4: 76.

Duncan, H., Beauchamp, G.K. and Yamazaki, K., 1992, Assessing odor generalization in the rat: A sensitivity technique. Physiol. Behav., in press.

Egid, K. and Brown, J.L., 1990, The Major Histocompatibility Complex and mating preferences in mice, Anim. Behav., 38:548.

Gilbert , A.N., Yamazaki, K., Beauchamp, G.K. and Thomas, L., 1986, Olfactory discrimination of mouse strains (Mus musculus) and major histocompatibility types by humans, J. Comp. Psychol., 100: 262.

Gubernick, D.J. and Alberts, J.R., 1985, Maternal licking by virgin and lactating rats: Water transfer from pups, Physiol. Behav., 34:501.

Klein, J., 1986, Natural History of the Major Histocompatibility Complex, John Wiley, New York.

Londei, T., Segala, P. and Leone, V.G., 1989, Mouse pup urine as an infant signal, Physiol. Behav., 45:579.

Ozato, K., Wan, Y. and Orrison, B.M., 1985, Mouse major histocompatibility Class I gene expression begins at midsomite stage and is inducible in earlier-stage embryos by interferon, Proc. Natl. Acad. Sci. USA, 82: 2427.

Potts, W.K., Manning, J.L. and Wakeland, E.K., 1991, MHC genotype influences mating patterns in semi-natural populations of Mus, Nature, 352: 619.

Singh, P.B., Brown, R.E. and Roser, B., 1987, MHC antigens in urine as olfactory recognition cues, Nature, 327:161.

Singh, P.B., Brown, R.E. and Roser, B., 1988, Class I transplantation antigens in solution in body fluids and in the urine: Individuality signals to the environment, J. Exp. Med., 168:195.

Singh, P.B., Herbert, J., Roser, B., Arnott, L., Tucker, D.K. and Brown, R.E., 1990, Rearing rats in a germ-free environment eliminates their odors of individuality, J. Chem. Ecol., 16:1667.

Thomas, L., 1975, Symbiosis as an immunologic problem: The immune system and infectious diseases. In: E. Neter and F. Milgrom, (eds.), Fourth International Congress of Immunology, S. Karger, Basel, p. 2.

Yamaguchi, M., Yamazaki, K., Beauchamp, G.K, Bard, J., Thomas, L. and Boyse, E.A., 1981, Distinctive urinary odors governed by the major histocompatability locus of the mouse, Proc. Natl. Acad. Sci. USA, 78:5817.

Yamazaki, K., Boyse, E.A., Mike, V., Thaler, H.T. Mathieson, B.J., Abbott, J., Boyse, J., Zayas, Z.A. and Thomas, L., 1976, Control of mating preferences in mice by genes in the Major Histocompatibility Complex, J. Exp. Med., 144:1324.

Yamazaki, K., Beauchamp , G.K., Kupniewski, D., Bard, J., Thomas, L. and Boyse, E.A., 1988, Familial imprinting determines H-2 selective mating preferences, Science, 240:1331.

Yamazaki, L. and Boyse, E.A. 1990a, Odortypes determined by the Major Histocompatability Complex in germfree mice, Proc. Natl. Acad. Sci. USA, 87:8413.

Yamazaki, K., Beauchamp, G.K., Bard, J. and Boyse, E.A., 1990b, Single MHC gene mutations alter urine odour constitution in mice. In: D.W. Macdonald et al. (eds), Chemical Signals in Vertebrates V, Oxford University Press, p. 255.

Yamzaki, K., Beauchamp, G.K., Bard, J. Thomas, L. and Boyse, E.A., 1991, Chemosensory identity and immune function in mice. In: C. Wysocki and M.R. Kare (eds.), Chemical Senses Genetics of Perception and Commmication, New York, NY: Marcel Dekker, p. 211.

MHC GENES, CHEMOSIGNALS, AND GENETIC ANALYSES OF MURINE SOCIAL BEHAVIORS

Stephen Clark Maxson

Biobehavioral Sciences Graduate Degree Program
Department of Psychology
The University of Connecticut
Storrs, CT   06269-4154   U.S.A.

INTRODUCTION

In many ways, behavior genetics is a dual discipline (Fuller and Thompson, 1978), and two types of behavior genetics are described in this section. These are labeled as Type I and Type II behavior genetics.

Type I is concerned primarily with the causes and adaptiveness of individual differences in the behaviors of a species (Plomin, 1986), and this type of behavior genetics traces its origins to the works of Francis Galton (1869, 1876). Individual differences in behaviors are conceived as being due to effects of variation in genotypes, environments, and their interactions (Plomin, 1986), and the methods of quantitative genetics are usually used to investigate the causes of individual differences in behaviors (Jinks and Broadhurst, 1974). The mapping of genes with effects on urinary odor types or chemosignatures to chromosomes X, Y, and 17 of mice (Yamazaki et al., 1990) obviously has relevance to this type of approach to variation and evolution in olfactory social signals in this species. Namely, variation in MHC haplotypes or genes is an identified cause of individual differences in at least one kind of chemosignal in mice. This is the focus of some of the articles presented at this symposium (Luszyk et al., 1991; Tsuchiya et al., 1991; Yamazaki et al., 1991).

Type II is concerned primarily with the chromosomal mapping of individual genes with effects on behavioral variation and development, and then determining each gene's function at all levels and each gene's action from DNA sequence to behavior (Ginsburg, 1958). The origins of this type of behavior genetics can be traced to proposals by Tolman (1924). Methods of chromosomal and molecular genetics are usually used to investigate gene mapping, function, and action for behavioral traits (Maxson, 1992a, b). The mapping of genes with effects on urinary odor types or chemosignatures to chromosomes X, Y, and 17 of mice obviously has relevance to research on the function and development of olfactory social signals in this species. Such research is a focus of some of the articles presented at this symposium Eklund et al, 1991; Potts et al., 1991; Schellinck and Brown, 1991; Wakeland, 1991; Yamazaki et al., 1991).

The genetics of social behaviors (Type I or II) is concerned not only with display of social signals (such as urinary odor type or chemosignature)

Chemical Signals in Vertebrates VI,   Edited by R.L. Doty and
D. Müller-Schwarze, Plenum Press, New York, 1992

but also with perception of social signals and the motivation for social behaviors. This article will focus on the relevance of MHC chemosignals for the chromosomal mapping of genes with effects on these aspects of murine social behaviors, with particular emphasis on offensive aggression in male mice. Thus, it is primarily in the tradition of Type II behavior genetics.

GENETICS OF SOCIAL BEHAVIORS OF MICE

The social behaviors of mice include the categories of urine marking, aggression, mating, parental or other adult care giving, and care soliciting by progeny or other young (Scott, 1983). Each of these social behaviors involves the interactions of at least two mice, with each mouse displaying stimuli to others and responding to stimuli from others. Three test paradigms for dyadic interactions have been used in research on the genetics of murine social behaviors (Maxson, 1992a). These are the homogeneous set test, the panel of testers, and the standard stimulus test.

In the homogeneous set test, all encounters or interactions are between pairs of mice of the same genotype (strain or Fl hybrid). Since in this paradigm, each individual of a pair must have the same genotype, chromosomal mapping of individual genes for a social behavior requires isogenic populations. For gene mapping with mice, these are either recombinant inbred strains or recombinant congenic strains (Maxson, 1992a). In the panel of testers, encounters or interactions are between pairs of mice of the same or different genotypes (strains or Fl hybrids), and each experimental group (strain or Fl hybrid) of genotypes is tested with mice from a panel of genotypes (strains or Fl hybrids). The panel of testees (experimental group) and the panel of testers (non-experimental group) may be of the same or different array of genotypes. Again, since the genotypes of both individuals in a dyadic encounter or interaction must be known in the panel of testers paradigm, only isogenic populations can be used for chromosomal mapping of the individual genes for a social behavior. For the testee (experimental) groups, this is limited again to genetic analyses with recombinant inbred strains or recombinant congenic strains (Maxson, 1992a). With the standard stimulus test, all encounters or interactions are between mice of one or another genotype and mice of a single, standard genotype (strain or Fl hybrid). Since only the genotype of the standard stimulus tester must be known in this paradigm, the standard stimulus test can be used in genetic analyses with not only isogenic populations, such as recombinant inbred or recombinant congenic strains, but also with heterogenic populations, such as F2 or F3 hybrids, backcrosses, heterogeneous stocks, and selected lines (Maxson, 1992a). This gives the standard stimulus test a major advantage in the chromosomal mapping of genes for social behaviors of mice. The advantage is that several types of genetic analyses can be used to determine the chromosomal map position of a gene with an effect on murine social behavior (Lander and Botstein, 1989; Paterson et al., 1991). For this and other reasons, I have recommended that the standard stimulus tester paradigm be used in the chromosomal mapping of genes for murine social behaviors (Maxson, 1992a).

Since MHC genes influence murine social signals (Boyse et al., 1990), the MHC genotype may be a relevant factor in the selection of the strain(s) to be used as the standard stimulus tester(s) in research on chromosomal mapping of genes with effects on male aggression and other social behaviors of mice. For this reason, the next sections of this article will review the effects of MHC social signals on aggression in male mice, and the following section will then consider the implications of these effects for selection of standard tester strain(s) to be used in chromosomal mapping of genes with effects on offensive aggression of male mice. Issues considered in these sections are also relevant to the selection of standard stimulus tester strains for chromosomal mapping of genes for other social behaviors of mice.

EFFECTS OF MHC SOCIAL SIGNALS ON OFFENSE OF MALE MICE

It has been known for some time that strain differences in offense of male mice, and therefore its genetics, often depend on the genotype not only of the experimental testee groups but also of the opponent tester group(s) (Maxson, 1992a). For example, NZB males are more aggressive (as measured by number of tail rattlings, number of attacks, and proportion of attacking males) when paired with BALB/c opponents than when paired with C57BL6 opponents (François et al., 1990). This difference may be due, at least in part, to differential effects of MHC haplotype on social signals. [BALB/c mice are $H-2^d$, and C57BL10 males are $H-2^b$.] This hypothesis was confirmed in a recent study by François (1990). F2's derived from crosses of BALB/c by BALB.C57BL10-H2^b were typed serologically for H-2 haplotype, and the homozygotes (bb or dd) were used as standard stimulus testers (standard opponents, hereafter) for NZB males. NZB males were more aggressive (as measured by latency to first tail rattle and to first attack) when the F2 opponent males had the homozygous $H-2^d$ haplotype than when they had the homozygous $H-2^b$ haplotype. F2's derived from crosses of C57BL10 and C57BL10.DBA2-H2^d were also serologically typed for H-2 haplotype, and the homozygotes (bb or dd) were also used as standard opponents for NZB males. On the C57BL10 background, there was no effect of haplotype on any measure of aggression. This may be due to the androgen insufficiency (as measured by serum levels of testosterone and target organ sensitivity to testosterone) of C57BL10 males, which can result in lower levels of testosterone-dependent urinary chemo-elicitors of offensive aggression in male mice (Maxson et al., 1983). Consistent with this proposal is the finding that for all measures of offense, males with the BALB/c background (regardless of H-2 haplotype) elicited higher levels of aggression from NZB males than did males with the C57BL10 background, regardless of H-2 haplotype (François, 1990).

An effect of the opponent's H-2 haplotype may also have been observed in our research on Y chromosomal genes and offense in male mice (Carlier, unpublished data; Didier-Erickson et al., 1989; Maxson et al., 1989). In these studies, DBA1 males are more aggressive than DBA1.C57BL10-Y males when the opponent has the $H-2^q$ or $H-2^a$ haplotype, whereas DBA1.C57BL10-Y males are more aggressive than DBA1 males when the opponent has the $H-2^d$ haplotype. Since $H-2^a$ and $H-2^d$ are identical in the C, S, and D and different in the K, A, B, J, E, and Tla regions of the MHC (Klein, 1981), it may be that the critical genes for this opponent effect map to the K to E or Tla regions of the MHC.

There are implications of these findings for the selection of the strain for standard opponent males in chromosomal mapping of genes for offensive aggression of males. Some aspects of this will be considered in the next section.

THE STANDARD OPPONENT AND CHROMOSOMAL MAPPING OF GENES FOR OFFENSIVE AGGRESSION OF MALE MICE

The MHC of mice may affect not only testosterone-independent urinary odor types or chemosignatures but also testosterone-dependent urinary chemo-elicitors of offense. There appears to be a gene in the MHC of mice with effects on serum levels of testosterone (Ivanyi et al., 1972), which may be the gene for 21 hydroxylase, a key enzyme in the synthesis of steroid hormones (Pla et al., 1987). There also appears to be another gene (Hom-1) in the MHC of mice with effects on target organ sensitivity to testosterone (Ivanyi et al., 1973), and it has been tentatively mapped to the K to S regions of the mouse MHC (Ivanyi et al., 1973). Variants in these loci may affect the levels of testosterone-dependent chemo-elicitors of offense in bladder urine (Ingersoll, 1986) and in preputial secretions (Ingersoll et al., 1986). Relevant to this suggestion is the observation that b and d alleles as well as q and d alleles of the S region have different patterns of restriction fragment length polymorphisms (RFLPs) for genomic DNA hybridizing to probes for the 21 hydroxylase gene (Pla et al., 1987). Since substitution in mice of b for d haplotypes (François, 1990) or of d for q haplotypes (Didier-Erickson et al., 1989) in opponent males changes the offensive aggression of testee males, this raises the issue of whether these MHC-dependent opponent effects are due to variation in H-2, class I antigen genes for testosterone-independent odor types or chemosignatures or to variation in other MHC genes for testosterone-dependent chemo-elicitors of offensive aggression.

I believe that this issue may be investigated by using a strain of transgenic mice with a null mutation for β2-microglobulin on the 129/Ola background (Koller et al., 1990). The homozygotes for this null mutation lack β2-microglobulin, and there are no H-2, Qa, or Tla class I MHC antigens on the cell surfaces of most, if not all tissues. For the proposed experiments, it will be necessary to transfer this null mutant by crossbreeding onto the BALB/c, BALB/c.C57BL6-H2^b, DBA1, and DBA1.BALB/c-H2^d strain backgrounds, and it will be necessary to show that urine from these congenic strain pairs, differing in H-2 haplotype, cannot be discriminated by mice in a Y maze or rats in an olfactometer. If such animals become available, the following experiment could be conducted to determine whether the MHC opponent effect is due to testosterone-independent or -dependent chemosignals.

Pairs of congenic strains with different MHC haplotypes and lacking β2-microglobulin would serve as standard opponents in tests of offensive aggression with either NZB or DBA1 males. The standard opponents would be either gonadally-intact or gonadectomized males. For the gonadally-intact standard opponent males, urine from gonadectomized males of either H-2 haplotype would be daubed on the anogenital area of the standard opponent prior to the test for offensive aggression. For example, urine from gonadectomized BALB/c or BALB/c.C57BL6-H2^b males would be daubed on the anogenital region of gonadally-intact BALB/c standard opponent males, and comparisons would be made for the aggressive responses of NZB males to urine of each haplotype daubed on the BALB/c standard opponent males. For the gonadectomized standard opponent males, urine from gonadally-intact males of either haplotype would be daubed on the anogenital area of the standard opponent prior to aggression testing. For example, urine from gonadally-intact BALB/c or BALB/c.C57BL6-H2^b would be daubed on the anogenital region of gonadectomized BALB/c standard opponent males; and comparisons would be made for aggressive responses of NZB males to urine of each haplotype daubed on the BALB/c standard opponent males.

If the differential response by NZB or DBA1 males to opponent haplotype were a function of MHC effects on testosterone-dependent urinary chemo-elicitors of offense, different levels of offensive aggression should be seen for testee males in response to urines of different haplotypes daubed on gonadectomized but not on gonadally-intact males. Also, gonadally-intact standard opponent males of different haplotypes should differ in levels of serum testosterone. If the differential response by NZB or DBA1 males to opponent haplotype were a function of MHC effects on testosterone-independent odor types or chemosignatures, different levels of offensive aggression should be seen by testee males in response to both gonadally-intact and gonadectomized standard opponents daubed with urine of the two haplotypes.

Also, the gonadally-intact standard opponent males of different haplotypes should have the same levels of serum testosterone.

If the MHC opponent effects on offensive aggression are mediated by testosterone-dependent urinary chemo-elicitors of offense, the selection of a standard opponent strain with regard to its MHC type becomes a simple matter. A standard opponent should elicit attack from testee males, but should not initiate attacks or respond with attacks (Denenberg et al., 1973). Thus, the level of serum testosterone should be sufficient to influence these accordingly. This is frequently accomplished by group housing (Carlier and Roubertoux, 1986) or repeatedly defeating (Didier-Erickson et al., 1989) the males to be used as standard opponents. In addition, preliminary experiments should be conducted to determine which treatment and which MHC haplotype maximizes the differences in offensive aggression between the parental inbred strains as well as reduces to a minimum the within-strain variability in offensive aggression. As is discussed by Lander and Botstein (1989) and by Maxson (1992a), accomplishing both of these can facilitate successful chromosomal mapping of genes for complex traits such as offensive aggression in male mice. Similar concerns should also be considered for other genes, such as that on the Y chromosome (Jutley and Stewart, 1985), which influence serum levels of testosterone, and which thereby influence the levels of urinary chemo-elicitors of offense in standard opponent males.

If the MHC opponent effects on offensive aggression are mediated by testosterone-independent odor types or chemosignatures, the selection of the MHC type for the standard opponent may be a more complex matter. More research would be needed to determine whether the effect of MHC odor type or chemosignature on offensive aggression is dependent on the MHC haplotype of the testee mice. If it is, it may be best to use parental inbred strains and a standard opponent strain with the same MHC haplotype. By doing this, differential response of testee mice to MHC haplotype of the standard opponent would not confound the genetic analysis. If it is not dependent on MHC haplotype of the testee mice, preliminary experiments should be done to determine which MHC haplotype for the standard opponent maximizes the difference in offensive aggression of the parental inbred strains and reduces to a minimum the within-strain variability in offensive aggression of testee mice. Since there are other genes (on the X, Y, and other autosomes) that influence odor types or chemosignatures (Yamazaki et al., 1990), there should be similar concern with regard to these in selecting the strain of the standard opponent.

In closing, this issue would be much clearer and simpler if we knew the chemical composition of MHC, testosterone-indpendent odor types or chemosignatures as well as of MHC, testosterone-dependent urinary chemo-elicitors of offensive aggression. As is indicated in presentations at this meeting and elsewhere, both of these types of urinary chemosignals appear to be mixtures of low molecular weight, volatile organic compounds (Lee and Ingersoll, 1983; Novotny et al., 1985; Schwende et al., 1984; Tsuchiya et al., 1991). Unfortunately, there remain great difficulties in identifying and characterizing the chemical constituents of these chemosignals. However, it should be obvious that knowledge about these would provide better control over stimulus parameters of the chemosignals for offensive aggression in mice, and that such control over the olfactory stimulus parameters for male mouse offensive aggression would greatly facilitate genetic or other analyses of this murine social behavior.

ACKNOWLEDGMENT

Supported in part by NATO Grant 0235/89.

REFERENCES

Boyse, E. A., Beauchamp, G. K., Bard, J., and K. Yamazaki, 1990, Behavior and the major histocompatibility complex of the mouse, in: "Psychoneuro-immunology, 2nd ed.," R. Ader, D. L. Felten, and N. Cohen, eds., Academic Press, New York.

Carlier, M. and Roubertoux, P., 1986, Differences between CBA/H and NZB mice on intermale aggression, in: "Genetic Approaches to Behaviour," J. Medioni and G. Vaysse, eds., Privat T.E.C., Toulouse.

Denenberg, V. H., Gaulin-Kremer, E., Gandelman, R., and Zarrow, M. X., 1973, The development of standard stimulus animals for mouse (Mus musculus) aggression testing by means of olfactory bulbectomy. Anim. Behav., 47:1.

Didier-Erickson, A., Maxson, S. C., and Ogawa, S., 1989, Differential effects of the DBA1 and C57BL10 Y chromosomes on the response to social or other stimuli for offense, Behav. Genet., 5:675.

Eklund, A., Egid, K., and Brown, J., 1991, The major histocompatibility complex and mate choice in male mice of two congenic strains, in: "Chemical Signals in Vertebrates VI," R.L. Doty and D. Müller-Schwarze, eds., Plenum, New York.

François, M.-H., 1990, Region chromosomique H-2 et aptitude a provoquer des comportements d'agression chez la souris male. These pour le Doctorat de l'Université Pierre et Marie Curie (Paris VI), Faculte de Medicine Pitie-Salpetriere, Paris, France.

François, M.-H., Nosten-Bertrand, M., Roubertoux, P. L., Kottler, M.-L., and Degrelle, H., 1990. Opponent strain effect on eliciting attacks in NZB mice: Physiological correlates, Physiol. Behav., 47:1181.

Fuller, J. L. and Thompson, W. R., 1987, Foundations of Behavior Genetics," C. V. Mosby Co., St. Louis.

Galton, F., 1869, "Hereditary Genius: An Inquiry into Its Laws and Conse-quences," Macmillan, London.

Galton, F., 1876, The history of twins as a criterion of the relative powers of nature and nurture, Roy. Anth. Inst. of Great Britain and Ireland, 6:391.

Ginsburg, B. E., 1958, Genetics as a tool in the study of behavior, Perspect. Biol. Med., 1:397.

Ingersoll, D. W., 1986, Latent aggression-promoting properties of mouse bladder urine activated by heat, Behav. Neurosci., 100:783.

Ingersoll, D. W., Morley, K. T., Benvenga, M., and Hands, C., 1986, An accessory sex gland aggression-promoting chemosignal in mice, Behav. Neurosci., 100:777.

Ivanyi, P., Hampl, R., Starka, L., and Mickova, M., 1972, Genetic association between H-2 gene and testosterone metabolism in mice, Nature, 238:280.

Ivanyi, P., Gregorova, S., Mickova, M., Hampl, R., and Starka, L., 1973, Genetic association between histocompatibility gene (H-2) and androgen metabolism in mice. Transplant Proc., V:189.

Jinks, J. L. and Broadhurst, P. L., 1974, How to analyse the inheritance of behaviour in animals -- the biometrical approach, in: The Genetics of Behaviour," J.H.F. van Abeelen, ed., North Holland, Amsterdam.

Jutley, J. K. and Stewart, A. D., 1985, Genetic analysis of the Y chromosome of the mouse: Evidence for two loci affecting androgen metabolism, Genet. Res. Camb., 47:29.

Klein, J., 1981, The histocompatibility-2 (H-2) complex, in: "The Mouse in Biomedical Research: Vol. I, History, Genetics, and Wild Mice," H. L. Foster, J. D. Small, and J. G. Fox, eds., Academic Press, New York.

Koller, B. H., Marrack, P., Kappler, J. W., and Smithies, O., 1990, Normal development of mice deficient in B2, MHC class I proteins and CD8+ T cells, Science, 248:1227.

Lander, E. S. and Botstein, D., 1989, Mapping Mendelian factors underlying quantitative traits using RFLP linkage maps, Genetics, 121:185.

Lee, C-T. and Ingersoll, D. W., 1983, Pheromonal influence on aggressive behavior, in: "Hormones and Aggressive Behavior," B. B. Svare, ed., Plenum, New York.

Luszyk, D., Eggeret, F., Uharek, L., Müller-Ruchholtz, W., and Ferstl, R., 1991, The influence of the homatopoietic system on the production of MHC-related odors in mice, in: "Chemical Signals in Vertebrates VI," R. L. Doty and D. Müller-Schwarze, eds., Plenum, New York.

Maxson, S. C., 1992a, Methodological issues in genetic analyses of an agonistic behavior (offense) in male mice, in: "Techniques for the Genetic Analysis of Brain and Behavior: Focus on the Mouse," D. Goldowitz, R. E. Wimer, and D. Wahlsten, eds., Elsevier Science Publishers, Amsterdam.

Maxson, S. C., 1992b, Potential genetic models of aggression and violence in males, in: "Genetically Defined Animal Models of Neurobehavioral Dysfunction," P. Driscoll, ed., Birkhäuser-Boston, Cambridge, MA.

Maxson, S. C., Shrenker, P., and Vigue, L. C., 1983, Genetics, hormones, and aggression, in: "Hormones and Aggressive Behavior," B. B. Svare, ed., Plenum, New York.

Maxson, S. C., Didier-Erickson, A., and Ogawa, S., 1989, The Y chromosome, social signals, and offense in mice, Behav. Neural Biol., 52:251.

Novotny, M., Harvey, S., Jemiolo, B., and Alberts, J., 1985, Synthetic pheromones that promote inter-male aggression in mice, Proc. Natl. Acad. Sci. USA, 82:2059.

Paterson, A. H., Damon, S Hewitt, J. D., Zamir, D., Rabinowitch, H. D., Lincoln, S. E., Lander, E. S., and Tanksley, S. D., 1991, Mendelian factors underlying quantitative traits in tomato: comparison across species, generation, and environments, Genetics, 127:181.

Pla, M., Rocca, A., Gillet, D., Villette, J.-M., Fiet, J., and Degos, L., 1987, Involvement of the H-2 complex in steroid hormone metabolism, in: "H-2 Antigens: Genes, Molecules, Function," C. S. Davis, ed., Plenum, New York.

Plomin, R., 1986, "Development, Genetics, and Psychology," Lawrence Erlbaum, Hillsdale, N.J.

Potts, W., Manning, C. J., and Wakeland, E. K., 1991, Strong MHC-based mating preferences in semi-natural populations of Mus: Evidence that they function primarily to avoid inbreeding, in: "Chemical Signals in Vertebrates VI," R. L. Doty and D. Müller-Schwarze, eds., Plenum, New York.

Schellinck, H., and Brown, R., 1991, Why does germ free rearing eliminate odours of individuality in rats but not in mice," in: "Chemical Signals in Vertebrates VI," R. L. Doty and D. Müller-Schwarz, eds., Plenum, New York.

Schwende, F. J., Jorgenson, J. W., and Novotny, M., 1984, Possible chemical basis for histocompatibility-related mating preference in mice, J. Chem. Ecol., 10:1603.

Scott, J.P., 1983, Genetics of social behavior in nonhuman animals, in: "Behavior Genetics: Principles and Application," J. L. Fuller and E. C. Simmel, eds., Lawrence Erlbaum Associates, Hillsdale, N.J.

Tolman, E. C., 1924, The inheritance of maze learning in rats, J. Comp. Psychol., 42:58.

Tsuchiya, H., Yamazaki, K., Singer, A. G., and Beauchamp, G. K., 1991, Chemical characterization of olfactory signals of MHC type, in: Chemical Signals in Vertebrates VI," R. L. Doty and D. Müller-Schwarze, eds., Plenum, New York.

Wakeland, E., 1991, MHC molecules and olfactory signalling: Structural constraints on simple explanations, in: "Chemical Signals in Vertebrates VI," R. L. Doty and D. Müller-Schwarz, eds., Plenum, New York.

Yamazaki, K., Beauchamp, G. K., Bard, J., and Boyse, E. A., 1990, Chemosensory identity and the Y chromosome, Behav. Genet., 20:157.

Yamazaki, K., Beauchamp, G. K., Bard, J., Thomas, L., and Boyse, E. A., 1991, MHC control of odortype in the mouse, in: "Chemical Signals in Vertebrates VI," R. L. Doty and D. Müller-Schwarze, eds., Plenum, New York.

MHC-RELATED ODORS IN HUMANS

Roman Ferstl[1], Frank Eggert[1], Eckhard Westphal[2],
Nicholaus Zavazava[2], and Wolfgang Muller-Ruchholtz[2]

[1]Department of Psychology and [2]Department of Immunology
Univeristy of Kiel, Kiel, Germany

INTRODUCTION

Up to now, MHC-related odors have been only described for rodents
Yamazaki et al., 1991; Roser et al., 1991). Nevertheless, it has been
speculated that a similar phenomenon may also occur in humans (Beauchamp et
al., 1985; Boyse et al., 1987). Individual specific body odors do indeed
play a role in human self-perception (Porter and Moore, 1981; Lord and
Kasprzak, 1989) and recognition of offspring (Porter et al., 1983; Kaitz et
al., 1987), but there is no information available on the biological basis
of these odors.

Since there exits no HLA-congenic groups to obtain urine or sweat
samples for further experimentation, a different approach has to be used to
identify possible MHC-related odors in humans. In this article, we report
on three studies. In the first and second studies, we tried to train rats
in a computer-controlled olfactometer to discriminate, by urine odors,
groups of human subjects according to their HLA class I type. In this
experiment, the hypothesis was tested that MHC-homogenous groups show a
correlated homogeneity of odor expression which allows trained animals to
group previously unknown stimuli according the MHC-type of the donors.
Groups of tissue typed subjects were used as donor panels for HLA class I
homogenous and heterogenous urine samples.

The second study was initiated by a report of a peculiar odor perception phenomenon which one of our female students had experienced for several months during her pregnancy; namely, among her fellows she named three
whose body scents were noticed by her as extremely aversive. Tissue-typing
her and the three students revealed that two of her fellows were identical
in one haplotype of their HLA class I loci A and B (A24, B62). The probability of identifying two such cases by random sampling (N=3, p=0.0079 base
rate in a representative German sample of 10,000 cases) is 0.00018. In our
third study, we tried to identify a larger sample of women who had experienced phenomena similar to that of our female student and to tissue type
the odor sensitive as well as the indexed persons and compare sample frequencies of HLA class I A and B loci with population distribution data. We
will report on the animal experiments first and consecutively outline the
design and results of our field research.

Chemical Signals in Vertebrates VI, Edited by R.L. Doty and
D. Müller-Schwarze, Plenum Press, New York, 1992

The first two studies were designed to test whether rats can be trained in an olfactory discrimination task to identify a group of subjects who share common HLA class I loci A and B by their urine odors. Class I locus C and Class II and III loci were not considered in these studies.

Animals, Materials and Methods

Three female rats of the inbred strain CAP, deprived of water for 20 hours before each session, were trained. They were maintained under standard conditions. For the first experiment, two rats (CAP, female) were trained in an odor concept formation task for HLA class I specificity A1B8. In the second experiment, one rat was confronted with the same type of task except that urine samples were obtained from females only (cf. lower part of Table I).

Table I. Urine Donor Panels

S+ Subjects					S- Subjects				
HLA-Class I				Sex	HLA-Class I				Sex
Panel 1									
A1	A1	B8	B8	male	A24	A28	B38	B56	male
A1	A1	B8	B8	female	A24	A24	B40	B51	female
A1	A1	B8	B8	female	A11	A24	B39	B61	female
A1	A2	B7	B8	female	A2	A26	A13	B55	male
A1	A2	B7	B8	female	A3	A24	B7	B57	male
A1	A2	B7	B8	female	A31	A25	B13	B39	female
Panel 2									
A1	A1	B7	B8	female	A1	A25	B37	B58	female
A1	A1	B7	B8	female	A1	A11	B35	B44	female
A1	A2	B7	B8	female	A25	A31	B13	B39	female
A1	A2	B7	B8	female	Ax	Ax	B35	B62	female

Stimulus material consisted of urine samples collected from four groups of tissue-typed subjects and sweat samples from one pair of S+/S- subjects (see Table 1). Urine was portioned and frozen until used for the experiment. Sweat was collected by means of axillary pads. For the donation of urine samples 12, subjects were recruited from the tissue type register at the University of Kiel. The upper part of Table 1 reveals A- and B-loci for the groups of S+ and S- urine samples. The S+ sample contained 3 homozygous and 3 heterozygous urine donors. Note that the S+ group contained one male donor and the S- group three male donors. In the first experiment, the sequence of introduction of pairs of S+ and S- were started for one animal with homozygous A1A1B8B8 samples and for the other with presentation of heterozygous A1A2B7B8 samples. In the second experiment, the sequence followed the order of panel 2 in Table I. Sweat was sampled from the last pair of subjects from panel 2.

The experimental set up is shown in Figure 1. Air was provided by a compressor (C), dehydrated through a column of silica gel (SG), and deodorized through a column of charcoal (CC). The filtered air was divided into four channels, one carrier stream and three odorant streams, and was partly rehydrated by being passed through distilled water in gas wash bottles (GWB). Each of the odorant streams was connected to an odor bottle (OB) containing one of the odorants used for training and testing. To allow presentation of more than three odors within one session, odor bottles (OB) were exchanged in a balanced order. Air flows in the four streams were controlled and monitored by flow meters (FM). The carrier stream passed through the test chamber (TC) from which air was continously exhausted (E). The air pulse from each odorant stream was controlled by a solenoid valve (SV).

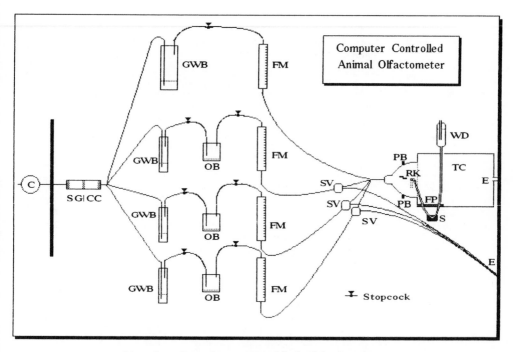

Fig. 1. Computer controlled olfactometer

When the rat interrupted a photobeam (PB) at the front of the test chamber, one of the odorant streams was added to the carrier stream. If the S+ had been presented (2 sec) and if the rat had made contact between the response key (RK) and a stainless steel floor plate (FP), a drop of water was delivered from a selenoid (S) controlled water dispenser (WD). Intertrial interval (1 sec), trial initiation and timing as well as order of odor presentation, were controlled and responses (correct/false) were recorded by a RHOTRON-ATARIcomputer. Each training session consisted of up to 200 single trials.

Urine contains several elements which are possible candidates for odor molecules (Albone 1984; Schwende et al., 1984). Considering this fact, as well as differences in the nutritional and hormonal status of the subjects, a stepwise odor concept formation task was designed. To test whether stimulus generalization occurs among the presented S+, a cumulative training and testing design was chosen. Training started with pairs of S+ and S- stimuli until a criterion of 70% correct was reached. Consequently, the

schedule of reinforcement (SOR) was reduced stepwise to 0% in four blocks
of 10 trials each in order to introduce a next pair of odors to test for
transfer of training (TTT). Two blocks were indexed as reference trials,
the other two as test trials.   After the first TTT, training continued
with four odors, i.e., two pairs of S+ and S- trials. A second TTT was
followed by two training phases with three different pairs of odors, each
of which was finished by a separate TTT. Due to technical reasons, no
training with four pairs of odors was possible.

Results

Table II-A reveals the results of reference and test trials based on
the two animals trained with urine samples from the first donor panel (cf.
Table I). The single and complex stimulus arrangement in every block of
reference trials, i.e., without any reinforcement for correct responses,
was correctly discriminated and the stimuli were correctly identified.
While the first and second introductions of a new pair of urine odors did

Table II. Transfer-of-training Tests with Urine Samples from
           Panels 1 and 2.

TTT after a training with	Discrimination S+/S-		Identification S+	
	RT	TT	RT	TT
A	(Urine Samples from Panel 1)			
1 pair of urines	87.3% *** N=150	49.5% N=140	71.3% *** N=80	20.0% N=60
2 pairs of urines	67.3% *** N=150	50.0% N=128	78.7% *** N=75	45.4% N=67
3 pairs of urines	64.4% *** N=160	64.8% *** N=122	85.0% *** N=80	83.9% N=62
3 pairs 2nd TTT	80.2% *** N=105	68.3% *** N=115	86.5% *** N=135	70.2% *** N=135
B	(Urine Samples from Panel 2)			
1 pair of urines	95.0% *** N=60	81.6% *** N=60	96.0% *** N=29	95.0% *** N=21
2 pairs of urines	86.6% *** N=60	83.3% *** N=60	84.0% *** N=32	83.3% *** N=30
3 pairs of urines	86.6% *** N=60	83.3% *** N=60	82.0% *** N=34	81.0% *** N=32
C	Transfer-of-Training Test to Sweat Samples			
3 pairs of urines	86.6% *** N=60	75.0% *** N=60	80.5% *** N=36	77.7% *** N=37

** p << 0.01, *** P << 0.001; binomial distribution. RT reference trials,
TT test trials

208

not reveal positive results for correct transfer of training after discrimination training with three pairs of odors, the forth new pair of S+ and S-stimuli was discriminable and the S+ was correctly identified. The results in the first and second transfer tests may be due to sex differences between the urine donors (cf. Table I, panel 1).

Table II-B shows the results of reference and test trials for the third animal, which was trained with urine probes from the second donor panel in which only females were included (cf. Table I, Panel 2). It also was able to discriminate the single and complex stimulus arrangement in every block of reference trials, i.e., without any reinforcement for correct responses. The first, second and third introduction of a new pair of urine odors revealed positive results for correct transfer of training.

The TTT with sweat samples of one pair of subjects revealed a correct discrimination, as well as identification, after training with three pairs of urine odors (cf. Table II-C).

HLA-SPECIFIC BODY SCENTS

About 400 women were interviewed on peculiar odor perception phenomena which occurred during their pregnancies and/or over their menstrual cycles. Among them we identified 19 (5%) women who fulfulled the following criteria: 1) perception of intensive body odors of at least two other people (male or female; relatives or partners being excluded from the study; effects of hygiene or fragrancies were ruled out), 2) commitment of all indexed subjects, including the interviewed woman, to participate in the study (i.e., to donate 20 ml of blood for HLA-classification). Overall 51 subjects (31 male, 20 female) with reportedly intense body odors were indexed. Additionally, 23 control subjects without noticeable body odors were named by the 19 women.

Results

The results revealed significantly higher frequencies for some HLA-alleles in our sample than in a representative unselected sample of tissue typed Germans (see Figure 2 for details). Ten out of the 19 women (55%) belonged to HLA class I A9(A23, A24) and/or B15(B62, B63). Twenty-nine of the 51 (60%) indexed subjects belonged to HLA class I A9(A23, A24) and/or B15(B62, B63). In the group of 23 control subjects, only six (23%) belonged to that tissue type.

GENERAL DISCUSSION

Our studies have shown that HLA class I specific odor phenotypes can be identified by rats and likely play a role in social communication among at least some humans. However, several questions and one shortcoming of our research need to be discussed. First, the data base for HLA- specific urine odors identified by rats is still small and should first be extended in two ways. First, more animals should be trained on the same task; second, homogenous urine donor panels should be recruited from other class I specificities. It also remains open whether restriction to a single class I locus (e.g. A1A1) still is sufficient for identifying a related odor concept. Up to now only compounds of A and B loci were tested. The main shortcoming of our field study was to select only those females who indexed at least two other persons for their intensive body odors. The reason for this strategy was to increase our group of subjects quickly. Further studies will include also females with one indexed subject.

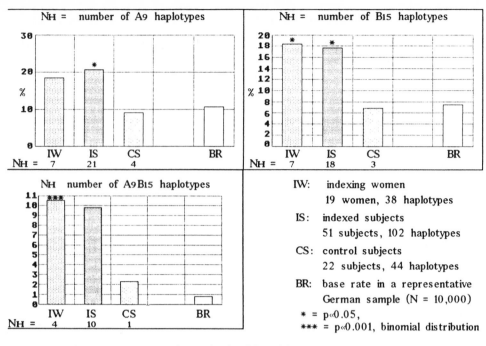

Fig. 2.   Frequencies of A9(A23, A24)-, B15(B62, B63)- and A9B15-MCH type.

As Zavazava et al., (1991; also this volume) have shown, serum concentrations of soluble class I HLA molecules of HLA-A9 and -B15 are significantly increased compared to other specificities. Moreover, SDS-PAGE analysis of both serum and sweat soluble class I antigens reveals that these molecules belong to the same 46 kD band. For 2 subjects with HLA-A24 (split of HLA-A9) sweat levels were >> 0.5 ug/ml. This supports the idea that HLA levels in sweat may be dependent on serum levels and could perhaps explain the odor intensity of HLA-A9, -B15 carriers.

ACKNOWLEDGEMENT

This work was supported by a grant from the Stiftung Volkswagenwerk.

REFERENCES

Albone, E.S., 1984, "Mammalian semiochemistry: The investigation of chemical signals between mammals", Wiley, New York.

Beauchamp, G.K., Yamazaki, K. and Boyse, E.A., 1985, The chemosensory recognition of genetic individuality, Sci. Am., 253:86.

Boyse, E.A., Beauchamp, G.K. and Yamazaki, K., 1987, The genetics of body scent, Tr. Genet., 3:97.

Kaitz, M., Good, A., Rokem, A.M. and Eidelman, A.I., 1987, Mothers' recognition of their newborns by olfactory cues, Developm. Psychobiol., 20:587.

Lord, T. and Kasprzak, M., 1989, Identification of self through olfaction, Percept. Mot. Skills, 69:219.

Porter, R.H. and Moore, J.D., 1981, Human kin recognition by olfactory cues, Physiol. Behav., 27:493.

Porter, R.H., Cernoch, J.M., and McLaughlin, F.J., 1983, Maternal recognition of neonates through olfactory cues, Physiol. Behav., 30:151.

Roser, B., Brown, R.E. and Singh, P.B., 1991, Excretion of transplantation antigens as signals of individuality, in: "Chemical Senses", Vol. 3, Genetics of Perception and Communication, C.J. Wysocki and M.R. Kare, eds., Marcel Dekker, New York.

Schwende, F.J., Jorgenson, W.J. and Novotny, M., 1984, Possible chemical basis for histocompatibility-related mating preference in mice, J. Chem. Ecol., 10:1603.

Yamazaki, K., Beauchamp, G.K., Bard, J., Boyse, E.A. and Thomas, L., 1991, Chemosensory identity and immune function in mice, in: "Chemical Senses", Vol. 3, Genetics of Perception and Communications, C.J. Wysocki and M. R. Kare, eds., Marcel Dekker, New York.

Zavazava, N., Westphal, E. and Muller-Ruchholtz, W., 1990, Characterization of soluble HLA molecules in sweat and quantitative HLA differences in serum of health individuals, J. Immunogenet., 17:387.

SEX DIFFERENCES IN THE USE OF THE MAJOR HISTOCOMPATIBILITY

COMPLEX FOR MATE SELECTION IN CONGENIC STRAINS OF MICE

Amy Eklund, Kathleen Egid, and Jerram L. Brown

Department of Biological Sciences
State University of New York
Albany, N.Y. 12222

Since the discoveries that mice can detect odors associated with genes of the major histocompatibility complex (MHC) and that these odors affect mate choice (Yamazaki et al., 1976, 1979), there has been speculation about the generality of these effects in other inbred strains of mice and in populations of wild mice. This interest in the effects of the MHC on behavior has arisen because it is possible that the observed disassortative mating could be maintaining, in part, the extensive genetic polymorphism found within this complex in natural populations (see Potts and Wakeland, 1990, for a more detailed discussion of hypotheses for MHC-disassortative mating). This polymorphism might also make the MHC an important tool in social contexts by providing unique labels for kin recognition and discrimination (Brown, 1983). It follows, then, that the significance of the MHC, particularly during mate selection, needs to be further investigated in diverse strains of mice, considering both intersexual and interstrain variation in strength and direction of preferences.

As a first step in comparing the behavior of mice of various strains, we tested the MHC-based odor preferences of females from three B6-congenic strains (two of these strains were used in the original work demonstrating male MHC-dependent mate preference). Here we report results from odor preference tests with these females and we compare these data to our previously reported results on the MHC-based odor and mate preferences of males and females of two B10-congenic strains of mice.

STRAINS OF MICE

Females from 3 strains of MHC-congenic mice, B6 (bb), B6-H-2^k (kk), and B6-H-2^{ER2} (bb, but see Table 1 notes), were tested for MHC-based odor preferences in a Y-maze (design supplied by K. Yamazaki). The first two strains were among those used by Yamazaki et al. (1976) in their initial demonstration of disassortative mating using MHC cues, and the third strain is identical to B6 (bb) except at the Tl region of the MHC.

We used two additional MHC-congenic strains of mice, B10.CHR51 and B10.GAA37, to test MHC-based mate preferences as well as odor preferences.

These strains, unlike conventional MHC-congenics, were produced by mating a wild mouse to a mouse from the inbred C57Bl/10Sn strain. After ten generations of backcrossing to the inbred strain with serological selection for the wild MHC-haplotype, the offspring were mated to produce the homozygous progeny needed to create a new inbred strain (CHR) with a recently-derived wild MHC-haplotype. The same procedure was repeated using a second wild mouse with a different MHC-haplotype to produce the MHC-congenic strain, GAA (Klein, 1973).

Using MHC-congenic strains allows us to investigate behavioral effects due solely to genes within the MHC, because these strains differ at the MHC but are genetically identical at all other loci. While other approaches, e.g., using wild or half-wild mice with known MHC-haplotypes, could provide more information regarding strength and frequency of preferences in wild populations, studying multiple inbred strains of mice provides a baseline to assess the potential effects of the MHC on behavior and population structure in house mice.

ODOR PREFERENCE EXPERIMENT

We tested the odor preferences of mice in a Y-maze or in a 3-chambered Plexiglas cage (see Egid and Brown, 1989 and Eklund et al., in press, for details). Briefly, a trial consisted of presenting a test animal with the odors of animals from two MHC-congenic strains. In each trial, one stimulus-odor came from a mouse of the same strain (same MHC-haplotype) as the test animal and the other from the congenic strain (different MHC-haplotype). Soiled bedding from the cages of stimulus mice served as the source for the odor. During the 15-minute trial we recorded the time a test animal spent in each arm of the Y-maze (or end-compartment of 3-chambered cage) sniffing an odor. We calculated a preference score for each animal as time spent near the different-strain odor minus time spent near the same-strain odor, divided by total time spent sniffing odors. Negative values therefore indicate more time near the MHC-similar odor.

Results demonstrated that strain, sex, and estrous condition affected the odor preferences of mice. Male CHR and GAA mice behaved randomly in response to the odors of females of either of these two MHC strains. When tested by strain or pooled, the mean preference scores of both groups of males were not significantly different from zero (Table 1). In contrast, female CHR and GAA mice did not behave randomly. Female behavior was affected by estrous condition: the combined preference scores of females in estrus were significantly different from females in metestrus or diestrus (ANOVA: $F = 13.27$, $p < .0004$). Females in estrus spent more time near the odor of the MHC-different-than-self male (Table 1).

These results for CHR and GAA mice were surprising since previous work with MHC-congenic strains indicated that while males exhibited preferences for MHC-dissimilar mates, females either mated randomly or showed a preference for a particular congenic strain regardless of their MHC-haplotype (Beauchamp et al., 1988). Since we tested different strains of mice and employed a different procedure from the original researchers, we decided to test the B6-congenic strains in our odor preference apparatus. When these strains (B6, and B6-H-2^k), and a third strain (B6-H-2^{ER2}), were tested for their odor preferences, all 3 strains behaved similarly. Specifically, females, regardless of their own MHC-type and estrous condition, showed weak preferences for the odors of B6 (bb) males (Table 1). Our results on odor preference for the B6-congenic strains are in agreement with the results from Beauchamp et al. when

Table 1. Comparison of preferences for different MHC-haplotypes among inbred mouse strains

Strain (MHC-type)	Sex[1]	Odor-choice test			Mate-choice test		
		Odor Score[2] (S.E.)	N	P	Percent different[3]	N	P
B10.CHR51 (w18)	F-E	.042[a] (.038)	25	p>.05	70[a]	20	p>.05
	F-N	-.081[a] (.038)	25	p>.05	n.a.	-	-
	M	.037[b] (.066)	20	p>.05	44[b]	18	p>.05
B10.GAA37 (w21)	F-E	.119[a] (.034)	25	p<.005	79[a]	19	p<.025
	F-N	-.020[a] (.034)	25	p>.05	n.a.	-	-
	M	.087[b] (.074)	20	p>.05	44[b]	9	p>.05
B6 (b)	F-E	-.124 (.096)	6	p>.05	36[c]	-	p<.01
	F-N	-.016 (.076)	10	p>.05	n.a.	-	-
	M	n.a.	-	-	67[c]	51	p<.05
B6-H-2[k] (k)	F-E	.073 (.155)	6	p>.05	64[c]	-	p<.01
	F-N	.081 (.042)	8	p>.05	n.a.	-	-
	M	n.a.	-	-	69[c]	61	p<.05
B6-H-2[ER2] (ER2[4])	F-E	-.054 (.034)	31	p>.05	n.a.	-	-
	F-N	-.044 (.039)	13	p>.05	n.a.	-	-

1) F-E: females in estrus, F-N: females in metestrus, M: males
2) mean odor preference score; see text for description; scores compared to random with two-tailed t-test
3) percentage mating with MHC-different mouse; frequencies compared to random with Chi-square test
4) ER2 MHC-haplotype: bbbbbbk (at K,IA,IE,D,L,Q,Tl)
a) data from Egid and Brown, 1989
b) data from Eklund et al., in press
c) data from Beauchamp et al., 1988; values for females are based on an approximate average of bb matings for both strains combined and P value is from Chi-square conducted on pooled frequencies.
n.a.- not available

they tested females of these same strains in a mate preference test. Considering the results from the CHR and GAA strains, along with the results from the B6 strains, it is clear that variability exists among inbred strains with respect to the sex which shows an MHC odor preference, and the direction of preference shown.

MATE PREFERENCE EXPERIMENT

Results from the odor preference tests with CHR and GAA mice provided a quick analysis of the ability of these strains to discriminate MHC-based odors and an indication of possible mating preferences. However, odor preference cannot be assumed to be the same as mate preference. Therefore we tested the preferences of sexually inexperienced male and female mice of these strains for actual mates. We used a test procedure which permitted us to discriminate between male and female preferences (described in Egid and Brown, 1989, and Eklund et al., in press). Briefly, the test animal, e.g. a male, was allowed to choose between two tethered females, one from the same strain as itself and one from the MHC-congenic strain, placed in opposite end chambers of a 3-chambered apparatus. The stimulus mice could not interact with each other, but the choosing animal could move into either of the two end-chambers. Only females determined to be in physiological estrus (by observing cell type from a vaginal lavage) were used as test or stimulus animals. Both amount of time spent with each stimulus animal and ejaculation with a stimulus animal were used as measures of preference.

Results from the mate choice study were comparable to results from the odor preference experiment: males mated randomly while females mated disassortatively (males compared to females: $X^2 = 4.86$, df = 1, p = .0274). Based on ejaculation data, females chose the male with an MHC-haplotype different from her own in 74% (29 out of 39) of the trials ($X^2 = 8.3$, df = 1, p < .01). In contrast, males mated with an MHC-different female in only 44% (12 out of 27) of the trials (Table 1).

For the male trials in the mate-preference experiment, we also calculated a time-preference score as time spent with the different-strain female minus time spent with the same-strain female, divided by total time spent with females. Time preference data from the 40 mating trials indicated that the amount of time both CHR and GAA males spent with females of the two strains was similar and not significantly different from random (CHR: X = 0.0595, t = 1.25, p = 0.239; GAA: X = -0.0877, t = 1.34, p = 0.196).

CONCLUSION

The impetus for conducting these experiments was to assess the degree of variation in the use of the MHC during mate choice in inbred mice, both between the sexes and among different inbred mouse strains. By combining the results of previous research with the results reported here, a broader representation emerges. Sexes appear to differ in their use of the MHC for mate choice. Surprisingly, the choosing sex may vary among different inbred strains. However, in several of the strains so far tested, at least one sex has shown disassortative mating preferences (6 out of 9). If occurring in wild populations, this pattern of mating would favor MHC heterozygosity. In addition, sexes of some strains exhibit preferences for specific MHC haplotypes, regardless of their own. It is possible that this behavior could result in some favored haplotypes increasing in frequency in wild populations. Given the diversity already discovered in inbred mouse strains, it will be interesting to elucidate the influence of the MHC on mating behavior of wild mice.

ACKNOWLEDGEMENTS

We thank Esther Brown for her help, Lorraine Flaherty for allowing us to test the B6 lines in her laboratory, and James Forman for providing breeding pairs of B10 mice. This research was supported in part by a Public Health Fellowship Award to K.E. and N.S.F. grant BSR8712242 to J.L.B.

REFERENCES

Beauchamp, G.K., Yamazaki, K., Bard, J. and Boyse, E.A., 1988, Preweaning experience in the control of mating preferences by genes in the major histocompatibility complex of the mouse, <u>Beh</u>. <u>Genet</u>., 18: 537.

Brown, J., 1983, Some paradoxical goals of cells and organisms: the role of the MHC, <u>in</u>: "Ethical Issues in Neural and Behavioral Sciences," D. Pfaff, ed., Academic Press, New York.

Egid, K. and Brown, J.L., 1989, The major histocompatibility complex and female mating preferences in mice, <u>Anim</u>. <u>Behav</u>., 38: 548.

Eklund, A., Egid, K., and Brown, J.L., In press, The major histocompatibility complex and mating preferences of male mice, <u>Anim</u>. <u>Behav</u>.

Klein, J., 1973, Polymorphism of the H-2 loci in wild mice, <u>in</u>: "International Symposium on Standardization of HL-A Reagents," R.H. Regamey and J.V. Sparck, eds., Karger, Basel.

Potts, W.K., and Wakeland, E.K., 1990, Evolution of diversity at the major histocompatibility complex, <u>TREE</u>, 5: 181.

Yamazaki, K., Yamaguchi, M., Baranoski, L., Bard, J., Boyse, E.A., and Thomas, L., 1979, Recognition among mice: evidence from the use of a Y-maze differentially scented by congenic mice of different major histocompatibility types, <u>J</u>. <u>Exp</u>. <u>Med</u>., 150: 755.

Yamazaki, K., Boyse, E.A., Mike, V., Thaler, H.T., Mathieson, B.J., Abbott, J., Boyse, J., Zayas, Z.A., and Thomas, L., 1976, Control of mating preferences in mice by genes in the major histocompatibility complex, <u>J</u>. <u>Exp</u>. <u>Med</u>., 144: 1324.

THE DISCRIMINATION OF HUMAN MALE URINE ODORS BY RATS IS NOT INFLUENCED

BY THE RELATEDNESS OF THE DONORS

Heather M. Schellinck and Richard E. Brown

Department of Psychology
Dalhousie University
Halifax, Nova Scotia, Canada B3H 4JI

INTRODUCTION

Urine odors provide rats with a unique individual identity. Rats could discriminate between the urine odors of two unrelated male rats as well as between the urine odors of congenic male rats which differed only at the Major Histocompatibility Complex (Brown, Singh, and Roser, 1987). Morever, they could retain the information they learned from session to session (Schellinck, Brown and Slotnick, 1991). Although discriminable odor cues were also present in the urine of genetically identical individuals, the rats could not remember these cues from session to session. These results support the hypothesis that genetic cues provide individual rats with a unique and unchanging odor.

Humans may also have a unique odor which has a genetic component. Hepper (1988) found that dogs could discriminate between the T-shirt odors of fraternal twins on the same diet and between identical twins on different diets. They were unable to discriminate between identical twins on the same diet. Furthermore, a computerized pattern matching analysis of gas chromatograms of axillary odors has shown that the chromotograms of identical twins are more similar than those of unrelated individuals (Sommerville, Green and Gee, 1990).

To determine if rats could detect consistent cues in human urine odors, we examined the ability of rats to discriminate between the urine odors of human males. If genetic cues provide a unique identifier in human urine, the urine odors of identical twins should be more difficult to discriminate consistently than the odors of fraternal twins, brothers or unrelated males.

METHOD

Subjects and Urine Donors

Five male Long Evans rats, which received food ad lib but were restricted to 15 ml of water each day, were used as subjects. Urine samples were collected from eight boys who ranged in age from 4 to 13 years. No attempt was made to control the diets of the urine donors. Urine samples were divided into 1 ml aliquots and frozen at -20° C until required.

Chemical Signals in Vertebrates VI, Edited by R.L. Doty and
D. Müller-Schwarze, Plenum Press, New York, 1992

Odor stimluli were delivered to the rat in its test chamber by a computer-controlled olfactometer, as described by Schellinck et al. (1991). Rats were trained to perform a go/no go sequential discrimination task in a series of steps which culminated in their learning to discriminate between the urine odors of two outbred rats. When an animal scored a minimum of 85 percent correct on the first block of 20 trials and a mean of 85 percent or higher on each of the remaining nine blocks of the 100 trial session, it was considered to have met the criterion for successfully acquiring the discrimination.

Following the completion of this preliminary training, all of the subjects were tested for their discrimination between the following four pairs of odors: 1) urine from two unrelated human males; 2) urine from two human male siblings; 3) urine from two male fraternal twins; 4) urine from two male identical twins. The odors were presented to the animals in a counterbalanced order and different samples from the S+ odor donor and from the S- odor donor were used for each session for all tasks. The criterion for learning was 85 percent correct, as described above.

RESULTS

The ability of the rats tò learn to discriminate between human urine odors was not significantly different over the four different tasks, F(3,21) < 1.0; (Figure 1). An examination of the data revealed a great deal of variability in the ability of the individual subjects to discriminate between the pairs of odors. All of the animals scored 85 percent correct on one or more blocks of trials within at least one session of each task; however, only one animal scored consistently above 85 percent from session to session in each of the four tasks. Figure 2 compares the performance of the best and worse rats on each task.

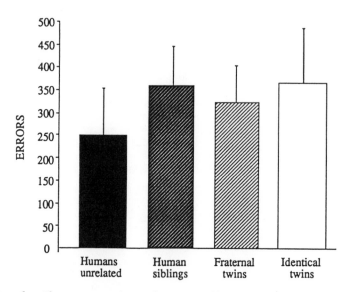

Fig. 1. The mean number of errors (± SEM) over eight sessions for five rats on each of four odor discrimination tasks.

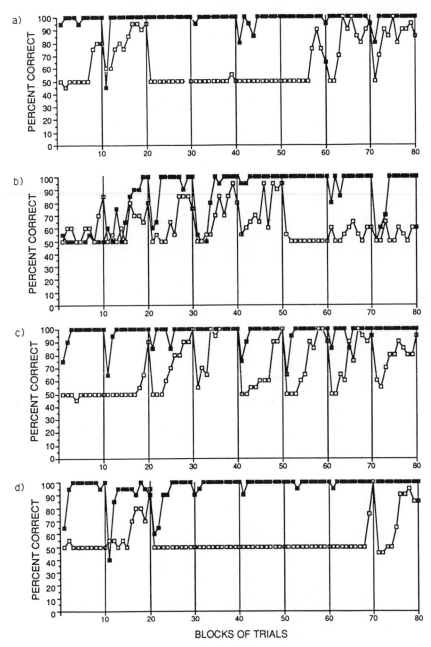

Fig. 2. The learning curves of the fastest (□) and slowest (■) of the five rats in four discrimination tasks in an operant olfactomer. The four tasks required the rats to discriminate between the urine odors of a) two unrelated human males, b) two brothers, c) two fraternal twins and d) two identical twins.

DISCUSSION

These results suggest that the discriminability of the odors of urine from boys ranging in age from 8 to 13 years is not influenced by the donors' degree of relatedness. Regardless of the level of performance of the individual subjects, there was little indication that any of the tasks were more difficult to learn than any of the others. The one subject which met criterion in all four tasks was able to retain this information from session to session, indicating that consistently discriminable cues can be found in the urine of humans. Moreover, since the urines from the identical twins were discriminable by this same subject, it would appear that genetic cues were not the basis for the discriminability. As the criterion for remembering the discrimination from session to session was set at a minimum of 85% correct on the first block of trials, one might argue that the acquisition of a learning set and subseuqent one trial learning could account for the performance of this individual. However, this explanation must be ruled out as the animal completed at least one error free session in each task.

Consequently, the question arises as to what cues enabled this subject to consistently remember the differences between these urines. Diet could be expected to provide discriminable differences between individual samples used in any given session. However, it is likely that diet varied within individuals from day to day, as well as between individuals, and therefore could not provide the difference necessary to discriminate consistenly between individuals. We have recently found that bacteria can influence the individual urinary odor of rats (Singh et al, 1990; Schellinck et al, 1991) and the odor of human urine may be determined by the interaction of diet, bacterial microflora, and MCH type.

As axillary and other skin gland odors are more prominent in humans, it may be more appropriate to examine the ability of rats to discriminate between human axillary odors rather than between human urine odors. Dogs were unable to discriminate between the T-shirt odors of identical twins on the same diet (Hepper, 1988); however, this may mean that the task is difficult rather than impossible. Differences between the odors of these individuals might have been detected if the dogs had been given more than ten trials to learn each of five pairs of identical twins. The rats in the present study were given 200 trials per session for eight sessions. Consequently, to further clarify the basis of the discriminability of human axillary odors, we plan to assess the ability of equally trained rats to discrimiinate between the axillary odors of humans of different degrees of relatedness.

The poor performance of the other subjects in this study is inexplicable, especially considering that all of the subjects were trained to an equal level of performance and like the "best" rat, they were able to discriminate between the urine odors of rats in the preliminary training. Evidence of such marked individual variability must be taken into account when designing future experiments.

ACKNOWLEDGEMENTS

This research was supported by NSERC of Canada grant A7441 to R.E. Brown.

REFERENCES

Brown, R.E., Singh, P., and Roser, B., 1987, The major histocompability complex and the chemosensory recognition of individuality in rats, Physiol. Behav., 40:65.

Hepper, P.G., 1988, The discrimination of human odor by the dog, Perception, 17:549.

Schellinck, H.M., Brown, R.E., and Slotnick, B., 1991, Training rats to discriminate between the odors of individual conspecifics, Anim Learn. Behav., 19, 2231-233.

Sommerville, B., Green, M., and Gee, D.J., 1990, Using chromatography and a dog to identify some of the compounds in human sweat which are under genetic influence, in: "Chemical Signals in Vertebrates 5," D.W. MacDonald, D. Müller-Schwartze, and S.E. Natynczuk, eds., Oxford University Press, Oxford.

Zlatkis, A., Bertsch, W., Lichtenstein, H.A., Tishbee, A., and Shunbo, F., Liebich, H.M., Coscia, A.M., and Fleischer, N., 1973, A profile of volatile metabolites in urine by gas chromatography - mass spectometry, Anal. Chem., 45:763.

THE INFLUENCE OF THE HEMATOPOIETIC SYSTEM ON THE PRODUCTION OF MHC-

RELATED ODORS IN MICE

Dagmar Luszyk[1], Frank Eggert[1], Lutz Uharek[2],
Wolfgang Muller-Ruchholtz[2], Roman Ferst[1]

[1]Department of Psychology and [2]Department of Immunology
University of Kiel, Kiel, Germany

INTRODUCTION

Yamazaki et al. (1985) were able to demonstrate an influence of the hematopoietic system on H-2-associated urine odors. H-2 is the designation of the major histocompatibility complex (MHC) of the mouse and is located on chromosome 17. The MHC was originally identified by its role in transplant rejection, but it is now recognized that proteins encoded in this region are involved in many aspects of immunological recognition. H-2 congenic imbred mice differ genetically only in the H-2 but not in the rest of their genomes, syngeneic mice share both haplotypes of the H-2, and semiallogeneic share one of the two haplotypes of the H-2.

Since Yamazaki et al. (1985) used semiallogeneic transplantations, it was not possible to assess the contribution of nonhematopoietic cells or tissues. Muller-Ruchholtz et al. (1980) developed a technique to prevent graft-versus-host disease, thus allowing a fully allogeneic bone marrow transplantation (BMT) in mice without any further medication. This technique enabled us to study the retention of odorant properties typical of the recipient in the case of transplantation between H-2 congenic strains and in the case of BMT between strains which share the same H-2 and differ only in the rest of their genomes. In the first case, we examined the influence of the hematopoietic system on the production of MHC-related odors, in the second case we tried to analyse the influence of the rest of the genome on odor production after BMT.

In reference to Yamazaki et al. (1985), four transfer tests were conducted in which the change of strain-specific urine odors after BMT between H-2 congenic inbred strains and strains which share the same H-2 and differ in the rest of their genomes was tested. In both cases we examined if there were any donor- and recipient-specific components in the chimera's urine odor.

MATERIALS AND METHODS

BMT's were made between (1) mice of H-2 congenic inbred strains; i.e., BALB/k (H-2k) and BALB/c (H-2d), (2) mice which share the same H-2 and differ in the rest of their genomes; i.e., BALB/k (H-2k) and C3H

Chemical Signals in Vertebrates VI, Edited by R.L. Doty and
D. Müller-Schwarze, Plenum Press, New York, 1992

(H-2k), and (3) mice of the same inbred strains; i.e., BALB/k (H-2k) and
C3H (H-2k) (syngeneic). The success of the reconstitution was determined by immunological tests for chimerism. Six weeks after BMT, urine
was collected from the transplanted mice by the application of slight
abdominal pressure and frozen until needed. Urine samples used for the
training were pooled from up to 20 animals and frozen until needed.

Table 1. Constitution and designations of transplantations

Donors (female)	Recipients (female)	Designation	Mice in each panel
1) H-2 congenic reconstituted			
BALB/c(H-2d)	BALB/k(H-2k)	BALB/c->>BALB/k	6
BALB/k(H-2k)	BALB/c(H-2d)	BALB/k->>BALB/c	6
2) Non-H-2 different reconstituted			
C3H(H-2k)	BALB/k(H-2k)	C3H->>BALB/k	12
BALB/k(H-2k)	C3H(H-2k)	BALB/k->>C3H	10
3) Syngeneic reconstituted			
C3H(H-2k)	C3H(H-2k)	C3H->>C3H	7
BALB/c(H-2d)	BALB/c(H-2d)	BALB/c>>BALB/c	3

Five female rats (CAP) were deprived of water for 20 hours before
each session and were trained in a computer controlled olfactometer (cf.
Ferstl et al., this volume) and reinforced for the identification of the
correct stimulus (S+) by a drop of water. Each training session consisted of 200 single trials. After reaching a criterion of 70% or more
correct responses for the S+, the schedule of reinforcement was reduced
stepwise to 0% in four blocks of 10 trials each to introduce the urine
samples of the chimeras to test for transfer of training (TTT). Two of
the four blocks were indexed as reference trials, the other two as test
trials. The purpose of the TTT is to test new panels of odor donors
entirely without reinforcement to exclude any new learning of adventi-
tious odor differences.

RESULTS

The first column of Table 2 reveals the discrimination rates and
the third column the precentage of correct identification of the S+. In
the first two tests we examined donor- and recipient-specific components
in the chimera's urine odor after BMT between H-2 congenic inbred
strains. Consequently, H-2-associated influences were investigated.

It is clearly shown that the trained rats were able to establish an
odor discrimination between two H-2 congenic inbred strains. Both
transfer tests demonstrate identification of donor- and recipient-specific components in the urine odors of these chimeras.

TTT's 3 and 4 examined donor- and recipient-specific components
after BMT between inbred strains which shared the same H-2 and differed
in the rest of their genomes. Table 2 shows the results of these two

TTT's. The trained rats were also able to discriminate between such inbred strains and they also identified both donor- and recipient-specific components in the urine odor of the chimeras. The hit-rates of these two tests were even higher than those of the first two TTT's.

Table 2. Results

Discrimination S+/S−	N	Identification S+	N
Training: BALB/k (S+) vs. BALB/c (S−)			
85.5%**	200	89.0%**	100
H-2-related donor-specific components			
TTT 1: BALB/k->>BALB/c vs. BALB/c->>BALB/c			
70.0%**	200	71.0%**	100
H-2-related recipient-specific components			
TTT 2: BALB/c->>BALB/k vs. BALB/c->>BALB/c			
69.0%**	200	81.0%**	100
Training: BALB/k (S+) vs. C3H (S−)			
86.0%**	450	91.0%**	225
H-2-unrelated donor-specific components			
TTT 3: BALB/k->>C3H vs. C3H->>C3H			
83.0%**	220	91.0%**	110
H-2-unrelated recipient-specific components			
TTT 4: CH3->>BALB/k vs. C3H->>C3H			
78.0%**	130	91.0%**	115

** $P \ll 0.01$ (binomial distribution, one-taled tests)

DISCUSSION

By training rats in a go - no go odor discrimination task we were able to show that H-2 congenic strains and inbred strains which differ in the rest of their genomes can be discriminated by their specific urine odors. Furthermore, we confirmed results of other experiments (Yamazaki et al., 1985; Ferstl et al., 1988; Luszyk et al., 1989; Luszyk et al., 1990; Eggert et al., 1989; Sobottka et al., 1990): BMT not only changes the immunologist but also the olfactory identity of the recipient. Both donor and recipient odor components contribute to the new chemosensory identity of the chimeras. This also could be demonstrated for the odor cues which are associated with the genetic background. Therefore, we have two conclusions: first, not only the hematopoietic system but also other tissues are involved in the production of H-2 specific odor cues; second, both the hematopoietic system and other tissues also contribute to the production of odor cues associated with loci outside the H-2.

ACKNOWLEDGEMENT

This work was supported by a grant from the Stiftung Volkswagenwerk.

REFERENCES

Eggert, F., Luszyk, D., Ferstl, R. and Muller-Ruchholtz, W., 1989, Changes in strain-specific urine odors of mice due to bone marrow transplantations, Neuropsychobiology, 22:57.

Ferstl, R., Welzel, C., Florian, M., Blank, M. and Muller-Ruchholtz, W., 1988, Ist das Knochenmark der einzige Ursprung korpereigener Duftkomponenten? Z. Exp. Angew. Psychol., 35:201.

Luszyk, D., Eggert, F., Ferstl, R., Blank, M., and Muller-Ruchholtz, W., 1989, Der Austausch des hamatopoetischen Systems verandert die chemosensorische Identitat, Z. Exp. Angew. Psychol., 36:239.

Luszyk, D., Eggert, F., Ferstl, R., Blank, M., and Muller-Ruchholtz, W., 1990, Exchange of the hematopoietic system alters the chemosensory identity, German J. Psychol., 14:27.

Muller-Ruchholtz,W., Wottage, H.-U. and Muller-Hermelink, H.K, 1980, Restitution potentials of allogeneically or xenogeneically grafted lymphocyte-free hematopoietic stem cells, in: "Immunobiology of Bone Marrow Transplantation", S. Thierfelder, H. Rodt, H.J. Kolb, eds., Springer, Berlin.

Sobottka, B., Eggert, F., Ferstl, R. and Muller-Ruchholtz, W., 1990, Altered chemosensory identity after experimental bone marrow transplantation: Recognition by another species, German J. Psychol., 14:213.

Yamazaki, K., Beauchamp, G.K., Thomas, L., Boyse, E.A., 1985, The hematopoietic system is a source of odorants that distinguish major histocompatibility types, J. Exp. Med., 162:1377.

WHAT'S WRONG WITH MHC MATE CHOICE EXPERIMENTS?

C. Jo Manning[1,2,3], Wayne K. Potts[2], Edward K. Wakeland[2] and
Donald A. Dewsbury[3]

[1]Department of Psychology, University of Washington
Seattle, WA; [2]Laboratory of Molecular Genetics, Department
of Pathology, University of Florida, Gainesville, FL
[3]Department of Psychology, University of Florida
Gainesville, FL

INTRODUCTION

It has been convincingly demonstrated that mice can discriminate
MHC-related odors among congenic strains under laboratory conditions
(see Yamaguchi et al. 1981). There is, however, debate about the use of
these MHC-related odors as mate choice cues (Partridge, 1988; Klein et
al., 1990; Nei and Hughes, 1991). MHC-based mate choice has been postulated to function either to produce presumably more fit MHC heterozygous offspring (Doherty and Zinkernagel, 1975) or to avoid close inbreeding (Brown, 1983). In either case, mating should be disassortative. Data from semi-natural half-wild enclosure studies (Potts et al.,
1991) suggest that disassortative matings produce an excess (from random
mating expectations) of MHC heterozygous progeny. Behavioral observations also suggest that these disassortative matings are due primarily
to female choice. In a subset of 305 pups of known maternity, those in
litters of mixed paternity (30 out of 58 litters) show a significant
excess of heterozygotes over what would have been produced if females
had mated only with their territorial males. Those litters sired only
by the female's territorial male did not deviate from the number of
heterozygotes expected. This suggests that an important factor in
producing more MHC heterozygous progeny is extra-territorial matings.
In 41 observations of mating pairs, all matings took place on the
territory of the male. All 13 extra-pair matings observed involved
females who had left their nesting territory to travel to the territory
of the male with which they were mating. No male was ever observed
mating off of his territory. For details see Potts et al. this volume.

LABORATORY MATE CHOICE EXPERIMENTS

Male Mate Choice

Despite our observations, current data from MHC-based mate choice
experiments in the laboratory would more strongly support male mate
choice. Female choice has been difficult to document. Male mate choice
has been observed in five of six strain combinations (Yamazaki et al.

Chemical Signals in Vertebrates VI, Edited by R.L. Doty and
D. Müller-Schwarze, Plenum Press, New York, 1992

1976; Yamazaki et al. 1978). Cross fostering experiments demonstrate
that these preferences are male and not female preferences because the
preference cue is learned from parents and the mating preferences of
males of one strain can be changed by rearing them with parents of a
different strain (Yamazaki et al. 1988). Originally, four of the six
strain combinations showed disassortative preferences, one mated randomly, but the sixth preferred to mate assortatively (Yamazaki et al.
1976). Four out of six strains is not significant. More strains need
to be tested in order to determine whether mating preferences are disassortative.

Female Mate Choice

Attempts to measure female preferences in some of the same strain
combinations have not been as successful (Beauchamp et al. 1988). Only
one female laboratory mate choice experiment has shown MHC-based disassortative mating preferences. Egid and Brown (1989) demonstrated disassortative mating preferences in B10.CHR and B10.GAA mice. Females of
these two strains, which differed only at MHC or closely linked loci,
preferred to mate with males of the opposite strain. In a test of males
of these two strains no significant preferences were detected (Eklund et
al., this volume).

Problems with Laboratory Mate Choice Experiments

In laboratory tests with inbred strains, mating preferences depend
on strain and male choice has been found more often than female choice.
The latter finding is not predicted by most evolutionary theory (but see
Dewsbury, 1982) or supported by observations made in our semi-natural
enclosures. Although there have been attempts to detect disassortative
mate preferences by females (Manning, unpublished data; Yamazaki et al.
1988; Egid and Brown, 1989), only one experiment using two inbred
strains has succeeded. Problems with Mus mate choice experiments have
not been confined to MHC-related mate choice experiments. Inconsistency
and confusion have been the hallmarks of other Mus mate choice experiments over the last 30 years (reviewed by D'Udine and Alleva, 1983;
Dewsbury, 1988b).

We believe there may be some fundamental problems with the exe-
cution of female Mus mate choice experiments. Of the three major components of mate choice experiments (the animals, the apparatus and the
test criteria), we have reservations about two, the animals and the test
criteria. The use of inbred mice for behavioral tests ignores the fact
that they have been artificially selected for specific traits, often
without regard to behavior, and that complex natural selection pressures
have been removed. As for the test criteria, these problems run the
gamut from obvious problems such as "mate choice" tests that do not employ copulation as part of the test procedure to unresolved problems
such as whether to score a single ejaculatory series as mate selection,
when the normal mating pattern of the animal may include multiple copulations (Dewsbury, 1988b). As we will show, there is good reason to re-
examine our test procedures.

The use of inbred strains: Inbred mouse strains seem, at first
glance, to be ideal for MHC-related mate choice experiments. The use of
congenic mice that differ only at specific MHC loci is often assumed to
control for variation at other genetic loci, so that resulting differences can be attributed directly to the MHC genetic differences between
the two strains. We believe this is a gross oversimplification, particularly for behavioral traits. Many genes become arbitrarily fixed
during the inbreeding process because many natural selection pressures

are removed while artificial selection pressures (often unwitting ones) are imposed. This could radically alter mate choice behavior. If mating preferences have a variable genetic component, then the trait is subject to selection. The process of inbreeding requires several generations of brother-sister matings. If either males or females prefer to avoid incestuous matings, their genes will be at a selective disadvantage in this artificial system. The animals exhibiting the most discrimination will fail to breed or breed less often, thereby eliminating their genes from the gene pool. MHC-related male mate choice may have survived the inbreeding process precisely because males are less choosy than females. A choosy male may not refuse to mate when his options are limited, but a choosy female may, thereby eliminating her genes from the gene pool. Is it wise to study disassortative mating in animals who have experienced strong selection against this trait?

Experimental tests: To test the hypothesis that inbred females may not exhibit mating preferences, we conducted mating preference tests with wild, half-wild and inbred (B10) females. Webster et al. (1982) devised a test of female choice for montane voles (Microtus montanus) in which they used several behavioral measures to determine whether females preferred one male over another. In their test a female vole was given a choice of mating with either or both of two males tethered at each end of the test apparatus. By comparing such measures as numbers of visits, mounts, intromissions and ejaculations allowed each male, they determined that individual females displayed preferences for one of the males over the other. This study employs similar methods.

Experiments were conducted with three different kinds of females. Wild females were F_{1-3} offspring of wild-caught animals from the Gainesville area. Half-wild females were F_1 offspring of wild-caught mice crossed with C57BL/6 (B6) mice. The inbred strain was C57BL/10 (B10). Sixteen half-wild females and eight each of the wild and B10 females were tested. Males for all three tests were B10 and B10.A congenics. The same six B10 and six B10.A males were used for testing the half-wild and the B10 females. A different set of four B10 and five B10.A males were used for testing the wild females. B10 and B10.A males differ only in the region of MHC. All of the half-wild females shared one MHC haplotype (b) with the B10 males; the MHC genotypes of the wild females is not known, and the B10 females shared the MHC genotype of the B10, but not the B10.A males.

Tests were conducted in a three-chambered apparatus consisting of three 48x27x13 cm polycarbonate cages connected by two 8 cm lengths of clear vinyl tubing, 3.2 cm in diameter. A plastic wire tie with a rachet closing, fitted with a 3.5 cm length of wire, was placed around the neck of the males. The protruding wire prevented the males from passing through the tubing to the other chambers. Females were given a two hour habituation period to familiarize themselves with all three chambers of the apparatus. Screens of 1/2-inch hardware cloth were placed over the opening of the tube. Collared males, along with the soiled bedding from their home cages were placed in the two male chambers. During a 15 minute male habituation period, we recorded the time the female spent sniffing and nosing at the opening of each male chamber. The hardware cloth screens were then removed and the female was allowed to visit and mate with the males at her discretion. Number and duration of visits to each male were recorded, as were all copulatory events. When one male ejaculated the test was ended.

B10 females, but not wild or half-wild females, mated indiscriminately with both males. A comparison of proportions of intromissions by successful males (males that achieved ejaculation) showed that wild and

half-wild females allowed successful males a significantly higher proportion of the intromissions, while B10 females divided intromissions
equally between them (Fig.1a). Another measure, time spent in the cage
of each male (Fig.1b), also indicated that wild and half-wild females
were exhibiting a preference while the B10 females were not. For two
measures more removed from copulatory events, number of visits to each
male (Fig.1c) and time spent investigating males through hardware cloth
barriers (Fig.1d), the differences between successful males (males that
ejaculated) and unsuccessful males was significant only for wild females, suggesting that at least on these measures, wild females are more
discriminating than either half-wild females or B10 females. This also
suggests that care must be exercised when interpreting data that may
more accurately represent "social" preferences than mating preferences.

We concluded that B10 females do not show a preference for any
traits that differ between B10 and B10.A males. However, since B10
males ejaculated with fewer intromissions than B10.A males, a first male
to ejaculate criterion would have led us to conclude that B10 females
preferred B10 males. Seven out of eight B10 males ejaculated first in
spite of B10 and B10.A males achieving the same numbers of intromissions. Random choice on the part of females, compounded by differential
sexual behavior on the part of males, would have given us an incorrect
result if we had not analysed the behavior of individual females. Although this finding should serve as a warning when attempting mate
choice experiments with inbred strains, it does not apply to all inbred
strains. If it applied to the two strains Egid and Brown (1989) tested,
either the females would have mated randomly or both would have shown a
"preference" for the same strain of male.

Wild and half-wild females did behave as if they were choosing one
male in preference to the other (Fig.1). However, even though their behavior indicated mate choice, half-wild females did not exhibit the MHC
disassortative mating preferences we predicted. Based on the MHC haplotype they received from their B6 parent (b), we predicted they would
mate disassortative with the B10.A (aa) male. Their choice between B10
(bb) and B10.A (aa), based on a criterion of first male to ejaculate,
was random with respect to MHC. First male to ejaculate is the usual
female mate choice criterion, but we would like to suggest that it may
be inappropriate.

Test criteria: We observed large amounts of multiple paternity in
litters in our semi-natural mouse enclosures, suggesting that when more
than one male is available, females often mate multiply. Using only MHC
for paternity exclusion (an underestimate), we found multiple sires in
28 out of 58 litters. Two other litters were sired entirely by extra-
territorial males. In addition, in a recent series of female mating
tests, females were given a choice of two males and allowed to mate ad
lib. for the duration of their receptive period (13:00h to 8:00h, reversed light cycle). Copulations were observed and recorded on a computer through the first ejaculation and then video-taped for the next
nine hours. In only 1 of 38 sessions filmed did the female fail to cop-
ulate again after the first male's ejaculation. In 7 cases, she copulated again with the first male and never with the second male. In the
other 30 cases she copulated with both males at least once, although in
5 cases the second male failed to ejaculate.

It would appear that when a female is confronted with two males,
she is likely to mate with both of them. This behavior may serve a
variety of functions, such as genetic variability in offspring or infan-
ticide insurance. This does not, however, mean that she doesn't
exercise preference. Egid and Brown (1989) have argued that since there

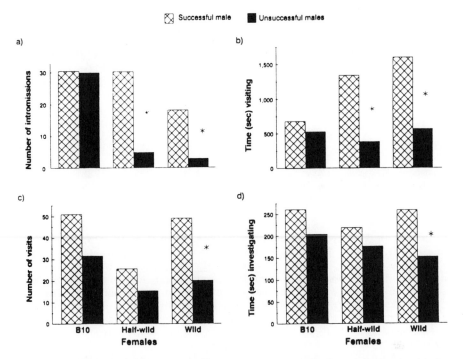

Fig.1. Comparisons of measures of behavior toward successful males (males that achieved ejaculation)(cross-hatch) and unsuccessful males (solid) by B10, half-wild and wild females. Measures include a) mean numbers of intromissions allowed to both categories of males, b) mean time spent visiting each category of male, c) mean number of visits to each category of male, and d) mean time (sec) spent investigating each category of male through hardware cloth barriers. *p<0.02, Wilcoxon test

is a first male advantage in sperm competition studies done by Levine (Levine, 1967), the first male mated should be the preferred male. We have confirmed this first male advantage under specific conditions (unpublished data), but neither Levine's or our sperm competition study manipulated timing of matings. The order and timing of the matings can influence the paternity patterns in the litter. In three species of rodents, golden hamsters (Huck et al. 1989), rats (Spinka, 1990) and deer mice (Dewsbury, 1988a), alteration of timing relative to ovulation has been shown to change the outcome of competitive matings. Since ovulation time varies with strain of mouse and with each individual mouse (Bingel and Schwartz, 1969; Braden, 1957), it would be difficult to do a definitive sperm competition study in Mus. Perhaps the only appropriate mating preference experiment in Mus is one that allows a female to mate at will with the stimulus males and then paternity types the offspring. Even under these conditions, we must know something about the behavior of our study animals. Does the female's behavior indicate she is making a choice? Are the males representative of the similarities and differences available in nature, or do they have arbitrarily fixed phenotypes that will artifactually influence the outcome of mate choice experiments?

CONCLUSIONS

We suggest that in the future a careful evaluation of all components of mate choice experiments is in order. The behavior of the
animals must be carefully monitored. In addition, test criteria must be
reevaluated in the light of the normal behavior of the subject animals.
After all of these problems are addressed and resolved, we may be able
to make some sense out of MHC mate choice experiments.

REFERENCES

Beauchamp, G.K., Yamazaki, K., Bard, J. and Boyse, E.A., 1988, Preweaning experience in the control of mating preferences by genes
in the major histocompatibility complex of the mouse, Behav.
Genet., 18:537.
Bingel, A.S. and Schwartz, N.B., 1969, Timing of LH release and ovu-
lation in the cyclic mouse, J. Reprod. Fertil., 19:223.
Braden, A.W.H., 1957, The relationship between the diurnal light cycle
and the time of ovulation in mice, J. Exp. Biol., 34:177.
Brown, J.L., 1983, Some paradoxical goals of cells and organisms: the
role of the MHC, in: "Ethical questions in brain and behavior",
D.W. Pfaff, ed., Springer Verlag, New York.
D'Udine, B. and Alleva, E., 1983, Early experience and sexual prefer-
ences in rodents, in: "Mate choice", P. Bateson, ed., Cambridge
University Press, Cambridge.
Dewsbury, D.A., 1982, Ejaculate cost and male choice, Am. Nat., 119:601.
Dewsbury, D.A., 1988a, Sperm competition in deer mice (Peromyscus
maniculatus bardi) Effects of cycling versus postpartum estrus and
delays between matings, Behav. Ecol. Sociobiol., 22:251.
Dewsbury, D.A., 1988b, Kin discrimination and reproductive behavior in
muroid rodents, Behav. Genet., 18:525.
Doherty, P.C. and Zinkernagel, R.M., 1975, Enhanced immunological
surveillance in mice heterozygous at the H-2 gene complex, Nature,
256:50.
Egid, K. and Brown, J.L., 1989, The major histocompatibility complex and
female mating preferences in mice, Anim. Behav. 38:548.
Huck, U.W., Tonias, B.A. and Lisk, R.D., 1989, The effectiveness of
competitive male inseminations in golden hamsters, Mesocricetus
auratus, depends on an interaction of mating order, time delay
between males, and the time of mating relative to ovulation, Anim.
Behav., 37:674.
Klein, J., Kasahara, M., Gutknecht, J. and Figueroa, F., 1990, Origin
and function of MHC polymorphism. Chem. Immunol., 49:35.
Levine, L., 1967, Sexual selection in mice. IV. Experimental demonstration of selective fertilization, Am. Nat., 101:289.
Nei, M. and Hughes, A.L., 1991, Polymorphism and evolution of the major
histocompatibility complex loci in mammals, in: "Evolution at the
molecular level", R.K. Selander, A.G. Clark, and T.S. Whittam,
eds., Sinauer, Sunderland.
Partridge, L., 1988, The rare-male effect: What is its evolutionary
significance?, Philos. Trans. R. Soc. Lond. B Biol. Sci., 319:525.
Potts, W.K., Manning, C.J. and Wakeland, E.K., 1991, Mating patterns in
seminatural populations of mice influenced by MHC genotype,
Nature, 352:619.
Spinka, M., 1990, The effect of time of day on sperm competition and
male reproductive success in laboratory rats, Physiol. Behav.,
47:483.
Webster, D.G., Williams, M.H. and Dewsbury, D.A., 1982, Female regu
lation and choice in the copulatory behavior of montane voles
(Microtus montanus), J. Comp. Physiol. Psych., 96:661.

Yamaguchi, M., Yamazaki, K., Beauchamp, G.K., Bard, J., Thomas, L. and
    Boyse, E.A., 1981, Distinctive urinary odors governed by the major
    histocompatibility locus of the mouse, Proc. Natl. Acad. Sci. USA,
    78:5817.
Yamazaki, K., Boyse, E.A., Mike, V., Thaler, H.T., Mathieson, B.J.,
    Abbott, J., Boyse, J., Zayas, Z.A. and Thomas, L., 1976, Control
    of mating preferences in mice by genes in the major histocompati-
    bility complex, J. Exp. Med., 144:1324.
Yamazaki, K., Yamaguchi, M., Andrews, P.W., Peake, B. and Boyse, E.A.,
    1978, Mating preferences of F 2 segregants of crosses between
    MHC-congenic mouse strains, Immunogenetics, 6:253.
Yamazaki, K., Beauchamp, G.K., Kupniewski, D., Bard, J., Thomas, L. and
    Boyse, E.A., 1988, Familial imprinting determines H-2 selective
    mating preferences, Science, 240:1331.

WHY DOES GERMFREE REARING ELIMINATE THE ODORS OF INDIVIDUALITY

IN RATS BUT NOT IN MICE?

Heather M. Schellinck and Richard E. Brown

Department of Psychology
Dalhousie University
Halifax, Nova Scotia, Canada B3H 4J1

INTRODUCTION

The Major Histocompatibility Complex (MHC) appears to provide the genetic basis for an individuality marker in the urine of rats and mice (Brown, Singh and Roser, 1987; Brown, Roser and Singh, 1990; Yamazaki et al., 1991). Recent findings suggest that commensal bacteria are also involved in the production of individually distinct urine odors in rats. Using a habituation-dishabituation task, Singh et al. (1990) determined that rats could not discriminate between the urine of germfree MHC congenic rats (PVG vs. PVG.R1). When donors were moved to conventional housing, their urines could be discriminated. Also, rats trained in a go-no go operant task in an olfactometer made more errors in learning to discriminate between urines of germfree MHC congenic rats than between urines of conventionally housed MHC congenic rats (Schellinck, Brown and Slotnick, 1991). Urine odors from conventionally housed MHC congenic rats were readily discriminated; the discrimination was remembered from session to session and no drop in performance was observed when the samples were changed to different donors of the same MHC type in midsession. Although the rats were able to discriminate between urine samples from individual germfree MHC congenic rats, they consistently performed at chance at the beginning of each session. One explanation of this inability to retain the information from session to session could be that constant cues were not being used to discriminate between the urine samples from germfree rats. Moreover, when the urine samples from germfree rats were changed to others of the same strain in midsession, performance was also disrupted.

The urine odors of germfree MHC congenic mice, however, can be discriminated as readily as those of conventionally housed animals, by mice which were trained in a Y-maze (Yamazaki et al., 1990). The test mice could also generalize their learning from urine samples of conventionally housed mice to samples from germfree mice. Such a finding would suggest that, in mice, the germfree status of the urine donors does not influence the MHC determined odor type. To determine if rats could also detect consistent cues in the urine odors of germfree mice, we examined the ability of rats to discriminate between the urine odors of conventionally housed and germfree B6 and B6-H2^k mice in an operant conditioning paradigm.

Chemical Signals in Vertebrates VI, Edited by R.L. Doty and
D. Müller-Schwarze, Plenum Press, New York, 1992

METHOD

Subjects and Urine Donors

Six male Long Evans rats which received food ad lib but which were restricted to 15 ml of water per day were used as subjects. All of the urine samples used in this experiment were supplied by Dr. Gary Beauchamp from the Monell Chemical Senses Center, Philadelphia. The urine from the conventionally housed B6 and B6-H-2^k donors was collected at Monell. The urine from the germfree donors was collected at Taconic Farms, Germantown, New York, as described by Yamazaki et al (1990). Urine samples were frozen at -20° C until used.

Apparatus and Procedure

Odor stimuli were delivered to the rat in a test chamber by a computer-controlled olfactometer as described by Schellinck et al. (1991). Rats were trained to perform a go/no go sequential discrimination task in a series of steps which culminated in their learning to discriminate between the urine odors of two outbred rats. When an animal scored a minimum of 80 percent correct on the first block of 20 trials and a mean of 80 percent or higher on each of the remaining nine blocks of the 200 trial session, it was considered to have met the criterion for successfully acquiring the discrimination.

Following the completion of this preliminary training, the six subjects were required to discriminate between the following pairs of odors: (1) pooled urine from conventionally housed B6-H-2^k and B6 mice and (2) pooled urine from germfree B6-H-2^k and B6 mice. One strain was designated the S+ from three of the subjects and the other strain was the S+ for the other three subjects. Three of the subjects were tested with the urines of conventionally housed mice first while the other three subjects learned to discriminate between germfree mice first. Each subject was tested for eight sessions (1600 trials) on each task.

RESULTS

The number of errors made by the rats in learning to discriminate between the urine odors from conventionally housed B6-H-2^k and B6 mice and between the urines from B6-H-2^k and B6 mice raised in germfree conditions was not significantly different (t (5) = 0.456, n.s.) (Figure 1). Four subjects reached the criterion in both tasks; the remaining two subjects did not meet criterion on either of the two tasks.

DISCUSSION

The results of this experiment replicate those of Yamazaki et al. (1990). The urine odors from germfree MHC congenic mice, which can be discriminated by mice in a Y-maze, can also be discriminated by rats in an operant olfactometer. We are thus left with the question as to why the urine of germfree MHC congenic mice can be discriminated while the urine of germfree MHC congenic rats cannot. If, as Yamazaki et al (1990) conclude, germfree status is irrelevant to the production of individual odors, then what caused the odors of individuality to be removed from the germfree rat samples used by Singh et al. (1990)? Conversely, if bacteria do contribute to the production of an individually unique urinary odor, what other variables created the discriminable differences found in the germfree mouse urine by Yamazaki et al. (1990)?

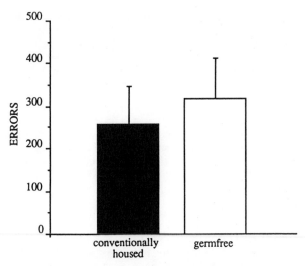

Fig. 1. The mean number of errors (+SEM) over eight sessions for
six rats in discriminating between urine from conventionally housed and germfree mice.

There are several methodological factors which might account for the
discrepancy between the findings of the rat and mouse studies. First,
there may be differences in the urine samples themselves. Both labs
confirmed the germfree status of their urine donors; however, the samples
differed in a number of other respects. The urine donors used by
Yamazaki et al. (1990) differed at all three MHC loci whereas the urine
donors used by Singh et al. (1990) and Schellinck et al. (1991) differed
only at the class IA locus. The H-2 system in mice may influence quantitative metabolic variations (Klein, 1986). For example, Ivanyi (1978)
has compiled a list of 35 different physiological and pathological traits
in which B10A and B10 mice may differ. These include differences in
hormone metabolism and differences in lymphoid cell populations. Whether
the MHC in rats produces such changes is unknown, nor is it known whether
differences at three MHC loci rather than one would differentially
influence the discriminability of the odor cues in the urine.

Second, the urine samples used by Singh et al. (1990) and Schellinck
et al. (1991) were from individual rats; the samples used by Yamazaki et
al. (1990) were pooled from a number of individual mice. While the use
of pools would eliminate spurious differences in the urine within a
strain, it would compound the effect if there are actual differences in
H-2 strains which are expressed in the urine. If the absence of bacteria
eliminates the most salient cues, it is possible that these other non-
antigenic related cues would be sufficient to produce constant discriminable differences in the pooled mouse urines.

Third, the diet supplied to the germfree rat urine donors used by
Singh et al. (1990) was irradiated; the diet given to the germfree mice
was sterilized (Beauchamp, 1991, personal communication). If irradiation
alters the structure of proteins in food and sterilization does not,
metabolic products may have been excreted in the urine of the germfree
mice which would not be present in the urines of the germfree rats.

Fourth, there may be extraneous differences in the urines. Because
of the absence of bacteria in germfree animals, macromolecules are
present which cause impaired water absorption in the gut; also, there are
no bacterial enzymes to degrade mucus secreted by the intestine. Consequently, animals kept in germfree conditions have more liquid contents in
the cecum and colon and are subject to chronic diarrhea (Van der Waaij,
1984). This makes it difficult to collect urine samples which are free
from fecal matter. In order for this to influence the discriminability
of the germfree mouse urines, the gut of the two different strains of
mice would have to be differentially affected by their germfree status,
thus producing consistently discriminable differences in the urines of
the two different strains. Although this is unlikely, until the urine
samples are analyzed for such differences, this possibility cannot be
dismissed.

Finally, the experimental procedures and criteria used to assess
learning may not be comparable. Three different procedures were used to
examine the discriminability of germfree urines. Rats were tested in a
habituation-dishabituation paradigm (Singh et al., 1990); they were also
trained in a go/no go operant task (Schellinck et al., 1991) as well as
in the results reported here. Yamazaki et al. (1990) used a Y-maze for
their studies with mice. While the latter two instrumental conditioning
procedures are comparable, they have both introduced an artificial
component into the discrimination process. The subjects were water
deprived and received water reinforcement for discriminating between the
samples. Thus, they were highly motivated and may have detected differences between urine odors that they would not normally attend to. The
habituation-dishabituation task examined the rats' ability to detect odor
differences in a more natural situation, and may be a more reliable
indicator of the animal's natural discrimination processes. In the
present study, rats were able to discriminate between the urine odors of
germfree mice in an operant task. An examination of the ability of rats
to discriminate between the urines from germfree mice in a habituation-
dishabituation paradigm should confirm whether procedural differences
could have influenced the outcome of the experiments.

Despite the possible differences between rat and mouse urine donors,
the urine samples collected, and the experimental procedures used, it may
be that bacteria are actually necessary for the production of individually unique urine odors in rats and not in mice and that MHC differences in
mice can be expressed without bacterial influence. We do know that mice,
in general, are more sensitive to the odors produced by genetic differences. For example, the odors from mice which differ only at the MHC
induce pregnancy block in mice (Yamazaki et al., 1983) whereas no stimuli
have yet been found which produce pregnancy block in rats (Brown, 1985).
In rats, when the MHC Class I antigens are removed, the urine odors are
still discriminable (Brown, Roser and Singh, 1987). The effect of
removing the Class I antigens in mice is unknown and should be investigated.

ACKNOWLEDGEMENTS

This research was supported by a National Sciences and Engineering
Research Council of Canada grant A7441 to R.E. Brown. We wish to thank
Dr. Gary Beauchamp from Monell Chemical Senses Center for providing the
samples used in this experiment.

REFERENCES

Brown, R.E., 1985, The rodents I: effects of odours on reproductive
    physiology (primer effects), in "Social Odours in Mammals," R.E.
    Brown and D.W. MacDonald, eds., Oxford University Press, England.
Brown, R.E., Roser B., and Singh, P.B., 1990, The MHC and individual
    odours in rats, in: "Chemical Signals in Vertebrates 5," D.W.
    MacDonald, D. Muller-Schwarze, and S.E. Natynczuk, eds., Oxford
    University Press, Oxford.
Brown, R.E., Singh, P., and Roser, B., 1987, The major histocompatibility
    complex and the chemosensory recognition of individuality in rats,
    Physiol. Behav. 40:65.
Ivanyi, P., 1978, Some aspects of the H-2 system, the major histocompati-
    bility system in the mouse, Proc. R. Soc. London, Ser B., 202:117.
Klein, J., 1986, "Natural history of the major histocompatibility com-
    plex," John Wiley and Sons, New York.
Schellinck, H.M., Brown, R.E., and Slotnick, B., 1991, Training rats to
    discriminate between the odors of individual conspecifics, Anim.
    Learn. Behav., 19, 223-233.
Singh, P.B., Herbert, J., Roser, B., Arnott, L., Tucker, D.K., and Brown,
    R.E., 1990, Rearing rats in a germ-free environment eliminates their
    odors of individuality, J. Chem. Ecol., 16:1667.
Van der Waaij, D., 1984, The immunoregulation of the intestinal flora:
    Experimental investigations on the development and the composition
    of the microflora in normal and thymusless mice, Microecol. Therapy,
    14:63.
Yamazaki, K., Beauchamp, G.K., Bard, J., Boyse, E.A., and Thomas, L.,
    1991, Chemosensory identity and immune function in mice, in, "Chemi-
    cal Senses. Volume 3," C.J. Wysocki and M.R. Kare, eds., Marcel
    Dekker, New York.
Yamazaki, K., Beauchamp, G.K., Imai, Y., Bard, J., Phelan, S., Thomas,
    L., and Boyse, E.A., 1990, Odortypes determined by the major histo-
    compatibility complex in germfree mice, Proc. Natl. Acad. Sci. USA,
    87:8413.
Yamazaki, K., Beauchamp, G.K., Wysocki, G.J., Bard, J., Thomas, L., and
    Boyse, E.A., 1983, Recognition of H-2 types in relation to the
    blocking of pregnancy in mice, Science, 221:186.

SECTION FIVE

SEMIOCHEMICALS AND ENDOCRINE PROCESSES

CHEMOSIGNALS AND REPRODUCTION IN ADULT FEMALE HOUSE MICE

Lee C. Drickamer

Department of Zoology
Southern Illinois University
Carbondale, IL 62901 U.S.A.

INTRODUCTION

Whitten (1958, 1959) first reported in detail that odors from male
and female mice could influence the timing of estrous cycles and
reproduction in adult conspecifics. Urinary chemosignals from male mice
enhance female cycles, whereas urinary chemosignals from other females
retard female cycles. Chemosignal modulation of ovarian cyclicity has
been documented for a broad range of mammals (see summary in McClintock,
1983). Vandenbergh (1969) first reported that urinary chemosignals from
conspecifics influenced the onset of puberty in young female mice. Several
decades of research have produced a rather thorough understanding of the
manner in which urine from male and female mice influences the sexual
development of young females (Vandenbergh, 1983; Vandenbergh and Coppola,
1986; Drickamer, 1986a). The release of these puberty-influencing urinary
chemosignals by wild mice has been documented in both the laboratory and
under field conditions (Drickamer, 1979; Massey and Vandenbergh, 1980,
1981). In recent tests conducted in outdoor field pens, we have shown
that urinary chemosignals affect adult reproduction, puberty in young
mice and the size of mouse populations (Drickamer and Mikesic, 1990).

Urinary chemosignals that influence puberty have been tested with
regard to types of urine that are effective in modulating development,
timing effects in terms of exposure duration and circadian and circannual
influences and hormonal involvement for production and release of
chemosignals, social factors such as density. They have also been tested
with regard to social status and degree of relatedness, dose levels and
using urine from several sources simultaneously, non-social environmental
effects on chemosignal release, and behavioral responses in terms of
urine deposition by donors and attraction to or avoidance of urine by
potential recipients (Vandenbergh, 1983; Drickamer, 1986a, 1989). Only
a few of these questions have, to this time, been tested with regard to
effects of urinary chemosignals on the timing of estrous cycles in adult
female mice.

It is my purpose in this paper to provide evidence concerning effects
of urinary chemosignals on adult female mice for several types of
questions that have been tested in young females (see McClintock, 1983;
Hoover and Drickamer, 1978; Drickamer, 1986b). In particular, I will

Chemical Signals in Vertebrates VI,   Edited by R.L. Doty and
D. Müller-Schwarze, Plenum Press, New York, 1992

report on (1) basic estrous cycle patterns for females treated with a variety of urine types and the possible effects of various experimental procedures on estrous cycles (Experimental Series I), and (2) the questions pertaining to timing and biological rhythm effects on adult female cycles (Experimental Series II).

GENERAL METHODS

The mice used were from two sources: laboratory house mice were of a heterogeneous ICR/Alb strain and wild mice were second and third generation laboratory mice bred from animals trapped in the vicinity of Williamstown, Massachusetts, U.S.A. All of the breeding stock and test mice were housed in a series of four rooms maintained at 21-25°C and 30-60% relative humidity on a 12 h light:12 h dark daily cycle with lights on from 0600 h to 1800 h. All mice were housed in polypropylene cages (15 x 28 x 15 cm deep) with opaque sides and fitted wire lids on a bedding of wood shavings that was changed weekly except as prescribed in the detailed methods for some experiments. Procedures for breeding mice for testing are outlined in detail in Drickamer (1977, 1989). For assaying the effects of urinary chemosignals, I used adult virgin females that were 100 to 160 days of age at the start of the test period.

The assay for ovarian cyclicity in my studies was the vaginal smear; wet-mount smears taken daily were evaluated immediately with a light microscope according to the criteria of Rugh (1968) and Vandenbergh (1969). The total number of days over which each experiment was conducted varied. Thus, two dependent measures were calculated from the vaginal smear information; (1) the number of estrous cycles/week, and (2) the number of days the mouse was in estrus/N days. Since both measures produced the same patterns of results, only data on (2) are presented here. All means are shown ± 1 standard error. Except for the first experiment in the first series, all tests were carried out using the domestic laboratory stock of house mice.

EXPERIMENTAL SERIES-I - BASIC PATTERNS

Laboratory versus Wild Stocks

The first experiment tested whether urine from 8 different sources would affect cycles of laboratory and wild stock adult females. Nine treatments were used: (1) no treatment; (2) water; and urine from (3) males; (4) singly-caged females; (5) females housed 8/cage; (6) pregnant females; (7) lactating females; (8) estrous females; and (9) diestrous females. Treatments, all involving fresh urine, were applied daily to the external nares with a small paintbrush. Test females were housed individually and smeared for 42 days. Sample sizes were 9 test mice/treatment for the wild stock and 16 mice/treatment for the laboratory stock. The urine types used for each stock of mice were collected from within the same stock.

One-way ANOVAs revealed significant treatment effects for domestic [$F (8,135) = 30.97$; $P < 0.001$] and wild [$F (8,135) = 21.76$; $P < 0.001$] stock mice. For each stock, treatment with urine from females housed 8/cage resulted in fewer days in estrus/30 days than the other treatments: 7.6±0.6 for domestic mice and 4.9±0.7 for wild mice. Four treatments produced intermediate number of estrous days (domestic stock mean first, followed by wild stock mean): no treatment (14.1±0.9/8.9±0.5); water (13.4±0.9/8.9±0.5); urine from singly-caged females (13.8±0.9/11.4±0.7); and urine from females in diestrus (12.6±1.0/12.0±0.9). Four treatments resulted in significantly more days in estrus: urine from adult males

(21/5±0.9/15.2±1.4); urine from pregnant females (19.9±0.8/14.6±0.9); urine from lactating females (20.1±0.8/17.4±0.8); and urine from females in estrus (20.2±1.1/16.3±1.0).

Effects of Doing Vaginal Smears

The second experiment tested the effects of doing vaginal smears; no chemosignal treatments were given to any of the mice. For group one treatment, ten individually caged females were smeared daily for 42 days. These data were compared with two other treatment regimes. In group two (N = 70 mice), 10 females were smeared each day, but each mouse was smeared only once per week on the same day. In the third comparison group (N = 70 mice), 10 females selected at random were smeared each day. Mice in treatments two and three were also housed individually smeared for 6 weeks. Thus, there were 3 treatments and 6 1-week blocks in the design.

Across all three treatment conditions, the mean values for numbers of days in estrus/7 days varied between 2.0 and 2.6. Neither the treatment effect [$F_{(2,27)} = 0.01$; $P > 0.50$], nor that for the weeks [blocks: $F_{(5,135)} = 0.12$; $P > 0.50$] were significant.

Are Control Treatments the Same?

The last experiment in the first series tested whether a variety of control treatments produced similar results. Five treatments were used: (1) no treatment; (2) water painted on the external nares; (3) exposure to clean bedding; (4) adding clean bedding to the cage; and (5) placing 0.1 cc of water on the nares with a syringe. Treatments (2) through (5) were performed daily. There were 12 individually housed mice in each treatment and the mice were smeared for 30 days.

The mean values for the 5 treatments varied from 9.5 to 10.6 days in estrus/30 days. There were no significant treatment differences [$F_{(4,55)} = 0.48$; $P > 0.50$].

EXPERIMENTAL SERIES-II - TIMING EFFECTS

Numbers of Days of Treatment Needed

How many days of treatment are necessary to produce effects on adult female cycles? The experimental design involved 6 treatment conditions with N = 12 mice/treatment: (1) water; (2) male urine; (3) urine from females in estrus; (4) urine from pregnant females; (5) urine from lactating females; and (6) urine from females housed 8/cage. Mice were housed individually and smeared for 60 days; the time period was divided into 6 10-day blocks.

A repeated measures ANOVA revealed significant treatment [$F_{(5,54)} = 22.26$; $P < 0.001$] and block [$F_{(5,342)} = 9.41$; $P < 0.001$] effects. Urine from females housed 8/cage (mean = 2.1±0.1) resulted in fewer average days in estrus/10 days than for water (3.3±0.2). The other 4 treatments all resulted in significantly more days in estrus/10 days; male urine (4.7±0.2), pregnant female urine (4.9±0.3); lactating female urine (4.8±0.2); and estrous female urine (4.5±0.2). Treatment during the first 10-day block resulted in 3.2±0.2 days in estrus across all 6 treatments. By the second 10-day block, this increased significantly to 3.9±0.2 days in estrus. The mean days in estrus per block continued to increase, reaching values of 4.3±0.2 and 4.4±0.2 by the fifth and sixth 10-day blocks, though there were no significant differences among the means for the second through sixth blocks.

Length of Daily Exposure Needed

How long must the daily exposure to the urinary chemosignal be in order for the female cycles to be affected? The same 6 treatments were used as in the previous experiment. However, in this experiment, exposure to the treatment conditions was accomplished by putting the test mouse into a cage with clean or soiled bedding for a prescribed period of time and then returning it to its home cage. Four different exposure durations were used: (1) 0.5 h/day; (2) 1.0 h/day; (3) 2.0 h/day; and (4) 4 h/day. The data were analyzed with a 2-way ANOVA.

Both the main treatment effects for urine source [F (5,312) = 101.64; P < 0.001] and duration of treatment [F (3,312) = 56.37; P < 0.001], as well as the interaction factor [F (15,312) = 17.48; P < 0.001] were significant. With regard to treatment sources, the results were similar to data presented above regarding which types of urine resulted in enhancement and retardation of estrous cycle. For clean bedding, there were no significant effects with regard to duration of exposure. For bedding soiled by males, the mean value for 0.5 h/day (12.8 days in estrus/40 days) exposure was significantly less than for 2 h/day (21.4) or 4 h/day (21.1); exposure for 1 h/day resulted in an intermediate mean value (16.0). For bedding soiled by pregnant females, the mean values for 0.5 h/day (13.8) and 1.0 h/day (16.2) exposure were significantly less than those for 2 h/day (20.3) or 4 h/day (20.9). For bedding soiled by lactating females, the mean values for 0.5 h/day (14.0) and 1.0 h/day (15.6) exposure were significantly less than those for 2 h/day (20.4) or 4 h/day (21.6). For bedding soiled by estrous females, the mean value for 0.5 h/day (14.1) exposure was significantly less than for 2 h/day (21.2) or 4 h/day (21/1); exposure for 1 h/day resulted in an intermediate mean value (17.4). For bedding from females housed 8/cage, the mean values for both 0.5 h/day (13.4) and 1.0 h/day (11.9) exposure were significantly more than the values for 2.0 h/day (8.2) and 4.0 h/day (8.5).

Effects of Intermittent Exposure

Does intermittent exposure have the same effect as continuous exposure? The same 6 treatment types were used as in the previous 2 experiments. Test mice were provided with 2, 4, 6, or 8 15 min exposures/day to soiled or clean bedding from the designated source. The exposures were provided during a single 4 h time block from 0800 h to 1200 h for all test mice. Treatments were continued for 40 days and there were 14 mice tested for each of the 24 different cells of the experimental design. The results were analyzed using a 2-way ANOVA.

Both the main treatment effect for urine source [F (5,312) = 54.81; P < 0.001] and number of exposures [F (3,312) = 25.73; P < 0.001], as well as the interaction factor [F (15,312) = 15.31; P < 0.001] were significant. With regard to treatment sources, the results were similar to data presented earlier concerning enhancement and retardation of estrous cycles. For clean bedding, there were no significant effects with regard to duration of exposure. For bedding soiled by males, the mean value for 2 exposures/day (14.5 days in estrus/40 days) was significantly less than for 4 exposures/day (17.9), 6 exposures/day (19.3) or 8 exposures/day (20.5). For bedding soiled by pregnant females, the mean values for 2 exposures/day (14.9) and 4 exposures/day (15.6) were significantly less than those for 6 exposures/day (19.1) or 8 exposures/day (20.9). For bedding soiled by lactating females, the mean values for 2 exposures/day (14.5) and 4 exposures/day (14.8) were significantly less than those for 6 exposures/day (19.5) or 8 exposures/day (20.1). For bedding soiled by estrous females, the mean values for 2 exposures/day (14.1) and 4

exposures/day (14.7) were significantly less than those for 6 exposures/day
(19.4) or 8 exposures/day (20.6). For bedding from females housed 8/cage,
the mean values for 2 exposures/day (15.3) and 4 exposures/day (13.6) were
significantly larger than for 6 exposures/day (10.4); the mean value for
8 exposures/day (7.5) was significantly less than all other means.

Circadian Effects - Donors and Recipients

Are there any circadian rhythms with regard to either chemosignal
release by donors or chemosignal effectiveness in recipient mice? The
design for this experiment was similar to that that I used previously
(Drickamer, 1982), with the difference that I used 6 intervals for
collecting urine (0000, 0400, 0800, 1200, 1600, and 2000 h) and the same
6 intervals for treating test females. Twelve mice were tested in each
of the 36 cells of the design and smears were conducted for 30 days.
This design was repeated 4 times, using male urine, urine from pregnant
females, urine from females in estrus and urine from females housed
8/cage. The results were analyzed using 2-way ANOVAs with collection
time and treatment time as main factors. Complete analyses are available
from the author; only descriptive findings will be provided here.

Male urine treatment resulted in more days in estrus for test
females when it was collected at 0400 or 0800 hours and male urine was
more effective in terms of inducing more estrous days in females when
the treatment was applied at 0400 h or 0800 h. There was no significant
effect with urine from pregnant females with regard to time of collection,
but his urine type was more effective when test females were treated at
0800 h, and also, to a lesser degree at 0400 h and 1200 h. Urine from
females in estrus was more effective when it was collected at 0400 h or
0800 h; there were no significant effects with regard to the time of day
for treating test females. For urine from females housed 8/cage there
were no significant effects for either the time of day of urine
collection or the time of day for treating test females.

Circannual Effects - Donors and Recipients

Are there any circannual rhythms with regard to either chemosignal
release or effectiveness in recipient mice? This question was tested
using a compilation of data from other experiments conducted during all
months of the year over a period of 10 years. Sample sizes for each
month varied depending on available data. Sufficient data were available
only for the control treatment condition and for females treated with
either adult male urine or urine from females housed 8/cage. Data from
different experiments were standardized so that the dependent variable
was the number of days in estrus/10 days. For each of the three test
conditions, separate 1-way ANOVAs were used to analyze the data.

For the control conditions [F (11,181) = 6.08; P < 0.001), female
mice exhibited significantly more days in estrus/10 days during the
period of June through September than at other times of the year. The
mice exhibited fewer days in estrus during December and January and the
remaining months were intermediate. For the male urine condition [F
(11,181) = 9.48; P < 0.001], female mice exhibited significantly more
estrous days/10 days during the months of May through August, fewer days
in estrus during December and January, with intermediate values for the
remaining months. Lastly, for the condition involving urine from
females housed 8/cage [F (11,168) = 2.73; P < 0.005], test females
exhibited significantly more estrous days/10 days from April through
October and fewer days in estrus from November to March, with no months
involving intermediate values.

DISCUSSION

Several conclusions emerge from the forgoing experiments: (1) Both laboratory and wild stocks of house mice exhibit similar effects with regard to the effects of 7 different types of urine on the estrous cycles of adult females. It is noteworthy, however, that for each of the urine types tested, the effects, measured as the number of days in estrus, were somewhat higher for the laboratory stock mice. (2) Neither conducting the vaginal smears on a daily basis nor the various types of treatment procedures used for control conditions affected the estrous cycles of the adult female mice. (3) With regard to the timing effects, 10 or more consecutive days of treatment, somewhere between 1 and 2 hours of daily exposure, and at least 4 exposures in a 4 h period are necessary to produce the enhancement or retardation effects of the urinary chemosignals. (4) There were many fewer effects of time of day for collection or time of day for treatment with the urinary chemosignals on adult female estrous cycles than have been reported previously with regard to effects on the timing of puberty in the same mice (see Drickamer, 1986a). (5) The patterns of seasonal effects for the male urine enhancement of estrus and grouped female urine retardation of estrus, measured here only as month-to-month changes in the numbers of days in estrus for recipient females, were similar to the patterns reported previously for effects on puberty in young mice (see Drickamer, 1986a); females are in estrus more often in the summer months and less often during winter months.

With the exception of the rather substantial differences with regard to circadian influences on chemosignal effects, the findings for adult female mice reported here are quite similar to those reported over the past 2 decades for effects on puberty in young females. I do not have an explanation for the differences found in the foregoing experiments concerning daily rhythms of either chemosignal production and release or chemosignal effectiveness in recipient females. Taken together, I suggest that these findings confirm that similar urinary chemosignals are involved in young and adult mice and that at least a major portion of the mechanism pathways in recipient mice are likely the same in young and adult mice. We have recently demonstrated that the cycles of adult females and the frequency with which they become pregnant can be influenced in mice living in 0.1 ha outdoor enclosures (Drickamer and Mikesic, 1990). Thus, it appears that under field conditions, both the frequency of estrus and the consequent reproduction of house mice can be influenced in a manner that is consistent with the types of findings reported here and elsewhere from laboratory settings for urinary chemosignal influences on adult female estrous cycles.

ACKNOWLEDGMENTS

I thank the many undergraduate students and Williams College who participated in the data collection. This research was supported in part by NIH Grant HD-08585 and by funds from Williams College.

REFERENCES

Drickamer, L. C., 1977, Delay of sexual maturation in female house mice by exposure to grouped females or urine from grouped females, J. Reprod. Fert., 51:77.

Drickamer, L. C., 1979, Acceleration of first vaginal estrus in wild Mus musculus, J. Mammal., 60:215.

Drickamer, L. C., 1982, Acceleration and delay of sexual maturation in female mice via chemosignals: circadian rhythm effects, Biol. Reprod., 27:596.

250

Drickamer, L. C., 1986a, Puberty-influencing chemosignals in mice: ecological and evolution considerations, in, "Chemical Signals in Vertebrates IV," D. Duvall, D. Muller-Schwarze and R. M. Silverstein, eds., Plenum, New York.

Drickamer, L. C., 1986b, Peripheral anosmia affects puberty-influencing chemosignals in mice: donors and recipients, Physiol. Behav., 37:741-746.

Drickamer, L. C., 1989, Patters of deposition of urine containing chemosignals that affect puberty and reproduction by wild stock male and female house mice (Mus domesticus), J. Chem. Ecol., 15:1407.

Drickamer, L. C. and Mikesic, D. G., 1990, Urinary chemosignals, reproduction and population size in wild house mice in enclosures, J. Chem. Ecol., 16:2955.

Hoover, J. E. and Drickamer, L. C., 1979, Effects of urine from pregnant and lactating female house mice on oestrous cycles of adult females, J. Reprod. Fert., 55:297-301.

Massey, A. and Vandenbergh, J. G., 1980, Puberty delay by a urinary cue from female mice in feral populations, Science, 209:821-822.

Massey, A. and Vandenbergh, J. G., 1981, Puberty acceleration by a urinary cue from male mice in feral populations, Biol. Reprod., 24:532-537.

McClintock, M. K., 1983, Pheromonal regulation of the ovarian cycle, in, "Pheromones and Mammalian Reproduction," J. G. Vandenbergh, ed., Academic Press, New York.

Rugh, R., 1968, "Biology of the Laboratory Mouse," Burgess, Minneapolis.

Vandenbergh, J. G., 1969, Male odor accelerates female sexual maturation in mice. Endocrinology, 84:658.

Vandenbergh, J. G., 1983, Pheromonal regulation of puberty, in, "Pheromones and Mammalian Reproduction," J. G. Vandenbergh, ed., Academic Press, New York.

Vandenbergh, J. G. and Coppola, D. M., 1986, The physiology and ecology of puberty modulation by primer pheromones, Adv. Stud. Behav. 16:71.

Whitten, W. K., 1958, Modification of the oestrous cycle of the mouse by external stimuli associated with the male. J. Endocrinol., 17:307.

Whitten, W. K. 1959, Occurrence of anoestrus in mice caged in groups. J. Endocrinol., 18:102.

INDUCTION OF ESTRUS AND OVULATION IN FEMALE GREY SHORT-TAILED OPOSSUMS,

MONODELPHIS DOMESTICA, INVOLVES THE MAIN OLFACTORY EPITHELIUM

S.A. Pelengaris, D.H. Abbott[*], J. Barrett, and H.D.M. Moore

Institute of Zoology, Zoological Society of London
Regent's Park, London NW1 4RY, England, UK

INTRODUCTION

Grey short-tailed opossums are South American didelphid marsupials. Females rarely exhibit behavioral estrus without direct contact with a male or male chemosignals (Fadem, 1985, 1989; Baggott et al., 1987; Baggott and Moore, 1990). Male pheromonal cues alone are quite sufficient to induce behavioral estrus (Fadem, 1989). The stimulus male can be removed after five days, once proestrus behavior is initiated, and ovulation will occur without any copulatory stimulus (Baggott et al., 1987). Thus, while the presence of a male or male chemosignals is important for the induction of estrus in female opossums, it has not been established whether (i) male chemosignals are the only stimuli involved, or (ii) which olfactory epithelium mediates the chemosensory information. This study represents an initial investigation of both these aims.

METHODS

All female opossums were at least 5 months old and were maintained in a breeding colony at the Institute of Zoology in London, as previously reported (Baggott et al., 1987). A combination of Ketamine (1 mg/kg) and Xylazine (0.6 mg/kg) anesthesia (0.2 ml, i.m. injection) was used in all ablation and control procedures. Local anesthesia [Lignocaine (0.1%), 0.1 ml] was injected into the upper or lower jaw during vomeronasal organ ablation or control procedures, respectively.

Olfactory ablations

Ten females received bilateral diathermic cautery ablations to the vomeronasal organ following an incision through the hard palate (VNX group). This incision was made 4 mm rostral to the upper incisor teeth and soft dental cement was used to cover the incision. Ten females received bilateral ablations to the main olfactory epithelium from

---

*Current address: Wisconsin Regional Primate Research Center, 1223 Capitol Court, Madison, WI 53715.

irrigation of the nasal passages with 0.1 ml 5% zinc sulphate solution per nostril (MOX group). Twelve females received bilateral diathermic cautery of the lower jaw (CONTROL group).

Behavioral tests

Two days following one of the above procedures, each female was presented in her "home cage" (56 x 38 x 18 cm) with an unfamiliar male in a small wire cage (18 x 13 x 10 cm). The male remained in the female's cage for 2h during each of the next 3 days. Following the third 2h exposure, the male was released into the female's cage until behavioral observations ceased. This procedure reliably induces estrus after about 8 days following the male's introduction (Baggott et al., 1987). To quantitatively assess development of pro-estrus behavior, females were observed during a 30 min daily test with an unfamiliar male in a clean "home cage". The tests were conducted 3-5h into the dark period of the light/dark cycle and from days 5-11 following introduction of a male into the female's "home cage" (or until copulation was observed during tests). The test methods and behaviors scored were similar to those developed by Baggott et al. (1987).

Assessment of reproductive tract and olfactory ablations

All females were sacrificed 16 (n=26) or 23 (n=6) days after the initial introduction of a male into the female's "home cage". The presence of corpora lutea and embryos and the size and weight of the uterus were determined. Histological examination of the olfactory ablations was performed after each head had been placed in 10% formalin for 2 weeks and 8% formic acid for 8-12 weeks.

RESULTS

Figure 1 illustrates the ablation of either the main olfactory epithelium or the vomeronasal organ, in randomly selected females from the MOX or VNX groups, respectively. A control female illustrates the intact epithelia. Figure 1(2) illustrates the ablation of the vomeronasal organ in a VNX female in comparison to the intact vomeronasal organ in the a normal CONTROL female (Figure 1[1]). Rostral portions of the vomeronasal organ in the VNX female remained containing degenerate epithelium. While figure 1(4) illustrates the degenerate main olfactory epithelium in a MOX female in comparison to a normal CONTROL (Figure 1[3]), approximately 20% of the mail olfactory epithelium appeared undamaged or had regenerated in the MOX female.

All CONTROL females and 70% of VNX females exhibited proestrus in the behavioral tests. These females displayed female-initiated circling, naso-cloacal contact (genital sniffing) and approaching of the male and acquiescence to male-initiated naso-cloacal contact. Only 10% of MOX females displayed these behaviors, suggesting that the MOX group remained almost entirely in anestrus. Estrus was displayed by all CONTROL females in the behavioral tests (Table 1) and occurred 7.6 ± 0.2 days (mean ± SEM) after the first introduction of the resident male to the "home cage". Estrus behavior was not observed in the MOX group of females and in only one female in the VNX group (Table 1).

However, when the occurrence of ovulation and pregnancy was assessed from examination of the female's reproductive tracts, ovulation was only impaired in the MOX group (Table 1). All those females assessed as post-ovulatory from the presence of ovarian corpora lutea were also found to

Table 1.   The Occurrence of Estrus and Ovulation in Female Grey Short-
Tailed Opossums Receiving Sham Procedures (CONTROL), or
Ablation of the Vomeronasal Organ (VNX) or Main Olfactory
Epithelium (MOX).

Treatment	No. of Females Displaying:	Estrus	Ovulation
CONTROL		11/11[a]	12/12
VNX		1/10[b]	8/10
MOX		0/10[b]	3/10[c]

a: one female was pregnant prior to testing
b: p < 0.01 vs. CONTROL group
c: p < 0.05 vs. CONTROL and VNX group
(Fisher Exact Probability Test, one-tailed)

be pregnant because of embryos recovered from their greatly enlarged
uterine horns. Thus, while the VNX group of females appeared to have a
similar inability to display estrus in the behavioral tests as the MOX
females, copulation (and presumably, estrus) was achieved with the male
resident in the VNX females' "home cages" outside the behavioral tests.

DISCUSSION

These preliminary results suggest that olfactory cues from males are
of paramount importance for the induction of estrus and ovulation in
female grey short-tailed opossums. The physical and behavioral presence
of the male appears to be insufficient. This finding supports and
extends those of Fadem (1985, 1989). Interestingly, the main olfactory
epithelium and not the vomeronasal organ may be the primary sensory
epithelium involved in mediating this olfactory effect. However, the
vomeronasal organ in females may mediate male olfactory cues which
enhance the progression of proestrus behavior because VNX females failed
to display estrus in the behavioral tests but were apparently induced
into estrus by prolonged exposure to a male resident in their "home
cage". VNX females seemed to have an impaired ability to display estrus,
unlike the inability of MOX females (which remained in anestrus). The
importance of the apparent olfactory deficits in VNX and MOX females is
further supported by the successful induction of estrus and ovulation in
CONTROL females in a similar time to that reported in a previous study by
Baggott et al. (1987).

In this marsupial, as in many eutherian mammals (Halpern, 1987;
Wysocki and Meredith, 1987; Signoret, 1990), olfactory cues play a
crucial role in controlling reproduction. However, unlike many eutherian
mammals, the main olfactory epithelium rather than the vomeronasal organ
may be the more important sensory epithelium mediating the olfactory
signals. Certainly, in the prairie vole, Microtus ochrogaster, a rodent
equivalent to the grey short-tailed opossum (in terms of male-induced
estrus and ovulation [Carter et al., 1981, 1987]), the vomeronasal organ
in females has been implicated as playing the major role in mediating
male olfactory cues stimulating estrus and ovarian activation (Lepri et
al., 1990). However, further experiments are necessary to delineate the

importance of non-vomeronasal sources of sensory information in both
marsupial and eutherian mammals because (i) specific ablations of the
main olfactory epithelium have not been tested in the prairie vole and
(ii) prolonged physical and behavioral contact with males partially
overcame the estrus deficits in VNX females and resulted in mating in 45%
of VNX female prairie voles (Lepri and Wysocki, 1987), at least 70% of
VNX female grey short-tailed opossums (Table 1), and 100% of VNX female
rats (Saito and Moltz, 1986). The main olfactory epithelium may play a
greater role in mediating olfactory cues which influence reproduction
than was previously thought. This latter point has been suggested for
the sheep, where ablation of the vomeronasal organ in ewes failed to
block the rise in luteinizing hormone (LH) secretion induced by fleece
extracts from rams (Cohen-Tannoudji et al., 1989).

Further work on estrus and ovulation inducing cues is necessary in
the female grey short-tailed opossum to determine (i) the neural mechanism mediating the main olfactory epithelium effect and (ii) whether this
olfactory mechanism is peculiar to this marsupial species or is typical
of olfactory mechanisms in marsupials generally. Such studies could
provide useful new information regarding olfactory control of physiological function.

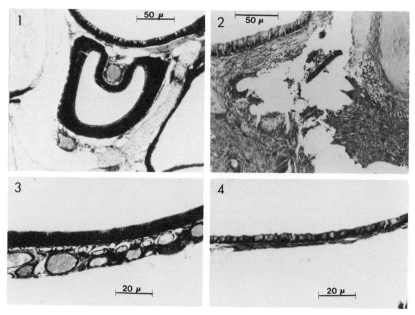

Fig. 1. Coronal histological sections of olfactory epithelia in
female grey short-tailed opossums illustrating (1) an
intact vomeronasal organ from a CONTROL female (magnification: x 160), (2) an ablated vomeronasal organ from a
VNX female (magnification: x 160), (3) a normal portion of
main olfactory epithelium in a CONTROL female (magnification: x 240) and (4) an ablated portion of main olfactory
epithelium from a MOX female (magnification: x 240).

ACKNOWLEDGEMENTS

We thank Professor A.P.F. Flint for criticism of this manuscript; D. Spratt for valuable technical assistance with the histology; T. Dennett and M.J. Walton for preparation of the plate; T. Noble and D. Stoula for valuable technical assistance and maintenance of the animals; and A. Richards McCormick for typing the manuscript.

REFERENCES

Baggott, L.M., Davis-Butler, S., and Moore, H.D.M., 1987, Characterization of estrus and timed collection of oocytes in the grey short-tailed opossum, Monodelphis domestica, J. Reprod. Fertil., 79:105-114.

Baggott, L.M. and Moore, H.D.M., 1990, Early embryonic development of the grey short-tailed opossum, Monodelphis domestica, in vivo and in vitro, J. Zool. Lond., 222:623-639.

Cohen-Tannoudji, J., Llavenet, C., Locatelli, A., Tillet, Y., and Signoret, J.P., 1989, Non-involvement of the accessory olfactory system in the LH response of anoestrous ewes to male odour, J. Reprod. Fertil., 86:135-144.

Fadem, B.H., 1985, Evidence for the activation of female reproduction by males in a marsupial, the grey short-tailed opossum (Monodelphis domestica), Biol. Reprod., 33:112-116.

Fadem, B.H., 1989, The effects of pheromonal stimuli on estrus and peripheral plasma estradiol in female grey short-tailed opossums (Monodelphis domestica), Biol. Reprod., 41:213-217.

Halpern, M., 1987, The organization and function of the vomeronasal system, Ann. Rev. Neurosci., 10:325-362.

Lepri, J.J. and Wysocki, C.J., 1987, Removal of the vomeronasal organ disrupts the activation of reproduction in female voles, Physiol. Behav., 40:349-355.

Lepri, J.J., Wysocki, C.J., Gerber, K., and Lisk, R.D., 1990, The vomeronasal organ of the prairie vole: role in chemosignal activation of female reproduction and observations of fine structure, in: "Chemical Signals in Vertebrates 5", D.W. Macdonald, D. Müller-Schwarze, and S.E. Natynczuk, eds., Oxford University Press, Oxford, pp. 147-153.

Saito, T.R. and Moltz, H., 1986, Sexual behavior in the female rat following removal of the vomeronasal organ, Physiol. Behav., 38:81-87.

Signoret, J.P., 1990, Chemical signals in domestic ungulates, in: "Chemical Signals in Vertebrates 5", D.W. Macdonald, D. Müller-Schwarze, and S.E. Natynczuk, eds., Oxford University Press, Oxford, pp. 610-626.

Wysocki, C.J. and Meredith, M., 1987, The vomeronasal system, in: "Neurobiology of Taste and Smell", T.E. Finger and W.L. Silver, eds., John Wiley and Sons, New York, pp. 125-150.

INFLUENCE OF SPECIFIC SKIN GLANDS ON THE SEXUAL MATURATION OF MALE

GOLDEN AND CAMPBELL'S HAMSTERS

Vladimir E. Sokolov, Nina Yu. Vasilieva
and Edvard P. Zinkevich

Institute of Animal Evolutionary Morphology and Ecology
Academy of Sciences, Leninsky Prospect 33
Moscow 117071, Russia

INTRODUCTION

Chemical signals play an important role in mammalian communication (Brown, 1979, 1985). Secretions of skin glands have been of particular interest. Glands used for chemical communication are widespread among mammals from different taxonomic groups (Quay, 1977; Adams, 1980). Although several researchers have studied specialized skin glands and associated marking behaviors (Eisenberg and Kleiman, 1972; Johnston, 1973, 1990; Thiessen and Rice, 1976), many questions regarding their function remain. It has been suggested that such glands may contribute to the regulation of population density by influencing reproduction (Naumov, 1973). Apparently this influence can occur through both behavioral (communicative) and physiological mechanisms. The communicative significance of skin glands has been studied intensively, but their physiological effects have not been investigated. The influence of stimuli from specialized skin glands on the rate of sexual maturation has been demonstrated only for primates (Dr. Jane Barrett, personal communication) and the Campbell's hamster (Sokolov et al., 1989 a,b).

In this study, we investigated the influence of different social factors, including specific skin gland secretions, on the sexual maturation of young male hamsters. Two species with contrasting types of social organization were used: a solitary species, the golden hamster Mesocricetus auratus, and a highly social species, the Campbell's hamster Phodopus campbell.

MATERIALS AND METHODS

Two different laboratory stocks of P. campbelli were used. One group of young males came from a highly inbred laboratory colony maintained since the early 1970's that was outbred once in 1984 with a wild male captured in the Altai (laboratory males), and another group came from a randomly-bred colony established in our laboratory from 5 pairs of hamsters captured in NE Mongolia in 1985 (Mongolian males). All M. aratus were from a colony maintained at the Institute since 1978. Breeding animals 3 to 6 months old were housed in 40 x 20 x 25 cm (P. campbelli) or 45 x 32 x 25 cm (M. auratus) glass cages at 20 ± 2 C^o, with a 14L:10D photoperiod, and provided with natural food ad libitum.

Chemical Signals in Vertebrates VI, Edited by R.L. Doty and
D. Müller-Schwarze, Plenum Press, New York, 1992

259

Young males were reared in their original litters of mixed sexes. Litter sizes were $3 \leq N \leq 6$ for P. campbelli and $4 \leq N \leq 7$ for M. auratus. From the day of birth until one month of age, P. campbelli were housed with their mother and either (1) their intact father, (2) their father with the midventral gland surgically removed, (3) no father, or in the case of Mongolian males only, (4) no father, but young males treated with sebum from the midventral gland of a strange male beginning on day 2 postpartum. In this last treatment group, sebum was applied around the nares and in the mouth every other day using a metal spatula. The sebum dose was varied according to the age of the young males (0.1 - 1.5 mg.). M. auratus juveniles were housed together with their mothers and treated by having either water or a water solution containing secretions from the flank glands of strange adult males painted directly on the external nares and dropped into the mouth every other day. At one month of age, all experimental animals were killed by an overdose of ether, and body weights and weights of reproductive organs obtained.

RESULTS AND DISCUSSION

There was a significant effect of the experimental treatment on the rate of sexual maturation of young P. campbelli males (Table 1). Males housed with fathers exhibited delayed development in comparison with males housed without fathers or with fathers that had the midventral glands removed. The differences between experimental groups appeared stronger in the Mongolian males than in the laboratory males. Young

Table 1. Body and reproductive organ weights in one-month-old P. campbelli males of Mongolian and laboratory stocks in different experimental groups (Mean $\pm$ SEM). Body weight is in grams; testes and epididymis weights are in milligrams.

Stock	Parameters	Experimental Exposure Groups			
		Intact Father	Operated Father	No Father	No Father + sebum
Mongolian	Body weight	22.2 $\pm$ 0.6	25.3 $\pm$ 0.5 **a	24.3 $\pm$ 0.9 *a	24.7 $\pm$ 0.5 *a
	Testes weight	284.7 $\pm$ 33.6	417.1 $\pm$ 27.2 **a,b	474.5 $\pm$ 29.1 ***a,b	298.0 $\pm$ 34.8
	Epididymis weight	4.4 $\pm$ 0.7	6.6 $\pm$ 0.5 **a,b	7.6 $\pm$ 0.8 ***a,b	3.9 $\pm$ 0.6
	n males/n litters	24/11	41/16	15/8	30/13
Laboratory	Body weight	24.3 $\pm$ 0.8	24.6 $\pm$ 0.8	22.2 $\pm$ 1.1	–
	Testes weight	332.9 $\pm$ 26.6	420.2 $\pm$ 20.2 *a	409.9 $\pm$ 19.7 *a	–
	Epididymis weight	5.6 $\pm$ 0.4	7.7 $\pm$ 0.6 **a	8.5 $\pm$ 0.7 **a	–
	n males/ n litters	28/9	13/6	15/6	–

* $P < 0.05$; ** $P < 0.01$; *** $P < 0.001$; a indicates significant difference from "Intact Father"; b indicates significant difference from "No Father + Sebum".

Table 2. Body weight and reproductive organ weight in one month old M. auratus males treated by water and male flank gland secretion (Mean ± SEM). Body weight is in grams; testes and epididymis weights are in milligrams.

| Parameters | Treatment Procedure | | Significance of difference (t-test) |
	Water	Secretion	
Body weight	52.9 ± 1.3	52.1 ± 1.2	NS
Testes weight	390.2 ± 11.1	345.5 ± 15.4	<0.05
Epididymis weight	24.2 ± 0.7	22.4 ± 0.4	<0.05
n males/n litters	27/9	29/9	

Mongolian males housed without fathers but treated with sebum from strange male midventral glands had reproductive organs with weights similar to those housed with intact fathers.

Although midventral gland secretions from adult males had a suppressing effect on the sexual maturation of young male Campbell's hamsters, the influence of intact fathers had a stronger suppressing effect on body weight than gland secretions alone. This indicates that additional factors such as tactile stimuli or urine may also affect growth and maturation, as documented for young female P. campbell (Reasner and Johnston, 1988).

Young male M. auratus treated with flank gland secretions from strange males did not differ from control males in body weight, but had significantly lighter reproductive organs (Table 2).

Overall, the results of the present set of studies clearly demonstrate the ability of specialized skin gland secretions to suppress the sexual maturation of young male hamsters in two species which differ in gregariousness.

REFERENCES

Adams, M.G., 1980, Odour producing organs of mammals, in: "Olfaction in Mammals", D.M. Stoddart, ed., Symp. Zool. Soc., London, 45:57.
Brown, R.E., 1979, Mammalian social odors: a crital review, Adv. Study Behav. 10:103.
Brown, R.E., 1985, The rodents II: Suborder Myomorpha, in: "Social Odors in Mammals 1", R.E. Brown and D.W. Macdonald, eds., Claredon Press, Oxford.
Eisenberg, G.F. and Kleiman, D., 1972, Olfactory communication in mammals, Ann. Rev. Ecol. Syst., 3:1.
Johnston, R.E., 1973, Scent marking in mammals, Anim. Behav., 21:521.
Johnston, R.E., 1990, Chemical communication in golden hamsters: from behavior to molecules and neural mechanisms, in: "Contemporary Issues in Comparative Psychology", D. Dewsbury, ed., Sinduer, Suderland, Massachusetts, USA.
Naumov, N.P., 1973, Biological (signaling) fields and their significance in mammals, Vestnik AN SSSR, 2:55, in Russian.
Quay, W.B., 1977, Structure and function of skin glands, in: "Chemical Signals in Vertebrates I", D. Muller-Schwarze and R.M. Silverstein, eds., Plenum Press, New York.

Reasner, D.S., and Johnston, R.E., 1988, Acceleration of reproductive development in female djungarian hamsters by adult males, <u>Physiol. Behav.</u>, 43:57.

Sokolov, V.E., Vasilieva, N. Yu. and Zinkevich, E.P., 1989a, Secretion of Djungarian hamster males' midventral gland contains the factor influencing maturation, <u>Doklady</u> <u>Acad.</u> <u>Sci.</u> <u>S.S.S.R.</u>, 308:1274 (in Russian).

Sokolov, V.E., Vasilieva, N. Yu., Voznesenkaya, V.V., Lebedev, V., Teplova, E.N., and Zinkevich, E.P., 1989b, The effect of chemical signals on reproductive state and behavior of females of some rodent species, in: "V Intern. Theriol. Congr. Abstr., Pap. 2", Roum.

Thiessen, D.D., and Rice, M., 1976, Mammalian scent gland marking and social behavior, <u>Psychol.</u> <u>Bull.</u>, 83:505.

MALE CHEMOSIGNALS INCREASE LITTER SIZE IN HOUSE MICE

Fan Zhiqin[1] and John G. Vandenbergh[2]

[1]Chinese Academy of Science, Beijing, China
[2]Department of Zoology, North Carolina State University
Raleigh, NC 27695-7617

INTRODUCTION

Reproduction in house mice, Mus musculus, and many other mammals is strongly influenced by urinary priming pheromones (Bronson, 1989; Vandenbergh, 1983). In house mice, such priming pheromones synchronize estrus and block pregnancy when exposure to the stimulus occurs prior to implantation. Priming pheromones also accelerate and inhibit the onset of puberty (Whitten, 1956; Bruce, 1959; Vandenbergh 1983). The acceleration of puberty by male urine is the most relevant to this report. Both puberty acceleration and the effect on litter size described in this report involve a positive influence of male stimulation on ovulation. The onset of puberty is hastened in female mice by the presence of adult males or their urine (Vandenbergh, 1967; 1969). A fraction of male urine at about 860 Daltons has been isolated that contains the biological activity (Vandenbergh et al., 1976), although another report suggests that the active pheromone may be a mixture of isobutylamine and isoamylamine (Nishimura et al., 1989). Recent work in our laboratory failed to replicate that latter finding (Price and Vandenbergh, submitted). The presence of the puberty accelerating pheromone in male urine is androgen dependent and is influenced by the social status of the donor (Lombardi and Vandenbergh, 1976). Since a male-produced pheromone could influence the first ovulation of a female, we reasoned that it might also influence the number of eggs ovulated by an adult, cycling female mouse.

Ovulation rate and litter size in rodents is known to be influenced by a number of environmental factors such as nutrition, parity, and social status (Bronson and Rissman, 1986; Hamilton and Bronson, 1985; Steiner et al., 1983; Huck et al., 1988). In addition, litter size is a trait that is subject to selection and, thus, has a strong genetic basis (Falconer, 1953; Eisen, 1978). Pituitary gonadotropins, regulated by gonadotropin hormone-releasing hormone (GnRH) from the hypothalamus, control ovulation rate (Lipner, 1988). Further, pheromonal stimulation from the male facilitates hormonally induced ovulation in the immature female house mouse and the effect of the male chemosignal varies among inbred strains (Zarrow et al., 1970; 1973). In the vole (Microtus agrestis), exposure to the male or male urine significantly increased the number of mature follicles in comparison to isolated females (Jemiolo et al., 1980). Thus, stimuli that increase gonadotropins or GnRH can potentially increase ovulation rate. For example, artificial selection for increased litter size leads to enhanced ovarian sensitivity to gonadotropins (Durrant et al., 1980). Treating outbred female mice with gonadotropins can also increase the number of ova shed (Greenwald and Terranova, 1988), though intrauterine embryo mortality reduces the chances that superovulation will produce supernormal litter sizes (Wilson and Edwards, 1963). A variety of environmental stimuli are also known to affect the release of hypothalamic factors regulating reproduction (Gilmore and Cook, 1981), providing the possibility that a male pheromone can influence the hormonal regulation of ovulation. To test the possibility that male pheromonal stimulation can influence litter size, the two following experiments were conducted.

EFFECT OF MALE-SOILED BEDDING ON LITTER SIZE

In the first experiment, cycling adult female mice of the ICR strain were exposed to male soiled bedding during the diestrus phase of the ovarian cycle and were subsequently paired with the same male that produced the soiled bedding. The male was removed from the female's cage when a copulatory plug was detected or after 8 days had elapsed. Pregnant females were checked daily, and at birth, the number, sex, and body weight of the pups were recorded.

The latency of females to show estrus, detected by vaginal lavages, was reduced by exposure to male soiled bedding, but the soiled bedding did not affect latency to insemination of the females (Table 1).

Table 1.   The latency to estrus and to mating in days as indicated by a copulatory plug in female mice exposed to either clean bedding or bedding soiled by an adult male.

Measurement	Clean Bedding Mean	$\pm$S.E.	Soiled Bedding Mean	$\pm$S.E.
Latency to estrus	6.9	0.87	4.6*	0.49
Latency to mating	3.0	0.26	3.0	0.36
n	29		25	

*$P<0.05$

Exposing females to male soiled bedding resulted in a significant increase in both the number of implantation scars (12.2 $\pm$ 0.47 compared to 14.2 $\pm$ 0.34) and litter size (11.6 $\pm$ 0.49 compared to 13.0 $\pm$ 0.50). The number of scars represents the number of embryos implanted and exceed litter size because of intrauterine wastage. The mean weights of the pups did not differ between groups and, thus, increased litter size following male stimulation was accompanied by a higher total weight of the litters. Therefore, exposure to male soiled bedding increased litter size without a reduction in the body weight of the pups.

Females responded to males differently as a function of prior exposure to the male's chemosignals. Latency for the female to sniff the male was shorter and the females remained closer in proximity to the males in the treatment than in the control group (P<0.025).

EFFECT OF MALE URINE ON OVULATION RATE IN FEMALES

The purpose of this experiment was to determine if the urine of the male was responsible for the enhancement of ovulation and to measure ovulation rate directly.

Females were reared and handled as described above; however, in this experiment, the females were directly exposed to male urine delivered to the filtrum rather than receiving male-soiled bedding delivered to the cage. These females (n=24) were treated with male urine from diestrus to proestrus; a control group of females (n=24) received saline. Urine or saline (0.03 ml) was delivered to the filtrum daily. At proestrus females were paired with the same male that donated the urine and saline treated females were paired with a strange male. When a female displayed a copulatory plug she was anesthetized. The oviducts along with the distal portion of the uterus were removed and examined microscopically (15-30 X). The ampulla of each oviduct was punctured, the eggs extruded and counted.

Females treated with male urine ovulated 13.3 ± 0.56 eggs compared to 11.5 ± 0.63 eggs from the saline treated females (P<0.04). This result extends the first experiment by demonstrating that male urine induces an increased ovulation rate.

DISCUSSION

These results demonstrate that litter size is subject to pheromonal stimulation. Bedding material soiled by a male and his urine contains a pheromone that induces increased litter size by increasing the number of eggs ovulated by the female. This study adds another effect to the list of priming pheromone effects known to be operating in the house mouse (Bronson, 1989; Vandenbergh 1983).

Increased litter size through male stimulation could have consequences for population dynamics in small, rapidly breeding rodents. Further studies are underway to determine the physiological mechanisms involved in this effect, whether the identity of the male is important, the significance of the effect in natural populations, and if the enhancement of litter size by male stimulation occurs in other chemosensitive mammals.

REFERENCES

Bronson, F. H., 1989, "Mammalian Reproductive Biology," Univ. of Chicago Press, Chicago.
Bronson, F. H., and Rissman, E., 1986, Biology of puberty, Biol. Rev. 61:157-195.
Bruce, H. M., 1959, An exteroceptive block to pregnancy in the mouse, Nature (London) 814:105.
Durrant, B. S., Eisen, E. J., and Ulberg, L. C., 1980, Ovulation rate, embryo survival and ovarian sensitivity to gonadotropins in mice selected for litter size and body weight, J. Reprod. Fert. 59:329-339.
Eisen, E. J., 1978, Single-trait and antagonistic index selection for litter size and body weight in mice, Genetics (Princeton) 88:781-811.
Falconer, D., 1953, Selection for large and small size in mice, J. Genet. 1:470-501.

Gilmore, D., and Cook, B., 1981, "Environmental Factors in Mammal Reproduction," Macmillan, London.

Greenwald, G. S., and Terranova, P. F., 1988, Follicular selection and its control, in: "The Physiology of Reproduction," E. Knobil and J. D. Neill, eds., Raven Press, New York, pages 387-446.

Hamilton, G. D, and Bronson, F. H., 1985, Food restriction and reproductive development: male and female mice and male rats, Am. J. Physiol. 250:R370-R376.

Huck, U. W., Pratt, N. C., Labov, J. B., and Lisk, R. D., 1988, Effects of age and parity on litter size and offspring sex ratio in golden hamsters (Mesocricetus auratus), J. Repro. Physiol. 83:209-214.

Jemiolo B., Marchlewska-Koj, A., and Buchalczyk, A., 1980, Acceleration of ovarian follicle maturation of female caused by male in Microtus agrestis and Clethrionomys glareolus, Folia Biol. (Krakow) 28:269-272.

Lipner, H., 1988, Mechanisms of mammalian ovulation, in: "The Physiology of Reproduction," E. Knobil and J. D. Neill, eds., Raven Press, New York, pages 447-488.

Lombardi, J. R., and Vandenbergh, J. G., 1977, Pheromonally-induced sexual maturation in females: regulation by the social environment of the male, Science 196:545-546.

Nishimura, K., Utsumi, K., Yuhara, M., Fujitani, Y., and Iritani, A., 1989, Identification of puberty-accelerating pheromones in male mouse urine, J. Exptl. Zool. 251:300-305.

Steiner, R. A., Cameron, J. L., McNeill, T. H., Clifton, D. K., and Bremner, W. J., 1983, Metabolic signals for the onset of puberty, in: "Neuroendocrine Aspects of Reproduction," R. L. Norman, ed., Academic Press, New York, pages 183-227.

Vandenbergh, J. G., 1967, Effect of the presence of a male on the sexual maturation of female mice, Endocrinol. 81:345-348.

Vandenbergh, J. G., 1969, Male odor accelerates female sexual maturation in mice, Endocrinol. 84:658-660.

Vandenbergh, J. G., 1983, Pheromonal control of puberty, in: "Pheromones and Mammalian Reproduction," J. G. Vandenbergh, ed., Academic Press, New York, pages 99-112.

Vandenbergh, J. G., Finlayson, J. S., Dobrogosz, W. J., Dills, S. S., and Kost, T. A., 1976, Chromatographic separaton of puberty accelerating pheromone from male mouse urine, Biol. Reprod. 15:260-265.

Whitten, W. K., 1956, Modifications of the oestrus cycle of the mouse by external stimuli associated with the male, J. Endocrinol.13:399-404.

Wilson, E. D., and Edwards, R. G., 1963, Parturition and increased litter size in mice after superovulation, J. Reprod. Fert. 5:179-186.

Zarrow, M. X., Eleftheriou, B. E., and Denenberg, V. H., 1973, Sex and strain involvement in pheromonal facilitation of gonadotropin-induced ovulation in the mouse, J. Reprod. Fert. 35:81-87.

Zarrow, M. X, Estes, S. A.. Denenberg, V. H., and Clark, J. H., 1970, Pheromonal facilitation of ovulation in the immature mouse. J. Reprod. Fert. 23:357-360.

OLFACTORY CUES AND OVARIAN CYCLES

V.E. Sokolov, V.V. Voznessenskaya and E.P. Zinkevich

A.N. Severtzov Institute of Evolutionary Animal Morphology
and Ecology, 33 Leninsky Prospekt, Moscow V-71, Russia

INTRODUCTION

Ovarian cyclicity is a sensitive indicator of the physiological status of the female. Some parameters of ovarian cycles (length, regularity, etc.) can be influenced by exogenous factors and the exact values of these parameters can be reliable criteria for action efficiency. Some potential effectors of ovarian cycle parameters have been identified in a number of mammalian species, including humans (McClintock, 1971, 1983, 1984; Russell, Switz and Thompson, 1980; Cutler, Preti, Huggins, Garcia and Lowrey, 1986; Preti, Cutler, Huggins, Garcia and Lowley, 1986). Apparently, ovarian cycle parameters can be drastically changed by chemical cues originating from both male and female conspecifics. Nevertheless, knowledge about the chemical signals that influence ovarian cyclicity is quite rare.

PURPOSE

When we were investigating the influence of the boar sex pheromone androstenone (5α-androst-16-en-3-one) on ovarian cyclicity and synchrony in the pig (Zinkevich, 1988), we unexpectedly observed that the menstrual cycles of some of the women who worked very closely with androstenone were altered. This observation was the starting point of the present series of experiments which were designed to answer the following questions: 1) Does exposure to androstenone and its odorous analogs influence ovarian cycle parameters in human females? 2) If it does, why is the boar sex pheromone effective in humans? 3) Does exposure cause identical alterations for every woman? 4) Is it possible to change the menstrual cycle length of women with specific anosmia, i.e. unconsciously? 5) Does sensitivity to boar sex pheromone vary during the different stages of the menstrual cycle? 6) How long-lasting are menstrual cycle disturbances that take place under the influence of the boar sex pheromone? 7) Does repeated exposure to the action of the boar sex pheromone cause habituation? 8) Is androstenone a unique effector or are other powerful odorants also modulators of menstrual cycle length?

PROCEDURE

Experiment 1: We tested 11 women (6 could smell the sex boar
pheromone whereas 5 denied that they could smell it). Each female was
exposed to saturated vapors of the boar sex pheromone once during one 3-30
minute period. Subjects had recorded the onset of menstruation for the 12
months before this exposure and continued monitoring their cycle for 3
months following exposure. All females had exhibited regular menstrual
cycle length during the 3 months prior to exposure to the pheromone.

Experiment 2: Test subjects were 6 women with regular menstrual
cycles; their ages ranged from 27 to 31 years of age. Samples of American
mink (Mustela vison) anal sac secretions were obtained from 1 to 2 year-old
standard mink males; collection occurred mostly during the autumn slaughter
on the Pushkino farm near Moscow. We used a mixture of 10 secretion
samples in our experiments. The mink secretion samples were stored in
sealed vials refrigerator at -1° C to +4° C for up to 10 years. Each woman
was exposed to one drop of mink secretion that was applied to cotton wool;
the exposure period lasted 20-25 minutes and took place inside a non-ven-
tillated box (about 8 sq.m.). Number of exposures ranged from 1 to 6 per
female. Each female recorded the onset of menstruation for 3-7 months
before initiating the experiment and for 2-3 months after exposure.

RESULTS

Experiment 1: Menstrual cycle length was altered by androstenone in 5
of the subjects (45.6%) tested; all were sensitive to the smell of the sex
boar pheromone and considered it unpleasant. We did not observe any
menstrual cycle alterations in the remaining female who was sensitive to
the smell. The menstrual cycles of the 5 women with the specific anosmia
(non-sensitive to the smell of boar sex pheromone) did not show any
alterations in menstrual cycle length.

We found that there were at least two periods in menstrual cycle
sensitivity to boar sex pheromone stimulation: the follicular and luteal
phases. Olfactory stimulation during the follicular phase caused a
shortening of the menstrual cycle for 2-3 days; stimulation during the
luteal phase caused lengthening of the menstrual cycle for 3-4 days. All
alterations in menstrual cycle length were observed for one cycle only.
The length of the next (after exposure) cycle was regular. Thus,
androstenone can be considered a menses modulator for subjects sensitive to
this effector. Actually androstenone is the first olfactory signal that is
a modulator of menstrual cycle length in which its chemical structure has
been characterized. Probably androstenone is one of the components of
physiologically active mixtures known to change menstrual cycle parameters.
Future investigations will be designed to determine the physiological
activity of androstenone analogs that are known to have lower olfactory
thresholds and higher concentrations in saturated vapors.

Experiment 2: American mink anal sac secretion caused a lengthening
of the rat estrous cycle (Voznessenskaya, Wysocki, C.J. and Zinkevich,
1992). It also altered human menstrual cycle length (see Fig. 1). This
modulator was more effective than that reported above for androstenone.
All women could smell mink secretion odor and considered it extremely
unpleasant. We observed a shortening of the menstrual cycle of length of
1-7 days when women were exposed to mink secretion during the follicular
phase (8 exposures in 6 test subjects). When women were exposed to mink
secretion during the luteal phase, menstrual cycle was lengthened for 2-14
days (10 exposures in 6 test subjects). The length of the menstrual cycle
following exposure was similar to the cycle length reported during the pre-
exposure period. We failed to obtain any alterations in menstrual cycle

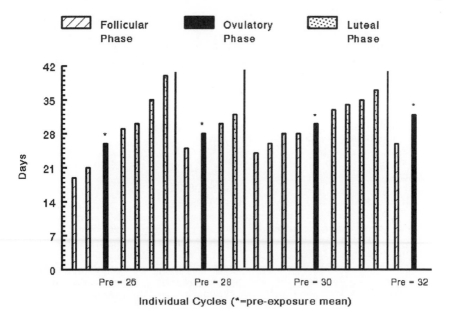

Fig.1.   Menstrual cycle lengths after exposure to mink anal sac
          secretion during the phase of the cycle indicated in the
          legend. Cycles are grouped according to the length of the pre-
          exposure cycle.

length (4 test subjects, 4 exposures) when women were exposed to mink
secretion during the peri-ovulatory period.

     Preliminary findings on the chemical composition of anal sac mink
secretion (Sokolov, Albone, Flood, Heap, Kagan, Vasilieva, Roznov and
Zinkevich, 1980) suggests that it consists of a complex mixture of sulfur
compounds, monoesters, fatty acids, amines etc. Nevertheless, in our
present experiments, samples of fresh mink secretion (kept in the refrigerator less than 2 years) were ineffective. Heating of these samples at +70
to +80Č for 30 minutes, after which its color darkened, induced physiological activity. Generally, we used aged mink secretion samples (kept up
to 10 years, -1 to +4Č in a closed vessel) and it was effective. We now
must solve the following problem: What chemical substances are responsible
for the physiological activity of aged mink anal gland secretion. These
experiments are now in progress.

CONCLUSIONS

     1. The boar sex pheromone, androstenone, is a menses modulator, but
only for women who are sensitive to its saturated vapors.

     2. Aged anal sac secretion from the American mink (<u>Mustela vison</u>) is
a menses modulator in 100% of tested subjects.

     3. There are at least two sensitive periods of olfactory stimulation
of the menstrual cycle: follicular, and luteal. Olfactory stimulation

during the follicular period caused a shortening of the menstrual cycle while during the luteal phase lengthening was observed. Peri-ovulatory stimulation was ineffective (or considerably less effective) than follicular and luteal phase olfactory stimulation by mink anal sac secretion.

ACKNOWLEDGEMENT

We thank Dr. J.A. Mennella for her comments, criticisms and assistance in the preparation of this contribution.

REFERENCES

Cutler, W.B., Preti, G., Huggins, G.R., Garcia, C.R., and Lowrey, H.J., 1986, Human axillary secretions influence women's menstrual cycles: The role of donor extract of men, Horm. Behav., 20:463.

McClintock, M.K., 1971, Menstrual synchrony and suppression, Nature, 229:244.

McClintock, M.K., 1983, Pheromonal regulation of the ovarian cycle: enhancement, suppression and synchrony, in: "Pheromones and reproduction in mammals," J.G. Vandenbergh, ed., Academic Press, New York .

McClintock, M.K., 1984, Estrous synchrony: modulation of ovarian cycle length by female pheromones, Physiol. Behav., 32:701.

Preti, G., Cutler, W.B., Huggins, G.R., Garcia, C.R., and Lowley, H.J., 1986, Human axillary secretions influence women's menstrual cycles: the role of donor extract from women, Horm. Behav., 20:474.

Russell, M.J., Switz, G.M., and Thompson, K., 1980, Olfactory influences on the human menstrual cycle, Pharmacol. Biochem. Behav., 13:737.

Sokolov, V.E., Albone, E.S., Flood, P.F., Heap, P.F., Kagan, M.Z., Vasilieva, V.S., Roznov, V.V., and Zinkevich, E.P., 1980, Secretion and secretory tissues of anal sac of the mink, Mustela vision, J. Chem. Ecol., 6:805.

Voznessenskaya, V.V., Wysocki, C.J., and Zinkevich, E.P., 1992, Regulation of the rat estrous cycle by predator odors: role of the vomeronasal organ, in: Chemical Signals in Vertebrates VI (this volume).

Zinkevich, E.P., 1988, Influence of olfactory stimuli on behavior and physiological state of domestic pig, in: Actual problems of morphology and ecology in higher vertebrates. Part II. Behavior and communication (in Russian).

OLFACTORY SIGNALS OF CONSPECIFICS STIMULATE ADRENAL FUNCTION

IN FEMALE MICE

A. Marchlewska-Koj, M. Kruczek and M. Zacharczuk-Kakietek*

Mammalian Reproduction Group, Institute of Zoology
Jagiellonian University, Cracow
*Institute of Sports, Warsaw, Poland

INTRODUCTION

The influence of conspecific chemosignals on the reproductive activity of female mice is a classic example of pheromonal interaction
between mammals (see Marchlewska-Koj, A., 1984, for review). Chemical
signals may be important in the regulation of reproductive activity in
wild populations; for example, specific olfactory cues released by
females may limit reproductive activity under high density conditions.
It has been observed that mice kept in such conditions show increased
activity of the adrenal glands and decreased circulating gonadotrophins
(Christian et al., 1965), effects which can be duplicated to some
degree by injection of ACTH (Christian, 1964). Nichols and Chevins
(1981a) found that female mice housed in groups had higher plasma
corticosterone levels than singly reared animals. However, Bronson
and Chapman (1968) observed heavier adrenal glands in individually
housed females than in grouped ones. The concept that the neurogenic
stressors associated with high density (social stress) increase the
release of adrenocorticotrophin and reduce circulating gonadotrophins
was reviewed by Bronson (1990).

We report here studies which examine the influence of male and
female chemosignals on adrenocortical activity in estrous female mice.
Also, the role of vomeronasal system in the transmission of these
chemosignals was investigated.

MATERIALS AND METHODS

The experiments were performed on 127 intact and 33 vomeronasale-
ctomized (VNX) virgin, 2- to 3- month old female and 20 male mice (Mus
musculus domesticus) from outbred stock bred in our laboratory. The
animals were kept in polyethylene cages (18x15x10 cm), with a 14:10 L:0
schedule (light on 7 am to 9 pm), and fed a standard pelleted diet of
Murigran (Bacutil, Motycz, Poland) and water. The sawdust bedding was
changed once a week.

Females were housed, one per cage, in a male-free room at least 20
days prior to the experiments. To establish estrous cycle phase, vaginal smears were taken every morning. The first day when the smear contained only cornified epithelial cells was regarded as estrus. Females

Chemical Signals in Vertebrates VI, Edited by R.L. Doty and
D. Müller-Schwarze, Plenum Press, New York, 1992

with at least two consecutive 4- or 5-day estrous cycles were used for further experiments.

The experiments were performed between 10 and 11 AM. To determine the basic level of plasma corticosterone in outbred females (CONTROL), blood samples were taken on the day of estrus. Other estrous females were randomly divided into two experimental groups. In the first experiment, the females were returned to their own cages after taking the vaginal smears and test samples were introduced. They were exposed either to 5 ml of clean bedding, bedding from a singly reared male, or bedding from 8 females that were housed together. The males were separated and the females were grouped 2 weeks before the bedding was used for the experiments. Blood samples were taken from tested females after 10 or 20 min of exposure.

In the second experiment, singly reared estrous females were transferred to either: a clean cage, a cage with bedding from 8 adult females, or a cage with bedding from a male, in each case for 20 min. After that period they were returned to their own cages and left undisturbed for the next 30, 90 or 180 min.

The third experiment was performed on VNX females in home cages. The vaginal smears were taken and females returned to their own cages. Approximately 5 ml of clean bedding, bedding from a single male, or bedding from 8 females was introduced to estrous female cages. Blood samples were taken from tested females 10 min after exposure. The basic level of plasma corticosterone in VNX females (CONTROL) was measured in blood samples taken immediately after the estrous phase has been determined.

The surgical procedure: Each mouse was anesthetized with sodium pentobarbitol and placed in a mouse stereotaxic unit. The vomer bone was fractioned and the entire vomeronasal organ extirpated from the nasal cavity (VNX). These animals were used for further experiments 2 weeks after the operation.

Blood samples from tested females were always collected by retro-orbital puncture under rapid ether anesthesia within 1-2 min. Blood was centrifuged for 5 min at 3000 rpm and plasma stored at -20°C until assayed.

Corticosterone was determined by a specific direct (non-extracted) radioimmunoassay, using rabbit antiserum against corticosterone-3-CMO-BSA.

RESULTS

The plasma corticosterone level in female mice was affected by the presence of the bedding of conspecifics during 20 min of exposure. As indicated in Table 1, 10 minutes of exposure of the individually-housed females to the bedding of singly reared males or grouped females raised the plasma corticosterone levels. The reaction was most pronounced to the male bedding. The concentration of adrenal hormone in the plasma decreased significantly during the next 10 min of exposure. Clean bedding applied to a cage did not affect the corticosterone level in singly-reared females.

The introduction of individually housed females to a cage previously occupied by a single male or 8 females evoked significant increase in the plasma corticosterone level during the first 10 min of exposure (Table 2). This was followed by a significant fall in the

Table 1. Plasma corticosterone level (ng/ml) of estrous female mice exposed to bedding of a single male or 8 females during 10 min or 20 min.
Mean $\pm$ S.E.M. Number of females shown in parentheses.

Treatment	Time of exposure (min)		ANOVA**
	10	20	
Control*	$130.8\pm3.6^{A,B}$ (6)		
Clean bedding	$205.2\pm41.8^{a,C}$ (5)	203.4 ± 35.8 (5)	$F_{(1,9)}=0.001$ NS
Male bedding	461.6 ± 52.5^{AC} (5)	272.2 ± 79.9 (5)	$F_{(1,9)}=3.938$ NS
8 female bedding	$405.0\pm76.5^{a,B}$ (5)	166.2 ± 32.9 (6)	$F_{(1,10)}=7.153$ P<0.05
ANOVA	$F_{(3,18)}=9.068$	$F_{(2,14)}=1.450$	

* For further detail see Materials and Methods.
** ANOVA for the means in horizontal row.
Those means in the vertical column marked by the same letter significantly differ: A,B,C, P<0.01; a, P<0.05. ANOVA and Duncan's a new multiple range test.

Table 2. Plasma corticosterone level (ng/ml) of estrous female mice transferred to a cage with bedding of a single male or of 8 females during 10 or 20 min.
Mean $\pm$ S.E.M. Number of females shown in parentheses.

Treatment	Time of exposure (min)		ANOVA*
	10	20	
Clean bedding	329.2 ± 78.2^{A} (5)	203.1 ± 52.6 (8)	$F_{(1,12)}=1.938$ NS
Male bedding	$716.0\pm83.6^{A,a}$ (5)	99.0 ± 41.3 (5)	$F_{(1,9)}=43.756$ P<0.01
8 female bedding	444.0 ± 29.9^{a} (5)	208.8 ± 19.0 (6)	$F_{(1,10)}=53.572$ P<0.01
ANOVA	$F_{(2,13)}=9.068$ NS	$F_{(2,17)}=1.24$ NS	

* ANOVA for the means in horizontal row. Those means in the vertical column marked by the same letter significantly differ from each other: A, P<0.01; a, P<0.05. ANOVA and Duncan's new multiple range test.

adrenal hormone level during the next 10 min. The adrenal reaction was significantly more pronounced when females were exposed to male bedding than to female bedding for the same period.

The adrenal function was also affected by stress connected with moving females from the home cage to a new one. The corticosterone level in females transferred to a clean cage for 10 min was significantly higher than in undisturbed control animals (Tables 1 and 2) ($F_{(1,10)}=5.527$; P<0.05).

Table 3. Plasma corticosterone level (ng/ml) of estrous female mice
exposed to bedding of a single male or of 8 females during 20
min and kept in the home cage for 30, 90 or 180 min.
Mean $\pm$ S.E.M. Number of females shown in parentheses.

Bedding Treatment	After exposure (min)				ANOVA[**] F
	0	30	90	180	
Clean	203.1$\pm$52.6* (8)	168.8$\pm$33.8 (5)	64.0$\pm$18.6 (5)	118.4$\pm$41.6 (5)	1.908
Male	99.0$\pm$41.3^{ac} (5)	10.4$\pm$3.9^{ab} (5)	17.6$\pm$7.4^{cd} (5)	98.2$\pm$28.2^{b} (5)	3.724
8 female	208.8$\pm$19.0^{ABC} (6)	61.4$\pm$21.0^{A} (5)	53.6$\pm$28.2^{B} (5)	51.6$\pm$8.5^{C} (5)	15.436

[*] These data correspond to the results presented in Table 2, column 2.
[**] These means in the horizontal row marked by the same letters
significantly differ from each other: A,B,C, $P<0.01$; a,b,c, $P<0.05$.
ANOVA and Duncan's new multiplerange test.

Exposure of individually – housed female mice to the bedding of
conspecifics for 20 min affected the plasma corticosterone level during
the next 180 min (Table 3). Those transferred to clean cages for 20 min
showed no statistically significant changes in adrenal function after
being returned to their own cages. However, individually – housed
females introduced for 20 min to a cage with bedding from a single
male or from 8 females showed significant decreases in corticosterone
levels 30 min following the return to their own cages. The concentration of this hormone in all experimental groups was significantly lower
than in females kept in clean cages ($F_{(2,13)}=19.409$; $P<0.01$).

As indicated in Table 4, 10 minutes of exposure to the bedding of
singly reared males raised the level of plasma corticosterone in individually housed VNX females. Similar exposure to clean bedding or to
the bedding of grouped females did not alter the adrenal hormone concentration in VNX females.

Table 4. Plasma corticosterone level (ng/ml) of VNX, estrous female
mice exposed for 10 minutes to the bedding of a single male
or 8 females.
Mean $\pm$ S.E.M. Number of females shown in parentheses.

Treatment	Time of exposure 10 min
Control[*]	189.6$\pm$22.5^{A} (5)
Clean bedding	227.4$\pm$37.6 (5)
Male bedding	345.5$\pm$26.5^{A,a} (6)
8 female bedding	187.4$\pm$38.5^{a} (5)
ANOVA	$F_{(3,17)}=6.072$

[*] For further details see Materials and Methods.
Means marked by the same letter significantly differ from each other:
A, $P<0.01$; a, $P<0.05$.

DISCUSSION

Social rank influences the adrenal function of grouped mice. Bronson (1973) reported that the plasma conticosterone level of males increased 5 to 6 fold during fighting, and returns to basal level in dominants during the next 1 to 3 days; in subordinates, this process took 6 days. Dominant female mice evidence lower plasma corticosterone levels than subdominant social partners within 30 min after grouping (Schuhr, 1987).

There is no doubt that other non-social stress factors, such as light-dark rhytms, and handling, contribute in part to the activity of the adrenal glands. Although, as shown by the present results and those of Nichols and Chevins (1981a,b), 2 weeks of daily vaginal smearing at the beginning of the light phase did not affect adrenal activity, the estrous cycle does alter adrenal function. This is manifested in mice by an increased plasma corticosterone level in proestrous and estrous phases.

An increase in plasma corticosterone has been observed in female mice after 30 min of exposure to male urine (Nichols and Chevins, 1981c). The authors suggested that the adrenals may be stimulated by an alarm pheromone present in the urine. The males were subjected to various stresses during the urine collection. The production of these types of social chemosignals by stressed animals has been described in mice (Carr et al., 1971, 1980). In the present experiments, bedding from single undisturbed males increased the corticosterone levels of females. The highest increase occurred in females transferred to a male cage. This may be the effect of two independent influences on the adrenal system: non-social stress evoked by transferring them to a new cage, and male chemosignals present in the bedding. The peak hormone level after 10 min was followed by a dramatic decrease during the next 90 min. This may reflect complex feed-back processes within the adrenal-pituitary axis.

There is evidence that in a few species of mammals, including pri-mates (Kendrich and Dixon, 1989) and the musk shrew (Suncus murinus) (Rissman and Bronson, 1989), adrenal steroids stimulate sexual behavior in females. It is possible that a male pheromone activates the adrenal glands and an acute increase in corticosterone additionally stimulates such behavior.

Our data show that exposure to the male bedding increased plasma corticosterone levels in VNX females, whereas exposure to female bedding did not. However, the male bedding evoked a more pronounced effect in non-operated females than in VNX females. This suggests that the effect is facilitated by the vomeronasal system.

REFERENCES

Bronson, F.H., 1973, Establishment of social rank among grouped male mice: relative effects on circulating FSH, LH, and corticosterone, Physiol. Behav., 10:947.
Bronson, F.H., 1990, "Mammalian Reproductive Biology," University of Chicago Press, Chicago.
Bronson, F.H., and Chapman, V.M., 1968, Adrenal-oestrous relationships in grouped or isolated female mice, Nature, 218:483.
Carr, W.J., Roth, P. and Amore, M., 1971, Responses of male mice to odors from stress vs nonstressed males and females, Psychonomic Science, 25:275.

Carr, W.J., Zunino, P.A. and Landauer, M.R., 1980, Responses by young
  house mice (<u>Mus</u> <u>musculus</u>) to odors from stress vs nonstressed
  adult conspecifics, <u>Bull. Psychonomic Soc.</u>, 15:419.
Christian, J.J., 1964, Effect of chronic ACTH treatment on maturation
  of intact female mice, <u>Endocrinology</u>, 74:669.
Christian, J.J., Lloyd, J.A. and Davis, E.D., 1965, The role of
  endocrines in self-regulation of mammalian populations, <u>Recent
  Prog.-Horm. Res.</u>, 21:501.
Kendrich, K.M. and Dixson, A.F., 1984, Ovariectomy does not abolish
  proceptive behavior cyclicity in the common marmoset (<u>Callithix
  jacchus</u>), <u>J. Endocr.</u> 101:155.
Marchlewska-Koj, A., 1984, Pheromones and mammalian reproduction <u>in</u>:
  "Oxford Reviews of Reproductive Biology" Vol. 6., J.R. Clarke ed.
  Oxford Univ. Press, Oxford, UK.
Nichols, D.J. and Chevins, P.F.D., 1981a, Effects of housing on
  corticosterone rhytm and stress responses in female mice, <u>Physiol.
  Behav.</u>, 27:1.
Nichols, D.J. and Chevins, P.F.D., 1981b, Plasma corticosterone
  fluctuations during the oestrous cycle of the house mouse,
  <u>Experientia</u>, 37:319.
Nichols, D.J. and Chevins, P.F.D., 1981c, Adrenocortical responses and
  changes during the oestrous cycle in mice: effects of male
  presence, male urine and housing conditions, <u>J. Endocr.</u>, 91:263.
Rissman, E. and Bronson, F.H., 1989, Role of the ovary and adrenal
  gland in sexual behavior of the musk shrew, <u>Suncus murinus,</u> <u>Biol.
  Reprod.</u>, 36:664.
Schuhr, B., 1987, Social structure and plasma corticosterone level in
  female albino mice, <u>Physiol. Behav.</u>, 40: 689.

REPRODUCTIVE ACTIVATION AND METABOLISM IN FEMALE VOLES

Rhonda R. Gardner, John J. Lepri and Robert E. Gatten, Jr.

Department of Biology
The University of North Carolina at Greensboro
Greensboro, NC 27412 USA

INTRODUCTION

Female prairie voles, <u>Microtus</u> <u>ochrogaster</u>, require stimulation by males to become reproductively activated and normally do not express "spontaneous" cycles of ovarian activity (Richmond and Conaway, 1969). Thirty-six to 48 hr after the onset of exposure to an unfamiliar male, virgin females become sexually receptive (Carter et al., 1980). In this period of time, follicular maturation takes place and uterine mass substantially increases. Ovulation does not occur until stimulated by copulation (Richmond and Conaway, 1969). The evolutionary factors that have resulted in this type of reproductive system have been mostly a matter of speculation. For example, it's unclear whether the voles' system is either primitive to or derived from that of rodents in which spontaneous ovulatory cycles are the norm.

The hormonal processes of reproductive activation in female voles can be initiated by the chemoreception of "male pheromones" by the vomeronasal organ (Lepri and Wysocki, 1987). Naso-nasal and naso-genital sniffing of unfamiliar males probably results in the delivery of stimulatory odors to the vomeronasal organ, initiating a neural cascade leading to the ovarian secretion of estradiol. The physiological and behavioral indicators of reproductive activation can be elicited in ovariectomized females by the administration of exogenous estradiol (Dluzen and Carter, 1979; Carter et al., 1987a). Therefore, estradiol appears to play a major role in reproductive activation.

It is not unusual for uterine mass to increase three-fold in the initial 48 hr of reproductive activation (Lepri and Wysocki, 1987; Wysocki et al., 1991). The magnitude of this change, plus the fact that spontaneous cycles of ovarian activity are rare, led us to the hypothesis that this system of reproductive activation helps females conserve metabolic energy until social and environmental conditions are favorable for the successful production and rearing of young. In an effort to characterize the energetic consequences of this reproductive pattern, we measured the metabolic rates of virgin females before and after they were exposed to males and/or male odors.

METHODS AND RESULTS

In resting, unfed mammals, basal metabolism (or basal metabolic rate, BMR) is the minimal energy expenditure necessary to sustain life (Hill and Wyse, 1989). The animal must be post-absorptive during testing to avoid metabolic costs associated with processing food. In addition, the animal must be tested in its thermo-neutral zone to assure that no energy is being used for thermoregulation. Furthermore, the animal should be resting and inactive but not sedated, drugged or asleep. Finally, the animal should be tested during the inactive phase of its daily cycle of behavioral activity and in a lightless environment (Hill and Wyse, 1989).

We measured the BMR of female voles as the average of the lowest level of oxygen consumption that was maintained for three minutes during each of four consecutive hours. Oxygen consumption was determined with an open flow system and a Beckman F3 oxygen analyzer following condition B of Hill (1972). Data were expressed in ml O_2 $g^{-1}h^{-1}$ consumed and were corrected to STP. Each female was tested in triplicate, on different days. The first test was to habituate each female to the apparatus; data from that test were not used. On the following day, each female was tested again and then placed into one of four treatment groups (n=8 females per treatment group). The four treatment groups were: i) physical contact with an unfamiliar male for 18 hr followed by 48 hr of exposure to his soiled cage-bedding (contact & odor), ii) physical contact followed by 48 hr in a clean cage (contact only), iii) 18 hr in a clean cage followed by 48 hr in a male-soiled cage (odor only) and iv) 18 hr in a clean cage followed by 48 hr in a clean cage (no exposure). All male-soiled cages had housed a male for at least one week prior to use. Following treatment, each female was tested again and then killed by an overdose of pentobarbital; the uterus was removed and weighed. All males and females were virgins less than seven months old at testing. All females weighed at least 30 g at time of testing and were housed individually for at least one week prior to being randomly placed in a treatment group.

The effect of contact and/or odor on oxygen consumption was expressed as the net change in BMR due to treatment (▲BMR), and was obtained by subtracting pre-treatment BMR from post-treatment BMR. Statistical analysis of ▲BMR was conducted using a two-way analysis of variance with a significance criterion of p=0.05. Uterine mass was standardized to mg (35 g body mass)$^{-1}$ and compared between treatment groups using t-tests (Steel and Torrie, 1980).

Females exposed to males (contact and/or odor) had higher standardized uterine mass than females with no exposure (Table 1, t30=2.06, p=0.049). The ▲BMR of females varied significantly with treatment group (Table 1, F3,28=7.44, p=0.001). The ▲BMR of females exposed to males and their odors was significantly greater than the ▲BMR of females with no such exposure (F1,27=13.99, p=0.001). Females exposed to males but without additional exposure to their odors also had a significantly higher ▲BMR than females without any male exposure (F1,27=16.15, p=0.001). However, females exposed only to male odors did not have a significantly higher ▲BMR than females with no exposure (F1,27=2.50, p= 0.126).

DISCUSSION

We found that male-induced reproductive activation in female prairie voles is associated with increased metabolic rates. These results support the hypothesis that non-activated female voles conserve a

Table 1. Mean[*] (± SEM) Uterine Mass and ▲BMR in Female Prairie Voles
Exposed or Not Exposed to Males (n=8 per group).

	TREATMENT GROUP			
	contact & odor	contact only	odor only	no contact
Uterine Mass mg (35 g body mass)$^{-1}$	$40.6^a(\pm7.1)$	$39.4^a(\pm8.4)$	$34.7^a(\pm7.1)$	$22.2^b(\pm4.4)$
▲BMR (ml O_2 g^{-1} hr^{-1})	$0.09^c(\pm0.02)$	$0.10^c(\pm0.02)$	$0.03^d(\pm0.02)$	$-0.02^d(\pm0.03)$

[*]*Means in the same row with the same super-*
script letter are not significantly different.

biologically significant amount of metabolic energy. This energy
conservation could benefit females by allowing them to channel energy into
growth, rather than unproductive estrous-cycles. Survival rates in times
of environmental stress should be higher for larger animals with more
stored energy. It is also possible that energy savings are expressed in
lower food requirements which would conserve resources, increasing the
number of animals supported by a specific foraging area. Some avenues for
further study include changes in food-intake associated with activation,
timing and total magnitude of the metabolic increase, hormone and
metabolic levels throughout activation, and studying the metabolic cost of
the entire reproductive process.

Notably, exposure to male odors without body contact did not alter
metabolic rates. These results corroborate earlier findings that odor
exposure, by itself, was less effective in stimulating lordosis than was
physical and odor contact with males (Carter et al., 1987b). Perhaps
physical contact with males initiated increased locomotor activity or some
other form of arousal that was not apparent to us because our testing
apparatus did not permit us to make behavioral observations. It seems
reasonable to hypothesize that cues from many sensory channels, such as
those experienced during physical contact with males, lead to greater
responses in the females.

There are at least three different ways to account for the increased
metabolic costs associated with reproductive activation. The first is by
increasing food intake to raise metabolism, the second is utilization of
fat stores to raise metabolism, the third involves shifting metabolism
from fat storage, cell repair, etc. to where it is required. Future work
is needed to identify the source of the energy used in activation. For
example, we plan to monitor food consumption during reproductive
activation to see if there is an increase resulting from exposure to
males. With this type of experimentation, we hope to gain more insight on

the evolution of this unique system of reproductive coordination, including its chemosensory mediation, and its neuroendocrine and metabolic mechanisms.

REFERENCES

Carter, C.S., Getz, L.L., Gavish, L., McDermott, J.L. and Arnold, P., 1980, Male-related pheromones and the activation of female voles (Microtus ochrogaster), Biol. Reprod., 23:1038-1045.

Carter, C.S., Witt, D.M., Auksi, T., and Casten, L., 1987a, Estrogen and the induction of lordosis in female and male prairie voles (Microtus ochrogaster). Horm. Behav., 21:65-73.

Carter C.S., Witt, D.M., Schneider, J., Harris, Z.L., and Volkening, D., 1987b, Male stimuli are necessary for female sexual behavior and uterine growth in prairie voles, (Microtus ochrogaster). Horm. Behav., 21:74-82.

Dluzen, D.E. and Carter, C.S., 1979, Ovarian hormones regulating sexual and social behaviors in female prairie voles, Microtus ochrogaster. Physiol. Behav., 23:597-600.

Hill, R.W., 1972, Determination of oxygen consumption by use of the paramagnetic oxygen analyzer. J. Appl. Physiol., 33:261-263.

Hill, R.W. and Wyse, G.A., 1989, "Animal Physiology", Second Edition, Harper & Row, New York.

Lepri, J.J. and Wysocki, C.J., 1987, Removal of the vomeronasal organ disrupts the activation of reproduction in female voles. Physiol. Behav. 40:349-355.

Richmond, M. and Conaway, C.H., 1969, Induced ovulation and oestrus in Microtus ochrogaster. J. Reprod. Fert., Suppl. 6:357-376.

Steel, R.G.D. and Torrie, J.H., 1980, "Principles and Procedures of Statistics: A Biometrical Approach", Second edition, McGraw-Hill Book Company.

Wysocki, C.J., M. Kruczek, L.M. Wysocki and J.J. Lepri. 1991. Activation of reproduction in nulliparous and primiparous voles is blocked by vomeronasal organ removal. Biol. Reprod., 45:611-616.

REGULATION OF THE RAT ESTROUS CYCLE BY PREDATOR ODORS: ROLE OF THE

VOMERONASAL ORGAN

V.V. Voznessenskaya[1], C.J. Wysocki[2] and E.P. Zinkevich[1]

[1]A.N. Severtzov Institute of Evolutionary Animal
Morphology and Ecology, 33 Leninsky Prospekt, Moscow V-71
U.S.S.R.; [2]Monell Chemical Senses Center, 3500 Market
Street, Philadelphia, PA 19104

INTRODUCTION

The behavior of prey can be changed dramatically by the presence of
predator odors (Stoddart, 1976; Dickman and Doncaster, 1984; Gorman, 1984).
Small mammals tend to avoid predator odors.   Faeces, urine or secretions
from some predators are effective in reducing feeding damage by snowshoe
hares, black-tailed deer (Odocoileus hemionus columbianus) and voles
(Microtus montanus and M. pennsylvanicus; Sullivan, Grump and Sullivan.,
1988; Sullivan, 1985, 1986).   2,2-Dimethylthietane from mink (Mustela
vison) causes a decrease in food consumption in snowshoe hares (Lepus
americanus) and rabbits (Oryctolagus cuniculus) (Robinson, 1990).

In the present work, the effects of anal sac secretions from mink (M.
vison) on the reproductive status of potential prey of mink was determined
under laboratory conditions.   Wistar rats with regular and short estrous
cycles were chosen as subjects.

MATERIALS AND METHODS

Test subjects were 120 Wistar rats, 4-8 months of age.   Animals were
kept under standard vivaria conditions in plastic cages, 4-5 per cage.
Lighting conditions were 14L:10D (light was switched off at 0700).   The
estrous cycle of females was determined by taking vaginal smears for 3-4
complete cycles or more.   Anal sac secretions were obtained from male mink
that were typically 1 to 2 years old at the time of an autumn kill on the
Pushkino farm near Moscow.   We used mixtures of 10 such samples.   These
mixtures were kept in cold storage (4°C) for periods ranging from 2-10
years.   We put 3 or 4 drops of mink anal sac secretion on the fresh bedding
in the cage of the females.   The duration of female-exposure was 24 hours.
After that, the females were housed in cleaned cages.   The interval between
exposures was 16-20 days.   We conducted observations after every exposure
for 16-20 days.

The vomeronasal organ was removed in a standard procedure via an oral
approach.   Removal was verified histologically.   For blockade of the main
olfactory system we used 5% $ZnSO_4$.   For control purposes, a sham vomeronasal surgery was performed or we applied 0.9% NaCl to the nasal cavity.
All animals were tested on the day after surgery or nasal irrigation.

Chemical Signals in Vertebrates VI,   Edited by R.L. Doty and
D. Müller-Schwarze, Plenum Press, New York, 1992

Table 1.   Influence of mink anal sac secretion on the length of the rat estrous cycle -- proestrous-estrous females.

Number of Females in Proestrus/Estrus	Number of Females with Extended Cycles	Percent of Females with Extended Cycles
11	1	9%
14	0	0%
4	0	0%
Σ = 29	1	3.5% (average)

RESULTS AND DISCUSSION

No alteration of the estrous cycle was noted when mink secretions were presented to females who were in proestrus/estrus (Table 1); however, mink secretion applied to the bedding of rat's in metestrus caused a delay of the next estrous by 4-12 days (Table 2). We also obtained shortening of estrous cycles by 1 day when mink secretion was applied on the bedding of females in diestrus-early proestrus. The estrous cycle length of these animals before application of mink secretion was 5 days.

We failed to reproduce the apparently robust effects of mink secretions on females in metestrus when we used samples that were stored for much less time than 10 years. As can be seen from the tables the percentage of females with extended estrous cycles under the influence of mink secretion is significantly ($p < 0.01$) higher in experiment 1 (Table 2;

Table 2.   Influence of mink anal sac secretion on the length of the rat estrous cycle -- metestrous females.

Number of Females in Metestrus	Number of Females w/ Extended Cycles	Percent of Females w/ Extended Cycles
14	12	85.7%
11	3	27.3%
21	15	71.4%
Σ = 46	30	65.2% (average)

Table 3. Influence of the vomeronasal organ on the effects of anal sac secretion on the length of the estrous cycle.

Number of Females	Exptl Group	Number of Females with Extended Cycles	Percent of Females with Extended Cycles
11	VNX	2	18%
9	VNX	0	0%
6	VNX	0	0%
Σ = 26	VNX	2	7.7% (average)
10	SHAM	3	30%
12	SHAM	7	58.3%
8	SHAM	4	50%
Σ = 30	SHAM	14	46.7% (average)

stored 10 years) than in experiments 2 and 3 (Tables 3 and 4). In experiment 2 we used secretions that were stored for 3 - 4 years and in experiment 3, they were stored less than 2 years.

Mink anal sac secretion is a complex mixture of mainly sulphur containing substances, other neutral components, and substances that are pH-dependent (Sokolov, Albone, Flood, Heap, Kagan, Vasilieva, Roznov and Zinkevich., 1980). During lengthy storage in the presence of oxygen (even at low temperatures) the production of S++ oxidation products -- powerful odorants -- is a distinct possibility. It appears that increasing the concentration of these substances may produce a physiologically active mixture.

Extirpation of vomeronasal organ reduced significantly ($p < 0.001$) the number of females that exhibited an extended estrous cycle upon exposure to anal sac secretions (Table 3). Relative to the effects of NaCl, the application of 5% $ZnSO_4$ in the nasal cavity did not eliminate this effect (Table 4). Thus, it appears that the vomeronasal organ may play a role in the effects produced on the rat estrous cycles by mink anal sac secretions.

CONCLUSIONS

Mink secretions apparently caused an extension of the estrous cycle for 4-12 days when they were applied to the bedding of rats in metestrus. However, the apparent expression of the effect may depend in part on the age of the secretions. Removal of the vomeronasal organ substantially reduced the numbers of females responding to the secretions.

Table 4. Influence of the main olfactory system on the effects of anal sac secretion on estrous cycle length.

Number of Females	Exptl Group	Number of Females w/ Extended Cycles	Percent of Females w/ Extended Cycles
13	ZnSO$_4$	5	38.5%
27	NaCl	8	29.6%

REFERENCES

Dickman, C. R., and Doncaster, C. P., 1984, Responses of small mammals to Red fox (Vulpes vulpes) odour, J. Zool., Lond., 204:521.

Gorman, M. L., 1984, The response of prey to stoat (Mustela erminea) scent, J. Zool., Lond., 202:419.

Sokolov, V. E., Albone, E. S., Flood, P. F., Heap, P. F., Kagan, M. Z., Vasilieva, V. S., Roznov, V. V., and Zinkevich, E. P., 1980, Secretion and secretory tissues of anal sac of the mink, Mustela vison, J. Chem. Ecol., 6:805.

Stoddart, D. M., 1976, Effect of the odour of weasels (Mustela nivalis) on trapped samples of their prey, Oecologia, 22:439.

Sullivan, T. P., 1985, Use of predator odours as repellents to reduce feeding damage by herbivores. ii Black-Tailed Deer (Odocoileus hemionus columbianus), J. Chem. Ecol., 11:921.

Sullivan, T. P., 1986, Influence of wolverine (Gulo gulo) odor on feeding behavior of snowshoe hares (Lepus americanus), J. Mammal., 67:385.

Sullivan, T. P., and Grump, D. R., 1984, Influence of mustelid scent-gland compounds on the suppression of feeding by snowshoe hares (Lepus americanus), J. Chem. Ecol., 10:1809.

Sullivan, T. P., Grump, D. R., and Sullivan, D. S., 1988, Use of predator odours as repellents to reduce feeding damage by herbivores. iii. Montane and Meadow Voles (Microtus montanus and Microtus pennsylvanicus), J. Chem. Ecol., 14:363.

Robinson, I., 1990, The effect of mink odour on rabbits and small mammals, in: "Chemical Signals in Vertebrates", D. W. Macdonald, D. Müller-Schwarze, S. E. Natynczuk, ed., Oxford University Press, Oxford-New York-Tokyo.

THE FUNCTIONAL PROPERTIES OF STEROID SEX PHEROMONES OF THE

MALE YELLOWFIN BAIKAL SCULPIN (COTTOCOMEPHORUS GREWINGKI)

T.M. Dmitrieva, P.L. Katsel, R.B. Valeyev,
V.A. Ostroumov, and Y.P. Kozlov

Institute of Biology, Irkutsk University
Irkutsk, Russia

INTRODUCTION

The native yellowfin Baikal sculpin develops, during the spawning period, a system of chemical communication which determines a complex of reproductive behaviors (Dmitrieva and Ostroumov, 1986). A source of sex pheromones associated with this complex of behaviors is the spawning male's urine. We have isolated and identified three sex pheromones from such urine. One of them appears to be of nonsteroidal origin -- polyene alcohol. The other two are the steroids testosterone (T) and 11b-hydroyx-testosterone (11HT)(Katsel, 1990).

A number of authors report that steroids and their conjugates can be fish sex pheromones (for review, see Stacey et al., 1987). In general, steroidal activity is mostly displayed in the induction of maturation of gonads and in the attraction of conspecifics to the source of the signal.

In the present studies, we determined how the influences of steroidal sex pheromones differ from those of nonsteroidal sex pheromones in the yellowfin Baikal sculpin and what kind of steroids can induce maturation of gonads and behavioral activity in females.

METHODS AND MATERIALS

Spawning yellowfin Baikal sculpins were caught in south Baikal. They were brought to the Irkutsk University biostation (Bay Bolshiye Koty, Lake Baikal) and put into 600 liter pools with a constant inflow of running water from the lake.

Bioassays

The behavioral influences of the steroids were studied in mature females during day light using two-choice preference test tanks (40 l; 1-3 l/minute flow). The water was pumped to the tanks from the lak and was at the same temperature as the lake. The fish were first adapted to the tanks for 48 hours. One hundred ml of aqueous solutions of steroids (-11 log M) were added to one of the two tank compartments and the sequence of behavioral reactions observed. These reactions included a whole complex of stereotypic elements of sexual behavior: female "testing motions"

Chemical Signals in Vertebrates VI, Edited by R.L. Doty and
D. Müller-Schwarze, Plenum Press, New York, 1992

(periodic wide opening of the mouth, rotational movements of the thoracic fins, and turns toward a stimulus compartment); movement toward a stimulus compartment; "courting dance" behaviors (snake-like body movements in the opposite direction, head bending towards the bottom); "nest entering-like movements" (head bending towards the bottom and short movements forward with short moments of standing still); and "egg laying-like" movements (female attempts to turn upside down)(Dmitrieva et al., 1988).

The experiments on stimulation of maturation in females were carried out in 40 1 tanks (0.2 1/min flow). The test solutions (100 ml, -11 log M) were added to tanks containing 6 nonmature females every four hours during a 24 hour period. After that all females were stripped and the number of maturing oocytes was determined using morphologic signs (Katsel, 1990). As a control, an aqueous solution of urine (-8 log % volume) was used.

RESULTS AND DISCUSSION

The sexual behavior of mature females was significantly altered by the androgens testosterone (T) and androstendione (AD)(ps < 0.01). The estrogens (17b-estradiol, estrone), corticosteroids (corticosterone, cortisone), and gestagens (progesterone, 17a-hydroxyprogesterone, 5b-pregnanetriol) had no statistically significant effects. We extended the series of steroids being tested and tried to find out which androgens can provoke female sexual behavior. 11b-hydroxytestosterone (11HT), adrenosterone (AT), 5a-dihydrotestosterone, androstan-3,16-diol and 5b-dihydroepitestosterone were evaluated. It was found that only Δ^4-3-keto C19-steroids such as T, 11HT, AD and AT possessed pheromonal activity (ps < 0.01).

It should be pointed out that the behavioral reaction of females to Δ^4-3-deto C19-steroids differed from the reactions to urine (U) and to non-steroidal pheromones (NP). After the introduction of U and NP, the duration of effects did not exceed five minutes. The impact of steroids was longer (more than 10 minutes) and elements of a "courting dance" and "nest entering-like movements" were observed. At the end of the behavioral sequence, female activity increased considerably. It should be pointed out that only the females who had ovulated showed this reaction at the end of the spawning period. Females who had not ovulated did not demonstrate these behaviors in response to the steroids. The effect of NP was mainly to attract females to the compartment to which the stimulus had been added, regardless of whether or not they had ovulated. NP was identified in the urine of males with immature testes even at the beginning of spawning (Katsel, 1990). Thus, NP likely functions as a sex attractant. Steroidal pheromones secreted by the male at the end of spermatogensis probably function as aphrodisiacs, facilitating spawning.

The male yellowfin Baikal sculpin is polygamous. Up to five females in consecutive order can spawn in a "nest" with one male. Therefore, the female's ovaries should not mature simultaneously but just before spawning; that is, there should be synchronoous maturation. Data in a number of papers show that steroids can function as primer pheromones (e.g., Stacey et al., 1987). We have also observed that, during the biotest of 11HT isolated from urine, that the females used in the biotest layed oocytes in 10 to 24 hours. We studied the effect of Δ^4-3-keto C19-steroids on the sexual maturing of female oocytes (Fig. 1). AT caused 90% maturation, whereas the other C19-steroids were not effective. Thus, 11HT can probably be considered a primer pheromone. AT was not identified in urine.

On the basis of these results, a model of the functioning of sex pheromones of the male yellowfin Baikal sculpin during the reproductive period can be developed. At the beginning of spawning, the males first come to

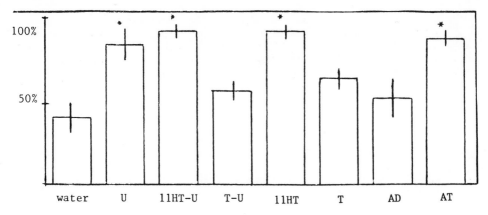

Fig. 1. The effect of Δ^4-3-keto C19-steroids on the maturing
oocytes of yellowfin Baikal sculpin females. Abbreviations: U indicates isolated from urine; 11HT -
11b-hydroxytestosterone; T - testosterone; AD - androstendione; AT - andrenosterone. *indicates significantly
different from water at 0.05 probability level.

the spawning grounds and occupy "nests". Females come a little bit later
and keep to the edge of the spawning grounds until the beginning of spawning. The males are not seen at this time, being hidden behind large stones
at a considerable distance from the females. To attract females, the males
secrete polyene alcohol which is present in urine in the early stages of
spawning. The females, attracted to a "nest", do not enter it but remain
close by. Probably at this moment, chemical signals from each fish help
to synchronize the reproductive maturation of both sexes. A primer pheromone, 11HT, secreted by the male at the end of spermatogensis, likely ini-
tiates the maturation of the oocytes. In turn, sex hormones, the synthesis
of which increases as a result of ovulation, also beome involved. They
help to alter the male's steroidal pheromones in a way that activates the
female's sexual behavior, culminating in "nest" entering and spawning.

REFERENCES

Dmitrieva, T.M., and Ostroumov, V.A., 1986, The role of chemocommunication
    in the organization of reproductive behavior of the yellowfin Baikal
    sculpin, Biol. Nauki, 10:38.
Dmitrieva, T.M., Katsel, P.L., Valeyev, R.B., Ostroumov, V.A., and Kozlov,
    Y.P., 1988, The separation of male sex pheromones of the yellowfin
    Baikal sculpin, Biol. Nauki, 6:39.
Katsel, P.L., 1990, The identification and functional properties of male
    sex pheromones in yellowfin Baikal sculpin, Sci.D. Thesis, Irkutsk.
Stacey, N.E., Sorensen, P.W., Dulka, J.G., Van Der Kraak, G.J., and Hara,
    T.J., 1987, Teleost sex pheromones: recent studies on identity and
    function, in: Proc. 3rd Internat. Symp. Reprod. Physiol. Fish, D.R.
    Idler, L.W. Crim, and J.M. Walsch, eds., Memorial Univ. Press, St.
    John's Newfoundland.

SECTION SIX

CHEMICAL REPELLENTS AND CHEMOSENSORY AVERSIONS

AVIAN CHEMICAL REPELLENCY: A STRUCTURE-ACTIVITY APPROACH AND

IMPLICATIONS

Pankaj S. Shah[1], J. Russell Mason[2] and Larry Clark[2]

[1] Monell Chemical Senses Center, 3500 Market Street, Philadelphia, PA 19104; [2]USDA/APHIS/S&T/DWRC, c/o Monell Chemical Senses Center, 3500 Market Street, Philadelphia, PA 19104

INTRODUCTION

Until recently, the discovery of avian sensory repellents has been empirical (Mason, Adams and Clark 1989). However, recent studies in our laboratory have shown that many avian repellents have similar perceptual and structural properties (Mason et al. 1989; Mason Clark and Shah 1991; Clark and Shah 1991; Clark, Shah and Mason 1991; Shah, Clark and Mason 1991). For example, methyl anthranilate, which has a grapy odor, is repellent to birds (Kare and Pick, 1960). Ortho-aminoacetophenone has an odor and structure similar to that of methyl anthranilate, differing only in the substitution of a ketone for an ester group (Mason et al. 1991). Behavioral tests of this aminoacetophenone isomer showed that it is at least an order of magnitude more repellent to birds than methyl anthranilate (Clark and Shah 1991). These similarities in structure and function prompted us to undertake a series of studies to elucidate a predictive model of chemical structure-activity (Clark and Shah 1991; Clark et al 1991; Shah et al. 1991). As a consequence of these studies we hypothesize a model where the following structural features appear to be important:

(1) A phenyl ring with an electron donating or a basic group is central to repellency;

(2) An electron withdrawing group in resonance with a basic group decreases the repellency (as well as the toxicity) of a substance. These effects are pronounced when the groups are ortho to one another;

(3) The presence of an acidic group decreases repellency;

(4) The presence of an H-bonded ring or a covalently bonded fused ring that possesses the required features (e.g., electron donating and withdrawing groups ortho to each other) can enhance repellency, but is not essential;

(5) Steric hindrance can overpower the features described above, and can weaken the effectiveness of potentially aversive substances.

Chemical Signals in Vertebrates VI, Edited by R.L. Doty and
D. Müller-Schwarze, Plenum Press, New York, 1992

This model was arrived at by examining analogues of benzoic acid, its esters (anthranilates) and acetophenones from various perspectives. In particular, we studied (a) the relationship of repellency to H-bonded rings fused or adjacent to the phenyl ring; (b) the effect of increased or decreased electron donation on the phenyl ring; (c) the effect of increased or decreased electron withdrawal; (d) the effects of a fused heterocyclic ring that incorporates polar groups; (e) the importance of steric effects; and (f) the importance of the phenyl ring, per se. Our studies were thus designed to elaborate the steric and electronic factors involved in behavioral aversiveness from a chemical point-of-view. The data do not address biochemical and/or physiological pathways that could mediate responding.

CHEMICAL PARAMETERS OF BEHAVIORAL AVOIDANCE

Relationship of repellency to an H-bonded ring fused or adjacent to the phenyl ring

We studied the importance of this parameter by examining positional isomers of aminoacetophenones [i.e., 2 (ortho), 3 (meta), 4 (para)] as well as alpha-aminoacetophenone (here, the amino group is on the side chain) (Figure 1). Our data showed that the ortho-isomer was more active than the para-isomer, and that the para-isomer was more active than the meta-isomer. Alpha-aminoacetophenone repellency was intermediate to ortho- and para-aminoacetophenone (Mason et al. 1991). Examination of methyl anthranilate analogues showed a similar trend (Clark et al. 1991). To further explore the possible role of the H-bonded ring, we next substituted the amino group with hydroxyl and methoxyl groups. Although slightly less active than the amino derivatives, the methoxy derivatives were superior when compared to their hydroxy counterparts (Clark and Shah 1991). Here too the 2-substituted isomers were more active.

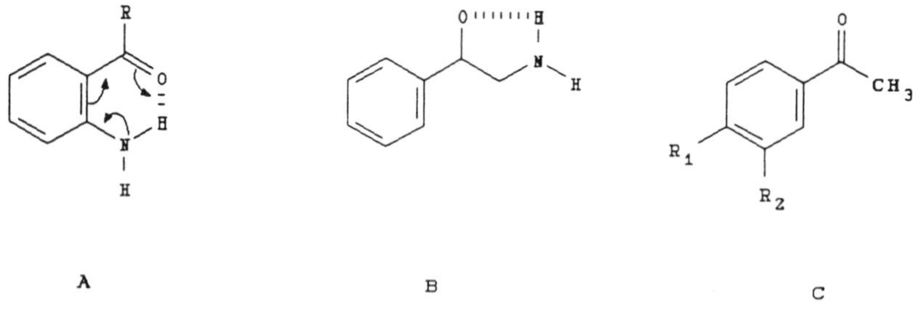

Figure 1. Positional isomers of aminoacetophenones. The arrows indicate electron withdrawal and donation paths. (A) When R = CH3, orthoaminoaceto-phenone; R = OCH3, methyl anthranilate. (B) Alpha-aminoacetophenone. (C) R1 = H and R2 = NH2, 3-aminoacetophenone; when R1 = NH2 and R2 = H, 4-aminoacetophenone.

Effect of increased or decreased electron density on the phenyl ring

Increased electron density on the phenyl ring can be obtained in two ways, namely through the addition of electron donating groups or by decreasing electron withdrawal. Conversely, electron density can be diminished by the addition of electron withdrawing groups or the removal of electron donating groups. Methyl groups donate electrons via inductive effect. When N-methyl-methylanthranilate (dimethyl anthranilate) was tested (Clark et al. 1991), it was slightly more repellent than methyl anthranilate. This finding,

292

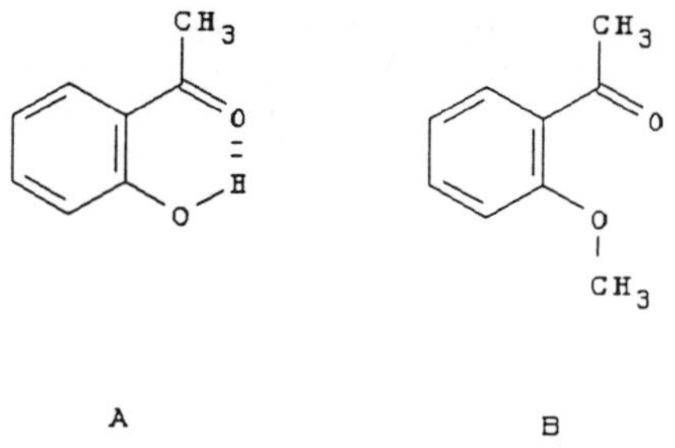

Figure 2. (A) 2-hydroxyacetophenone. (B) 2-methoxyacetophenone.
If an H-bonded ring (i.e., an intramolecular H-bond) was an essential
feature for repellency, then hydroxy derivatives would have been more
effective than their methoxy counterparts (Figure 2). Since this result
was not obtained, we examined the role played by the basic nature of the
substituent groups.

together with the data for methoxy and hydroxy aminoacetophenone derivatives, showed that electron donation (or basicity) of the substituent was
important. But still, the meta isomers were almost inactive, suggesting
that the resonance of this basic group with the carbonyl or carboxyl moiety
was affecting repellency. Also, the ortho-isomer was still superior to the
para-isomer, suggesting a possible ancillary role played by the H-bonded
ring. When electron donation by the amino group, was compensated by a
nitro group, in resonance (Clark et al. 1991), the compound (2-amino-5-
nitro benzoic acid) was not repellent to birds.

Similarly, when the electron withdrawing sulfonyl group was placed
between the amino group and the phenyl ring, as in the case of orthocarbo-
ethoxy-benzene sulfonamide, no aversiveness was observed (Clark et al.
1991). Alternatively, when 2'-amino-4',5'-dimethoxy-acetophenone was
examined, no substantial increase in the repellency was observed. Thus,
increased electron richness by dimethoxy substitution did not influence
repellency. However, increased electron withdrawal decreased aversiveness.
Together, these results suggest an upper limit to enhanced repellency
associated with electron enrichment of the phenyl ring. When the electron
donating group was removed as in the case of acetophenone or methyl benzoate, the compounds were found to be ineffective (Clark et al. 1991; Shah et
al. 1991). When an amino group was reintroduced at a different place (away
from the phenyl ring), as in alpha-aminoacetophenone, repellency was partially restored (Mason et al. 1991). This reintroduction of the amino
group also had a limited effect since benzamide was ineffective as a repellent although theoretically, the amino group in that position should have
compensated the electron withdrawal by the carbonyl group.

Interestingly, when the electron withdrawing carbonyl group was removed from the phenyl ring, toxicity was observed. Tests of 2-amino-benzyl
alcohol showed that the compound was moderately repellent at low concentrations, but lethal at higher levels. Thus the electron withdrawing group may
be associated with decreased repellency, as well as decreased toxicity.

Effect of fused heterocyclic rings

Since the fused H-bonded ring structure was found to play at least an
ancillary role, it was of interest to find the effect of covalently bonded
rings. Two compounds with covalently bonded rings were initially tried,

viz. isatoic anhydride and 4-ketobenztriazene (Clark et al. 1991). Both of these substances have the required electron donating amino and electron withdrawing carbonyl functionalities ortho with respect to each other (Figure 3). The difference between the two lies in the interlinking atoms that make the covalently bonded heterocyclic ring. In the case of isatoic anhydride, the amino group is connected to a carbonyl group and thus reduces the electron donation to the phenyl ring, whereas in 4-ketobenztriazene the amino group is connected to the carbonyl group via two nitrogen atoms. Apart from keeping the possible electron donation by the amino group the same, the nitrogen atom at position 3 can potentially reduce the electron withdrawal by the carbonyl group via internal compensation. As predicted, isatoic anhydride was found to be ineffective and 4-ketobenztriazene was as active as ortho-aminoacetophenone.

Anthraquinone has been registered as an effective avian repellent. If this substance was truly effective, then the result would contradict our understanding of avian repellency. Anthraquinone contains two electron withdrawing carbonyl groups ortho to each other, linked via two carbon atoms which form a part of another phenyl ring. When we evaluated the repellency of anthraquinone, however, we found it to be entirely ineffective. The fact that a registered bird repellent would be ineffective is not surprising (e.g., Dolbeer, Link and Woronecki 1988). However, we deduce that although the effectiveness of a stimulus cannot be increased (compared to orthoaminoacetophenone) by a covalently bonded fused ring structure, it can be decreased by placing electron withdrawing groups in the fused ring.

Steric effects

Steric effects were studied using a variety of anthranilate derivatives (4). We observed that increasing the bulk on either the amino group or the ester group decreased repellency if the group involved could potentially decrease the pi-orbital overlap by forcing them partially out of plane. Thus N,N-diisobutyl-methyl anthranilate, as well as phenethyl anthranilate, were ineffective repellents, whereas methyl anthranilate, dimethyl anthranilate, and ethyl anthranilate showed good activity. One compound in this series, namely linalyl anthranilate, showed good repellency, although the group itself was quite large. This finding might be due to planar positioning of the initial sp2 hybridized atoms of the linalyl group which would keep its chain away from the amino group.

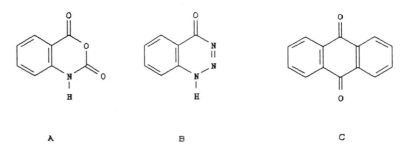

Figure 3. Structures for (A) isatoic anhydride, (B) 4-ketobenztriazine, (C) anthraquinone.

The importance of the phenyl ring

When the phenyl ring from methyl anthranilate is removed, the resulting compound is an ester derivative of beta-alanine. Hence, we tried
beta-alanine itself as a potential repellent. It had no apparent aversive
quality. These data suggest two possibilities. The first possibility is
that a phenyl ring may be required for repellency. Alternatively, it may
be that the position of the two polar groups must be rigid. In the latter
case, no phenyl ring need be present. We are currently exploring these two
possibilities.

ON-GOING RESEARCH

We are applying our model to the GRAS list of substances that are
approved by the U.S. Food and Drug Administration for human consumption.
First, this provides opportunities to evaluate the predictive capacities of
our model. Also, from a practical perspective, our expectation is that
substances will be more easily registered by the U.S. Environmental Protection Agency as avian repellents if they are already approved for use in
human and/or animal feeds. To this end, we tested vanillin, a widely used
flavoring agent, and two derivatives, vanillyl alcohol and veratryl alcohol
(5). These substances were chosen as stimuli because they permitted us to
examine the roles played by electron donating and withdrawing (polar)
groups and their contribution to resonance in the absence of a confounding
H-bonded ring at the ortho position (Figure 4).

In vanillin, the electron-withdrawing group is in resonance with the
hydroxyl group. This hydroxyl group and the methoxyl group form a 5-mem-
bered H-bonded ring. Vanillyl alcohol entirely eliminates the electron
withdrawal by resonance. Thus, comparison of data for vanillin and vanillyl alcohol would show the importance of the electron-withdrawing group.
Examination of veratryl alcohol would show the role of electron-donating
groups and the importance of the H-bonded ring for repellency because it
has neither an electron-withdrawing carbonyl group, nor an H-bonded ring.
Instead, it has a methoxyl group (rather than hydroxyl group), resulting in
increased electron-donation to the phenyl ring. Overall, then, we predicted that veratryl alcohol should be a more effective repellent than either
vanillyl alcohol or vanillin. The results showed that veratryl alcohol was
significantly more repellent (Shah et al. 1991).

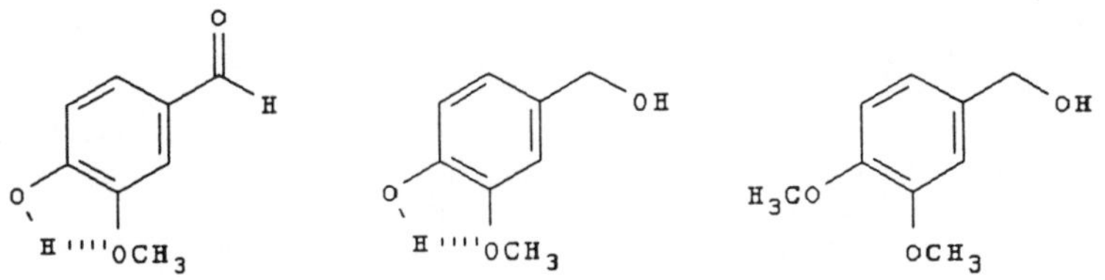

Figure 4. Structures for (A) vanillin, (B) vanillyl alcohol, (C) veratryl
alcohol.

We also are applying our model to GRAS listed substances such as
derivatives and analogs of coniferyl benzoate, a naturally produced material known to deter the feeding activity of ruffed grouse (Bonasa umbellus)
and other birds (Jakubas, Mason Shah and Norman 1991). Coniferyl benzoate

may be regarded as a derivative of cinnamyl alcohol. Although these compounds are very dissimilar to the compounds described above (Figure 5), we speculated that the electronic effects could still be used to predict repellency. Seven different compounds were tested in this series of experiments. The results obtained were broadly consistent with our hypothesis: dimethoxy cinnamyl alcohol and its benzoate were the best stimuli whereas coniferyl alcohol and its benzoate were the worst, showing the negative effects of the phenolic hydroxyl group.

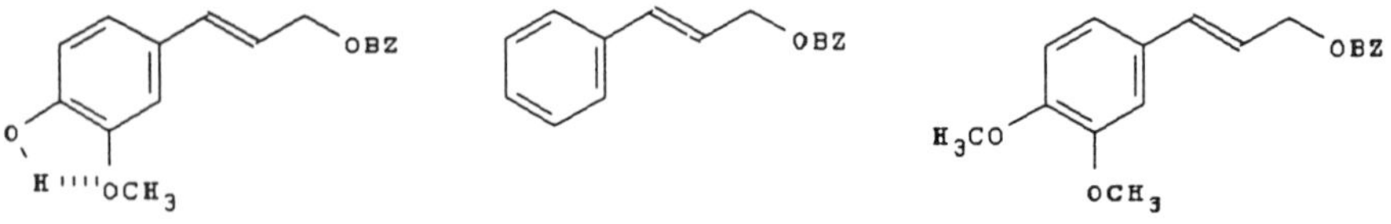

Figure 5. Structures for (A) coniferyl benzoate, (B) cinnamyl benzoate, (C) 3,4-dimethoxy cinnamyl benzoate.

CONCLUSIONS

From our experiments, it appears that potential repellents can be predicted on the basis of a few molecular attributes. Although this model provides a good qualitative correlation of aversiveness to basic chemical structure, it is unable to predict aversiveness in quantitative terms, e.g., shape of dose-response curves. For this reason, we have begun to use molecular modelling to examine the structure of compounds more closely. Also, we are testing different types of chemical substances to evaluate and refine the accuracy of the present model.

REFERENCES

Clark, L. and Shah, P.S. 1991, Nonlethal bird repellents: In search of a general model relating repellency and chemical structure. J. Wildl. Manage. 55:539-546.

Clark, L., Shah, P.S. and Mason, J.R. 1991, Chemical repellency in birds: relationship between structure of anthranilate and benzoic acid derivatives and avoidance response. J. Exptl. Zool. 260:in press.

Dolbeer, R.A., Link, M.A. and Woronecki, P.P. 1988, Napthelene shows no repellency for starlings. Wildl. Soc. Bull. 16:62-64.

Kare, M. R. and Pick, H.L. 1960, The influence of the sense of taste on feed and fluid consumption. Poult. Sci. 39:697-706.

Jakubas, W.J., Mason, J.R., Shah, P.S. and Norman, N. 1991, Avian feeding deterrence by coniferyl benzoate: a structure-activity approach. Ecol. Appl., in press.

Mason, J.R., Adams, M.J. and L. Clark. 1989, Anthranilate repellency to starlings: Chemical correlates and sensory perception. J. Wildl. Manage. 53:55-64.

Mason, J.R., Clark, L. and Shah, P.S. 1991, Ortho-aminoacetophenone repellency to birds: perceptual and chemical similarities to methyl anthranilate. J. Wildl. Manage. 55:334-340.

Shah, P.S., Clark, L. and Mason, J.R. 1991, Prediction of avian repellency from chemical structure: The aversiveness of vanillin, vanillyl alcohol, and veratryl alcohol. Pest. Biochem. Physiol. 40:169-175.

EFFECTS OF NERVE GROWTH FACTOR ON THE RECOVERY OF CONDITIONED TASTE

AVERSION IN THE INSULAR CORTEX LESIONED RATS

Federico Bermúdez-Rattoni, Martha L.Escobar, Ana Luisa Piña,
Ricardo Tapia, Juan Carlos López-García and Marcia Hiriart

Instituto de Fisiología Celular
Universidad Nacional Autónoma de México
Apdo. Postal 70-600
México, D.F.

INTRODUCTION

Recent research in our laboratory has focused on the influence of
brain grafts on the recovery of learning ability in cortical-lesioned
animals. The findings suggest that graft maturation and/or cholinergic
activity may play a role in the graft-mediated behavioral recovery
following brain lesions.

THE BEHAVIORAL MODEL: CONDITIONED TASTE AVERSION

Our studies have used conditioned taste aversion paradigms (CTA) to
examine the effects of brain grafts on learning in cortically lesioned
rats. Animals can acquire aversions to flavors if the taste stimulus (CS)
is followed by gastrointestinal illness (Garcia et al., 1985) as an unconditioned stimulus (US). Taste is readily associated with illness and can
be observed after a single taste-illness experience. Flavor-illness
association has been demonstrated in several laboratories and in different
animal species. A major advantage of this model in the study of the
neurobiology of learning and memory is the knowledge of the neural pathways involved. The pathways for CTA have been established with the use of
anatomical, electrophysiological and behavioral methods (for review see
Garcia et al., 1985; Kiefer, 1985).

THE INSULAR CORTEX

The insular cortex (IC) has been referred to as gustatory or visceral
cortex, since it receives taste and visceral information from the ventromedial nucleus of the thalamus that, in turn, receives afferences from the
pontine parabrachial nucleus, which is a second-order relay for taste and
visceral information (for review see Norgren, 1989; Kiefer, 1985). The
anatomical connections of the IC clearly suggest that this brain region
may play a role in integrating and possibly storing visceral information
(see Garcia et al., 1985). Moreover, it has been postulated that the IC
receives convergence of limbic afferences and primary sensory inputs that
is not seen within any other sensory area in the cortex (Krushel and van
der Kooy, 1988). Among the IC connections that may be important for
memory processing are those with the limbic system; i.e., amygdala, dorsomedial nucleus of the thalamus and prefrontal cortex (Krushel and van der

Chemical Signals in Vertebrates VI, Edited by R.L. Doty and
D. Müller-Schwarze, Plenum Press, New York, 1992

Kooy, 1988; Bermúdez-Rattoni and McGaugh, 1991; Escobar et al., 1989; van der Koy, 1984).

Although the IC connectivity is fairly well known, the neurotransmitters of the IC have not been extensively studied. In this regard, we have recently demonstrated that slices taken from the IC are able to release labeled GABA, ACh and glutamate but not dopamine. Additionally, there are significant glutamic acid decarboxylase, choline acetyl transferase and acetylcholinesterase activities in IC homogenates (Lopez-Garcia et al., 1990a).

Several studies have shown that the insular cortex area is involved in mediating the associative aspects of taste response, but is not involved in the hedonic responses to taste. Rats lacking the IC are impaired in acquiring and retaining taste aversion. That is, when the IC lesions are made either before or after acquisition of the CTA animals do not show taste aversion. However, the hedonic responses of lesioned IC rats appear to be normal: like normal rats, IC-lesioned animals prefer sucrose as well as low concentrations of sodium chloride over water and reject quinine and acid solutions. Also, it is known that taste responsiveness remains intact even in decerebrate rats (Kiefer, 1985; Braun, et al., 1982; Grill & Norgren, 1978).

RECOVERY OF FUNCTIONS BY FETAL BRAIN GRAFTS

Functional behavioral recovery from brain injury has recently been demonstrated using the fetal brain transplant technique in adult mammalian brains. It has been established that transplanted neurons differentiate and make connections with the host brain (Bjorklund and Stenevi, 1985). We recently showed that the fetal brain transplants produced a significant recovery in the ability of IC-lesioned rats to acquire a CTA. The possibility of spontaneous recovery was excluded, because the animals with IC lesions that did not receive transplants were unable to acquire the CTA 8 weeks after the transplantation, even with two series of acquisition trials (Bermúdez-Rattoni et al., 1987). In contrast, animals with lesions in the amygdala showed spontaneous recovery eight weeks after the lesion was induced. Similar spontaneous recovery of performance in an alternation task has been found in animals tested 6 weeks after having received large cortical ablations (Dunnett et al., 1987).

Elsewhere we have discussed in detail these functional differences between the amygdala and IC (Bermúdez-Rattoni et al., 1987). One possible explanation is that amygdala lesions may have resulted in reorganization of other elements in the neuronal network. Another plausible explanation is that for taste-aversion learning the IC may be a permanent memory store, whereas the amygdala may only serve to influence an initial step in the storage of CTA (Bermúdez-Rattoni et al., 1987; Bermúdez-Rattoni et al., 1989).

We have further shown that the degree of functional recovery induced by fetal brain tissue grafts depends on the place from which graft tissue was taken. Animals which received homotopic but not occipital cortical tissue recovered the CTA. Further, the animals that received either tectal heterotopic tissue or no transplant showed no behavioral recovery. Results based on horseradish peroxidase (HRP) histochemistry revealed that cortical, but not brain-stem grafts, established connections with amygdala and with the ventromedial nucleus of the thalamus (Escobar et al., 1989). Biochemical analyses revealed that IC fetal grafts released GABA, ACh and glutamate in response to K+ depolarization. In contrast, occipital grafts released labeled GABA and glutamate, but not ACh (Lopez-Garcia et al., 1990b). These results suggest that cholinergic transmission is important for CTA and that ACh may play a role in graft-mediated behavioral recovery.

The results discussed above indicate that some morphological recovery is necessary for the acquisition of taste aversion. However, it has been suggested that structural and morphological integrity of fetal brain grafts may not be essential for behavioral recovery after cortical brain injury (Dunnett at al., 1987; Kesslak et al., 1986). These investigators have speculated that brain injury or brain grafts induce a release of neurotrophic factors that can reactivate neural function and/or prevent injury-induced degeneration in the damaged host brain (Dunnett et al., 1987: Kesslak et al., 1986). Labbe and coworkers (1983) reported that rats lesioned in the frontal cortex and given cortical transplants were able to learn a spatial alternation task in fewer trials than lesioned controls or rats with cerebellar implants. The recovery effects were seen just one week after transplantation. In this regard, Dunnett and coworkers (1987) found that neocortical grafts produced short-lasting improvement in the T-maze alternation performance. They concluded that the short-lasting effects were attributable to diffused influences of the embryonic tissue on the lesioned host brain instead of a reconnection of the damage circuits. In contrast, findings from our laboratory clearly suggest that, with CTA, recovery of function increases with time after transplant.

TIME-DEPENDENT RECOVERY

In a series of experiments in our laboratory, rats with lesions of IC showing disrupted taste aversion received neocortical transplants and were retrained at 15, 30, 45 and 60 days after transplantation. The behavioral results showed almost complete functional recovery at 60 days, slight recovery at 45 and 30 days and a poor recovery at 15 days post-graft. HRP histochemistry revealed that at 15 days there were no HRP labeled cells in the ventromedial nucleus or into the amygdala. At 30, 45 and 60 days post-graft, there were increasing connections, almost as many as those seen in the controls, with the thalamus and with the amygdala. The behavioral recovery was correlated with increased acetylcholinesterase activity, detected histochemically, and with morphological maturation, revealed by Golgi staining (Fernández-Ruiz et al., 1991). The possibility that neuro-trophic factors alone may be involved in the functional recovery is unlikely, because it is necessary to wait at least 30 days to see any recuperation. Therefore, such findings suggest that if neurotrophic factors are involved, they need to be associated with cortical homotopic grafts and/or with some time dependent factor essential for producing functional recovery. The implication of these series of experiments is that, for the IC and CTA, functional recovery is related to the morpholo-gical maturation of the graft.

THE ROLE OF THE NERVE GROWTH FACTOR IN FUNCTIONAL RECOVERY

Several lines of evidence have demonstrated that the nerve growth factor (NGF) produces a significant regeneration, regrowth and penetration of cholinergic axons in the hippocampal formation (Gage 1990; Hagg et al., 1990). Chronic perfusion of NGF, in fimbria-fornix lesioned animals with severe learning impairments, produce functional and anatomical recovery (Varon et al., 1989). It has also been demonstrated that chronic intrace-rebral infusion of NGF improves memory performance in cognitively impaired aged rats (Gage and Bjorklund, 1986). Nevertheless, long-term impairments by application of NGF in combination with intrahippocampal septal grafts have been observed (Pallage et al., 1986).

In another series of experiments, we have assessed the role of NGF on the recovery of acquired conditioned taste aversions in cortical lesioned rats. For CTA, the animals were habituated for 10 days to drink tap water daily during 5 min sessions. After surgery, the animals were given one acquisition trial and two testing trials conducted every fourth day with baseline access to distilled water on the intervining days. The acquisition day involved the presentation of 0.1 M LiCl instead of water. It has been demonstrated that the taste of LiCl can readily be aversively conditioned to its gastric aftereffects, since it is also the agent inducing illness (Nachmann, 1963). The tests consisted of the presentation of 0.1 M of NaCl instead of the LiCl with three water intake baseline measures in between (see Bermúdez-Rattoni et al., 1987; Fernández-Ruiz et al., 1991). Rats cannot discriminate between the flavors of NaCl and LiCl (Nachmann, 1963). Three groups of rats showing disrupted taste aversions due to IC electrolytic lesions were grafted as follows: the first group received fetal (15E) cortical graft + NGF (ICNGF), the second group received gelfoam + NGF (NGF), and the third group received fetal cortical graft alone (IC). Unoperated animals were used as a control group (CON). The three grafted and control groups were subdivided in three subgroups (ICN, ICNGF, NGF and CON) that were retrained for CTA at 15-, 30- and 60-days post-graft respectively. As expected, the control groups showed strong taste aversions in all the post-graft times (see Fig. 1). The IC group showed a disrupted taste aversion at 15 days post-graft. During the 30 and 60 post-transplantation day, the IC groups showed recovered taste aversion when compared with the CON group. The ICNGF group showed a significantly recovered ability to acquire the taste aversions at the three post-graft times when compared with the NGF alone group. These results indicate that the application of NGF with the cortical graft accelerate the functional recovery up to 15 days after grafting.

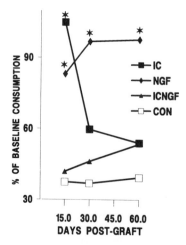

Fig. 1. The amount of NaCl consumed by control (CON), Insular Cortex (IC), Insular Cortex + NGF (ICNGF), and NGF with gelfoam grafted groups at 15-, 30-, and 60-days postgraft. * $p < 0.05$ (Newman-Keuls' test). For comparison the group CI was redrawn from Fernández-Ruiz et al., 1991.

The possibility that combinations of NGF with any other brain tissue could produce functional recovery in IC lesioned animals is currently being evaluated in our laboratory. Briefly, four groups of IC-lesioned rats showing disrupted taste aversions received either cortical graft + NGF, cortical graft + DMEM (Dulbecco's Modified Eagle Medium), mesencephalic tissue with NGF, or mesencephalic tissue alone. All the groups were retrained 15 days after transplantation. The results clearly indicate that the combination cortical grafts with NGF produce significant recovery in the lesioned animals to acquire taste aversions. Those animals that received mesencephalic grafted tissue in combination with NGF, mesencephalic tissue alone, or the vehicle with cortical grafts were unable to acquire the taste aversions after 15 days postgraft. The histochemical results show the presence of positive immunoreaction for NGF and many AChE positive labeled fibers and somas into the grafts of the cortical with NGF group. The grafts from the same group showed a noticeable growth and integration with the host tissue. In addition, preliminary biochemical experiments showed that choline-acetyltransferase (ChAT) activity in this group was very similar to that of the unoperated controls.

CONCLUSIONS

In our model, the application of NGF alone did not produce significant functional recovery in any of the post-graft times tested. In contrast, other authors using different learning models and different brain regions have found, after few days post-graft, recovery following acute application of NGF or other trophic factors (Kesslak, et al., 1986; Will and Hefti, 1985; Hefti et al., 1984; Hefti, 1986). Varon and coworkers have found functional recovery by chronic administration of NGF in fornix lesioned animals that had evidenced severe impairments in a Morris spatial task (Varon et al., 1989). In our study, the best functional recovery was seen when the NGF was associated with homotopical cortical grafts but not with heterotopical mesencephalic grafts or the application of NGF alone. These behavioral results appear to be related with the integration and maturity of the grafted tissue. Preliminary results using the Golgi staining technique indicate that the cortical grafts with NGF showed more neuronal maturation compared to the other groups. Therefore, as mentioned, if neurotrophic factors are involved, they need to be associated with cortical homotopic grafts and/or certain cortical factors essential for producing functional recovery.

ACKNOWLEDGMENTS

This research was supported by Grant from DGAPA-UNAM IN-204689. We thank Mrs. María Teresa Torres for preparing the manuscript.

REFERENCES

Bermúdez-Rattoni, F., Fernández, J., Sánchez, M.A., Aguilar-Roblero, R. and Drucker-Colín, R., 1987, Fetal brain transplants induce recuperation of taste aversion learning. Brain Res., 416: 147-152.
Bermúdez-Rattoni, F., Fernández, J. and Escobar, M.L., 1989, Fetal brain transplants induce recovery of morphological and learning deficits of cortical lesioned rats. In: Cell function and disease. L.E. Cañedo, L.E. Todd, L. Packer and J. Jaz (Eds.) Plenum Publishing Corporation.
Bermúdez-Rattoni, F., and McGaugh, J.L., 1991, Insular cortex and amygdala lesions differentially affect acquisition of inhibitory avoidance and conditioned taste aversion. Brain Res. 549: 165-170.

Bjorklund, A. and Stenevi, U., 1985, Intracerebral neural implants: Neuronal replacement and reconstruction of damaged circuitries. Ann. Rev. Neurosci. 7: 279-308.

Braun, J.J., Lasiter, P.S. and Kiefer, S.W., 1982, The gustatory neocortex of the rat. Physiol. Psychol. 10: 13-45.

Dunnett, S.B., Ryan, C.N., Levin, P.D., Reynolds, M. and Bjorklund, A., 1987, Functional consequences of embryonic neocortex transplanted to rats with prefrontal cortex lesions. Behav. Neurosc. 101: 489-503.

Escobar, M., Fernández-Ruiz, J., Guevara-Aguilar, R. and Bermúdez-Rattoni, F., 1989, Fetal brain grafts induce recovery of learning deficits and connectivity in rats with gustatory neocortex lesion. Brain Res., 478: 368-374.

Fernández-Ruiz, J., Escobar, M.L., Piña, A.L., Díaz-Cintra, S., Cintra-McGlone, F.L. and Bermúdez-Rattoni, F., 1991, Time-dependent recovery of taste aversion learning by fetal brain transplants in gustatory neocortex-lesioned rats. Behav. and Neural Biol., 55: 179-193.

Gage, F.H., 1990, NGF-dependent sprouting and regeneration in the hippocampus, Progress in Brain Res. 83: 357-370.

Gage, F.H. and Bjorklund, A., 1986, Enhanced graft survival in the hippocampus following selective denervation. Neurosc. 17: 89-98.

García, J., Lasiter, P.S., Bermúdez-Rattoni, F. and Deems, D.A., 1985, General theory of aversion learning. Ann. N.Y. Acad. Sci. 443: 8-20.

Grill, H.J. and Norgren, R., 1978, The taste reactivity test. I. Mimetic responses to gustatory stimuli in neurologically normal rats, Brain Res., 143: 263-279.

Hagg, T., Vahlsing, H.L., Manthorpe, M. and Varon, S., 1990, Nerve growth factor infusion into denervated adult rat hippocampal formation promotes its cholinergic reinnervation. J. of Neurosc. 10(9): 3087-3092.

Hefti, F., David, A., Hartikka, J., 1984, Chronic intraventricular injections of nerve growth factor elevate hippocampal choline-acetyltransferase activity in adult rats with partial septo-hippocampal lesions. Brain Res. 293: 305-311.

Hefti, F., 1986, Nerve growth factor promotes survival of septal cholinergic neurons after transections. J. Neurosci., 6: 2155-2162.

Kesslak, J.P., Brown, L., Steichen, C. and Cotman, C.W., 1986, Adult and embryonic frontal cortex transplants after frontal cortex ablation enhance recovery on a reinforced alternation task. Exp. Neurol. 94: 615-656.

Kiefer, S.W., 1985, Neural mediation of conditioned food aversions. J. Ann. N.Y. Acad. Sci., 443: 100-109.

Krushel, L.A. and van der Kooy, D., 1988, Visceral cortex: Integration of the mucosal senses with limbic information in the rat agranular insular cortex. The J. of Comp. Neurol. 270: 34-54.

Labbe, R., Firl, A., Mufson, E. and Stein, D., 1983, Fetal brain transplants reduction of cognitive deficit in rats with frontal cortex lesions. Science 221: 470-472.

López-García, J.C., Bermúdez-Rattoni, F. and Tapia, R., 1990a, Release of acetylcholine, G-aminobutirate, dopamine acid glutamate, and activity of some related enzymes, in rat gustatory neocortex. Brain Res., 523: 100-104.

López-García, J.C., Fernández-Ruiz, J., Bermúdez-Rattoni, F. and Tapia, R., 1990b, Correlation between acetylcholine release and recovery of conditioned taste aversion induced by fetal neocortex grafts. Brain Res., 523: 105-110.

Nachman, M., 1963, Learned aversion to the taste of lithium chloride and generalization to other salts. J. Comp. Physiol. Psychol. 56: 343-349.

Norgren, R., Nishijo, H. and Travers, S., 1989. Taste Responses from the Entire Gustatory Apparatus. In: L.H. Schneider, S.J. Cooper and K. A. Halami (Eds.) The Psychobiology of Human Eating Disorders. <u>Ann. N. Y. Acad. Sci.</u>, 575: 246-263.

Pallage, V., Toniolo, G., Will, B. and Helti, F., 1986, Long-term effects of nerve growth factor and neural transplants on behavior of rats with medial septal lesions. <u>Brain Res.</u> 386: 197-208.

van der Kooy, D.L., Koda, L.Y., McGinty, J.F., Gerfen, C.R. and Bloor, F.E., 1984, The organization of projections from the cortex to amygdala, and hypothalamus to the nucleus of the solitary tract in the rat. <u>J. Comp. Neurol.</u>, 224: 1-24.

Varon, S., Hagg, T., Vahlsing, L. and Manthorpe, M., 1989, Nerve growth factor in vivo actions on cholinergic neurons in the adult rat CNS. In: Cell function and disease. Edited by L.E. Cañedo, L.E. Todd, L. Packer and J. Jaz. <u>Plenum Press</u>: 235-248.

Will, B. and Hefti, F., 1985, Behavioral and neurochemical effects of chronic intraventricular injections of nerve growth factor in adult rats with fimbria lesions. <u>Behav. Brain Res.</u> 17: 17-24.

REPELLENT EFFECT OF TRIMETHYL THIAZOLINE IN THE WILD RAT

RATTUS NORVEGICUS BERKENHOUT

E. Vernet-Maury, B. Constant and J. Chanel

Physiologie neurosensorielle
CNRS-Université Claude Bernard/Lyon
F-69622, Villeurbanne cedex, France

INTRODUCTION

In 1968, Vernet-Maury et al. established that predator odor induced avoidance behavior in rats. Ethological analysis (Vernet-Maury et al., 1968), biochemical bioassays (Vernet-Maury, 1970) and bioelectrical recordings (Vernet-Maury and Chanel, 1967) led to the same conclusion: the predator odor increased emotional reactivity and induced freezing behavior in the albino rat. Recordings of olfactory bulb mitral cell electrical activity evidenced an inhibitory phenomenon, this inhibition being centrally controlled (Cattarelli et al., 1974). A pertinent index was recently added: under predator odor, ultrasonic vocalizations decreased (Vernet-Maury et al., 1992). Moreover, ontogenic analysis of this particular odor led to the conclusion that its frightening power is innate: pups,that regularly responded to this odor from birth to adulthood, did not habituate and showed freezing behavior in the open field test undertaken 3 months later (Duveau et al., 1981).

From GC-MC analysis, the active compound from the whole pentane extract was identified as 2,5,dihydro, 2,4,5, trimethyl thiazole or trimethyl thiazoline (TMT) (Vernet-Maury, 1980).This active compound was compared to other sulfur molecules to establish a structure/function relationship (Vernet-Maury and Polak, 1984) of these sulfur compounds. Once again TMT induced the higher emotional reactivity quantified through the open fiel test and corticosterone release.

The purpose of this study was to try to generalize the observations from laboratory rats to wild rats and, if this was possible, then to render the TMT molecule more efficient and thus to use it as an area repellent.

METHODS

Animals/Terrarium

The subjects were wild brown rats (*Rattus norvegicus* Berkenhout) in a pen population of about 100 animals living in semi-natural conditions in the terrarium of the National Veterinary School of Lyon. This population was started from three males and three females trapped from the center of France.

Chemical Signals in Vertebrates VI, Edited by R.L. Doty and
D. Müller-Schwarze, Plenum Press, New York, 1992

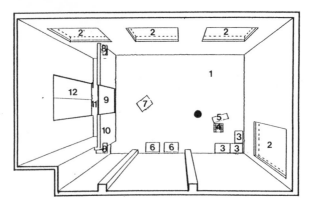

Fig. 1. Terrarium : Top view – 1 : soil (light
earth) – 2 : windows – 3 : truss of
straw – 4 : crate – 5 : board – 6 :drin-
king troughs – 7 : feeding troughs –
8 : side board –9 : entrance door – 10 :
partition – 11 : observation point –
12 : outside door – ● : holding device.

The concrete floor of a rectangular terrarium (5.5 m x 6.9 m) was covered with an earth layer, 50 cm deep. During the day, light came from 4
windows. At night, a 40–watt bulb was on continuously, but the rats' activity was not modified (see Fig. 1). Food troughs and four drinking troughs
were placed symmetrically. Food and water was available *ad libitum*.

Procedure

An odor was introduced in the terrarium, as discussed below. Observations took place at the beginning of the dark period corresponding to the
first peak of these rats activity   (from 21:00 to 22:30 hrs)(Rampaud,1981).
During the observation period, the number of visible rats was counted every
five minutes. About 80 rats were involved during the different parts of
these experiments.

The main natural behavior patterns we quantified were aggressive encounters between male rats, boxing postures, mating, social encounters and
grooming after copulation. "Confinement" was defined when one male prevented
another from coming out of the burrow. Maternal behavior was not observable
as nests were built underground and only pups were visible.

A preliminary experiment was undertaken to determine the appropriate
place of the beaker containing the odor to be tested, hanging on a metal
tripod. The question to be answered was : Are rats equally distributed
around the terrarium ? To solve this problem, the length of the terrarium
was roughly, and in imagination, divided into 3 parts in which the number
of rats was determined by four experimenters (see Fig. 1, number 11) at 2
given times: 20:45 and 21:15 hrs on 5 consecutive nights. For one minute,
each visible rat was indicated in a diagram. Thus a kind of snap shot of
the rats' array was obtained for a given minute. The resulting diagrams
were combined and 4 zones were thus delimited in which the rats were counted;
a total number was obtained for each night (Fig. 2).

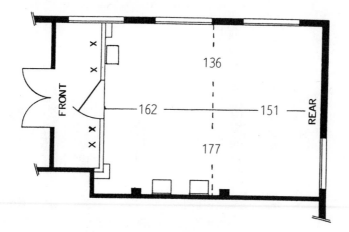

LEFT SIDE

RIGHT SIDE

Fig. 2. Terrarium divided into four zones according to
ventral and horizontal axes.
Mean number of observed rats in the four zones.

X : position of the four experimenters.

The main experiment lasted 11 days ; 2 days of habituation preceded
the first three experimental days: 1-3 control days when the holding device
was present without any odor, followed by 4-6 treatment days with 30 mg of
pure solution of trimethyl thiazoline in the holding device. The last 3 days
were control days without any odor.

A 50 cm diameter circle around the holding device was demarcated by
small stones ; the behavior of rats reaching this area was observed and
recorded.

This experimental procedure was preferred to the conventional one with
another control odor tested, in order to preserve this pen population of
wild rats from too many "manipulations" of their surroundings. Indeed, too
many disturbance days are forbidden in this semi-natural area. Three days
odor (plus odor in only a part of the terrarium) would allow us to discuss
a neophobia phenomenon.

RESULTS

a) Preliminary experiment: For each of the 4 zones of the terrarium
a Chi2 test was made and null hypothesis (no symmetry) accepted or rejected
(see Fig. 2). The total number of rats is identically counted for all experiments. Totalized results from the different night counts led to the conclusion that null hypothesis is accepted only with front-rear symmetry.
Chi2, calculated from the observed vs. theoretical values, is respectively
5.37 for right-left symmetry and 0.39 for front-rear symmetry.

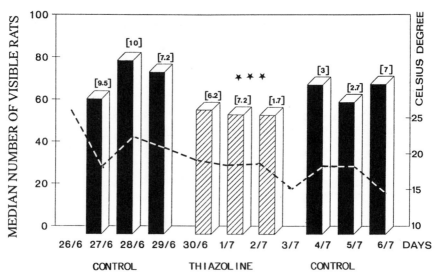

Fig. 3. Median number of visible rats under the Control and Trimethyl
Thiazoline treatment days – Semi-interquartile ranges indicated in brackets – The broken line represents temperature variations during the experiment.

The animals were not equally dispersed troughout the terrarium ; many more
rats were observed in the right area. Based on the finding, the decision
was made to locate the holding device to the right and away from observers,
as shown in Fig. 1.

   b) Main experiment with Trimethyl Thiazoline (TMT): With a view to statistical analysis results from different days were classified in 3 groups :
2 control – 3 days before and 3 after treatment – and 1 corresponding to
the 3 days with odor treatment. Friedman's analysis of variance was significant at threshold $p < .001$, evidencing a difference between the 3 groups.
A Wilcoxon test revealed that the difference was significant ($p < .01$)
between each control group and the TMT group, but not between the controls.
TMT brought about a decrease in the number of rats visible in the terrarium.
The median (semi-interquartile range) of the 30 counts is respectively
77(7) and 71(7) for the control and 56.5(7) for the TMT odor days. Results
are summarized in Fig. 3. In addition to this decrease in the number of visible rats, the aversion was exteriorized by characteristic behavioral patterns or by avoidance of the immediate surroundings ofthe TMT odor, as shown
in Fig. 4 obtained from six "snap shots" under TMT compared to control recorded in an equivalent temporal situation, i.e. after each count of rats
on the 2nd day (control or experimental).

   Animals kept to peripheral areas of the terrarium avoiding the preferential paths, around the holding device, clearly witnessed during preliminary experiments as the odor position was determined in the right part of
the terrarium, with frequently observed passing of rats. An ambivalent approach of the holding device was common during the first treatment days.
The number of rats recorded, in the circle defined around the holding device,
was respectively : 9 on day 1, 1 on day 2 and 0 on day 3 of treatment, compared to 14, the mean number on control days. On day 1, when approaching
the odor area, a significant number of rats typically came in groups of 2
or 3, not by themselves as was always observed on control days. During these
greagarious periods, rats sniffed one another around their noses with
numerous vibrissae movements.

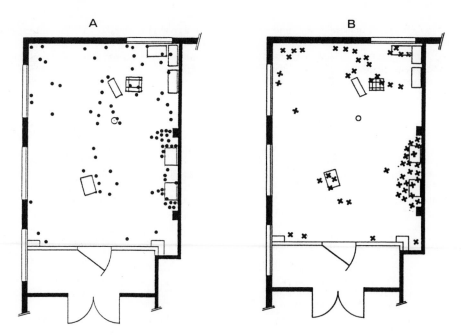

Fig. 4. Dispersion of rats in the terrarium - A : Control day
B : 30 min. after TMT odor exposure.

DISCUSSION

The main result of this experiment was an avoidance behavior of rats
induced by TMT odor. Is this effect specifically induced by TMT ? The approach behavior toward the holding device observed on day 1 of treatment
(and nor on day 2 or 3) at the TMT odor period,permits to reject a neophobia hypothesis that might explain the obtained results in response to
the odorant (Mitchell, 1976). Moreover, during the first treatment day,
the observed approach to the holding device was ambivalent and near the
container numerous withdrawals were recorded, followed by backward jumps.
The gregarious phenomenon noted during this approach is under analysis,
using concomitant recordings of ultrasonic calls, which are main way of
intraspecific communication of *Rattus norvegicus* (Joubert-Seguin, 1987).

The innate response to TMT odor described in laboratory rats (Duveau
et al., 1981) is demonstrated with wild rats:as far as on days 2 and 3 of
TMT treatment no pups were observed outside. Another result also agree
with those obtained from laboratory strains : no habituation to this odor
was evidenced since the number of visible rats was not statistically different in the TMT treatment days.

The decreased percentage of visible rats (24%) is identical to that
observed during field research by Sullivan et al. (1988) concerning the
number of orchard trees protected by TMT against wild rodents (Meadow
voles).

The repellent effect of TMT is thus evidenced in the wild rats, confirming observations on Wistar rats but with forms of response that are
different from those observed in limited spaces.

309

As TMT is the isolated active compound from fox predator, an adaptative value has to be postulated for this avoidance response and a kairomone power postulated. Results from two different wild rodent species having fox as a common predator, suggests that this kairomone is not specific of *Rattus norvegicus* species but could take an active part in the relationship between other rodents and fox.

The innate response added to the observed non-habituation lead us to try to modify the TMT molecule in order to enhance its repellent effect.

ACKNOWLEDGMENTS

Special thanks to Professor Lorgue(*Ecole Nationale Vétérinaire de Lyon*) for providing terrarium facilities for our experiments.

REFERENCES

Cattarelli, M., Vernet-Maury, E. and Chanel J., 1974,Olfactory bulb and integration of some odorous signals in the rat ; behavioral and electro-physiological study, Olfaction and Taste, V:235.

Duveau, A. Vernet-Maury, E. et Chanel, J., 1981, Ontogenèse des communications perceptuelles : vocalisations ultrasonores à point de départ olfactif chez le Rat, Journal de Physiologie, 77:52.

Joubert-Seguin, C., 1987, La communication sonore chez *Rattus norvegicus* ; étude d'une colonie de rats sauvages en semi-liberté et des effets d'une émission d'ultra-sons sur le comportement et la physiologie de ces animaux, Thèse de l'Ecole Nationale Vétérinaire de Lyon soutenue devant l'Université Claude Bernard de Lyon.

Mitchell, D., 1976, Experiments on neophobia in wild and laboratory rats, a reevaluation, J. Comp. Physiol. Psychol., 90:190.

Rampaud, M., 1981, Le comportement du Rat (*Rattus norvegicus*) en semi-liberté, Thèse de l'Ecole Nationale Vétérinaire de Lyon soutenue devant l'Université Claude Bernard de Lyon.

Sullivan, T.P., Crump, D.R. and Sullivan D.S., 1988, Use of predator odors as repellents to reduce feeding damage by herbivores, III- Montane and Meadow voles (Microtus montanus and Microtus pennsylvanicus), J. Chem. Ecol., 14:379.

Vernet-Maury, E. et Chanel, J., 1967, De quelques méthodes d'évaluation de l'émotion chez le Rat, Journal de Physiologie, 59:526.

Vernet-Maury, E., Le Magnen, J. et Chanel, J. 1968, Comportement émotif chez le Rat; influence de l'odeur d'un prédateur et d'un non-prédateur, Comptes Rendus de l'Académie des Sciences, 267:331.

Vernet-Maury, E., 1970, Excrétion urinaire des catécholamines chez le Rat en fonction de l'ambiance olfactive, Journal de Physiologie, 62:461.

Vernet-Maury, E., 1980, Trimethyl Thiazoline in Fox feces : A natural alarming substance, Olfaction and Taste V:407.

Vernet-Maury, E. and Polak, E., 1984, Structure-Activity relationship of stress-inducing odorants in the rat, J. Chem. Ecol.,10:1007.

Vernet-Maury, E., Vigouroux, M. and Chanel, J., 1992, Ultrasonic emission in response to predator odor in Wistar rats,Anim. Behav., submitted.

TAXONOMIC DIFFERENCES BETWEEN BIRDS AND MAMMALS IN THEIR RESPONSES

TO CHEMICAL IRRITANTS

J. Russell Mason[1], Larry Clark[1], and Pankaj S. Shah[2]

[1]USDA/APHIS/S&T/Denver Wildlife Research Center, c/o
Monell Chemical Senses Center, 3500 Market Street
Philadelphia, PA 19104; [2]Monell Chemical Senses Center
3500 Market Street, Philadelphia, PA 19104

INTRODUCTION

Ninety-five products are registered with the U.S. Environmental Protection Agency as bird damage control chemicals, but 38 (40%) are nonlethal chemical repellents (Eschen and Schafer, 1986). Of these products,
the active ingredients in 27 (71%) are methiocarb (a physiologic repellent
that acts through food avoidance learning) or polybutene (a tactile repellent). In general, chemical repellents are effective either because of
aversive sensory effects (irritation), or because of post-ingestional malaise (sickness). If the former, then chemicals are usually stimulants of
trigeminal pain receptors (i.e., undifferentiated free nerve endings) in
the nose, mouth, and eyes (Mason and Otis, 1990). Although many birds
possess adequate olfactory and gustatory capabilities (e.g., Berkhoudt,
1985, Kare and Mason, 1986) smell and taste, per se, are rarely of consequence for bird damage control. Here, we address chemosensory repellents
only.

Trigeminal chemoreception is a component of the common chemical sense,
and its biological function appears to be the initiation and mediation of
protective reflexes (Green et al., 1990). While the morphological organization of the peripheral trigeminal system in birds and mammals is similar,
there are broad functional discrepancies (Kare and Mason 1986). Thus, the
avian trigeminal system is essentially unresponsive to strong mammalian
irritants, such as capsaicin (Szolcsanyi et al., 1986).

Explanation of this taxonomic difference is of fundamental interest.
For example, it could reflect phylogenetic constraints present at the time
of divergence for each group, or an evolutionary response to selective
pressures relating to the chemical ecology prevailing at the time of divergence. Explanation of the taxonomic difference also is of practical interest, because it might lead to the systematic identification of new avian
repellents. To this end, we are examining derivatives of a basic phenyl
ring structure to develop a chemical model of avian repellency. Our presumption is that repellency and irritation are isomorphic (Mason et al.,
1989).

Chemical Signals in Vertebrates VI, Edited by R.L. Doty and
D. Müller-Schwarze, Plenum Press, New York, 1992

CHEMISTRY

Both methyl and dimethyl anthranilate are repellent to birds at concentrations that are accepted by mammals (Glahn, 1989). Avoidance of these
ester derivatives of anthranilic acid is based on odor quality and irritation (Mason et al., 1989). To humans, both have a grape-like or fruity
odor, and a slightly bitter, pungent taste (Furia and Bellanca, 1975).
Methyl anthranilate is a commonly used grape flavoring in human food preparations (Furia and Bellanca, 1975), and is the chemical traditionally
blamed for the "foxy" quality of red wines produced from Vitis lambrusca
grapes (Amerine and Singleton, 1966; Broadbent, 1970). The term "foxy"
presumably is derived from the colloquial name of the Concord grape (i.e.,
fox grape), and there are suggestions that wines from V. lambrusca have an
"animal-den" odor (Amerine and Singleton, 1966).

There is at least one other chemical responsible for the "foxy" off-
flavor of some red wines. Ortho-aminoacetophenone has an odor similar to
that of methyl anthranilate, and is structurally similar, differing only in
the substitution of a ketone for an ester group (Acree et al., 1990).
Coincidentally, ortho-aminoacetophenone is present in the scent glands of a
variety of mammalian species (Novotny et al., 1990; Hall, 1990), including
mustelids that prey on birds (Acree et al., 1990).

Our attempts at chemical modelling have focused on isomers of aminoacetophenone and anthranilate derivatives, and we have used behavioral
tests of European starlings (Sturnus vulgaris) to measure aversiveness.
This bird is used as the test species for several reasons. First, starlings exhibit good chemical sensing abilities (Mason and Silver, 1983;
Clark and Mason, 1987). Also, data on the responsiveness of starlings to
irritants is available (e.g., Mason et al., 1989; Glahn et al., 1989).
Finally, starlings are considered agricultural pests (Besser et al., 1968).

Initially, we found that the strength of repellency was related to
resonance of lone electron pairs and intramolecular hydrogen bonding (Mason
et al., 1991). Subsequent studies showed that hydrogen bonding is not
required for repellency, though it may play an ancillary role (Clark and
Shah, 1991). Also, we verified that increased electron donation and/or
decreased electron withdrawal to the phenyl ring enhances repellency (Shah
et al., 1991). At present, our investigations are designed to more precisely determine the influence of basicity, pi cloud planarity, and electron donating and withdrawing groups on avian repellency. A detailed
discussion of on-going attempts to relate chemical structure with behavioral activity is provided by Shah et al. in the present volume.

APPLICATIONS

Agriculture

Birds damage ripening grain (Dolbeer, et al., 1982; Bullard et al.,
1981; Dolbeer et al. 1978; Henne et al., 1979) and fruit crops (Hothem et
al., 1981; Tobin et al., 1989a,b; Avery et al., 1991; Hobbs and Leon,
1988). Although bird depredation on vegetables, nut crops, and legumes is
less publicized, it is a common complaint among growers (Mott et al.,
1972). In addition to depredation losses, per se, damage results in higher
levels of insect damage and spoilage (Woronecki et al., 1980). The econom-
ics of damage varies greatly. For example, a 1972 survey of sunflower
fields in North Dakota and Minnesota showed that the mean loss to birds was
only 13 kg/ha (Besser, 1978). Because 174,500 ha were planted in sunflower
during that year, we estimate that the national loss was 2,270 metric tons
(Woronecki et al., 1980). At an average value of $230 per metric ton
(Cobia, 1978), bird damage cost growers $522,100. On the other hand, Avery

et al. (1991) estimate that birds destroyed nearly 11% of the national blueberry crop in 1989. Because total blueberry production in 1989 was 158 million pounds, and the average price was $0.50/pound, bird damage may have cost growers as much as $8.5 million from a total market size of $77.3 million.

Non-food crops also are damaged. Turf (Laycock, 1982), flowers [e.g., orchids and anthurium (Cummings et al., 1990)], and cover crops are lost. Because some non-food crops remain in the field for years, damage can be cumulative and costly. Estimates of annual bird damage to orchids grown in the Hawaiian Islands are as high as 75% of the total crop; the 1985 market value of Hawaiian orchids exceeded $12 million (Kefford et al. 1987), representing a potential loss of $9 million.

Apart from field crops, bird damage has been documented in feedlots and at grain storage operations (Twedt and Glahn, 1982). Birds are a vector for transmissible gastroenteritis (Pilchard, 1965), tuberculosis (Bickford et al., 1966), and avian influenza. As predators, birds prey on livestock (Phillips and Blom, 1988), and take fish from pound nets (Craven and Lev, 1988) and fish-culture ponds (Parkhurst and Brooks, 1988). Estimates of bird damage to catfish operations in the Mississippi delta exceed $5 million annually (Stickley and Andrews, 1989).

A potentially more important problem than agricultural loss is the hazard that modern agricultural practices present to birds. Pelleted chemicals and chemically treated seeds are essential to no-till conservation farming, a practice that will be used on 60% of the nation's cropland by the year 2010 (Crosson, 1982). This technique generally benefits wildlife by providing cover and food (Castrale, 1987), and is environmentally safe (Greig-Smith, 1987). However, pelleted chemicals and treated seeds present dangers to birds that accidentally ingest them (Best and Gionfriddo, 1991); most granular insecticides are highly toxic to birds (Schafer et al., 1983).

Chemosensory repellents might be used to resolve many of the problems above. For example, methyl anthranilate or a similar substance (e.g., orthoaminoacetophenone) could be added to livestock feeds to repel birds without affecting consumption by livestock (Glahn et al., 1989). Not only is bird consumption of feed essentially eliminated, but bird numbers at treated sites are significantly reduced (Figure 1). Similarly, chemosensory repellents are effective when applied to nonfood crops like orchids.

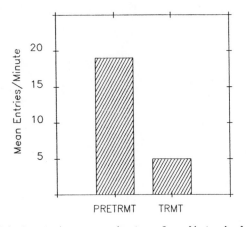

Fig. 1.   Mean bird entries per minute of undisturbed pretreatment and treatment (1.0% dimethyl anthranilate) observation at test sites in Kentucky, 1988

Finally, formulated agricultural chemicals can be treated with a chemosensory repellent to reduce nontarget hazards to birds (T. Miller, pers. commun.). For any application, an especially promising strategy may be to combine chemosensory repellents with other cue sources. Ecologically, the superiority of redundant cues is predictable from studies of predator-prey interactions. Toxic prey often use multiple aposematic signals to advertise unpalatability to predators (Wickler, 1968). Redundancy may decrease ambiguity, or might affect different types of predators (Mason, 1989).

Conservation

Industrial by-products and mine effluvia often are stored in open outdoor impoundments. Although the impoundments meet federal and state regulations for the protection of ground water, they pose serious risks to wildlife (Allen, 1990). Waterfowl, shorebirds, and other species are attracted to the freestanding water and risk both acute and chronic poisoning (Ohlendorf et al., 1989).

Birds also are a problem at airport (Blokpoel, 1976). In 1989, the economic loss to U.S. Military operations caused by bird strikes was about $80 million, while civilian losses were about $100 million (R.A. Dolbeer, pers. commun.). In many instances, birds are attracted to airports because of free-standing water that accumulates on paved surfaces. As for mining operations, traditional hazing methods, are ineffective because birds habituate to the harassment or simply move from one location to another.

Chemosensory repellents can be used to reduce consumption and use of free-standing water (Figure 2) (Clark et al. 1991). It is possible that these substances could be used as aversive additives to waste water or fresh water puddles at airports. At presant, the greatest obstacle to use is the lack of delivery technologies that insure an even distribution of repellent over the pond surface.

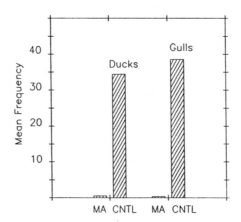

Fig. 2. Mean numbers of swimming bouts by mallards (Anas platyrhynchos) and ring-billed gulls (Larus delawarensis) in
treated (0.5% methyl anthranilate, MA) or control (CNTL)
pools; Ohio, 1990.

CONCLUSIONS

There is a clear taxonomic difference between birds and mammals in their responsiveness to chemosensory irritants. To date, we have not identified a single mammalian substance also avoided by birds. This observation has diverse practical implications. There are basic implications as well. Perhaps, avian insensitivity to mammalian irritants reflects phylogenetic constraints or an evolutionary response to some unknown selective pressure. One possible explanation might lie with the plants that produce many chemical irritants. Capsaicin illustrates this possibility. Although there is information about the repellency of this material to mammals, its gastronomic significance, and its importance as a neurotoxin, there is little if anything published concerning the biological role of this material. Maybe capsaicin was selected to act as a deterrent to seed predators (i.e., mammals), without influence on seed dispersal agents (i.e., birds). This explanation is especially interesting to us, because many plant manufactured compounds, including piperine, zingerone, gingerol, and mustard oil are repellent to mammals but not birds. Striking though it seems, there are to our knowledge no experimental tests of this proposition.

LITERATURE

Acree, T. E., Lavin, E. H., Nishida, R., and Watanabe, S., 1990, The serendipitous discovery of ortho-aminoacetophenone as the 'foxy' smelling component of Lambruscana grapes, Chem. Eng. News, 9:80.

Allen, C., 1990, Mitigating impacts to wildlife at FMC Gold Company's Paradise Peak Mine. in: "Proceedings of the Nevada Wildlife Mining Workshop", R. McQuivey, ed., Nevada Mining Assoc., Nevada Dept. Minerals and Nevada Dept. Wildl. Reno.

Amerine, M. A., and Singleton, V. L., 1966, "Wine: An introduction for Americans", Univ. Calif. Press, Los Angeles.

Avery, M. L., Nelson, J. W., and Cone, M. A., 1991, Survey of bird damage to blueberries in North America - 1989, Denver Wildlife Research Center Bird Section Research Report No. 468.

Berkhoudt, H., 1985, Structure and function of avian taste receptors, in: "Form and Function in Birds", A. S. King, and J. McLelland, eds., Academic Press, New York.

Besser, J. F., 1978, Birds and sunflower. in: "Sunflower Science and Technology", J. F. Carter, ed., American Society of Agronomy, Crop Science Society of America, Soil Science Society of America, Inc., Madison.

Besser, J. F., DeGrazio, J. W., and Guarino, J. L., 1968, Costs of wintering starlings and red-winged blackbirds at feedlots, J. Wildl. Manage., 32:179.

Best, L. B., and Gionfriddo, J. P., 1991, Characterization of grit use by cornfield birds, Wilson Bull., 103:68.

Bickford, A. A., Ellis, G. H., and Moses, H. E., 1966, Epizootiology of tuberculosis in starlings, J. Am. Vet. Med. Assoc., 149:312.

Blokpoel, H., 1976, Bird hazards to aircraft, Canad. Wildl. Serv., Ottawa, Canada, Unpubl. report.

Broadbent, J. M., 1970, "Wine tasting", Wine and Spirit Publ., Ltd., London.

Bullard, R. W., York, J. O., and Kilburn, S. R., 1981, Polyphenolic changes in ripening bird-resistant sorghums, J. Agr. Food Chem., 29:973.

Castrale, J. S., 1987, Pesticide use in no-till row-crop fields relative to wildlife, Indiana Acad. Sci., 96:215.

Clark, L., and Mason, J. R., 1987, Olfactory discrimination of plant volatiles by the European starling, Anim. Behav., 35:227.

Clark, L., Mason, J. R., Shah, P. S., and Dolbeer, R. A., 1991, Bird repellents: Water additive, waste water additive, edible additive, and methods for using same, U. S. Patent Application No. 679432.

Clark, L., and Shah, P. S., 1991, Nonlethal bird repellents: In search of a general model relating repellency and chemical structure, J. Wildl. Manage., 55:539.

Cobia, D. W., 1978, Production costs and marketing. in: "Sunflower Science and Technology", J. F. Carter, ed., American Society of Agronomy, Society of America, Soil Science Society of America, Inc., Madison.

Craven, S. R., and Lev, E., 1987, Double-crested cormorants in the Apostle Islands, Wisconsin, USA: Population trends, food habits and fishery depredations, Colonial Waterbirds, 10:64.

Crosson, W. F., 1982, Conservation tillage and conventional tillage: a comparative assessment, U. S. Environ. Protec. Agency Report No. 600/3-82-027, Washington, D.C.

Cummings, J. L., Mason, J. R., Otis, D. L., and Davis, J. E., 1990, Evaluation of methiocarb, zinc dimethyldithiocarbamate, and methyl anthranilate for reducing bird damage to dendrobium orchids. Interim Report to the Hawaiian Farm Bureau, unpublished report.

Dolbeer, R. A., Stickley, A. R., Jr., and Woronecki, P. P., 1978, Starling, Sturnus vulgaris, damage to sprouting wheat in Tennessee and Kentucky, U.S.A., Protec. Ecol., 1:159.

Dolbeer, R. A., Woronecki, P. P., and Stehn, R. A., 1982, Effects of husk and ear characteristics on resistance of maize to blackbird (Agelaius phoeniceus) damage in Ohio, USA, Protec. Ecol., 4:127.

Eschen, M. L. and E. W. Schafer, E. W., Jr., 1986, Registered bird damage chemical controls--1985, Denver Wildlife Research Center Bird Damage Research Report No. 356.

Furia, T. E., and Bellanca, N., 1975, "Handbook of Flavor Ingredients", CRC Press, Cleveland, Ohio.

Glahn, J. F., Mason, J. R., and Woods, D. R., 1989, Effects of dimethyl anthranilate as a bird repellent livestock feed additive, Wildl. Soc. Bull., 17:313.

Green, B. G., Mason, J. R., and Kare, M. R., 1990, "Chemical Senses: Irritation", Marcel Dekker, Inc., New York.

Greig-Smith, P. W., 1987, Hazards to wildlife from pesticide seed treatments. British Crop Protection Council Monograph 39: Application to seeds and soil.

Hall, D. H., 1990, Host odour attractants for Tsetse flies, in: "Chemical Signals in Vertebrates V", D. W. MacDonald", D. Muller-Schwarze, and S. E. Natynczuk, eds., Oxford University Publications, Oxford.

Henne, D. R., Pfeifer, W. D., and Besser, J. R., 1979, Bird damage to sunflower in North Dakota in 1978. in: "Proceedings of the Third Sunflower Forum", D. Lilliboe, ed., National Sunflower Association, Fargo.

Hobbs, J., and Leon, F. G., 1988, Great-tailed grackle predation on south Texas citrus (identifying a unique problem), in: "Proceedings of the Third Eastern Wildlife Damage Control Conference", N. R. Holler, ed., Univ. Alabama Press, Gulf Shores.

Hothem, R. L., Mott, D. F., DeHaven, R. W., and Guarino, J. L., 1981, Mesurol as a bird repellent on wine grapes in Oregon and California, Am. J. Enology Viticulture, 32:150.

Kare, M. R., and Mason, J. R., 1986, The chemical senses in birds. in "Avian Physiology", P. D. Sturkie, ed., Springer-Verlag, New York.

Kefford, N. P., Leonhardt, K. W., Tanaka, T., and Rohrbach, K. G., 1987, Orchid Industry Analysis. College of Tropical Agriculture and Human Resources, University of Hawaii, Report No. 3.

Laycock, G., 1982, The urban goose, Audubon, 84:44.

Mason, J. R., 1989, Avoidance of methiocarb-poisoned apples by red-winged blackbirds. J. Wildl. Manage., 53:836.

Mason, J. R., Adams, M. A., and Clark, L., 1989, Anthranilate repellency to starlings: Chemical correlates and sensory perception, J. Wildl. Manage. 53:55.

Mason, J. R., Clark, L. and Shah, P. S., 1991, Ortho-aminoacetophenone repellency to birds: perceptual and chemical similarities to methyl anthranilate, J. Wildl. Manage., 55:334.

Mason, J. R., and Otis, D. L., 1990, Effectiveness of six potential irritants on consumption by red-winged blackbirds (*Agelaius phoeniceus*) and starlings (*Sturnus vulgaris*), in: "Chemical Senses: Irritation", B. G. Green, J. R. Mason, and M. R. Kare, eds., Marcel Dekker, Inc., New York.

Mason, J. R., and Silver, W. L., 1983, Conditioned odor aversions in starlings (*Sturnus vulgaris*) possibly mediated by nasotrigeminal cues, Brain Res., 269:196.

Mott, D. F., Besser, J. F., West, R. R., and DeGrazio, J. W., 1972, Bird damage to peanuts and methods for alleviating the problem, Proc. Vertebr. Pest Conf., 5:118.

Novotny, M., Jemiolo, B., and S. Harvey, S., 1990, Chemistry of rodent pheromones: molecular insights into chemical signalling in mammals, in: "Chemical Signals in Vertebrates V", D. W. MacDonald, D. Muller-Schwarze, and S. E. Natynczuk, eds., Oxford University Publications, Oxford.

Ohlendorf, H. M., Hothem, R. L., and Welsh, D., 1989, Nest success, cause-specific nest failure, and hatchability of aquatic birds at selenium-contaminated Kesterson Reservoir and a reference site, Condor, 91:787.

Parkhurst, J. A., and Brooks., R. P., 1988, American kestrels eat trout fingerlings, J. Field Ornithol., 59:286.

Phillips, R. L., and Blom, S., 1988, Distribution and management of eagle/livestock conflicts in the western United States, Proc. Vertebr. Pest Conf., 13:241.

Pilchard, E. I., 1965, Experimental transmission of transmissible gastroenteritis virus by starlings, Am. J. Vet. Res., 26:1177.

Putt, E. D., 1978, History and present world status. in: "Sunflower Science and Technology", J. F. Carter, ed., American Society of Agronomy, Crop Science Society of America, Soil Science Society of America, Inc., Madison.

Schafer, E. W., Jr., Bowles, W. A., Jr., and Hurlbut, J., 1983, The acute oral toxicity, repellency, and hazard potential of 988 chemicals to one or more species of wild and domestic birds, Arch. Environ. Contam. Toxicol., 12:355.

Shah, P. S., Clark, L., and Mason, J. R., 1991, Prediction of avian repellency from chemical structure: The aversiveness of vanillin, vanillyl alcohol, and veratryl alcohol, Pest Biochem. Physiol., 40:169.

Stickley, A. R., and Andrews, K. J., 1989, Survey of Mississippi catfish farmers on means, effort, and costs to repel fish-eating birds from ponds, Proc. Eastern Wildl. Damage Cont. Conf., 4:105.

Szolcsanyi, J., Sann, H., and Pierau, F-K., 1986, Nociception in pigeons is not impaired by capsaicin. Pain, 27:247.

Tobin, M. E., Dolbeer, R. A., and Webster, C. M., 1989, Alternate-row treatment with the repellent methiocarb to protect cherry orchards from birds, Crop Protec., 8:461.

Tobin, M. E., Dolbeer, R. A., and Woronecki, P. P., 1989, Bird damage to apples in the MidHudson Valley of New York, Hort. Sci., 24:859.

Twedt, D. J., and Glahn, J. F., 1982, Reducing starling depredations at livestock feeding operations through changes in management practices, Proc. Vertebr. Pest Conf., 10:159.

Wickler, W., 1968, Mimicry in plants and animals. McGraw-Hill Publ. Co., New York, N.Y. 255pp.

Woronecki, P. P., Stehn, R. A., and Dolbeer, R. A., 1980, Compensatory response of maturing corn kernels following simulated damage by birds, J. Appl. Ecology, 17:7.

USE OF A TRIGEMINAL IRRITANT FOR WILDLIFE MANAGEMENT

Michael L. Avery, David G. Decker, and Curtis O. Nelms

USDA/APHIS, Denver Wildlife Research Center
Florida Field Station
2820 E. University Ave.
Gainesville, FL 32601

Aversive chemicals have been used for many years to manage wildlife in certain problem situations. A group of compounds that hold particular promise for the management of wild bird species is composed of derivatives of anthranilic acid. These esters, especially methyl anthranilate and dimethyl anthranilate, are very effective avian feeding deterrents and appear to act primarily through nasal trigeminal irritation (Mason et al. 1989). Methyl anthranilate (MA) is attractive as a wildlife management tool because (1) it is relatively inexpensive, (2) it is generally regarded as safe (GRAS-listed) for humans by the U. S. Food and Drug Administration (MA is used as a fruit flavoring in many foods and cosmetics), and (3) birds do not seem to habituate to it. The use of MA and related compounds represents a humane, nonlethal approach to the control of problem bird species.

In this paper, we briefly describe experiments with MA using captive wild birds to evaluate the potential usefulness of MA as a bird repellent in 3 diverse applications: (1) reducing damage to small fruit crops, (2) discouraging use of small bodies of water, and (3) protecting eggs of endangered species from avian predators.

MA AS A FEEDING DETERRENT TO FRUGIVOROUS BIRDS

Introduction

Each year, cedar waxwings (_Bombycilla cedrorum_) and other frugivorous bird species cause considerable damage to blueberry, cherry, and other fruit crops in North America. To evaluate the potential of MA as a bird repellent for fruit crops, we tested the repellency of technical grade MA to caged cedar waxwings.

Methods

We captured cedar waxwings in mistnets in blueberry fields near Gainesville, FL and transported them to the Florida Field Station of the Denver Wildlife Research Center. Birds were maintained and tested in 1.4 x 1.4 x 1.8-m cages within a roofed outdoor aviary. Banana mash (Denslow et al. 1987), whole bananas, and water were available _ad libitum_.

Chemical Signals in Vertebrates VI, Edited by R.L. Doty and
D. Müller-Schwarze, Plenum Press, New York, 1992

During the 5-day trials maintenance food was removed at 0800 and trials began at 0900. Each day, we offered the birds 4 berries dipped in propylene glycol and 4 berries dipped in a mixture of MA and propylene glycol. We tested 3 treatment levels: 0.25%, 0.5%, and 1.0% MA (g/g). Each berry was dipped immediately prior to placement on a metal feeder next to a perch readily accessible to the birds. We recorded when each berry was picked up, eaten, or dropped. Berries remaining in the feeders were removed after 10 min. The number of berries eaten and the number of berries dropped were compared among treatment levels with 1-way ANOVAs.

Results

The total number of MA-treated berries eaten (out of 80 presented) was 8, 10 and 25 for the 1.0%, 0.5%, and 0.25% groups, respectively. The totals for berries without MA were 56, 40, and 44. Thus, berry consumption was reduced 86%, 75%, and 43% for the 3 treatments. Significantly ($P = 0.05$) more treated berries were consumed by the 0.25% MA group than by higher treatment level groups. Also, significantly ($P = 0.03$) more treated berries were dropped at the 1.0% MA level than at the lower MA levels. There were no differences ($P > 0.4$) among groups in their responses to propylene glycol berries.

Two birds regurgitated MA-treated berries on each of 2 separate occasions. In the 0.5% group, an individual regurgitated a berry 125 sec after ingestion on test day 2 and regurgitated another berry 2 sec after eating it on test day 4. On each day, the bird ate untreated berries. In the 1.0% group, a bird regurgitated MA-treated berries 90 sec after ingestion on test day 3, and 128 sec after ingestion on day 4. Six minutes after regurgitation on day 4, this bird ate another MA-treated berry and kept it down.

Discussion

The persistence of the cedar waxwings in trying to eat MA-treated berries suggests either that birds were unable to detect MA on the berries without picking them up or that the MA was not sufficiently aversive. The latter possibility seems remote because (1) MA did reduce berry consumption, and (2) at least 2 birds regurgitated treated berries. Thus, repeated sampling of MA-treated berries was probably due to the birds' inability to detect the MA distally. This contention is supported by anatomy of the avian olfactory system. The external nostrils and outer nasal chambers contain no olfactory receptors. In contrast, the epithelium of the relatively large inner chamber is lined with olfactory receptor cells, and the chamber is connected to the mouth by a pair of large ducts, or choanae. The choanae probably carry odors from the mouth directly to the olfactory receptors enabling a bird to smell the food while holding it in the mouth (Welty 1975). Thus, the waxwings had to pick up the berries in order to detect the MA.

DISCOURAGING BIRD USE OF SMALL BODIES OF WATER BY MA TREATMENT

Introduction

The use of certain small bodies of water creates problems for birds as well as for humans. Examples of the former include agricultural drainwater evaporation ponds and gold mine tailing ponds each of which contain toxins detrimental to birds. Human safety is endangered when birds are attracted to temporary puddles on airport runways. Nuisance problems develop when waterfowl are attracted to the ponds in parks, lawns, and golf courses. In this study, we investigated the feasibility of applying MA to discourage birds use of small bodies of water.

Methods

Seven American coots (<u>Fulica</u> <u>americana</u>) were captured near Gainesville, Florida, and maintained in the 0.2 ha flight pen at the field station. The birds were allowed to forage freely within the flight pen, and we provided them with free access to water, except during the MA trial. After several weeks of acclimation to the flight pen, we presented the coots with 2 20-1 pans of water. To one pan we added 100 ml of MA plus 50 ml of isopropyl alcohol as a dispersant. The other pan received just the alcohol. The behavior of the coots were monitored from an observation blind, approximately 25 m from the water pans.

We recorded the birds' behavior on video tape to determine the number of times the birds dipped their bills in the pans to drink and the number of times they entered the pans to bathe. We also measured the water level in the pans each morning and afternoon. After 7 days we removed the untreated pan and monitored activity and water levels as before. We placed a pan of untreated water under bird-proof netting to record moisture gained or lost to the environment. During the 1-pan test the coots' only source of untreated water was seepage from the concrete bases of an irrigation pipe.

Results

During the taped observation periods, the coots used the MA-treated pan on the initial test day but not thereafter (Table 1). Water loss from the MA-treated pan averaged 0.5 cm/day compared to 4.8 cm/day from the untreated pan. The coots continued to avoid the MA-treated water during the 1-pan test. Video-taped observations on days 1, 2, and 4 (total time = 376 min) showed no bill dips or entries into the MA-treated water. Daily water loss from the treated pan averaged 0.3 cm compared to 0.2 cm from the control water pan.

When added to water, the MA quickly separated and sank to the bottom of the pan. Initially clear, the water in the treated pan soon acquired an orange tinge and became progressively darker. Eventually, the treated water was dark reddish-brown and murky. After 2 wks, the MA odor was very noticeable several m from the pan.

Table 1. Coot use of MA-treated and plain water pans.

Test day	Observation period (min)	Bill dips		Entries	
		MA	Plain	MA	Plain
1	95	16	62	7	25
2	120	0	63	0	19
5	109	0	50	0	15
6	122	0	82	0	5
7	120	0	83	0	24

Discussion

This limited evaluation indicates that MA can greatly reduce bird use of small bodies of water. The coots in this study were not immediately deterred by the treated water, but after limited use on the first day of exposure, they avoided it completely. Eventual application of this method to actual management situations requires additional tests with different dose rates, species, water depths, and MA formulations.

PROTECTION OF EGGS FROM AVIAN PREDATORS

Introduction

Crows and ravens regularly prey on the eggs of other species (e.g., Post 1990). Control of such depredations usually has involved lethal means -- shooting, poisoning, trapping. Egg predation control methods based on food avoidance learning have been tested (Nicolaus 1987), and here we present results of feeding experiments using MA to help produce a conditioned avoidance response in captive crows.

Methods

Fish crows (Corvus ossifragus) were trapped locally and housed individually in 1.3 x 9.3 x 1.8 m test enclosures. For 5 consecutive days, during 0800-1200, each bird was offered 2 treated quail eggs. We evaluated 2 treatments: methiocarb only and methiocarb-MA combined. Water and commercial dry dog food were available throughout. An egg was considered lost if after 4 h it had been moved from its original location even if the crow did not eat it.

To prepare the treated eggs, we injected an emetic dose (30 mg/egg) of methiocarb into raw quail eggs, then boiled them and sealed the openings with glue. Technical grade MA was mixed 1:1 with propylene glycol, and 0.15 ml of this mixture was applied with a syringe to each egg immediately prior to the daily trial.

Results and Discussion

Crows were not deterred from removing the quail eggs by the methiocarb treatment. Virtually every egg was taken on each of the test days ($\bar{x}$ = 1.7 egg/day, SD = 0.7). In contrast, methiocarb-MA-treated eggs were mostly avoided ($\bar{x}$ = 0.4, SD = 0.7). Some birds sampled the eggs throughout the test, so total deterrence was not achieved.

Protection of eggs is an all-or-nothing proposition. Eggs taken from a nest are lost whether they are eaten or not. Thus, to be effective, the repellent must cause total avoidance of the eggs. It appears from our preliminary results that an emetic compound inside the egg will not by itself produce an avoidance reponse. Rather, a topical treatment using MA or a similarly irritating compound paired with the emetic is more likely to elicit the desired response by the egg predator.

REFERENCES

Denslow, J. S., Levey, D.J., Moermond, T.C., and Wentworth, B. C., 1987, A synthetic diet for fruit-eating birds, Wilson Bull., 99:131-134.
Mason, J. R., Adams, M. A., and Clark, L., 1989, Anthranilate repellency to starlings: chemical correlates and sensory perception, J. Wildl. Manage., 53:55-64.
Nicolaus, L., 1987, Conditioned aversions in a guild of egg predators: implications for aposematism and prey defense mimicry, Am. Midl. Nat., 117:405-419.
Post, W., 1990, Nest survival in a large ibis-heron colony during a three-year decline to extinction, Colon. Waterbirds, 13:50-61.
Welty, J. C., 1975, "The Life of Birds", 2nd ed., W. B. Saunders, Philadelphia.

VISUAL CUE FAILS TO ENHANCE BIRD REPELLENCY OF METHIOCARB IN

RIPENING SORGHUM

Richard A. Dolbeer, Paul P. Woronecki, and Roger W. Bullard

U.S. Department of Agriculture, Denver Wildlife Research
Center, 6100 Columbus Avenue, Sandusky, OH 44870 (RAD, PPW)
P. O. Box 25266, Denver, CO 80225 (RWB)

INTRODUCTION

Methiocarb (Mesurol®) produces a conditioned aversion which birds
associate with a particular treated food and then avoid (Rogers 1974).
Numerous field tests have shown that methiocarb is generally effective
in reducing bird damage to ripening fruits when applied at rates of 1 to
2 kg/ha. Methiocarb was registered in the United States by the
Environmental Protection Agency for use in cherries and blueberries
during the 1980's (Dolbeer et al. 1988). These registrations were
withdrawn by the proprietary company in 1989 because of additional
studies required by EPA related to methiocarb residues and environmental
effects.

These registrations could perhaps be reinstated and additional
registrations for other crops obtained if application rates of
methiocarb could be reduced to lower levels without compromising
efficacy. Previous studies (Bullard et al. 1983) indicate that aversion
to methiocarb often occurs at levels lower than birds can discriminate
due to taste alone. To increase discrimination, inexpensive visual,
olfactory or tactile cues have been added to methiocarb in laboratory
tests to enhance detection and substantially reduce efficacious levels
(Bullard et al. 1983, Mason and Reidinger 1982, 1983, Avery 1984, Mason
1989, Avery and Nelms 1990). The cue, in theory, enhances the birds'
ability to recognize that a familiar food has been altered and to
associate this alteration with illness.

The study reported here represents an effort to reduce the amount
of methiocarb used in field applications by treating a crop (ripening
sorghum) with methiocarb in association with a visual cue. The specific
hypothesis tested was that methiocarb, applied to fields of ripening
sorghum at 1.1 kg/ha in association with a visual cue [calcium carbonate
(Elmahdi, et al. 1985)], would result in less bird damage than would
occur in fields with methiocarb alone. Previous field studies have
indicated that application rates of 2 to 3 kg of methiocarb (without

Chemical Signals in Vertebrates VI, Edited by R.L. Doty and
D. Müller-Schwarze, Plenum Press, New York, 1992

visual cue)/ha are necessary to protect small grain crops such as sorghum from birds (DeHaven et al. 1971, Crase and DeHaven 1976, Bruggers et al. 1981, 1984).

METHODS

Nine 60-row by 171-m (0.8 ha) plots of sorghum (Jacques 377-w) were planted on 17 June 1987 in a 30-ha field at Ottawa National Wildlife Refuge, Lucas County, Ohio. Plant spacing averaged 5 cm and row spacing was 0.75 m within plots. Plots were 60 to 90-m apart (Fig. 1). The field was 2 km S of Metzger Marsh, a traditional late-summer roosting site for blackbirds on the Lake Erie shoreline (Dolbeer 1980).

Starting 15 August and at 2-day intervals until 10 September, an observer walked through each plot and examined 100 arbitrarily selected plants to determine the date when 50% of the plants had infloresced (head or panicle emerged from boot) and when ripening grains received initial bird damage. Maturity differed among plots because of variable field moisture following planting; therefore, the plots were grouped into three maturity groupings or blocks (Fig. 1). One of three treatments (methiocarb at 1.1 kg/ha A.I.; methiocarb at 1.1 kg/ha A.I. plus visual cue at 15.7 kg of calcium carbonate/ha; and untreated) was assigned randomly to each of the three plots within each maturity grouping.

Methiocarb was obtained from Mobay Chemical Corporation (Mesurol[R] 75% WP). The visual cue was an economy interior flat latex paint with the white pigment replaced with calcium carbonate at 0.88 kg/L. For application, visual cue was mixed with water in ratio of 1:1.6. The treatments were applied by a licensed aerial applicator using a Pawnee C agricultural spray aircraft. On treatment days, the applicator first applied visual cue to the three designated plots. Then, the applicator landed, rinsed the tank, and refilled the tank with Mesurol[R] which was promptly applied to the six designated plots.

The first application was on 10 September, the day after initial bird damage had been detected in the plots. Two subsequent scheduled applications were made on 21 and 30 September and a supplemental application was made on 14 September following a 3.0-cm rainfall on 11 September. No other methods to repel birds were employed in the plots.

On six dates from 14 September to 8 October, a total of 14 upper leaves from randomly selected sorghum plants in the methiocarb-plus-cue plots were photogaphed with a reference measurement scale. The number of white spots was counted on four randomly located areas of the leaf in each photograph. The diameters of a sample of spots were also measured.

Bird damage was assessed in the plots at 7, 14, 21, 28 and 67 days after the first application (10 Sep). In each plot, eight rows were randomly selected (two each from rows 1-15, 16-30, 31-45 and 46-60). In each row, a subplot was randomly located within the first 24 m from the plot edge and five subsequent subplots were at 24-m intervals. At each subplot the lengths of four heads (one plant in each of four adjacent rows) were measured to nearest cm and the percent of grain removed by birds was visually estimated (Manikowski and DaCamara-Smeets 1979, Seamans and Dolbeer 1989).

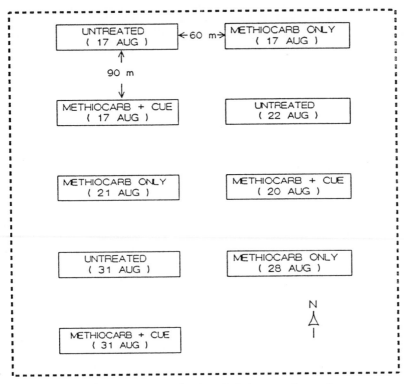

Fig. 1. Schematic map of nine 0.8-ha sorghum plots at
         Ottawa National Wildlife Refuge, Lucas County,
         Ohio. Dates in parentheses are when 50% of the
         plants had flowered.

The population size of the blackbird roost at Metzger Marsh was
estimated on 1, 5, 9 and 17 September. Two or four observers estimated
numbers of blackbirds [Red-winged blackbirds (*Agelaius phoeniceus*),
common grackles (*Quiscala quiscula*), brown-headed cowbirds (*Molothrus
ater*)] and starlings (*Sturnus vulgaris*) entering the roost in the
evening by "block counting" flightlines of birds (Meanley 1965).

From 27 August until 8 October, bird observations were made in
plots between 0800 and 1800 h, 1 to 4 times per day on 3 to 4 days per
week. Each plot was observed for 2 minutes from an elevated spot 60 to
90 m from the plot edge. The number of blackbirds in the plot was
estimated. The sequence of plots observed was randomly chosen each
time, beginning with either the south or north end of the test area.

A repeated measures, randomized block analysis of variance was
used to test for differences in mean daily percent loss among the three
treatment groups and between two time periods. Time periods were 10
September to 8 October (date of first treatment application to 8 days
after last application) and 9 October to 16 November (posttreatment
damage period). In addition, a randomized block analysis of variance
was used to test for differences in final percent loss among the three
treatment groups. An arc sine transformation was performed on the
percent damage estimates for individual heads. Bird numbers were

compared among the three treatment groups by a chi-square analysis of the proportion of observations in which a flock of ≥ 100 blackbirds was observed.

RESULTS

Visual-cue Characteristics

The visual cue distinctly speckled the upper leaves of plants with white spots 1 to 3 mm in diameter at a density of 8.6 ± 4.1 spots/cm^2 ($\bar{x}$ $\pm$ SD, $\underline{n}$ = 14). Persistence of spots was generally excellent. For example, on 8 October, 9 days after the last application of material, plants in the methiocarb-plus-cue plots were still distinctly marked. Plants in methiocarb-only treated fields had a very light coating of white spots that were generally <1 mm in diameter and much less distinctive and persistent than in the visual cue plots. The visual appearance of the two groups of treated fields was obviously different to human observers.

Bird Numbers in Roost and in Plots

The estimated numbers of blackbirds and starlings entering the roost in the evening were 50,000, 50,600, 15,900 and 55,200 on 1, 5, 9 and 17 September, respectively. Species composition was not determined but our subjective observations indicated that red-winged blackbirds predominated with lesser numbers of brown-headed cowbirds, common grackles and starlings.

Blackbird numbers were generally low in all plots during the pretreatment observation period (27 Aug - 9 Sep) with three of 45 plot observations recording a flock of ≥ 100 birds. During the treatment period, 26 of 216 plots observations had ≥ 100 birds with the proportion being significantly ($\underline{P}$ < 0.01) different among the three treatment groups: 17 in untreated plots, 5 in methiocarb-only plots and 4 in methiocarb-plus-cue plots (Table 1). Overall, the respective treatment groups averaged 30, 0 and 117 birds per plot during observations in the pretreatment period and 103, 35 and 22 during the treatment period. No sick or dead birds were noted during bird observations or during the six damage assessments conducted in the plots.

Blackbird Damage

There were significant differences among the three treatment groups in the mean daily percent loss and in the final percent loss of grain (Table 2). Both groups of methiocarb-treated plots averaged significantly less final loss (29%) than did the untreated plots (48.3%). The temporal pattern of loss was almost identical for the methiocarb-only and methiocarb-plus-cue plots (Fig. 2), indicating that although the birds responded negatively to methiocarb, the visual cue had no apparent additional impact.

There was a significant time x treatment group interaction for the mean daily percent loss (Table 2). Daily percent loss in both groups of methiocarb-treated plots increased during the posttreatment period (9 Oct - 16 Nov) to levels similar to that in control plots. However, because losses had been substantially reduced in the methiocarb-treated plots during the treatment period, final losses in these plots were still significantly less than in the untreated plots (Fig. 2).

DISCUSSION

This study demonstrated that methiocarb, applied to ripening sorghum at the rate of 1.1 kg/ha, can significantly suppress bird damage for at least an 8- to 10-day period following application. Thus, the study confirms previous work (cited in introduction) showing methiocarb to be an effective bird repellent for small grain crops and indicates that rates as high as 2 to 3 kg/ha may not always be necessary. However, the hypothesis that a visual cue would enhance the effectiveness of methiocarb was not supported.

The failure of the visual cue to enhance repellency could be because the methiocarb-only treatment also left a light coating of white spots on the leaves. Although not nearly as prominent or persistent as the calcium carbonate spotting, the birds may have quickly learned to associate any type of spotting with methiocarb. Furthermore, the relatively close proximity (60 to 90 m) of the plots may have aided the birds in quickly discerning treated from untreated plots, regardless of the degree of visual cue present.

Table 1. Number of 2-minute observation periods in which a flock of ≥ 100 and <100 blackbirds were recorded in 9 sorghum plots, Ottawa National Wildlife Refuge, Lucas County, Ohio during the pretreatment period (PTP) (27 Aug to 9 Sep) and the treatment period (TP) (10 Sep to 9 Oct).

No. of observations	Untreated plots (n=3) PTP	TP	Methiocarb-only plots (n=3) PTP	TP	Methiocarb-plus-cue plots (n=3) PTP	TP
with ≥ 100 birds	2[a]	17[b]	0[a]	4[b]	1[a]	5[b]
with <100 birds	13	55	15	68	14	67
Total	15	72	15	72	15	72

[a]The proportion of observations with ≥ 100 birds was not significantly (P = 0.40) different among the 3 treatment levels during the pretreatment period ($\underline{X}^2$ = 2.14, 2 df).
[b]The proportion of observations with ≥ 100 birds was significantly (P < 0.01) different among the 3 treatment levels during the treatment period ($\underline{X}^2$ = 13.7, 2 df).

The fact that bird damage was not suppressed in the methiocarb-treated plots during the posttreatment period (9 to 47 days following the final application) may be related to the rapid turnover in blackbird populations at this time. Previous studies (Dolbeer 1978) have demonstrated that late October and November is the peak period of autumn migration for blackbirds in the Great Lakes region. Thus, most blackbirds feeding in the plots during the posttreatment period were probably migrants with no prior conditioning to the methiocarb-treated grain.

Table 2. Estimated mean daily and final percent loss of grain to blackbirds in nine plots of sorghum at Ottawa National Wildlife Refuge, Lucas County, Ohio, 1987.

Damage measurment	Untreated plots (n = 3)		Methiocarb only plots (n = 3)		Methiocarb-plus-cue plots (n = 3)	
	x̄	SD	x̄	SD	x̄	SD
Daily % loss during treatment period (10 Sep-8 Oct)	0.98	0.77	0.20	0.20	0.25	0.22
Daily % loss during posttreatment period (9 Oct-16 Nov)	0.54	0.03	0.60	0.25	0.56	0.24
Daily % loss[ac] (10 Sep-16 Nov)	0.72A	0.33	0.43B	0.23	0.43B	0.24
Final % loss[bc] (16 Nov)	48.3A	22.3	29.0B	15.1	29.0B	16.1

[a]There is a significant difference among treatment group means (F = 16.02, 2 and 4 df, $P < 0.01$) and a significant treatment group x time period interaction (F = 7.24, 2 and 6 df, $P = 0.03$).
[b]There is a significant difference among treatment group means (F = 23.1, 2 and 4 df, $P < 0.01$).
[c]Means with different letters are significantly ($P < 0.01$) different, Duncan's multiple range test.

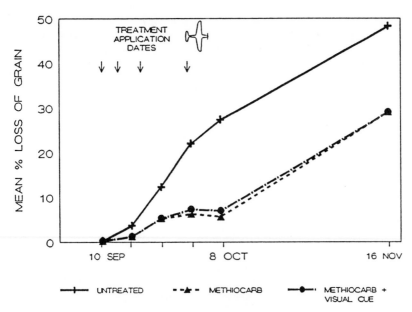

Fig. 2. Blackbird damage to three 0.8-ha plots of sorghum
in each of three treatment groups, Ottawa National
Wildlife Refuge, Lucas County, Ohio. Plots were
assessed for damage on six dates from 10 September
to 16 November 1987.

If further field tests of the sensory-cue hypothesis are
conducted, a more distinctive cue should be used. Perhaps prominently
marking the boundaries of repellent-treated plots with, for example,
mylar reflective ribbons (Dolbeer et al. 1986), would be a more
practical and effective cue than attempting to spray all plants with
particles of a distinctive color. Alternatively, the use of a
distinctive gustatory cue, such as tannin or acetic acid (Bullard et al.
1983, Shumake et al. 1976), mixed with the repellent formulation, might
be further explored. The fact that controlled tests with birds in cages
have repeatedly shown sensory cues to enhance chemical repellents
indicates that this concept should not yet be abandoned for field
situations.

Acknowledgements: We thank G. E. Bernhardt, E. J. Bly, E. C. Cleary, J.
R. Mason, E. Rodriguez, T. W. Seamans, and J. A. Shieldcastle for
assistance with field work. W. and J. Sharlow planted the crop, Gibbs
Aerospray, Inc., Fremont, Ohio applied the treatments, Ottawa National
Wildlife Refuge provided use of the farmland, and American Colors, Inc.,
Sandusky, Ohio provided material and mixed the visual-cue formulation.

REFERENCES

Avery, M. L., 1984, Relative importance of taste and vision in reducing
    bird damage to crops with methiocarb, a chemical repellent,
    Agric. Ecol. and Environ., 11:299-308.
Avery, M. L., and Nelms, C. O., 1990, Food avoidance by red-winged
    blackbirds conditioned with a pyrazine odor, Auk, 107:544-549.

Bruggers, R., Matee, J., Miskell, J., Erickson, W., Jaeger, M., Jackson, W. B., and Juimale, Y., 1981, Reduction of bird damage to field crops in eastern Africa with methiocarb, Trop. Pest Manage., 27:230-241.

Bruggers, R. L., Bohl, W. H., El Bashir, S., Hamza, M., Ali, B., Besser, J. F., DeGrazio, J. W., and Jackson, J. J., 1984, Bird damage to agriculture and crop protection in the Sudan, FAO Plant Prot. Bull., 32:2-16.

Bullard, R. W., Bruggers, R. L., Kilburn, S. R., and Fiedler, L. A., 1983, Sensory-cue enhancement of the bird repellency of methiocarb, Crop Prot., 2:387-398.

Crase, F. T., and DeHaven, R. W., 1976, Methiocarb: its current status as a bird repellent, Proc. Vertebr. Pest Conf., 7:46-50.

DeHaven, R. W., Guarino, J. J., Crase, F. T., and Schafer, E. W., Jr., 1971, Methiocarb for repelling blackbirds from ripening rice, Internat. Rice Comm. Newsl., 20(4):26-30.

Dolbeer, R. A., 1978, Movement and migration patterns for red-winged blackbirds: a continental overview, Bird-Banding, 49:17-34.

Dolbeer, R. A., 1980, Blackbirds and corn in Ohio, U.S. Fish and Wildl. Serv., Resour. Publ. 38, 18 pp.

Dolbeer, R. A., Woronecki, P. P., and Bruggers, R. L., 1986, Reflecting tapes repel blackbirds from millet, sunflowers, and sweet corn, Wildl. Soc. Bull., 14:418-425.

Dolbeer, R. A., Avery, M. L., and Tobin, M. E., 1988, Assessment of field hazards to birds and mammals from methiocarb applications to fruit crops, Denver Wildl. Res. Cent. Bird Sect. Res. Rep. No. 421, 43 pp.

Elmahdi, E. M., Bullard, R. W., and Jackson, W. B., 1985, Calcium carbonate enhancement of methiocarb repellency for quelea, Trop. Pest Manage., 31:67-72.

Manikowski, S., and DaCamara-Smeets, M., 1979, Estimating bird damage to sorghum and millet in Chad, J. Wildl. Manage., 43:540-544.

Mason, J. R., 1989, Avoidance of methiocarb-poisoned apples by red-winged blackbirds, J. Wildl. Manage., 53:836-840.

Mason, J. R., and Reidinger, R. F., 1982, Observational learning of food aversions in red-winged blackbird (Agelaius phoeniceus), Auk, 99:548-554.

Mason, J. R., and Reidinger, R. F., 1983, Importance of color for methiocarb food aversions in red-winged blackbirds, J. Wildl. Manage., 47:383-393.

Meanley, B., 1965, The roosting behavior of the red-winged blackbird in the southern United States, Wilson Bull., 77:217-228.

Rogers, J. G., Jr., 1974, Responses of caged red-winged blackbirds to two types of repellents, J. Wildl. Manage., 38:418-424.

Seamans, T. W., and Dolbeer, R. A., 1989, A comparison of sorghum damage assessment methods, Denver Wildl. Res. Cent. Bird Sect. Res. Rep. No. 433, 11 pp.

Shumake, S. A., Gaddis, S. E., and Schafer, E. W., Jr., 1976, Behavioral response of quelea to methiocarb (Mesurol), Proc. Bird Control Semin., 7:250-254.

SECTION SEVEN

BEHAVIOR AND CHEMICALLY-MEDICATED SOCIAL COMMUNICATION

PART 1: Fish

ARE SHARKS CHEMICALLY AWARE OF CROCODILES?

L.E.L. Rasmussen[1] and Michael J. Schmidt[2]

[1]Department of Chemical and Biological Sciences, Oregon
Graduate Institute of Science and Technology, Beaverton, OR
[2]Washington Park Zoo, Portland, OR

INTRODUCTION

The American crocodile, Crocodylus acutus, inhabits a coastal,
estuarine environment; its Atlantic distribution has included all the
major islands of the West Indies, the Atlantic coastal lowlands from the
southern coasts of Florida, and from Tamaulipas in Mexico to Venezuela
(Campbell and Winterbotham, 1985; Schmidt and Ingor, 1957; Minton and
Minton, 1973). The lemon shark, Negaprion brevirostris, frequents this
same coastal, estuarine environment (Springer, 1938, 1940; Gruber, 1982).

As apex predators, sharks have few natural enemies. Overlooked in
previous studies of interspecies interactions of these elasmobranchs
(Baldridge, personal communication) has been the substantial evidence for
predation of crocodiles on sharks in the estuarine environment (Cott,
1961; Daniel, 1983; Hutton, 1987; Messel et al., 1981; Worrel, 1964; Webb
and Manolis, 1988). The intriguing possibility that sharks may be aware
of the presence of potential danger by sensing chemical exudates from
crocodiles prompted this investigation of a possible semiochemical
interaction between C. acutus and N. brevirostris.

Results obtained during this investigation demonstrate that several
substances that exude from the crocodile, probably from feces or possibly
from the chin glands, the cloacal glands (Wright and Moffat, 1985),
and/or the urine, are present in the water of holding tanks containing
crocodiles. These crocodile-derived compounds, when presented to sharks,
elicit a visually discernible response; namely, the reversal of tonic
immobility, a temporarily quiescent state exhibited by a number of
vertebrates (Burghardt and Greene, 1988; Gallup, 1974; Gallup and Maser,
1977; Gruber, 1983; Pivik et al., 1981).

METHODS

Three-liter water samples were collected from separate habitat tanks
(20 by 30 feet by several feet deep) containing C. acutus, American
crocodile, at Busch Gardens, Tampa, Florida. Water flow had been stopped
for 72 hours prior to sample collection. At Gatorama, Florida, similar
water samples were obtained from ponds containing A. mississippiensis,
American alligator, and C. acutus. First, the water samples were bio-

assayed at the collected concentration. Subsequently, the water was concentrated 10- to 1000-fold by vacuum centrifugation prior to bioassay.

The animals utilized for the bioassay were immature lemon sharks, N. brevirostris, about equal numbers of males and females. The bioassays were conducted over a period of five years using six different sets of lemon sharks in three locations. The reversal of induced tonic immobility, an established test system for chemical awareness, was the response monitored for the bioassay. Tonic immobility has been documented in the immature lemon shark (Watsky and Gruber, 1990). Following inversion and restraint, the lemon sharks rapidly achieved tonic immobility. Normal tonic immobility duration for this species is 10 to 15 minutes. Accordingly, each animal was in the test situation less than 10 minutes. Test substances at various concentrations were dissolved in 5 ml of shark saline maintained at ambient habitat temperature. During the 1- to 2-minute test period (per sample), the 5 ml solution was slowly released (42 ul/sec) into the seawater within 0.5 cm of an incurrent olfactory nare of the test shark recently placed in tonic immobility. Eye blinking was monitored to indicate possible mechanical disturbances. For substances testing negative, each nostril in an individual shark was tested successively. In initial studies, because of the constraints of availability of test animals, each shark was used for 2 to 3 tests of randomly chosen samples, with 2- to 5-minute intervals between tests. Subsequently, more rigorous assays subjected an individual shark to only 1 test substance per 6 hour period. Testing was randomized by chance selection of all test samples and sharks. Single substances were tested 5 times using different sharks. All experiments were conducted as double blind tests. Reversals of tonic immobility were scored as positive responses. Absence of a visible response was scored as a negative. During the clearly positive response, the test shark demonstrated increased twitching and successfully righted itself by struggling out of the inverted position.

Concentrated water samples that elicited a positive response in the test system of reversal of tonic immobility were first extracted with diethylether and then suspended in dichloromethane and separated by high performance liquid chromatography (HPLC). A normal phase silica column (25 cm x 4 mm) was used. During the 45-minute separation, a linear gradient elution from 100% dichloromethane to 90% dichloromethane: 10% acetonitrile at a flow rate of 2 ml/minute and 2000 psi was used. The solvents used during extraction and fractionation were also tested to rule out false positive responses.

Subsequently, analyses of the fractions eliciting reversal of tonic immobility were performed on a VG 7070 E-HF double-focusing mass spectrometer to identify specific chemical compounds. The system used electron impact and/or chemical ionization at 70 eV and was equipped with a 11/250 data system. The NBS library of chemical compounds was utilized in computer searches for compound identification. After initial identification, the retention time of standards of authentic compounds were compared by capillary column gas chromatography and confirmed by duplicate runs on gc/ms.

Additional confirmation of the identifications of the 3 crocodile-water-derived compounds was provided by both ultraviolet/visible absorption spectral data and infrared spectral data comparison with their synthetic counterparts. A Perkin Elmer Lambda 9 recording spectrometer was used for ultraviolet/visible spectral analyses of samples suspended in methanol. Samples on sodium chloride circular crystal windows were analyzed using a Perkin Elmer 1800 Fourier Transform Infrared Spectrometer with a 7500 computer system.

RESULTS

Samples of native water gathered from C. acutus ponds tested positive, i.e., reversed tonic immobility in juvenile lemon sharks in 53% of the trials (Table 1). In contrast, the similar water collections from the ponds of A. mississippiensis tested 100% negative (Table 1). After 500- and 1000-fold concentration, the C. acutus waters elicited reversal of tonic immobility in 67% and 80% of the tests, respectively (Table 1).

Bioactivity, as represented by tonic immobility reversal and indicating substances eliciting the bioresponse, was located specifically in HPLC fraction-3 (Fr-3, eluting 10-15 minutes after the initiation of the gradient). In 10-15 tests using 3-5 sharks, evaporated equivalents (0.01-1%) or dilution equivalents of the solvents (acetonitrile, ethanol and dichloromethane) and dilutants (seawater, shark saline and pond water) all tested negative. Controls included extracts of land carnivore feces and fish tissues and blood (Table 1). In certain instances these samples were positive (see Discussion).

Table 1. Natural Extracts Tested

Substance	Test Sharks[2]	Total Tests[2]	Positive	Negative	Reversal
Crocodilian Substances					
Habitat Water[1]					
native	5	15	8	7	53%
500 XC[3]	3	6	4	2	67%
1000 XC[3]	2	5	4	1	80%
HPLC Fr.3[4]	5	10	8	2	80%
Controls					
Alligator water[1]	3	6	0	6	0%
Carnivore feces (Bengal tiger)					
water extract	3	6	0	6	0%
organic extracts	9	18	11[5]	7	50-67%[5]
Smelt muscle					
water extract	3	6	0	6	0%
CH2Cl2 extract	3	4	2[5]	2	50%[5]
acetone extract	3	5	5	0	100%
Teleost blood					
fresh	2	4	1	3	25%
aged (2 days)	2	4	2	2	50%

[1] 72 hours of non-circulation in habitat tank. Alligator water assayed both at native concentration and 1000 XC^3. [2] Each shark used for 3 non-successive tests. [3] XC = X-fold concentration by vacuum centrifugation. [4] Pre-gradient fraction and 0 to 10 minutes gradient were inactive; Fr-3 (active) eluted between 10 and 15 minutes gradient. [5] slow response (occurred at least 30 seconds from start of test).

Gas chromatographic separation of the active HPLC fraction (Fig. 1) was combined with mass spectrometric analysis to identify three specific compounds. By electron impact (EI) mass spectrometry the primary substance had a molecular weight of 141. Its EI spectrum contained the following mass/charge (m/z) {abundance}: 141 {25}, 113 {100}, 98 {26}, 85 {15}, 70 {64}, 55 {50}, and 42 {24}. By chemical ionization mass spectrometry there was an M+H at m/z 142. This compound (a) was identified as 2-ethyl-3-methylsuccinimide (3-ethyl-4-methylpyrrolidine-2,5-dione) (Figure 2). Two other related compounds, 2-ethyl-3-methylmaleimide (3-ethyl-4-methyl-1H-pyrrole-2,5-dione)(b) and 2-ethylidene-3-methylsuccinimide (c), were also identified.

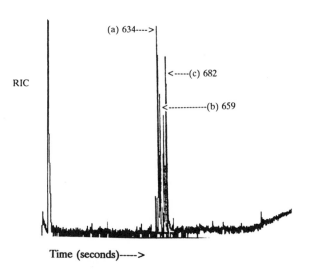

Fig. 1. Conditions: 60 meter DB-1 fused silica column; helium carrier gas; temperature programmed 6° C/minute, 35° C to 350° C; x-axis, retention time in seconds; y-axis, the relative intensity (RIC). At 634 sec compound (a), at 682 sec compound (b), and at 700 sec compound (c) eluted.

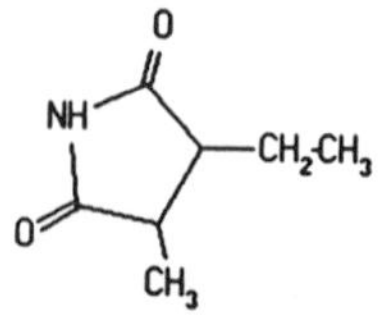

Fig. 2. Compound (a), 2-ethyl-3-methylsuccinimide. In compound c the ethyl group at position 2 is replaced by an ethylidene group and in compound b (2-ethyl-3-methylmaleimide) a double bond is present between positions 2 and 3.

All three compounds were first identified by gc/ms and confirmed by comparison of mass spectra and gc retention times (using both internal and external standards) with those of pure synthetic compounds, both synthesized in our laboratory (Fuhrhop and Smith, 1975; Killilea and O'Carra, 1978) and commercially obtained. The mass spectra of the naturally derived compounds and prepared synthetics were similar. For 2-ethyl-3-methylmaleimide the m/z (abundance) were 140 {2}; 139 {99}; 124 {63}; 11 {16}; 110 {23}; 96 {43}; 95 {16}; 94 {16}; 81 {14}; 68 {75}; 67 {99}; 53 {100}. The crocodile-derived compound (a) 2-ethyl-3-methylsuccinimide and its synthetic (Petterson, 1980) had both identical EI and CI mass spectra.

Bioassays of the synthetic compounds also supported our identifications. Synthetic 2-ethyl-3-methyl succinimide, at least in part an isomeric mixture, elicited a strong reversal of tonic immobility in a slow fashion at concentrations between 10^{-9}-10^{-7} M in 65% of the tests (Table 2). At similar concentrations, 2-ethyl-3-methylmaleimide elicited rapid reversal of tonic immobility in 80% of the tests. The results from these two compounds confirmed that this material accounted, at least in part, for the results observed using water from the crocodile ponds. The third compound, identified as 3-ethylidene-4-methylpyrrolidine-2,5-dione, was available for only a single bioassay during this study.

Table 2. Single Compounds[1] Eliciting Reversal of Tonic Immobility

Substance	Concentrations (1 x 10^{-7} to 1 x 10^{-9} M) in Shark Saline					
	Test Sharks	Number of Tests	Positive Slow	Fast	Mixed	Negative
succinimide	5	10	4		0	6
succinic acid	5	10		1	0	9
maleimide	5	7		5	1	1
n-ethylmaleimide	4	8		6	0	2
maleic acid	4	8		4	1	3
2-ethyl-3-methyl maleimide[1]	5	10	8		0	2
2-ethyl-3-methyl succinimide[1]	5	10		6	1	3
2-ethylidene-3-methyl succinimide[1,2]	1	1		1		

[1]Identified in crocodile waters. [2]One test only. [3]The following substances tested negative in all tests; 10 total tests utilizing at least 3 different sharks: D-serine, L-alanine, L-ethionine, L-glutamic acid, L-methionine, L-norleucine, L-glutamic acid δ-methyl ester, L-ornithine, pyrrole, purine, urea, holothurin, stereobilin, 4-butyl-1-phenyl-3,5 pyrazolidinedione, 1-phenyl-3-pyrazolidinone, phenylpropiolic acid, β-citronellol. Sodium dodecyl-sulfate, L-serine, and DL-farnesol were negative 80-90%. Slow responses refer to response times greater than 30 seconds.

Ultraviolet/visible absorption spectral data confirmed the identity of the three compounds. The naturally derived compounds had spectra similar to the spectra of the respective synthetics. The absorption maxima of 5 mM solutions in methanol were: 221 nm (0.49A) and 247 nm (0.23A) for compound (b); 221 nm (0.5A) and 227 nm (0.45A) for compound (b); and 221 nm (0.2A) and 252 nm (0.5A) for compound (c).

Infrared spectral data also confirmed the suggested identities as the spectra of the crocodile-derived compounds were identical to the spectrum of the synthetics. For example, the stretching carbonyls in the 1769-1712 CM^{-1} region varied slightly among the three compounds but were identical respectively to the appropriate synthetic.

DISCUSSION

Co-existence of some sharks and crocodiles in overlapping habitats has occurred since Mesozoic times. In estaurine habitats such overlap persists today in selected geographical locales, such as in Australian environs (salt water crocodile, Crocodylus porosus, and various shark species such as Sphryna lewini and Carcharhinus spallazini) and in northern hemispheric waters between the American crocodile and the lemon shark. Some crocodiles, " ... the top carnivores in many tropical wetlands..." (King, 1987), such as C. acutus, and some sharks, such as N. brevirostris, (Gruber et al, 1988; Nixon and Gruber, 1988) are apex predators. During their cohabitation and during food competition these two species probably are aware of each other. Usually the crocodile is the dominant species, but size probably is a deciding factor; smaller sharks may avoid areas where bigger crocodiles dwell. Whether such avoidance is learned or innate, the sensory perception of danger is a major factor in survival. Although other senses than olfaction certainly play a role, our study specifically focused on possible interspecies semiochemical communication.

Do crocodiles give off specific compounds into water that enables sharks to detect their presence chemically? We have identified 2-ethyl-3-methylsuccinimide, 2-ethyl-3-methylmaleimide, and 3-ethylidene-4-methylpyrrolidine-2,5-dione as the principal components in a fraction producing tonic immobility reversal. This fraction resulted from HPLC of concentrated pen waters of American crocodile. It is probable that such compounds have their origin in the feces. Possibly, there may be release from chin gland secretions. A possible biological origin of these succinimides and imides could be blood hemoglobin, presumably degraded via bilirubin. This is supported by some positive responses to other carnivore feces and aged blood and suggests that the anecdotal information from fishermen that sharks are aware both of blood and of decaying elasmobranch flesh may have some biochemical validity.

Synthetic 2-ethyl-3-succinimide also reversed tonic immobility in lemon sharks. At the present time we have not separated the several possible geometric (cis-trans) isomers, and we have not attempted to differentiate between these either in our native preparations or in synthetic material.

Lemon sharks may derive chemical information from these two compounds via the olfactory system, the gustatory system, or the common chemical sense systems. It is noteworthy that reversal of tonic immobility occurred in response to physiological concentrations (ppm) in water of these compounds placed in proximity of the incurrent nares of the olfactory system.

Could it be that such perception by lemon sharks of chemicals, evidencing the nearby presence of predatory crocodiles, might convey to those sharks timely information of such strategic value as to improve significantly their chances of survival, both as individuals and as a species? We think our preliminary observations strongly suggest this as a possibility. It remains to be studied whether such chemical interactions and any behavioral manifestations of them, if real, are learned or are perhaps inherited. The ability of the immature sharks to be chemically aware of such substances suggests the latter. We fully recognize the possibilities presented by our findings, and semiochemistry in general, for future studies on chemical mediation of behaviors, including repellency (Baldridge, 1990), by sharks and crocodilians in the natural environment.

ACKNOWLEDGEMENTS

We thank Dr. J. Olsen (Busch Garden Park, FL) and Mr. D. Thielen (Gatorama, FL) for their collection of the crocodilian habitat waters, Dr. S. Gruber (University of Miami and The Bimini Biological Station), Dr. C. Manire, Dr. R. Heuter (Mote Marine Laboratory, Sarasota, FL), Mr. Brad Wetherbee and Mr. Michael Braun for their assistance with the bioassay aspects of this study. We also thank Dr. H.D. Baldridge, Dr. G. Doyle Daves, and Dr. M. Silverstein for their comments on the chemical aspects of this study.

REFERENCES

Baldridge, H.D., 1990, Shark repellent: not yet, maybe never, _Milit_. _Med_., 155:358-361.

Burghardt, G.M., and Greene, H.W., 1988, Predator similation and duration of death feigning in neonate hognose snakes, _Anim_. _Behav_., 36:1842-1845.

Campbell, G.R., and Winterbotham, A.L., 1985, " The Natural History of Crocodilians with Emphasis on Sanibel Island's Alligators", Sutherland Press, Ft. Myers, FL, 154-156.

Cott, H., 1961, Scientific results of an inquiry into the ecology and economic status of the Nile crocodile (_Crocodilus_ _niloticus_) in Uganda and Northern Rhodesia, _Trans_. _Zool_. _Soc_. _London_., 4:295.

Daniel, J.C., 1983, "The Book of Indian Reptiles", Bombay Natural History Society, Bombay.

Fuhrhop, J., and Smith, K., 1975, _in:_ "Porphyrins and Metalloporphyrins", K. Smith, ed., Elsevier, Amsterdam.

Gallup, G.G., 1974, Animal hypnosis: Factual status of a fictional concept, _Psychol_. _Bull_., 21:836-853.

Gallup, G.G., and Maser, J.D. 1977, Tonic immobility: evolutionary underpinnings of human catalepsy, _in_ "Psychopathology: Experimental Models", J.B. Maser and M.E. Seligman, eds., Freeman, San Francisco, 334-357.

Gruber, S.H., 1982, Role of the lemon shark, _Negaprion brevirostris_ (Poey) as a predator in the tropical marine environment: A multidisciplinary study, _Fla_. _Scient_., 45:46-75.

Gruber, S.H., 1983, Protocols for a behavioral bioassay in "Shark Repellents from the Sea," B. Zahuranec, ed., AAAS Westview Press Co, 91-115.

Gruber, S.H., Nelson, D.R. and Morrissey, J.F., 1988, Patterns of activity and space utilization of lemon sharks, _Negaprion brevirostris_, in a shallow bahamian lagoon, _Bull_. _Mar_. _Sci_., 43:61-76.

Hutton, J., 1987, Growth and feeding ecology of the Nile crocodile, _J_. _Anim_. _Ecol_., 56:25-38.

Killilea, S.D. and O'Carra, P., 1978, Improved unidimensional thin-layer chromatographic system for the identification of bilin degradation products, _J_. _Chromatogr_., 166:338-340.

King, W., 1987, Changing outlook for crocodiles, _Species_ (Newsletter SSC), 10:38-39.

Messel, H., Vorlicek, G.C., Wells, A.J., and Green, W.J., 1981, "Survey of the Tidal River Systems in the Northern Territory of Australia and their Crocodile Populations", Pergamon Press, Brisbane.

Minton, S. and M. Minton., 1973, "Giant Reptiles", Scribners, New York, 50-65.

Nixon, A. and Gruber, S.H., 1988, Diel metabolic activity patterns of the lemon shark, Negaprion brevirostris, J. Exp. Zool., 248:1-6.

Pettersen, J., 1980, Adv. Mass Spectro., 8B:1291-7.

Pivik, R.T., Sircar, S. and Braun, C., 1981, Nuchal muscle tonus during sleep, wakefulness and tonic immobility in rabbits, Phys. Behav., 26:13-20.

Schmidt, K. and Ingor, R., 1957, "Living Reptiles of the World", Hamish-Hamilton, London, 47-68.

Springer, S., 1938, Notes on the sharks of Florida, Proc. Fla. Acad. Sci., 3:9-41.

Springer, S., 1940, The sex ratio and seasonal distribution of some Florida sharks, Copeia, 3:188-94.

Watsky, M. A., and Gruber, S.H., 1990, Induction and tonic immobility in the lemon shark, Negaprion brevirostris, Fish Physiol. Biochem., 8:207-210.

Webb, G.J. and Manolis, G., 1988, "Australian Saltwater Crocodiles", C. Webb, Australia, 33.

Worrell, E., 1964, "Reptiles of Australia", Angus and Robertson, Sydney.

Wright, D.E. and Moffat, L.A., 1985, Morphology and ultrastructure of the chin and cloacal glands of juvenile Crocodylus porosus (Reptilia, Crocodilia), in:"Biology of Australasia Frogs and Reptiles", G. Gigg, R. Shine, and H. Ehmann, eds., Surrey Beatty and Sons, Australia.

CHARACTERIZATION OF A SPAWNING PHEROMONE FROM PACIFIC HERRING

Joachim Carolsfeld[1], Nancy M. Sherwood[1], Ann L. Kyle[2],
Timothy H. Magnus[2], Steven Pleasance[3], and Henrik Kreiberg[4]

[1]Biology Department, University of Victoria,
Victoria, B.C., Canada, V8W 2Y2
[2]Department of Anatomy, University of Calgary
Calgary, Alberta, Canada, T2N 1N4
[3]Institute for Marine Biosciences, Halifax
N.S., Canada, B3H 3Z1
[4]Pacific Biological Station, Nanaimo, B.C., Canada, V9R 5K6

PACIFIC HERRING: A "SCHOOL" SPAWNING FISH

Pacific herring (Clupea harengus pallasi) spawn synchronously in
large schools of several million fish in the near-shore environment,
markedly discolouring the surrounding water with suspended milt. This
spawning activity has a rapid onset as a school moves inshore. Males
appear to initiate spawning in small pockets from which the activity
spreads to the rest of the school (Hay, 1985). Spawning continues for
several hours in small schools (Hourston and Rosenthal, 1976) and
several days in large schools, after which the fish move back into
deeper waters (Haegele and Schweigert, 1985). Without overt behavioural
interaction, the spawning female and male herring deposit gametes on
submerged vegetation or other surfaces as trails of sticky eggs or
viscous milt (Schaeffer, 1937). Over 20 layers of eggs result in some
cases (Haegele and Schweigert, 1985). The deposited milt, together with
milt released mid-water, gradually dissipates, resulting in the
milkiness typical of herring spawning sites. Hay (1985) suggests that
high concentrations of milt may inhibit spawning and hence regulate the
density of egg deposition.

Herring have been variously characterised as open substrate,
benthic spawners (Balon 1975), group spawners (Keenleyside 1979) or
promiscuous spawners (Turner, 1986). However, none of these
classifications adequately describes the most striking characteristic of
spawning in Pacific herring: it occurs in a large social grouping with
no courtship behaviour, competition, leadership, pair bonds, or
dominance relationships between individuals. This has been termed
"mass" spawning (Aneer et al., 1983; Haegele and Schweigert, 1985;
Hay, 1985), but is better described as "school" spawning because it
closely fits the definition of schooling behaviours (Keenleyside, 1979).

EVIDENCE FOR A SPAWNING PHEROMONE

A potent spawning pheromone contained in herring milt induces
spawning in ripe individuals of both sexes (Stacey & Hourston, 1982;

Chemical Signals in Vertebrates VI, Edited by R.L. Doty and
D. Müller-Schwarze, Plenum Press, New York, 1992

Sherwood et al., 1991) and is probably responsible for the spread of spawning activity throughout a school. Stacey and Hourston (1982) were able to elicit spawning behaviour in both spermiating and ovulated captive herring within a few minutes of exposure to herring milt or ripe herring testes filtrate. The response was not evoked by herring eggs or filtrates of testes of other fish species, and could not be elicited in herring of other maturational states.

We demonstrated that the spawning behaviour can also be elicited in individual herring in small aquaria. In a full response in these aquaria, the fish extends its gonadal papilla and spawns within one to three minutes of pheromonal stimulation. We have used this response as a bioassay for purification of the spawning pheromone (Sherwood et al., 1991).

STEROIDAL PHEROMONE CANDIDATES IN HERRING

Sorensen and Stacey suggest that reproductive pheromones of fish, as exemplified by the goldfish model, are largely excretory products of sexual hormones. Steroids and conjugated steroids of this kind have been the principal reproductive pheromones found to date (see Sorensen and Stacey, 1989).

A variety of reproductive steroids have been detected by radioimmunoassay in the milt of ripe Pacific herring (Scott et al., 1991). Of these, the principal components were conjugated and free 17α,20β-dihydroxyprogesterone. Also present in smaller quantities were conjugated and free 17α,21-dihydroxyprogesterone, 17α,20β,21-trihydroxy-progesterone, testosterone, 11- ketotestosterone and several 5β(3α)-reduced steroids. The 17α,20β di- and 17α,20β,21-trihydroxy-progesterones, respectively, have been implicated as pheromones in the goldfish and Atlantic croaker, in which they act as the final egg maturation hormone (Stacey and Sorensen, 1986; Trant et al., 1986). The 17α,20β-dihydroxyprogesterone may also be a male pheromone in the goldfish (Sorensen and Stacey, 1989).

Two steroidal compounds, not generally considered to be reproductive hormones, have also been found in expressed herring milt. Conjugated and free cortisol were the major components detected by radioimmunoassay (Scott et al., 1991), and taurocholic acid was the major component found by HPLC linked to mass spectrometry (unpublished results). Both may have originated as contamination from blood, urine, or bile.

HERRING PHEROMONE IS NOVEL

Our results suggest that the spawning pheromone of herring is not one of the currently recognised fish pheromones. Sixty compounds failed to mimic the native pheromone in the herring spawning bioassay, even though all are pheromones, pheromone candidates, or potent odourants in other fish or vertebrates, and/or elute in the area of interest from the HPLC (Fig. 1).

Several steroidal compounds eluted on the HPLC in positions identical to that of bioactive material (Fig. 1). These included compounds found to be pheromones in other fish species, compounds related to these pheromones, and cortisol. Taurocholic acid also co-eluted with the pheromonal substances (unpublished results). Other steroids and steroid conjugates (including all steroid sulfates tested)

eluted later than the bioactive substances. All of the steroidal
compounds tested in the spawning pheromone bioassay were inactive.

Prostaglandins are prime candidates for female spawning pheromones
in goldfish (Sorensen et al., 1988). All male fish pheromones
identified so far are steroidal (see Sorensen and Stacey, 1989), but
prostaglandins should not be ruled out. However, we do not yet have
clear evidence whether prostaglandins act as herring spawning
pheromones. Prostaglandin $F_{2\alpha}$ co-eluted with herring pheromonal
substances on the HPLC, whereas prostaglandins E_1 and E_2 eluted after the
bioactive material (Fig. 1). Prostaglandin $F_{2\alpha}$, Cloprostenol (an
analogue of prostaglandin $F_{2\alpha}$), prostaglandin E_2, and 6-keto
prostaglandin $F_{1\alpha}$ were all inactive in the bioassay (Sherwood et al.,
1991; unpublished results). Prostaglandin $F_{1\alpha}$ eluted within the active
region (unpublished results) and resulted in spawning in herring in the
one year it was tested (Sherwood et al., 1991). However, the delay
between stimulus and response was much longer than that seen with the
natural pheromone, and so far has not been replicated. Thus, this
result is questionable.

L-amino acids and N-acetylglucosamine-3-sulfate were inactive as
well in the pheromone bioassay (Sherwood et al., 1991).

CHEMICAL NATURE OF PHEROMONE

Herring spawning pheromone activity is present in the aqueous phase
of both petroleum ether and dichloromethane extractions of milt and
testes. It may also be present to a lesser extent in the non-polar
phase of the dichloromethane extraction. This suggests that the
pheromone is relatively polar, and if steroidal, it is primarily
conjugated. Also, the bioactivity is neither precipitated by acetone,
nor sensitive to digestion by a variety of peptidases, indicating that
it is not a large protein or a peptide. Treatment with ß-glucuronidase
reduced bioactivity, but did not eliminate it; digestion with a
combination of ß-glucuronidase and aryl-sulfatase did destroy the
pheromonal activity. Thus, the pheromone appears to be conjugated
(Sherwood et al., 1991).

The relatively broad elution of the bioactivity on the HPLC (Fig.
1) suggests that pheromonal activity may consist of several factors,
some of which may be glucuronidated while others are sulfated. It is of
interest that $17\alpha,20\beta$-dihydroxyprogesterone eluted from HPLC in the
bioactive area only if glucuronidated. The unconjugated form eluted
later and the sulfated form eluted more than 30 minutes after the
bioactive material. None of these three compounds was active in the
bioassay, but Fig. 1 demonstrates that the pheromone must be quite
polar. Indeed, none of the synthetic standards were as polar as the
early-eluting bioactive material. However, other unidentified compounds
sensitive to glucuronidase or sulfatase may elute in this area. For
example, taurocholic acid, while not active in the pheromone bioassay,
was present in expressed herring milt, sensitive to sulfatase digestion,
and sufficiently polar to elute in the bioactive region (unpublished
results). Clearly, the search for the herring pheromone should not be
limited to the common reproductive steroids and prostaglandins.

SITE OF SECRETION

We have confirmed the observation by Stacey and Hourston (1982)
that testes extracts contain pheromone activity. Even the testicular
tissue remaining after rinsing, maceration and centrifugation in salt

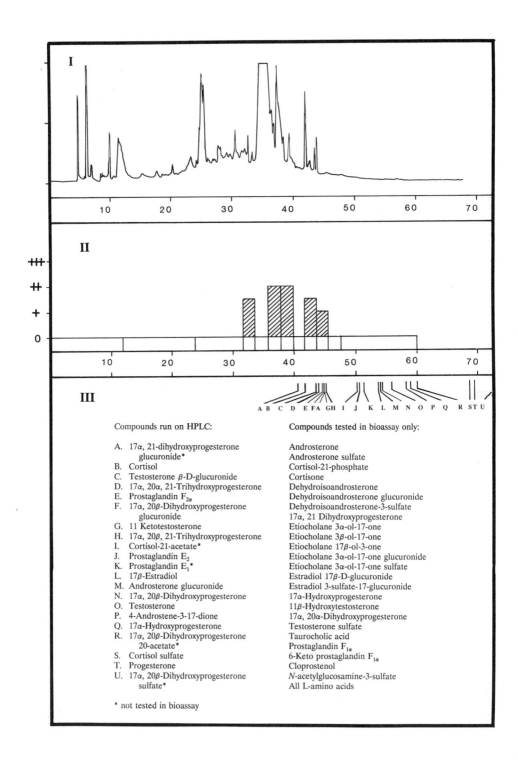

Compounds run on HPLC:

A. 17α, 21-dihydroxyprogesterone
   glucuronide*
B. Cortisol
C. Testosterone β-D-glucuronide
D. 17α, 20α, 21-Trihydroxyprogesterone
E. Prostaglandin $F_{2\alpha}$
F. 17α, 20β-Dihydroxyprogesterone
   glucuronide
G. 11 Ketotestosterone
H. 17α, 20β, 21-Trihydroxyprogesterone
I. Cortisol-21-acetate*
J. Prostaglandin E_2
K. Prostaglandin E_1*
L. 17β-Estradiol
M. Androsterone glucuronide
N. 17α, 20β-Dihydroxyprogesterone
O. Testosterone
P. 4-Androstene-3-17-dione
Q. 17α-Hydroxyprogesterone
R. 17α, 20β-Dihydroxyprogesterone
   20-acetate*
S. Cortisol sulfate
T. Progesterone
U. 17α, 20β-Dihydroxyprogesterone
   sulfate*

* not tested in bioassay

Compounds tested in bioassay only:

Androsterone
Androsterone sulfate
Cortisol-21-phosphate
Cortisone
Dehydroisoandrosterone
Dehydroisoandrosterone glucuronide
Dehydroisoandrosterone-3-sulfate
17α, 21 Dihydroxyprogesterone
Etiocholane 3α-ol-17-one
Etiocholane 3β-ol-17-one
Etiocholane 17β-ol-3-one
Etiocholane 3α-ol-17-one glucuronide
Etiocholane 3α-ol-17-one sulfate
Estradiol 17β-D-glucuronide
Estradiol 3-sulfate-17-glucuronide
17α-Hydroxyprogesterone
11β-Hydroxytestosterone
17α, 20α-Dihydroxyprogesterone
Testosterone sulfate
Taurocholic acid
Prostaglandin $F_{1\alpha}$
6-Keto prostaglandin $F_{1\alpha}$
Cloprostenol
N-acetylglucosamine-3-sulfate
All L-amino acids

water retained pheromonal activity (unpublished results). The testes are thus likely to be the prime site of pheromone production in the Pacific herring, but we have not shown definitively the preferred route of excretion.

HERRING AS A PHEROMONE MODEL

The response of herring to their spawning pheromone is a particularly marked behaviour, well suited for investigation. In addition, it is unique among reproductive behaviours so far studied in that no courtship or interactive behaviour occurs between sexes. The pheromone is also unique among known fish reproductive pheromones in that it is a releasing pheromone for both sexes, but is produced by only the male. Turner (1986) suggests that the ancestral method of reproduction in teleosts is "a promiscuous mating system with no courtship or mate choice" much like the "school" spawning of herring. The spawning pheromone of the Pacific herring may thus be an ancestral type. However, "school" spawning, to our knowledge, has not yet been clearly demonstrated in any other fish species and may not be typical of all herring (Rounsefell, 1930; Ewart, 1884). "School" spawning may, in fact, be a specialized type of spawning behaviour facilitated by long-lived sperm. Maximum fertilisation in herring occurs only 0.5 to 2.5 hours after gamete release (Hourston and Rosenthal, 1976), and sperm motility is still observable five days after activation (see Ginzberg, 1972). Nevertheless, a spawning strategy similar to that of herring may occur in other schooling pelagic fish (Balon, 1975). Reproductive behaviour in most of these species has not yet been reported, probably largely because it is difficult to observe. The Pacific herring may be an excellent pheromone model for such fish.

ACKNOWLEDGMENTS

We thank Carol Warby for excellent assistance throughout this study. This research has been funded by the Natural Sciences and Engineering Research Council of Canada (NSERC), the Department of Fisheries and Oceans, and Archipelago Marine Research, Victoria, B.C. The Science Council of British Columbia provided a graduate fellowship to J.C.

REFERENCES

Aneer, G., Florell, G., Kautsky, U., Nellbring, S. and Sjöstedt, L., 1983, In-situ observations of Baltic herring (Clupea harengus membras) spawning behaviour in the Askö-Landsort area, northern Baltic proper, Marine Biology, 74: 105.
Balon, E.K., 1975, Reproductive guilds of fishes: a proposal and definition, J. Fish. Res. Board Can., 32: 821.

Fig. 1. (I) the uv absorbance at 254nm of an extract of herring milt during HPLC. HPLC fraction numbers are listed on the abscissa. (II) bioactivity of the HPLC fractions of the milt in (I). The hatched bars indicate positive responses by pooled pairs of fractions. Bioassay scores represent: 0, no papilla extension; +, partial papilla extension; ++, full papilla extension, +++, spawn. (III) pheromone candidates tested in the herring bioassay and/or by HPLC. Compounds listed on the left were injected onto the HPLC to determine their elution in relation to that of the bioactive fractions. Compounds listed on the right were tested only in the bioassay. All compounds were tested at a concentration of 10^{-7} M. (Adapted from Sherwood et al., 1991).

Ewart, J.C., 1884, Natural history of the herring, Rep. Fish. Board Scotl., 2: 61.

Ginzberg, A.S., 1972, "Fertilization in fishes and the problem of polyspermy", B. Golek, ed., Z. Blake, translator, Israel Program for Scientific Translations Ltd., Jerusalem.

Haegele, C.W. and Schweigert, J.F., 1985, Distribution and characteristics of herring spawning grounds and description of spawning behaviour, Can. J. Aquat. Sci., 42. (Suppl 1): 39.

Hay, D.E., 1985, Reproductive biology of Pacific herring (Clupea harengus pallasi), Can. J. Fish. Aquat. Sci., 42 (Suppl. 1): 111.

Hourston, A.S. and Rosenthal, H., 1976, Sperm density during active spawning of Pacific herring, J. Fish. Res. Board Can. 33: 1788.

Keenleyside, M.H.A., 1979, "Diversity and Adaptation in Fish Behaviour", Springer-Verlag, Heidelberg.

Rounsefell, G.A., 1930, Contribution to the biology of the Pacific herring, Clupea pallasi, and the condition of the fishery in Alaska, Bull. Bur. Fish. (U.S. Dept. Commerce), 45: 227.

Schaeffer, M.B., 1937, Notes on the spawning of the Pacific Herring, Clupea pallasi, Copeia, 1937: 57.

Scott, A.P., Sherwood, N.M., Canario, A.V.M. and Warby, C.M., 1991, Identification of free and conjugated steroids, including cortisol and 17α,20ß-dihydroxy-4-pregnen-3-one, in the milt of Pacific herring, Clupea harengus pallasi, Can. J. Zool., 69: 104.

Sherwood, N.M., Kyle, A.L., Kreiberg, H., Warby, C.M., Magnus, T.H., Carolsfeld, J. and Price, W.S., 1991, Partial characterisation of a spawning pheromone in the herring Clupea harengus pallasi, Can. J. Zool., 69: 91.

Sorensen, P.W., Hara, T.J., Stacey, N.E. and Goetz, F.W., 1988, F-prostaglandins function as potent olfactory stimulants that comprise the post-ovulatory female sex pheromone in goldfish, Biol. Reprod., 39: 1039.

Sorensen, P.W. and Stacey, N.E., 1990, Identified hormonal pheromones in the goldfish: the basis for a model of sex pheromone function in teleost fish, pp. 302 - 311, in: "Chemical Signals in Vertebrates V.", MacDonald, D., Müller-Schwarze, D. and Natynczuk, S.E., eds., Oxford University Press, Oxford.

Stacey, N.E. and Hourston, A.S., 1982, Spawning and feeding behavior of captive Pacific herring, Clupea harengus pallasi. Can. J. Fish. Aquat. Sci., 39: 489.

Stacey, N.E. and Sorensen, P.W., 1986, 17α,20ß-dihydroxy-4-pregnen-3-one: a steroidal primer pheromone increasing milt volume in the goldfish, Carassius auratus, Can. J. Zool., 64: 2412.

Trant, J.M., Thomas, P. and Shackleton, C.H.L., 1986, Identification of 17α,20ß,21-trihydroxy-4-pregnen-3-one as the major ovarian steroid produced by the teleost Micropogonias undulatus during final oocyte maturation, Steroids, 47: 89.

Turner, G., 1986, Teleost mating systems and strategies, pp. 253-274, in: "The Behavior of Teleost Fishes", Pitcher, T.J., ed., Croom Helm, London.

CHEMOSENSORY ORIENTATION TO CONSPECIFICS AND RAINBOW TROUT IN ADULT LAKE

WHITEFISH COREGONUS CLUPEAFORMIS (MITCHILL)

Gunnar Bertmar

Department of Animal Ecology
Umeå University
S-901 87 Umeå, Sweden

INTRODUCTION

Functions subserved by chemosensory orientation (scanning the area for chemical cues) have been known for many years (Stoddart, 1980; Kleerekoper, 1982). Pheromones (chemical cues for communication within a species) have been implicated in a number of fish behaviours, including schooling, aggression, sexual attraction, individual recognition, recognition of offspring and parents, homing, recognition of predators, and alarming conspecifics (Pfeiffer, 1977, 1982). Both behavioural (Höglund and Åstrand, 1973; Newcombe and Hartman, 1973; Groot et al., 1986; Quinn and Tolson, 1986; Stabell, 1987) and physiological (Döving et al., 1974; Hara and McDonald, 1976; Olsén, 1989) experiments have demonstrated the existence of species- or population-specific chemical signals in salmonids. Reactions to pollutants have also been reported in salmonids (Bertmar, 1979, 1982; Brown et al., 1982). However, experiments have usually been performed on young and subadult fishes, and no chemical orientation between salmonid species (allomones) has been described.

Preliminary experiments have indicated that lake whitefish are either attracted or repelled by skin mucous collected from conspecifics, suggesting a role in individual or sexual attraction (Hara et al., 1984). From a sensory ecology perspective (Bertmar, 1985), the working hypotheses for the present study were, therefore, that ripe lake whitefish have not only orientation by sight, as is expected in a schooling species, but also chemosensory orientation towards chemical cues of conspecifics of the same population and of a population with older conspecifics. They also might be able to detect nonconspecific chemical signals from a population of potential predators like rainbow trout (Oncorhynchus mykiss).

Test fish

One hundred twelve ripe 4-year-old fish (total length ranging from 26 to 35 cm) were tested. The fish were from an Ontario wild stock and were reared at the Biology Department, York University, Toronto. They were kept in a 600 liter fiberglass tank with running dechlorinated tap water. This water was also used in the experiments. The fish were fed in the afternoon with food pellets. One to five fish from this home tank were also put in a 200 liter home aquarium when used for special experi-

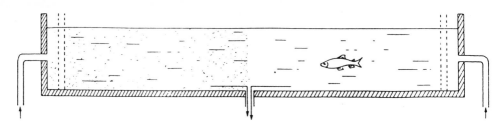

Fig. 1. Schematic illustration of avoidance/preference trough.
Water flows into each end and out the center of the
trough, with a test solution introduced at one end or the
other. As there were no stream variations, the fish
moving back and forth along the trough were able to con-
tinually scan the environment by repeatedly sampling the
contents of the water. A perforated plate of equal size
as the inlet side was placed 3 cm from each end. Incoming
water was mixed and dye tests with Methylene Blue showed
the horizontal gradient and stream velocity. Modified
from Brown et al. (1982).

ments. All the test fish were taken at random and showed natural
behaviour when they were put back after the experiments.

Behavioural bioassay

The test apparatus consisted of a 120 x 32 x 22 cm. trough (Fig. 1)
in the same room as the tank and home aquarium. Water was pumped from
the tank and aquarium to the trough. Before the start of an experiment,
the inflow/outflow balance of the trough was adjusted to 900 ml./min. for
the first two tests and to 600 ml./min. for the other tests. The water
temperature was 12° C. There was a light panel about 2 m. above the
trough, and natural light came from windows located about 3 m. from one
side of the trough. The test fishes were continuously observed through a
one-way mirror in a screen. The test fishes were adapted to the trough
and only fishes with stable swimming patterns were used. They were not
fed during the experiments. Studies were usually run all day with one to
five fish at a time taken from the home tank or home aquarium.

Testing procedure

The positions of the test fish were recorded every 15 sec. The
tests were usually run in 2 x 10 min. At no time during or immediately
following a test session was the behaviour of a fish reinforced. Test
water from a tank with ten 8-year-old lake whitefish (total length 35-45
cm. and of same stock as the test fish) and from a tank with eight ripe
rainbow trout (3 years old, total length 30-45 cm., from an Ontario
stock) was also used. These fish were kept in another room and had the
same kind of running tap water as the test fish.

Natural light

During the total 1233 min. that the fish were observed, they showed
some area preference for the part of the trough that was closest to the
windows. The fish were observed 7424 times in that part versus 6808 in
the other half of the trough. This did not influence the chemosensory
results, however, as these were based on reversal of inflow of test
water. The present results certainly indicate that lake whitefish do

Table 1.   Orientation of 4-year-old lake whitefish.

| Fish nr | Type of chem. cues | Orientation towards | | X^2-type p=0.05 |
		home water	tap water	
1	Home tank with ca 50 fish	45	15	Signif.
2	"	86	34	"
3	"	75	31	"
3	Home aquarium with 4 fish	45	15	"
4	Home tank	54	26	"
5	"	66	14	"
6	Home aquarium with 4 fish	52	28	"
7	Home aquarium with 3 fish	46	34	Insignif.
8	Home tank	64	16	Signif.
9	Home aquarium with 3 fish	47	33	Insignif.
10	Home aquarium with 6 fish	51	29	Signif.
11	Home tank	77	3	"
12	Home tank	39	1	"
13-17	"	245	155	"
18-22	Home aquarium with 5 fish	271	129	"
23-27	Home tank	261	139	"
$\sum$ 27 fish	Home water	1396	670	"

have visual orientation. But the results also show that their orientation drive towards chemical cues was stronger. Orientation to temperature, visual and chemical cues in lake whitefish is therefore another example of Integrated Hierarchial Orientation earlier suggested to guide homing salmonid fishes (Bertmar, 1982).

Intraspecies chemical cues compared with tapwater

The fishes were diurnal and nonaggressive and were therefore very suitable for orientation studies of the present type. The experiments started with testing twelve single fishes for various durations (10-60 min.). Ten fish showed preference responses for water either from the home tank, or the home aquarium containing four to six fishes (Table 1). If there were only three fish in the home aquarium there was still a preference for home water, but this was not statistically significant. It can therefore only be hypothesized that there is a lower limit in the number of fishes that should be present in a water volume in order to give a response. Three groups of five fish each also showed preference

Table 2. Orientation of 4-year-old lake whitefish to tap water and tank water from ripe 3-year-old rainbow trout.

Fish nr	Type of chem. cues	Orientation towards		X^2-type p=0.05
		trout water	tap water	
38-42	Rainbow water	202	198	Insignif.
43-47	"	227	173	"
48-52	"	206	194	"
53-57	"	233	167	Signif.
58-62	"	245	155	"
63	"	38	42	Insignif.
64	"	56	24	Signif.
65	"	75	5	"
66	"	56	24	"
67	"	62	18	"
$\sum$ 30 fish	"	1401	999	"

Table 3. Orientation of 4-year-old lake whitefish to home water and tank water from rainbow trout.

Fish nr	Type of chem. cues	Orientation towards		X^2-type p=0.05
		home water	trout water	
88-92	Home/rainbow water	120	80	Signif.
93-97	"	118	82	"
98-102	"	132	68	"
103-107	"	122	78	"
108-112	"	123	77	"
$\sum$ 25 fish	"	615	385	"

Table 4.   Orientation of 4-year-old lake whitefish to tap water, home
            water and tank water from 3-year-old rainbow trout and 8-
            year-old whitefish.

Fish nr	Type of chem. cues	Orientation towards		X^2-type p=0.05
		chem. cues	tap water	
28-32	Home tank water	246	154	Signif.
"	Rainbow water	278	122	"
"	Whitefish 8 year	321	79	"
33-37	"	201	199	Insignif.
"	Rainbow water	200	200	"
"	Home tank water	270	130	Signif.
68-72	Rainbow water	276	124	"
"	Whitefish 8 year	293	107	"
"	Home tank water	254	146	"
73-77	Rainbow water	216	184	Insignif.
"	Home tank water	268	132	Signif.
"	Whitefish 8 year	240	160	"
78-82	"	227	173	"
"	Home tank water	281	119	"
"	Rainbow water	195	205	Insignif.
83-87	Home tank water	253	147	Signif.
"	Whitefish 8 year	229	171	"
"	Rainbow water	196	204	Insignif.
$\sum$ 30 fish	Chemical cues	4472	2756	Signif.

responses to home water (Table 1). Usually, three to four fish schooled together and often assembled at the inlet of the home water and even showed increased swimming activity.

Interspecies chemical cues compared with tapwater

Thirty lake whitefish were tested in groups or individually to see if they could discriminate between tap water and rainbow water. Some schooling and most single lake whitefish displayed preference responses to rainbow water (Table 2). The summary of these tests show that lake whitefish may trace chemical cues (allomones) from a population of a potential predator like ripe rainbow trout.

Interspecies cues compared with intraspecies chemical signals

Five groups of five fish each were offered the opportunity to discriminate between water from the home tank and from the rainbow tank. But the behaviour now changed and they showed a very strong preference to home water (Table 3). Later on, six different groups of five fish each were offered both home water and rainbow water, as well as water from a tank with 8-year-old lake whitefish and in different sequences and combinations. These experiments showed that 4-year-old lake whitefish seem to be able to discriminate between home water and water from lake whitefish of another age, as well as water from a potential predator species (Table 4). It did not matter in which order the three types of water were offered. The fish were alert most of the time. The results also indicate that lake whitefish show the most preference for chemical cues of the home water (pheromones) and the least preference, or even avoidance, for rainbow water (allomones).

DISCUSSION

Sample behavioural responses of two lake whitefishes, one to an attractant (food extract) and one to a repellant (quinine), were used to demonstrate a new behavioural bioassay (Jones and Hara, 1985). The present study of 112 mature lake whitefish strengthens this finding and the hypothesis that these fish can use chemical orientation in a similar manner to many other salmonid species. Indeed, they may even prefer these cues to visual ones. This strategy may be more effective and energy saving for a schooling species. As lake whitefish spawn in November and December, and the experiments were performed in May and June, the chemical cues were probably not used for sex attraction but for schooling. The chemical recognition of a lake whitefish population of another age, as well as predators (rainbow trout), may also be valuable strategies for a schooling species like Coregonus clupeaformis.

REFERENCES

Bertmar, G. 1978, Home range, migrations and orientation mechanisms of the River Indalsälven trout, Salmo trutta, Report Inst. Freshwater Research, Drottningh, 58:5-26.

Bertmar, G., 1982, Structure and function of the olfactory mucosa of migrating Baltic trout under environmental stresses, with special reference to water pollution, in "Chemoreception in Fishes", Vol. 8, T.J. Hara, ed., Elsevier, Amsterdam.

Bertmar, G., 1985, General ecology of primitive fishes, in "Evolutionary Biology of Primitive Fishes," R.E. Foreman, A. Gorbman, J.M. Dodd and R. Olsson, eds., Plenum Press, London.

Brown, S.B., Evans, R.E., Thompson, B.E. and Hara, T.J., 1982, Chemoreception and aquatic pollutants, in "Chemoreception in Fishes", Vol. 8, T.J. Hara, ed., Elsevier, Amsterdam.

Döving, K.B., Nordeng, H. and Oakley, B. 1974, Single unit discrimination of fish odours released by char (Salmo alpinus L.) populations, Comp. Biochem. Physiol., 47A:1051-1063.

Groot, C., Quinn, T.P., and Hara, T.J., 1986, Responses of migrating adult sockey salmon (Onchorhynchus nerka) to population specific odours, Can. J. Zool., 64:926-932.

Hara, T.J. and McDonald, S., 1976, Olfactory responses to skin mucous-substances in rainbow trout Salmo gairdneri, Comp. Biochem. Physiol., 54A:41-44.

Hara, T.J., McDonald, S., Evans, R.E., Mauri, T. and Arai, S., 1984, Morpholine, bile acids and skin mucus as possible chemical cues in salmonid homing: electrophysiological re-evaluation, in "Mechanisms of Migration in Fishes", J.D. McCleave, G.P. Arnold, J.J. Dodson and W.H. Neill, eds., Plenum Press, London.

Höglund, L.B. and Åstrand, M., 1973, Preference among juvenile charr (Salvelinus alpinus) to intraspecific odours and water currents studied with the fluviarium technique, Report Inst. Freshw. Res Drottningholm, 53:21-30.

Jones, K.A. and Hara, T.J., 1985, Behavioural responses of fishes to chemical cues: results from a new bioassay, J. Fish Biol., 27:495-504.

Kleerekoper, H. 1982, The role of olfaction in the orientation of fishes, in "Chemoreception in Fishes," Vol. 8, T.J. Hara, ed., Elsevier, Amsterdam.

Olsen, K.H., 1989, Sibling recognition in juvenile Arctic charr, Salvelinus alpinus (L.), J. Fish Biol., 34:571-581.

Newcombe, C. and Hartman, G., 1973, Some chemical signals in the spawning behavior of rainbow trout (Salmo gairdneri). J. Fresh Res. Board Can., 30:995-997.

Pfeiffer, W. 1977, The distribution of fright reaction and alarm substance cells in fishes, Copeia, 77:653-665.

Pfeiffer, W. 1982, Chemical signals in communication, in "Chemoreception in Fishes," Volume 8, T.J. Hara, ed., Elsevier, Amsterdam.

Quinn, T.P. and Tolson, G.M., 1986, Evidence of chemically mediated population recognition in coho salmon (Oncorhynchus kisutch), Can. J. Zool., 64:84-87.

Stabell, O.B. 1987, Intraspecific pheromone discrimination and substrate marking by Atlantic salmon parr, J. Chem. Ecol., 13:1625-1643.

Stoddart, D.M., 1980, "The Ecology of Vertebrate Olfaction," Chapman and Hall, London.

ELECTROPHYSIOLOGICAL MEASURES OF OLFACTORY SENSITIVITY SUGGEST THAT

GOLDFISH AND OTHER FISH USE SPECIES-SPECIFIC MIXTURES OF HORMONES AND

THEIR METABOLITES AS PHEROMONES

P.W. Sorensen[1], I.A.S. Irvine[1], A.P. Scott[2] and N.E. Stacey[3]

[1]Department of Fisheries and Wildlife, University of Minnesota, St. Paul, MN 55108 U.S.A., [2]Directorate of Fisheries Research, Lowestoft, Suffolk NR33 0HT U.K., [3]Department of Zoology, University of Alberta, Edmonton, Alberta T6G 2E9 Canada

INTRODUCTION

Twenty years ago Kittredge et al. (1971) speculated that aquatic organisms may have evolved to use hormonal metabolites as sex pheromones. Their reasoning was simple: these compounds are pre-adapted for this function because they are naturally released to the environment in synchrony with important reproductive (endocrinological) events and their recognition might only require a simple mutation by which endocrine receptors are expressed on chemosensory tissue. Ten years later, Colombo et al. (1980) found that a fish, the male black goby (<u>Gobius jozo</u>), produces and releases a conjugated reduced androgen to water where it attracts ovulated females. Soon thereafter, Van den Hurk and Lambert (1983) reported that a mixture of two conjugated steroids, testosterone glucuronide and estradiol glucuronide, attracts male zebra danio (<u>Brachydanio</u> <u>rerio</u>). These exciting findings prompted us to test whether goldfish (<u>Carassius</u> <u>auratus</u>) also use hormonal metabolites as pheromones. Using electrophysiological measures of olfactory sensitivity (the electro-olfactogram or EOG) as a bioassay we discovered that goldfish are extremely sensitive to the steroidal maturation hormone $17\alpha,20\beta$-dihydroxy-4-pregnen-3-one ($17,20\beta$P) (see Sorensen, 1992). Subsequently we discovered that $17,20\beta$P is released to the water where if functions as a preovulatory "primer" pheromone, increasing male gonadotropin (GtH) and milt (sperm and seminal fluid) production. Knowing that ovulated goldfish also release a pheromone which stimulates male behavior, we continued our pursuit of hormonal pheromones using EOG recording to screen putative hormonal pheromones. We subsequently discovered that metabolites of prostaglandin $F_{2\alpha}$ ($PGF_{2\alpha}$) are likely released by recently ovulated fish to function as a postovulatory 'releaser' pheromone. A preliminary account of these findings was presented at Chemical Signals in Vertebrates V (Sorensen and Stacey, 1990), along with predictions that many other fish use hormonal metabolites as pheromones.

The past three years have witnessed significant advances in the study of fish hormonal pheromones. Our understanding of the chemical nature of goldfish sex pheromones and the neural mechanisms responsible

Chemical Signals in Vertebrates VI, Edited by R.L. Doty and
D. Müller-Schwarze, Plenum Press, New York, 1992

for their detection has improved considerably. Also, we have screened the olfactory sensitivity of over 15 species of fish and now have preliminary evidence that many fish detect hormonal derivatives (unpublished). In addition, several other groups of fish endocrinologists have made significant headway understanding hormone metabolism and have implicated the use of hormonal pheromones in a variety of other fish. Three recent reviews summarize much of this information. Sorensen (1992) reviews the literature on goldfish pheromones, the influence of hormones on pheromonal function, and the history of the term "pheromone' and its applicability to fish hormonal pheromones. Stacey and Sorensen (1991) provide a comprehensive review of recent information on fish hormonal pheromones and discuss the evolutionary process by which hormones may have come to be used as chemical signals. More recently, Stacey et al. (1992) review the literature on fish pheromones and speculate about how they might be applied to aquaculture. Rather than review these data yet again, we will update our article from Chemical Signals in Vertebrates V by presenting new data on the composition of goldfish hormonal pheromones and the olfactory sensitivity of several cypriniform fish to hormonal compounds.

RECENT PROGRESS IN UNDERSTANDING GOLDFISH HORMONAL SEX PHEROMONES

Goldfish ovulate in the spring in response to a surge in GtH triggered by rising temperature, aquatic vegetation, and pheromones. Like other oviparous teleosts, goldfish become sexually receptive at the time of ovulation. Because females spawn (release eggs) within a few hours, male-female reproductive physiology and behavior must be tightly synchronized. This synchrony is mediated by at least two classes of pheromones with different actions and identities. Although we initially believed that these pheromones were simple mixtures of largely unmodified compounds, we now believe that the opposite is the case. We will briefly review some of this evidence to provide a background for our cross-species comparisons.

Like many fish, goldfish synthesize the gonadal steroid 17,20βP in response to surging GtH levels to induce final oocyte maturation. In goldfish, circulating 17,20β increases 12 h prior to ovulation, after which it decreases to basal levels (Stacey et al., 1989). Using radioimmunoassay, we discovered that this steroid, or a compound closely resembling it, is released by ovulatory females to function as a pheromone which stimulates GtH release in males. Recent evidence strongly suggests that 17,20βP is only one of several components in a complex mixture of steroids. Some components of this mixture appear to have independent actions and some do not. For instance, although we now know that ovulatory goldfish release immunoreactive(IR)-17α-hydroxyprogesterone, and IR-17,20βP glucuronide (17,20βP-G) together with IR-17,20βP (Van Der Kraak et al., 1989), these steroids are much less stimulatory than 17,20βP and their actions are attributable to the same olfactory receptor mechanism responsible for responsiveness to 17,20βP; if they have a role, it is to amplify the 17,20βP signal (Sorensen et al., 1990). On the other hand, we (Sorensen et al. 1991a) have recently found that another conjugate of 17,20βP, 17,20βP-sulphate (17,20βP-S), stimulates the goldfish olfactory system as effectively as 17,20βP and that it is independently recognized. Similarly, we now know that ovulatory goldfish also release IR-androstenedione (A) and the goldfish olfactory system detects A with great sensitivity and distinguishes it from 17,20βP-like compounds. The goldfish steroidal pheromone is likely comprised of at least three compounds with independent actions and perhaps many minor 'amplifying' components.

The functions of the preovulatory steroid pheromone appear more complex than initially supposed. For instance, A appears to have inhibitory actions; males exposed to $17,20\beta P$ and A have smaller GtH and milt increases than males exposed to $17,20\beta P$ alone (Stacey, unpublished). Because peak release of A probably occurs before peak $17,20\beta P$ release, it is possible that A restricts male GtH responses to the time of peak $17,20\beta P$ release. Alternatively, it is possible that A is released by other species or male goldfish and has the effect of giving this steroidal pheromone a gender- or species-specific character. The function and release of $17,20\beta$-S remains to be determined. As predicted (Sorensen and Stacey, 1990), pheromone release patterns appear important: the $17,20\beta P$-induced rise in male GtH is transient (Stacey, unpublished). The actions of $17,20\beta P$ do not appear to be simply to increase milt; males exposed to $17,20\beta P$ for several hours become more active and compete for females more successfully (deFraipont and Sorensen, 1992). Finally, females also detect $17,20\beta P$ and respond to it with an increased ovulation rate (see Sorensen, 1992); $17,20\beta P$ may be a bisexual pheromone.

The PGF postovulatory pheromone also appears to be more complex than first supposed. Originally we thought that the PGF pheromone was comprised of two components, one closely resembling or identical to $PGF_{2\alpha}$, the other its metabolite 15-keto-prostaglandin $F_{2\alpha}$ ($15K$-$PGF_{2\alpha}$) (Sorensen et al., 1988, 1989). We now know that this cannot be the case. Although ovulated females release ten times more IR-PGF than non-ovulated fish, only 1% of the $PGF_{2\alpha}$ injected into females is released as immunoreactive $PGF_{2\alpha}$, far too little to explain the olfactory potency of $PGF_{2\alpha}$-injected fish water (Sorensen et al., 1988). We have also traced PGF metabolism in ovulated and $PGF_{2\alpha}$-injected fish using radiolabeled $PGF_{2\alpha}$ and high pressure liquid chromatography and found that three unknown PGF metabolites are released (Sorensen and Goetz, unpublished).

Several recent experiments indicate that responsiveness to hormonal sex pheromones in goldfish is exclusively mediated by the olfactory nerve (Cranial Nerve 1). First, severing the medial olfactory tract totally eliminates whole-animal responsiveness to $17,20\beta P$ and PGFs (see Sorenson, 1992). Second, central responsiveness to $17,20\beta P$ has been verified by recording from the olfactory tracts (Sorensen et al., 1991b) and because we have been unable to measure electrical responses from the terminal nerve (Cranial Nerve 0; which runs within the olfactory tracts), it seems likely that the olfactory system alone is responsible for pheromonal responsiveness (Fujita et al., 1992). Third, recording from the goldfish gustatory system has found it to be unresponsive to $17,20\beta P$ and PGFs (Sorensen and Hara, unpublished). It is now clear that the EOG is a valid index of chemosensitivity to hormonal pheromones. Exciting questions about the precise chemical composition of hormonal mixtures, their relationship to the endocrine system, and their functions remain to be answered.

SCREENING THE OLFACTORY RESPONSIVENESS OF MISSISSIPPI RIVER FISHES:
A TEST OF THE APPLICABILITY OF THE GOLDFISH HORMONAL PHEROMONE MODEL

To test whether fish might commonly be using hormonally-derived compounds as sex pheromones and whether differences in sensitivity to these compounds might lend them a species-specific character, the olfactory sensitivities of four species of cypriniform fishes from the Mississippi River drainage were measured and compared to the goldfish (another member of the minnow family [Cyprinidae]). The hornyhead chub (Nocomis biguttatus) is a stream-dwelling minnow which represents an interesting contrast to the 'gang-spawning' goldfish; males of this species build large nests which they vigorously defend against other males but permit females to enter for spawning. Sperm competition is not likely to be a factor in this species and male gonadosomatic indices are small (<1.0%). Three sympatric species common to the Mississippi River were also captured for testing: the carp (<u>Cyprinus carpio</u>; a minnow), the smallmouth buffalo (<u>Ictiobus bubalus</u>; a sucker [family <u>Catastomidae</u>]), and the shorthead redhorse (<u>Moxostoma macrolepidotum</u>: another sucker). The carp, an exotic, is interesting because its reproductive physiology and behavior appear similar to the goldfish, and goldfish-carp hybrids have been reported (Taylor and Mahon, 1977); these species might 'share' pheromone systems. The smallmouth buffalo is native to North America, and like the carp, has drab coloration and spawns in the spring over flooded vegetation (Becker, 1983); because hybrids have not been reported, one might predict that it has a pheromone system different form the carp's. The shorthead redhorse is an interesting contrast to the carp and buffalo because, although it also spawns in the spring, it spawns over rocky substrate and has distinctive coloration (a bright red tail)(Becker, 1983); it need not have a different pheromone system.

Olfactory sensitivity was assayed using EOG standard recording techniques during which electrode position was determined by the location which elicited the largest response to our standard, 10^{-5}M L-serine, a component of food odor (see Sorensen et al., 1990). Males and females of these four species were captured by electroshocking or netting during their spawning season (spring). In addition to testing L-serine, responses were recorded to 10^{-7}M concentrations of: A, etiocholanolone, etiocholanolone glucuronide, testosterone, testosterone glucuronide, estradiol, 5α-pregnen-3β,17α,20β-trihydroxy-4-pregnen-3-one, $17,20\beta$P, $17,20\beta$P-G, $17,20\beta$P-S, prostaglandin E_2, $PGF_{2\alpha}$, 15-K-$PGF_{2\alpha}$, 13,14-15K-$PGF_{2\alpha}$ (13,14-dihydro-15-keto-prostaglandin $F_{2\alpha}$; a metabolite of $PGF_{2\alpha}$). Dose responses were performed for stimulatory compounds and responses plotted.

Species-specific differences in olfactory sensitivity to steroids and PGFs were evident among these fishes (Fig. 1). Only members of the minnow family (carp, goldfish, chub) detected any of the steroids. The olfactory systems of the carp and goldfish detected the same steroids and PGFs and the dose response-relationships for these compounds were the same with only two seemingly minor exceptions: $17,20\beta$P was slightly more stimulatory at lower concentrations for carp, and 13,14-15K-$PGF_{2\alpha}$ was a more potent olfactory stimulant at higher concentrations for carp. However, olfactory responsiveness of the chub differed dramatically from that of the goldfish and carp. This species was anosmic to A and responded with much greater sensitivity to glucuronated $17,20\beta$P than free $17,20\beta$P ($17,20\beta$P-S was intermediate); its pattern of sensitivity is the opposite of goldfish and carp which barely detect $17,20\beta$P-G. The sensitivity of the chub to PGFs was also markedly different from that of goldfish and carp; it detected all three PGFs with equivalent sensitivities, suggesting that it has only one olfactory receptor mechanism for these compounds. Similarly, the olfactory sensitivities of buffalo and redhorse to PGFs appeared to be different from each other. Like the chub, the buffalo detected all three PGFs with equivalent sensitivities. On the other hand, redhorse only detected $PGF_{2\alpha}$ with any real certainty and even then not well. This appears to reflect a difference in specific sensitivity because we recorded the largest responses to L-serine from this species. Conclusions must be drawn very carefully from this study. For instance, although we have yet to encounter a gender difference in olfactory function, this topic is not understood, and we were testing fish of mixed sex and maturity. Also, we were testing synthetic compounds which, at least in the case of the PGFs, are unlikely to be the actual pheromones. Still, at the very least, these data describe the

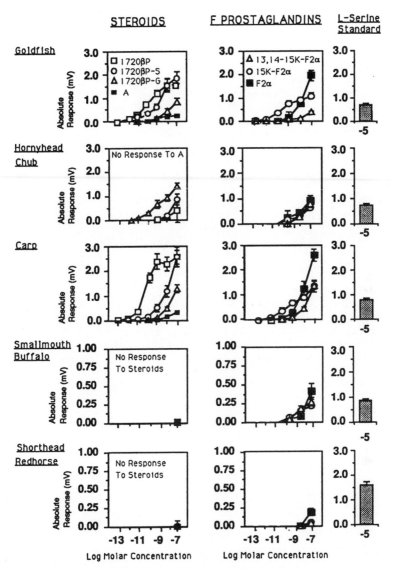

Fig. 1. Mean EOG responses (+/- standard error) of various
cypriniform fish to concentrations of steroids and F
prostaglandins ranging from 10^{-14} Molar to 10^{-7} Molar.
Abbreviations for steroids and prostaglandins are as in
text. Goldfish (n=7 mature males); Hornyhead chub (n=7
mature males); carp (n=7 mature males); Smallmouth buffalo
(n=6; 3 mature and 3 immature males); Shorthead redhorse
(n=4; 2 mature males, 1 mature and 1 immature female).

first clear species-specific differences in fish olfactory sensitivity
and suggest that hormones and their metabolites can function as species-
specific cues.

GOLDFISH-CARP HYBRIDIZATION: HOW MUCH CAN WE EXTRAPOLATE ABOUT FISH
REPRODUCTIVE BIOLOGY FROM EOG PROFILES TO SYNTHETIC COMPOUNDS?

EOG responses of carp and goldfish to synthetic PGFs are remarkably
similar and these species have similar reproductive physiologies and
behaviors. The fact that anosmic goldfish do not spawn is generally
interpreted to mean that these fish rely heavily on pheromones to reproduce, and goldfish-carp hybrids have been collected in the Great Lakes.
Can similar EOG patterns to synthetic hormones be interpreted to reflect
a lack of pheromonal specificity? Further, if this is really the case,
and these species are as dependent on olfactory cues to spawn as we
presently suppose, might it not be reasonable to suggest that they should
freely hybridize? To address this question we studied behavioral interactions between mature male goldfish and goldfish and carp injected with
$PGF_{2\alpha}$, a treatment which induces female sexual receptivity and pheromone
release in goldfish. Although the effects of $PGF_{2\alpha}$-injection on carp have
yet to be tested, it is reasonable to believe that its actions should be
similar, for $PGF_{2\alpha}$ stimulates male-female reproductive behavior in at
least 6 other species of cyprinids (see Stacey and Sorensen, 1991).

To first test whether male goldfish will interact with carp and
whether $PGF_{2\alpha}$-injection influences the nature of these interactions (i.e.
does it elicit the release of pheromone-like substances?), male goldfish
were exposed to carp before and after $PGF_{2\alpha}$-injection and their responses
compared to those elicited by exposing the same group of males to goldfish treated in the same manner. Briefly, male goldfish were held in 70
L aquaria (18°C) containing floating and submerged vegetation. During
the morning of day one, either a carp or a goldfish was injected with
saline, placed into a male's aquarium, and, after a 20-min acclimatiza-
tion period, the resident male's behavior noted for 20 min. This procedure was repeated on day two using the same fish but this time injecting
it with 15µg $PGF_{2\alpha}$. Two behaviors were noted: nudging by males (physical
contact, generally in the vicinity of urogenital aperture, an activity
characteristic of arousal) and spawning. In a second experiment, a $PGF_{2\alpha}$-
injected carp and a $PGF_{2\alpha}$-injected goldfish were placed together into

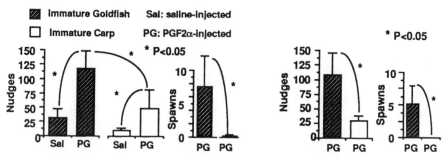

Fig. 2.   (A) Behavioral responses of male goldfish to saline-in-
          jected (Sal) or $PGF_{2\alpha}$-injected (PG) immature goldfish and
          carp. (B) Responses of male goldfish simultaneously ex-
          posed to a $PGF_{2\alpha}$-injected goldfish and a $PGF_{2\alpha}$-injected
          carp. Mean values for 20 min. test periods +/- standard
          error.

aquaria to determine directly which species male goldfish prefer. Equivalent-sized and colored (black) immature goldfish and carp were used in these experiments.

The results of these simple studies were clear: goldfish readily distinguish goldfish from carp and prefer to interact with the former (Fig. 2). For instance, although goldfish nudged $PGF_{2\alpha}$-injected carp three times more than saline-injected carp, they nudged $PGF_{2\alpha}$-injected goldfish 2-3 times more often than $PGF_{2\alpha}$-injected carp, and spawned more often with goldfish than carp. This did not appear to be attributable to the fact that carp are faster swimmers because immobile carp were frequently ignored, particularly in Experiment 2 when male goldfish consistently chased the other goldfish. To the human eye, carp and black goldfish are similar in appearance and behavior; could different chemical cues have been involved? Obviously, this experiment is very far from proving this point, but it emphasizes how little we know abut the sensory cues which mediate reproductive synchrony. EOG sensitivities to synthetic compounds must be cautiously applied to questions about chemical signaling and its functions.

GENERAL CONDITIONS

Electrophysiological measures of olfactory activity strongly indicate that cypriniform fish are likely to commonly use species-specific mixtures of steroids, F prostaglandins and their metabolites as sex pheromones. However, although EOG sensitivity to synthetic compounds provides rapid insight into the general nature of these systems, conclusions must be drawn cautiously given our poor understanding of both hormone metabolism and the likelihood that many pheromones may be novel hormonal metabolites.

REFERENCES

Becker, C.L., 1983, "Fishes of Wisconsin," University of Wisconsin Press. Madison, Wisconsin.

Colombo, L., Marconato, A.,. Belvedere, P.C. and Frisco, C. 1980, Endocrinology of teleost reproduction: a testicular steroid pheromone in the black goby, Gobius jozo. L. Boll. Zool., 47:355.

DeFraipont, M., and Sorensen, P.W., 1992, The steroidal primer pheromone, $17\alpha,20\beta$-dihydroxy-4-pregnen-3-one, enhances behavioral spawning success and sperm quantity and quality in male goldfish, Anim. Behav. (submitted).

Fujita, I., Sorensen, P.W., Stacey, N.E., and Hara, T.J., 1992, The olfactory system, not the terminal nerve, functions as the primary chemosensory pathway mediating responses to sex pheromones in goldfish, Brain. Behav. Evol. (in press).

Kittredge, J.S., Terry, M., and Takahashi, F.J., 1971, Sex pheromone activity of the moulting hormone, crustecdysone, on male crabs (Pachygrapsus crassipes, Cancer antennarius, and C. anthonyi). Fish. Bull., 69:117.

Sorensen, P.W., 1992, Hormones, pheromones, and chemoreception. In: "Fish Chemoreception", T.J. Hara, ed., Chapman and Hall, London., (in press).

Sorensen, P.W., and Stacey, N.E., 1990, Identified hormonal pheromones in the goldfish: The basis for a model of sex pheromones function in teleost fish. In: "Chemical Signals in Vertebrates 5", D.W. MacDonald, D. Muller-Schwarze, and S.E. Natynczuk, eds., Oxford University Press, New York.

Sorensen, P.W., Hara T.J., and Stacey, N.E., 1991b, Sex pheromones selectively mediate olfactory tracts of male goldfish, Brain Res. 558:343.

Sorensen, P.W., Stacey, N.E. and Chamberlain, K.J., 1989, Differing behavioral and endocrinological effects to two female sex pheromones on male goldfish, Horm. Behav., 23:317.

Sorensen, P.W., Goetz, F.W., Scott, A.P., and Stacey, N.E., 1991a, Recent studies of the goldfish indicate both unmodified and modified hormones function as sex pheromones. In: "Proceedings of the Fourth International Symposium on the Reproductive Physiology of Fish," A.P. Scott, J.P. Sumpter, D.E. Kime, and M.S. Rolfe, eds., University of East Anglia Press, United Kingdom.

Sorensen, P.W., Hara, T.J., Stacey, N.E. and Dulka, J.G., 1990, Extreme olfactory specificity of male goldfish to the preovulatory steroidal pheromone 17α,20β-dihydroxy-4-pregnen-3-one, J. Comp. Physiol. A., 166:373.

Sorensen, P.W., Hara, T.J., Stacey, N.E. and Goetz, F.Wm., 1988, F prostaglandins function as potent olfactory stimulants comprising the postovulatory female sex pheromone in goldfish, Biol. Reprod., 39:1030.

Stacey, N.E., and Sorensen, P.W., 1991, Function and evolution of fish hormonal pheromones. In: "Biochemistry and Molecular Biology of Fishes," P.L. Hochachka and T.P. Mommsen, eds, Elsevier, Toronto.

Stacey, N.E., Sorensen, P.W., Dulka, J.G., Van Der Kraak, G.J., 1989, Direct evidence that 17α,20β-dihydroxy-4-pregnen-3-one functions as the preovulatory pheromone in goldfish, Gen. Comp. Endocrinol., 75:62.

Stacey, N.E., Sorensen, P.W., Dulka, J.G., Cardwell, J.R., and Irvine, A.S., 1992, Fish sex pheromones: current status and potential applications, Bull. Instit. Zool. Academia Sinica, (in press).

Taylor, J., and R. Mahon, R., 1977, Hybridization of Cyprinus carpio and Carassius auratus, the first two exotic species in the lower Laurentian Great Lakes, Env. Biol Fish., 1:205.

Van den Hurk, R. and Lambert, J.G.D., 1983, Ovarian steroid glucuronides function as sex pheromones for male zebrafish, Brachydanio rerio, Can J. Zool., 61:2381.

Van Der Kraak, G.J., Sorensen, P.W., Stacey, N.E., and Dulka, J.G., 1989, Periovulatory female goldfish release three potential pheromones: 17α,20β-dihydroxyprogesterone, 17α,20β-dihydroxyprogesterone glucuronide and 17α-hydroxyprogesterone, Gen. Comp. Endocrinol., 73, 452.

Contribution #19,146 from the Minnesota Agricultural Experiment Station

GUSTATORY BEHAVIOR OF CHANNEL CATFISH TO AMINO ACIDS

T. Valentincic[1] and J. Caprio[2]

[1]Department of Biology, University of Ljubljana
Askerceva 12, 61000 Ljubljana, Slovenia
[2]Department of Zoology and Physiology, Louisiana
State University, Baton Rouge, Louisiana 70803. U.S.A.

INTRODUCTION

Electrophysiological recordings from both nerve twigs (Caprio,1978; Davenport and Caprio, 1982; Kanwal and Caprio, 1983) and single taste fibers (Kohbara et al., 1990) provided evidence that the taste system of the channel catfish is highly responsive to the amino acids, L-alanine, L-arginine and L-proline. Dose-response relations determined electrophysiologically indicate that the taste system of the channel catfish is most sensitive to L-alanine and L-arginine (Caprio, 1975), whereas L-proline becomes highly stimulatory at stimulus concentrations >10^{-3}M (Kanwal and Caprio, 1983; Wegert and Caprio, 1991). Electrophysiological cross-adaptation (Davenport and Caprio, 1982; Kanwal and Caprio, 1983; Wegert and Caprio, 1991) and biochemical competition (Cagan, 1986) experiments indicated relatively independent receptor sites for these three amino acids. Recent patch clamp experiments (Teeter et al., 1990; Brand et al. 1991) directly confirmed the existence of independent receptor sites for the L-isomers of alanine, arginine and proline and additionally provided insight into the respective transduction mechanisms associated with these receptor sites. Current evidence indicates that the receptor sites for L-arginine and L-proline are direct ligand-operated, whereas those for L-alanine involve GTP-binding proteins and the generation of second messengers. Interestingly, the majority of amino acid taste information from the direct ligand-operated receptor sites is transmitted to the central nervous system by narrowly-tuned taste fibers that are highly responsive to L-arginine, whereas the majority of amino acid taste information involving second messenger generation is transmitted centrally by relatively broadly-tuned taste fibers that are most stimulated by L-alanine (Kohbara et al., 1990).

Although much information is presently accruing concerning the neurophysiology and biochemistry of amino acid taste in ictalurid catfish, little specific information was known concerning the taste-mediated behavior of these organisms to individual amino acids. Channel catfish were aversively conditioned with lithium chloride to avoid food gels containing specific amino acids; this behavior was indicated to be based on taste alone (Little, 1977). Studies using a heart-rate conditioning paradigm to investigate the detection of amino acids in channel catfish were in conflict as to whether olfaction (Little, 1981) or taste (Holland and Teeter, 1981) was the sensory system responsible for the conditioning. Except for brown bullhead catfish, <u>Ictalurus nebulosus</u>, turning towards a higher concentration of two stimuli that were simultaneously presented to the right and left maxillary barbels (Johnsen and Teeter, 1980), no published data exist describing the precise nature of behavioral responses that are specifically mediated through the taste system of ictalurid catfish. Preliminary reports on conditioned food searching (swimming) responses demonstrated that brown bullhead catfish discriminated L-arginine, L-alanine, L-norleucine and L-cysteine from each other and from other amino acids (Valentincic et al., 1990). Channel catfish conditioned to single chemicals responded with searching swims to most of the amino acids and the response to conditioned amino acids lasted nearly twice as long as responses to non-conditioned stimuli (Valentincic et al., 1991). In the present report, the experimental protocol was directed at answering the following questions: 1. Do channel catfish respond behaviorally to the amino acids determined electrophysiologically to be the most stimulatory taste stimuli? 2. Are behavioral responses of anosmic channel catfish to amino acids the same as those of intact channel catfish? 3. Are different dose-response relations for L-alanine, L-arginine and L-proline determined electrophysiologically reflected in the behavior of the organism? 4. Do L-alanine, L-arginine and L-proline release the same patterns of feeding behavior?

MATERIALS AND METHODS

Channel catfish obtained from a local catfish farm were housed individually in 80 liter aquaria, which were extensively aerated and biologically filtered. Amino acid stimuli were delivered hydraulically from a distance of three meters from Pasteur pipettes which were hung above each aquarium. To prevent mechanical noise of stimulus injection from alerting the fish, the stimulus was directed over the turbulent area created by the aquarium aeration system. To determine the amino acid stimulus distributions and concentrations at different times after stimulus injection into the aquaria, an electrochemical dopamine probe (Moore et al. 1989) was used in a subset of experiments. The electrochemical probe was positioned at the bottom directly below the stimulus pipet, at the bottom at the opposite end of the aquarium and 5 cm directly below the stimulus delivery pipet. All stimulus concentrations reported an estimated contact concentration unless otherwise stated. Behavioral responses of channel catfish were video-recorded with a s-VHS video recorder. Bites and movements of the hyoidal region were counted in slow motion.

RESULTS AND DISCUSSION

The stimulus concentration in the aquaria after delivery as determined with the electrochemical probe indicated that the dopamine distribution outside the initial cloud was patchy and that the dopamine was distributed irregularly in the stimulus plumes carried by microcurrents (Fig. 1). For an injected concentration of 10mM dopamine, a 1000 to 10,000-fold dilution occurred for the highest dopamine peak which passed the electrode positioned at the bottom of the aquarium (Fig. 1a, b and c). Following injections of 1 ml of a mixed solution of 10mM dopamine and 1M L-arginine to increase the density of the liquid, the initial dilution by the time the substances reached the bottom under the electrode was generally >300-fold (Fig. 1d). At a 10mM initial concentration, a single dopamine plume was sensed by the electrode (Fig. 1a, b and c), whereas the injection of a mixture of 1M L-arginine and 10mM dopamine resulted in several stimulus peaks following the initial peak, which was most concentrated (Fig. 1d). The dopamine concentration at the sensor increased to its maximum value

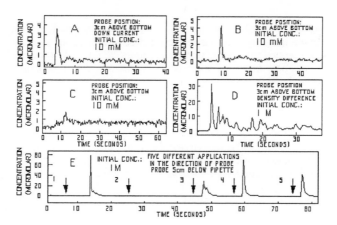

Fig. 1. Dopamine distribution in aquaria after delivery of 1 ml of dopamine solution

within 2 seconds, irrespective of the stimulus concentration within the center of plume. If the electrode was positioned 5 cm below the delivery point, the initial peak concentration detected by the electrode was different, depending on the position of the electrode relative to the initial cloud formed by the delivered solution (Fig. 1e, compare stimulus applications 1 and 2). The observed dilutions were in most cases >100-fold and in some cases >3000-fold.

The single L-amino acids alanine, arginine and proline released the complete set of behavior patterns (orienting, appetitive and consummatory phases of the feeding response) both in intact and anosmic channel catfish. Other amino acids were far less effective and frequently did not release the consummatory behavior. L-proline at probable contact concentrations > 10^{-3} M was the most effective amino acid stimulus at releasing biting behavior (Fig. 2); however, L-arginine (Fig. 3) and L-alanine were the most effective stimuli releasing biting behavior at lower amino acid concentrations (10^{-6} to 10^{-4} molar) . This difference in sensitivity is reflected in the electrophysiological dose-response relations determined for these stimuli (Kanwal and Caprio, 1983). None of the stimulus peaks lasted longer than 5 seconds and most of the snapping behavior occurred within this same time frame. The biting responses to L-alanine and L-arginine saturated at contact concentrations of 3×10^{-6}M with 2 - 4 median number of bites per stimulation. A further increase in the number of biting responses occurred in response to multiple peaks of stimuli following an 80-liter aquarium of >100mM aminoacids (Fig 1d). Other physiologically less

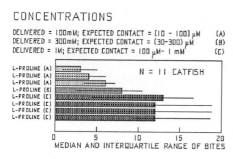

Fig. 2. Biting to L-proline
stimulation

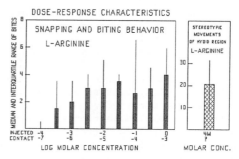

Fig. 3. Biting and hyoid movements
to L-arginine stimulation

effective amino acids released the biting behavior at high concentrations, but were far less effective than L-alanine, L-arginine and L-proline. Thus, current evidence suggests that under natural conditions, L-arginine and L-alanine most probably release gustatory behavior at a greater distance from the source and at lower concentrations than does L-proline; however, L-proline is the major single amino acid stimulus that releases consummatory biting behavior at high stimulus concentrations. In addition, only >10mM L-arginine released stereotypic hyoid movements that were probably related to mastication. Interestingly, of all the amino acids tested on the oropharyngeal cavity in an electrophysiological study of glossopharyngeal and vagal taste responses in the channel catfish (Kanwal and Caprio, 1983) only >10mM L-arginine triggered reflexive motor activity that was possibly related to the observed movements of the hyoid in the present behavioral experiments; however, the biological role of this behavior is presently unknown.

SUMMARY

Single amino acids, L-alanine, L-arginine and L-proline, released in intact and anosmic channel catfish, Ictalurus punctatus, the complete set of feeding behavior patterns. The consummatory patterns, turning, biting, snapping and masticating, were controlled primarily by the taste system. A single stimulus plume of either L-alanine or L-arginine with a peak concentration in the plume >10^{-6}M passing over the fish caused biting responses with a median number of bites between two and four. L-proline released biting behavior at contact concentrations >10^{-4}M. A median number of snapping responses > 5 occurred when several successive plumes of high stimulus concentration passed over the fish. If the concentration of the injected amino acid was >100mM, the density of the delivered stimulus was greater than the density of water. Thus, the high stimulus density slowed the speed of mixing of the stimulus with the aquarium water, resulting in multiple high concentration peaks. As many as 25 snaps occurred per test in response to high concentrations of L-proline and L-alanine. Contact concentrations of >100mM L-arginine released stereotypic hyoid movements probably related to mastication. All other amino acids were less effective stimuli for the consummatory biting behavior.

Acknowledgements: Supported by NSF BNS-8819772, the Visiting Scientist Program of the College of Basic Sciences, Louisiana State University and a travel grant of the Ministry of Science, Research and Technology of Slovenia.

REFERENCES

Brand, J.G., Teeter, J.H., Kumazawa, T., Huque, T. and Bayley, D.L. 1991. Transduction mechanisms for the taste of amino acids. Physiol. Behav. 49:899-904

Cagan, R.H. 1986. Biochemical studies of taste sensation. XII. Specificity of binding of taste ligands to sedimentable fraction from catfish taste tissue. Comp. Biochem. Physiol., 85A:355-358

Caprio, J. 1975. High sensitivity of catfish taste receptors to amino acids. Comp. Biochem. Physiol. 52A:247-251

Caprio, J. 1978. Olfaction and taste in the channel catfish: An electrophysiological study of responses to amino acids and their derivatives. J. Comp. Physiol., 132:357-371

Davenport, C.J. and J. Caprio. 1982. Taste and tactile recordings from the ramus recurrens facialis innervating flank taste buds in the catfish. J. Comp. Physiol. 147:217-229

Holland, K.N. and Teeter, J. H. 1981. Behavioral and cardiac reflex assay of the chemosensory acuity of channel catfish to amino acids. _Physiol. Behav._ 27:699-707

Johnsen, P.B. and Teeter, J.H. 1980. Spatial gradient detection of chemical cues by catfish. _J. Comp. Physiol._ 140:95-99

Kanwal, J.S. and Caprio, J., 1983. An electrophysiological investigation of the oropharyngeal (IX-X) taste system in the channel catfish, _Ictalurus punctatus_. _J. Comp. Physiol._ 150:345-357

Kohbara, J., Wegert, S., Caprio, J. 1990. L-proline information is transmitted primarily by a portion of arginine-best taste fibers in the channel catfish. _Chem. Senses_ 15(5):601

Little, E.E. 1977. Conditioned aversion to amino acid flavors in the catfish _Ictalurus punctatus_. _Physiol. Behav._ 19:743-747

Little, E.E. 1981. Conditioned cardiac response to the olfactory stimuli of amino acids in the channel catfish, _Ictalurus punctatus_. _Physiol. Behav._ 27:691-697

Moore, P.A., Gerhardt, G.A., Atema, J., 1989. High resolution spatio-temporal analysis of aquatic chemical signals using microelectrochemical electrodes. _Chem. Senses_ 14(6):829-840

Teeter, J.H., Brand, J.G. and Kumazawa, T. 1990 A stimulus-activated conductance in isolated taste epithelial membranes. _Biophys. J._ 58:253-259

Valentincic, T., Ota. D., Blejec, A. and Metelko, J. 1990. Chemosensory similarity of amino acids and other low molecular compounds by catfish. _Chem. Senses_ 15(5):648

Valentincic, T., Wegert, S. and Caprio, J. 1991. Behavioral responses of channel catfish to amino acids. AChemS, abs. no. 26

Wegert, S. and Caprio, J. 1991. Receptor sites for amino acids in the facial taste system of the channel catfish. _J. Comp. Physiol._ 168A:201-211

INTERSPECIFIC EFFECTS OF SEX PHEROMONES IN FISH

Vitaly A. Ostroumov

Institute of Biology
Lenin Street 3
Irkutsk State University, Irkutsk, Russia

INTRODUCTION

Goldfish and other teleosts use released hormones (steroids, prostaglandins) as sex pheromones (Sorensen and Stacey, 1990; Sorenson, Irvine,
Scott and Stacey, 1992; Stacey and Sorenson, 1991) raising the question of
whether such pheromones can be species-specific, given the limited chemical
diversity of hormones. This complex issue involves consideration of the
specific nature of reproductive sympatry, of non-hormonal components of the
pheromone which could create species specificity, and of non-olfactory cues
which might obviate the need for a species-specific odor. This paper presents experiments which address the problem of species-specific sex pheromones in fish by examining responses to heterospecific odors prepared from
gonadal fluids and urine of several teleost species.

SCULPINS (COTTIDAE)

Typically, male sculpins defend a cryptic crevice to which females
are attracted for spawning. We have shown that in several Lake Baikal
cottids (yellowfin sculpin, Cottocomephorus grewingki; sand sculpin,
Paracottus kessleri; stone sculpin, P. kneri) females are attracted to
odor of conspecific males, and that odor of C. grewingki males attracts
females and induces ovulation (Dmitrieva and Ostroumov, 1986; Ostroumov,
1987). The present studies indicate that cottid sex pheromones are present in urine and ovarian fluid and are not species-specific.

Methods

Male and female P. kessleri and P. kneri were captured in spawning
areas and held at the Bolshiya Koti research station in running Baikal
water. Untreated male sand sculpins, and males made anosmic by plugging
the nares with petroleum jelly-soaked tissue paper, were placed individually in 20 liter aquaria with a stone nest-site which they occupied
within several days. Occurence of the distinctive courtship behavior
normally displayed to ovulated females was observed during exposure to
the following test solutions (0.5 liters per min for 30 min): (a) holding
water: water (non-running) from 20 liter aquaria in which either 10 mature
male P. kessleri, 10 ovulated P. kessleri or 10 ovulated P. kneri, had
been held for 6 hours; (b) egg wash: ovulated eggs from one P. kessleri
or P. kneri stripped into 100 µl water diluted to 100 liters.

Chemical Signals in Vertebrates VI, Edited by R.L. Doty and
D. Müller-Schwarze, Plenum Press, New York, 1992

Table 1. Behavioral response of male and female P. kessleri to conspecific and heterospecific odors.

Response of males			Response of females		
Test solution	% displaying courtship		Test solution	% positive response	
holding water			holding water		
conspecific female	100		conspecific male	90	
conspecific male	0		conspecific female	0	
P. kneri female	75		P. kneri male	75	
egg wash			urine		
conspecific female			conspecific male		
test male untreated	100		mature	90	
test male anosmic	0**		immature	20	
P. kneri female	100		mature P. kneri	90	

*N = 20 in all but one treatment: **N = 10

To examine female responses to male odor, ovulated P. kessleri were placed individually in a 35 liter Y-maze in which water flowed continuously (5-6 liters per min) into each choice arm. After several hours adaptation, the female was observed following addition of a bolus (100 µl) of the following test solutions: (a) holding water: water (non-running) from 30 liter aquaria in which 10 mature male P. kessleri or P. kneri, or 10 ovulated P. kessleri, had been held for 6 hours; (b) male urine; urine (1 ml) collected by syringe from the bladder of mature or immature P. kessleri, or mature P. kneri, and diluted in 100 ml water.

Results and Discussion

Male P. kessleri did not respond to holding water from male conspecifics, but exhibited a suite of courtship responses (color change, pectoral fin vibration, body undulations) to holding water from conspecific and heterospecific females. The fact that diluted ovarian fluid of conspecifics and heterospecifics induced courtship in all test males indicates an ovarian source for the active components in holding water.

When exposed to holding water from conspecific or heterospecific males, female P. kessleri left the neutral area of the maze and entered the test arm, often increasing ventilation rate and exhibiting search behaviors. Although similar responses occurred during exposure to urine from mature male conspecifics and heterospecifics, urine from immature male conspecifics induced only weak responses in a few females.

EUROPEAN MINNOW (CYPRINIDAE)

The European minnow (Phoxinus phoxinus) mating system is typical of many cyprinids, groups of males attending an ovulated female and competing for position at the time of oviposition. Because of the intense sperm competition in this mating system, we examined whether males increase milt (sperm and seminal fluid) volume in response to female odors.

Methods

Phoxinus in spawning condition were captured in a tributary of the Angara River which drains Lake Baikal and males were placed in groups of

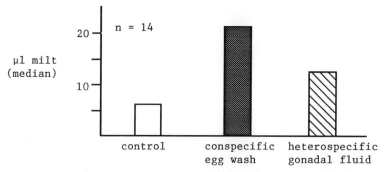

Figure 1. Volume of milt stripped from Phoxinus 12 hours after exposure
to blank water, Phoxinus egg wash, or the seminal and ovarian
fluids of Arctic cisco (Coregonus autumnalis migratorius).

7 in 60 liter aquaria. The volume of milt which could be stripped was
measured 12 hours after exposure to 100 µl of the following test substances:
(a) water control; (b) conspecific egg wash: ovulated eggs from one
Phoxinus were stripped into 100 µl water, which then was diluted in 100
liters; (c) heterospecific gonadal stimulus: fluid from dishes in which the
stripped milt and eggs of Arctic cisco (Coregonus autumnalis migratorius;
Salmonidae) had been placed for artificial fertilization was lyophilized
and reconstituted in water (1 µg per liter).

Results and Discussion

Male Phoxinus exhibited no obvious behavioral reaction to the conspecific or heterospecific test substances. However, exposure to both
these substances caused a significant ($p < 0.05$) increase in the volume of
milt which could be stripped 12 hours later (Fig. 1).

Although it appears that, as in goldfish Carassius auratus (Stacey
and Sorenson, 1991), male Phoxinus increase milt volume in response to an
ovulatory pheromone, this study examined neither the identity nor the mode
of action of the gonadal factors increasing Phoxinus milt volume. The
primer pheromone of goldfish has been proposed to be 17a,20β-dihydroxy-4-
pregnen-3-one (17,20β-P), an ovarian steroid which acts as a horomone to
induce oocyte maturation, and as an olfactory phermone to increase milt
volume (Sorensen and Stacey, 1990). It seems unlikely that present milt
response of Phoxinus is due to 17,20β-P; 17,20β-P levels in cyprinids are
low after ovulation (goldfish, Stacey and Sorensen, 1991) and milt and egg
wash preparation from Coregonus contains no immunoreactive 17,20β-P
(G. Van Der Kraak, personal communication).

MULLUS BARBATUS (MULLIDAE)

Unlike the cottids and Phoxinus, in which the shallow, benthic spawning is readily investigated, Mullus are pelagic spawners of deeper water
(>10 m) where observation is more difficult. Mullus in the Black Sea
have an extended summer spawning; females may ovulate every day, which
makes this species a good model for primer pheromone studies.

Methods

Mullus were captured during spawning and held in monosex groups (7-
10 fish) in 100 liter aquaria at the Karadag Field Station, Sevastopol.
The weight of ovulated eggs and volume of milt were measured 12 hours
after exposure to 100 µl of: (a) water control; (b) conspecific egg wash;
prepared as for the sculpin and Phoxinus experiments; (c) conspecific

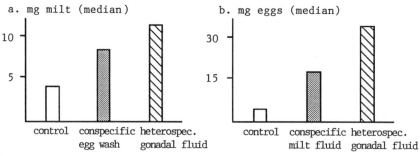

Figure 2. Milt and ovulated eggs stripped from <u>Mullus barbatus</u> 12 hours after exposure to blank water or to conspecific and heterospecific gonadal fluids.

<u>milt fluid</u>; stripped <u>Mullus</u> milt (1 ml) stirred in 100 ml water, allowed to settle, and the supernatant diluted in 100 liters water; (d) <u>heterospecific gonadal stimulus</u>; prepared as in the <u>Phoxinus</u> experiments.

Results and Discussion

Volume of ovulated eggs and milt both were increased by exposure to conspecific and heterospecific gonadal fluids (Fig. 2). These primer effects were similar to those in <u>Phoxinus</u>, in that no behavioral effects were observed in response to odor additions.

CONCLUSIONS

All species in these studies exhibited either releaser (sculpins) or primer (<u>Phoxinus</u>, <u>Mullus</u>) responses to aqueous extracts of heterospecific gonadal products. Because the active components of these extracts have not been identified, it is not known whether these heterospecific responses indicate the use of similar chemical signals in diverse fish species.

ACKNOWLEDGEMENTS: This manuscript was kindly prepared by N.E. Stacey.

REFERENCES

Dmitrieva, T.M. and Ostroumov, V.A., 1986, Role of chemical communication in spawning behavior of the yellowfin Baikal sculpin (<u>Cottocomephorus grewingki</u>), <u>Biol. Nauki.</u>, 10:38.
Ostroumov, V.A., 1987, Ethophysiological analysis of sex pheromone systems in Cottid fish. Ph.D. Thesis, Irkutsk State University.
Ostroumov, V.A., and Dmitrieva, T.M., 1990, The influence of sex pheromones on the gonad maturation and behavior in Arctic cisco (<u>Coregonus autumnalis migratorius</u>), <u>J</u>. <u>Ichthyol.</u>, 30:497.
Sorensen, P.W., and Stacey, N.E., 1990, Identified horomonal pheromones in the goldfish: the basis for a model of sex pheromone function in teleost fish, <u>in</u> "Chemical Signals in Vertebrates, Volume V," D. MacDonald, D. Muller-Schwarze, and S.E. Natynczuk, eds., Oxford University Press, Oxford.
Sorensen, P.W., Irvine, I.A.S., Scott, A.P. and Stacey, N.E., 1992, Electrophysiological measures of olfactory sensitivity suggest that goldfish and other fish uses species-specific mixtures of hormones and their metabolites as sex pheromones, <u>in</u>: "Chemical Signals in Vertebrates VI: R. Doty and D. Muller-Schwarze, eds., Plenum Press, New York.
Stacey, N.E., and Sorenson, P.W., 1991, Function and evolution of fish hormonal pheromones, <u>in</u> "The Biochemistry and Molecular Biology of Fishes, Volume I," P.W. Hochachka and T.P. Mommsen, eds., Elsevier Press, Amsterdam.

LEARNED PREDATOR AVOIDANCE BEHAVIOR AND A TWO-LEVEL SYSTEM FOR CHEMOSENSORY

RECOGNITION OF PREDATORY FISHES IN JUVENILE BROOK TROUT

MaryLouise Keefe*, Timothy A. Whitesel*, and Howard E. Winn

Graduate School of Oceanography
University of Rhode Island
Narragansett, RI   02882

*Current address:
Oregon Department of Fish and Wildlife
211 Inlow Hall
Eastern Oregon State College
La Grande, OR   97850

INTRODUCTION

The ability of a fish to recognize and respond appropriately to chemical cues can affect its growth, survival, and reproductive success (Liley, 1982). The chemical detection of predators by a potential prey is one chemosensory response with potential impact on individual survival. An abundance of research with prey fishes has shown that many species can detect and behaviorally respond to chemical cues emanating from predators (reviewed by Weldon, 1990), including some nonsympatric and exotic predatory fishes (Reed, 1969).

As early as 1941 there was evidence suggesting that predator avoidance in fishes was an acquired behavior (Göz, 1941, cited in Kleerekoper, 1969). Göz demonstrated that several species of minnow (Phoxinus spp.) exhibited anti-predator behaviors to chemical cues of pike (Esox) only after they had been exposed to a pike preying on other minnows. More typically, cyprinids exhibit innate anti-predator behaviors to an alarm chemical, Schreckstoff, which is given off by damaged skin cells of ostariophysans (reviewed by Pfeiffer, 1982; Smith, 1986). The ostariophysan alarm reaction varies somewhat among species (Hara, 1986). However, in general, the alarm reaction appears to require the integration of a learned component, that of predator recognition, with the innate behavioral reaction (Magurran, 1989; Suboski et al., 1990).

Relatively little information is available concerning chemically mediated anti-predator responses in fishes that do not exhibit a Schreck-reaktion, such as salmonids. Several salmonids have been shown to avoid the chemical washes of mammalian (Brett and MacKinnon, 1954) and piscine (Rehnberg and Schreck, 1986; Keefe, 1990) predators. Keefe (1990) also found that brook trout (Salvelinus fontinalis) avoided an unfamiliar predatory fish, Atlantic salmon (Salmo salar), and that this response was related to the piscivorous diet of the salmon. Thus, the chemical avoidance exhibited by juvenile brook trout was presumably initiated by chemicals

Chemical Signals in Vertebrates VI, Edited by R.L. Doty and
D. Müller-Schwarze, Plenum Press, New York, 1992

released by the salmon that reflect its recent diet. Since the brook trout had no prior exposure to piscivorous salmon, Keefe (1990) suggested that these fish either innately respond or learn to avoid some general cue(s) emanating from piscivorous fishes.

Learning seems to play a role in the predator avoidance behavior of salmonids (Kanayama, 1968; Patten, 1977; Olla and Davis, 1989). These studies reported on increased rates of survival in salmon that had prior exposure to predators when compared to "predator naive" controls. The increase in survival of the exposed fish may have resulted from learning to avoid predation or may have reflected an increased adaptation to the general stress associated with the experimental protocol. For instance, sockeye salmon (Oncorhynchus nerka) fry that experienced enumeration without injury survived subsequent predation equally well as fry that had previously been exposed to predators (Ginetz and Larkin, 1976). Thus, the two objectives of the present study were (1) to determine whether the chemosensory avoidance of predators exhibited by juvenile brook trout is a learned or innate behavior. (2) We also wanted to establish whether the anti-predator cues that brook trout avoid represent the chemical signature of a predatory species or are a reflection of chemical by-products associated with the predator's recent feeding regime.

METHODS

Fish collection and maintenance

Brook trout juveniles were reared in the laboratory (Department of Zoology, University of Rhode Island) from wild stock. They were reared in tanks receiving flow-through municipal water. Fry were moved to East Farm Aquaculture Facility at the University of Rhode Island. They were held in 250 L fiberglass bioassay tanks supplied with artesian well water (7.5 L/min) under a simulated natural photoperiod. Trout were fed daily a diet of commercial trout feed. Since these trout had no contact with any other fish species, they were considered to be naive.

Red fin pickerel (Esox americanus) and Atlantic salmon were used to provide chemical cues of predatory fish. Six adult red fin pickerel were obtained by electrofishing in Paris Brook, Exeter, Rhode Island. Three sub-adult Atlantic salmon were obtained from the Aquaculture Department at the University of Rhode Island. These stimulus fish were held at the East Farm Aquaculture Facility under similar conditions as those described for trout. The pickerel were divided into two groups, which were held in separate tanks. One group of pickerel, as well as the salmon, were maintained on diets of live goldfish (Carassius auratus) (pickerel-goldfish), while the other group of pickerel were fed mealworms (Tenebrio spp.) (pickerel-mealworms).

General testing procedures

Choice experiments with individual brook trout were conducted in 120 x 30 x 25 cm, fiberglass coated, plywood troughs. Eighty percent of the length of each trough was separated into two equal arms by a watertight divider. The remaining length of each trough was isolated by a removable gate, constituting a holding area for the test fish. During each trial, trout were tested individually for preference between two water sources following the protocol described by Keefe and Winn (1991).

Predator-conditioned waters were produced by siphoning water from the 250 L tanks holding the respective stimulus fish. The predators were fed 24 hours prior to collecting conditioned waters. To make certain food

odors did not contaminate the conditioned water, the water in the predator's tank was drained and refilled with clean well water after the
predator had eaten. The conditioned water was siphoned 30 min prior to the
start of the choice experiment.

Choice experiments

The study included six choice experiments conducted between September
1990 and February 1991. Forty-eight naive trout were randomly assigned to
three experimental groups. Trout in the first group were tested for
avoidance of water conditioned by pickerel-goldfish over unconditioned well
water. The second group of trout were tested for avoidance of water
conditioned by Atlantic salmon over unconditioned well water. The third
group of naive trout were subject to five training episodes and then were
tested for avoidance of pickerel-goldfish conditioned water when paired
with unconditioned well water. In an effort to determine whether these
fish had been trained to avoid pickerel-specific or dietary-specific cues,
choice tests were also conducted pairing water conditioned by Atlantic
salmon fed goldfish with unconditioned well water and pairing water conditioned by pickerel-goldfish with water conditioned by pickerel-mealworms.

Training episodes

The training episodes were conducted with individual trout in the
choice troughs following a modified version of the protocol used during
choice experiments. During training, water conditioned by pickerel-
goldfish was paired with unconditioned well water. Each time a trout
entered the arm of the choice trough containing pickerel-goldfish conditioned water it received 25V electric shocks from an electric stimulator
(Grass, Model S-44). The trout experienced 0.5 second shocks every 1.0
second for as long as it remained in the pickerel-goldfish arm of the
trough. Shocking was continued each time the test fish entered the pickerel arm for either a 10-minute training period or until the fish stayed
in the well water arm for three minutes.

Sixteen brook trout were used in each choice experiment. Invariably,
some individuals did not move within the first half of the test period and
thus, were eliminated from the analysis. Sample sizes for each choice experiment varied from 10 to 15. To make certain there was no directional
bias in the experimental apparatus nor inherent to the sample populations,
a control test, pitting well water against well water, was conducted prior
to any experimental trials.

Analyses

The time of an individual's entry into and exit from each water source
was recorded. The total time each fish spent in each water source was then
calculated. These data were analyzed using a Wilcoxon matched-pairs signed-
ranks test for each choice experiment. A binomial test was conducted for
the choice experiments pairing pickerel-goldfish with unconditioned well
water to analyze whether the number of trout entering control water first,
having their longest foray in control water, and spending the greatest
total time in control water was different from random expectations. Significant avoidance was determined at the alpha = 0.05 level.

RESULTS

Choices of the two arms in the control trials were not significantly
different from random (Table 1.). No directional bias associated with the
test apparatus, protocol, or trout population was detected. In initial

Table 1. Results of choice experiments with naive and trained juvenile brook trout. N equals the number of fish run in each experiment.

Experimental group	Stimulus Choice (*)	N	Median total time spent (seconds)	Wilcoxon T	P
NAIVE 1	CW	12	202	36	> 0.05
	P-GF		191		
NAIVE 2	CW	10	305	27	> 0.05
	ATS-GF		204		
TRAINED	CW	14	237	24	< 0.05
	P-GF		105		
TRAINED	CW	12	204	28	> 0.05
	ATS-GF		93		
TRAINED	P-MW	15	164	42	> 0.05
	P-GF		210		
CONTROL	CW	12	200	29	> 0.05
	CW		162		

(*) CW, unconditioned well water; P-GF, red fin pickerel maintained on goldfish; ATS-GF, Atlantic salmon maintained on goldfish; P-MW, red fin pickerel maintained on mealworms

choice experiments, naive juvenile brook trout did not avoid either pickerel-goldfish or Atlantic salmon conditioned waters when these source waters were paired with unconditioned well water (Table 1). However, after five training episodes, brook trout juveniles spent significantly more total time in control water than pickerel-goldfish water (Table 1). In contrast to naive trout, significantly more of the trained trout entered the control water first, spent a greater amount of time in control water, and made their longest foray into control water (Table 2). Trained fish did not avoid water conditioned by Atlantic salmon, nor did they behaviorally discriminate between pickerel-goldfish and pickerel-mealworm conditioned waters (Table 1).

DISCUSSION

The avoidance behavior of brook trout to the chemical cues associated with piscivory does not appear to be an innate response. The trout used in the present experiment were naive to the existence of other fish species and did not avoid the chemical milieu of either piscivorous red fin pickerel or Atlantic salmon. This is in contrast to the hypothesis of Keefe (1990) who demonstrated that wild brook trout juveniles avoided an unfamiliar piscivore and suggested, therefore, that this behavior may be an innate response to cues associated with piscivory. It is probable that the wild trout used by Keefe (1990) had prior exposure to predation on brook trout by other species or adult brook trout. Thus, these fish may have developed a response to chemical cues associated with piscivory before being brought into the laboratory.

Table 2. Binomial analyses of choice experiments pairing pickerel-
goldfish water with unconditioned well water. Twelve naive
and 13 trained brook trout juveniles were tested.

Response variable	Experimental group	P-value
First water type entered	NAIVE	> 0.050
	TRAINED	< 0.001
Water type of longest foray	NAIVE	> 0.050
	TRAINED	= 0.035
Water type in which the greatest total time was spent	NAIVE	> 0.050
	TRAINED	= 0.022

Juvenile brook trout are capable of learning to recognize and avoid the chemical cues given off by red fin pickerel. This finding is in agreement with research done on cyprinids (Magurran, 1989; Soboski et al., 1990) which showed that they learn to recognize novel chemical cues that are introduced in concert with either alarm substance or the alarm reaction of conspecifics as predator cues. Furthermore, as previously suggested from circumstantial evidence (Kanayama, 1968; Patten, 1972; Olla and Davis, 1989), our data demonstrate that salmonids are capable of learning to recognize a predator. This directs attention to the importance of learned anti-predator behaviors in fishes, such as salmonids, which do not exhibit an ostariophysan Schreckreaktion.

The brook trout in these experiments learned to recognize chemical cues specific to red fin pickerel. This became apparent when the trout, which were trained to avoid the chemical traces of red fin pickerel main-tained on goldfish, did not avoid chemical washes of Atlantic salmon which were fed goldfish or behaviorally discriminate between pickerel fed gold-fish and those fed mealworms. The lack of effect of predator diet on the avoidance response does not clearly support the hypothesis proposed by Keefe (1990) that juvenile brook trout respond to chemical cues associated with piscivory. However, the duration of the training that the brook trout received while in the laboratory was not very extensive and did not provide potentially important reinforcement in the form of, for example, visual or tactile cues. As Magurran (1989) cited, animals are best at learning those skills that are basic to their survival. It seems reasonable to presume that the intensity of learning in the wild is greater than that resulting from the electric shock conditioning used in our experiments. Thus, we believe it is probable that wild juvenile brook trout learn to make dis-tinctions between the chemical cues emanating from piscivorous and non-piscivorous fishes.

When considered together with the findings of Keefe (1990), this study illustrates that brook trout are capable of chemosensory assessment of other fishes on two levels: 1) recognition of species-specific chemical cues and; 2) detection of the recent feeding regime of individual fish. A two-level system for the recognition of predatory fishes has clear ecolog-ical implications for juvenile brook trout. Early detection is considered an essential step in predator defense (Bertram, 1979) and may allow prey to avoid encounters with predators. The suite of predators encountered by juvenile brook trout may change frequently due to ontogenetic and seasonal shifts in the habitat of both the trout and their predators. A two-, or perhaps multi-level chemosensory recognition system for the detection of predators not only permits juvenile brook trout to avoid the chemical signatures of proven predators but, at the same time, may increase their ability to avoid unfamiliar piscivorous species. At this point in time it is difficult to assess whether this phenomenon occurs in other fishes because studies on fish chemosensory behavior generally do not report the diet of species used to generate chemical cues. Future studies are needed to determine if these findings have broader implications for teleosts.

ACKNOWLEDGEMENTS

This research was supported by an Electric Power Research Institute fellowship in population dynamics to M. Keefe.

REFERENCES

Bertram, B.C.R., 1978, Living in groups: predators and prey, in: "Behavioral Ecology: An Evolutionary Approach," J.R. Krebs and N.B. Davies, eds., Blackwell, Oxford.

Brett, J.R., and Mackinnon, D., 1954, Some aspects of olfactory perception in migratory adult coho and spring salmon, J. Fish. Res. Bd. Can., 11:310-318.

Ginetz, R.M., and Larkin, P.A., 1976, Factors affecting rainbow trout (Salmo gairdneri) predation on migrant fry of sockeye salmon (Oncorhynchus nerka), J. Fish. Res. Bd. Can.,33:19-24.

Hara, T.J., 1986, Role of olfaction in fish behavior, in: "The Behavior of Teleost Fishes,"T.J. Pitcher, ed., The Johns Hopkins University Press, Baltimore.

Kanayama, Y., 1968, Studies of the conditioned reflex in lower invertebrates X: Defensive conditioned reflex of chum salmon fry in groups, Mar. Biol, 2:77-87.

Keefe, M., 1990, Chemosensory behavior in wild brook trout, Salvelinus fontinalis, Doctoral dissertation, University of Rhode Island, Kingston.

Keefe, M., and Winn, H.E., 1991, Chemosensory attraction to home stream water and conspecifics by native brook trout, Salvelinus fontinalis, from two southern New England streams, Can. J Fish. Aquat. Sci., 48: 938-944.

Kleerekoper, H., 1969, "Olfaction in fishes," Indiana University Press, Bloomington.

Liley, N., 1982, Chemical communication in fish, Can. J. Fish. Aquat. Sci., 39:22-35.

Magurran, A., 1989, Acquired recognition of predator odors in the European minnow (Phoxinus phoxinus), Ethology, 82:216-223.

Olla, B.L., and Davis, M.W., 1989, The role of learning and stress in predator avoidance of hatchery-reared coho salmon (Oncorhynchus kisutch) juveniles, Aquaculture, 76:209-214.

Patten, B.G., 1977, Body size and learned avoidance as factors affecting predation on coho salmon, Oncorhynchus kisutch, fry by torrent sculpin, Cottus rhotheus, Fish. Bull., 75:457-459.

Pfeiffer, W., 1982, Chemical signals in communication, in: "Chemoreception in Fishes,".T.J. Hara, ed., Elsevier, Amsterdam.

Reed, J.R., 1969, Alarm substances and fright reaction in some fishes from the southeastern United States, Trans. Am. Fish. Soc., 4:664-668.

Rehnberg, B.G., and Schreck, C.B., 1986, Chemosensory detection of
     predators by coho salmon and the physiological stress response, <u>Can.
     J. Zool.</u>, 65:481-485.
Smith, R.J.F., 1986, The evolution of chemical alarm signals in fishes, <u>in</u>:
     "Chemical Signals in Vertebrates IV," D. Duvall, D. Muller-Schwarze,
     and R.M. Silverstein, eds., Plenum Press, New York.
Suboski, M.D., Bain, S., Carty, A.E., McQuoid, L.M., Seelen, M.I., and
     Seifert, M., 1990, Alarm reaction in acquisition and social
     transmission of simulated predator recognition by zebra danio fish
     (<u>Brachydanio</u> <u>rerio</u>), <u>J. Comp. Psychol.</u>, 104(1):101-112.
Weldon, P.J., 1990, Responses by vertebrates to chemicals from predators,
     <u>in</u>: "Chemical Signals in Vertebrates V," D.W. Macdonald, D.
     Muller-Schwarze, and S.E. Natynczuk, eds., Oxford University Press,
     Oxford.

PREY ODORS AS CHEMICAL STIMULANTS OF FEEDING BEHAVIOR IN THE RED SEA

MORAY EEL SIDEREA GRISEA

Rhonda Tannenbaum[1], Lev Fishelson[1], Sheenan Harpaz[2]

[1]Tel-Aviv University, Ramat Aviv, Israel
[2]Agricultural Research Organization, Bet Dagan, Israel

INTRODUCTION

Moray eels of the family Muraenidae are carnivorous fishes which inhabit, in the tropics, both the coral reef itself and the sandy areas with scattered boulders and corals. There are three genera of moray eels in the Gulf of Aqaba, Red Sea: Echidna, Gymnothorax and Siderea. The moray eel Siderea grisea is the most common species of morays in the Northern Red Sea inhabiting the coral and rocky subtidal.

Lateral extensions of chemosensory organs, such as the moray eel's olfactory organ, can aid in detection of an odor gradient produced by various prey, thus eliciting orientation. Fish possessing a well de-veloped sense of smell (such as the moray eel) have numerous olfactory folds in the nasal cavity. In addition to the olfactory apparatus, the moray eel has a vast array of taste buds lining its lower jaw and the sides of its face (Bardach et al., 1959). As observed, moray eels rely heavily upon chemoreception during their daily activities.

Very little work has been carried out on the feeding behavior of moray eels and virtually no information is available regarding Siderea grisea. The aim of this study was to investigate the foraging behavior and ecological habits of Siderea grisea and to examine the importance olfaction and taste play in its feeding behavior sequence.

MATERIALS AND METHODS

Behavioral observations were carried out at coral reefs located off the H. Steinitz Marine Laboratory in the Gulf of Aqaba, Elat, Israel. Observations consisted of approximately 100 dives, lasting one to two hours, at various times of the day.

Atherinid fishes, cut into small pieces or blended forming a fish slurry, were presented separately at selected sites of the coral reef. The arrival of different species of moray eels and the number of indi-viduals were noted.

For behavioral analyses in the laboratory, juvenile Siderea grisea were collected and were then transported to the Zoology Department of

Chemical Signals in Vertebrates VI, Edited by R.L. Doty and
D. Müller-Schwarze, Plenum Press, New York, 1992

Tel-Aviv University. Each eel was given time to acclimate and placed in a separate 100 litre aquarium equipped with a biological filter. The full feeding behavior sequence from food presentation (either as intact prey or a fish homogenate) until ingestion or rejection was analyzed using a video camera and recorder system.

To determine if the moray showed a marked preference for the smell of fish over that of sea water, a rubber tube was attached to a separatory funnel at one end and a feeding tube at the other end. A stopclock regulated the flow of the solutions and the number of entries into each tube were recorded.

Experiments testing prey preference in moray eels used three different food sources: fish (Sparidae), shrimp and squid. Three opaque containers with plankton netting covering their openings were attached equidistantly to a plexiglass slate. One of the three food sources was placed randomly in one container while the other two were left empty. The test-unit was then presented to a single moray and a lapse of five minutes was allowed for observations. Time of approach to the first container by the moray as well as the number of touches applied to each container were noted. Six trials were performed on each of five different morays. Subsequent investigations used two of the three food sources in different permutations and combinations, and arrival time was then measured.

RESULTS

Experiments done under natural conditions showed a marked difference between responses to the intact prey and fish slurry. More species and more individuals of moray eel arrived at the feeding site when a fish slurry was presented.

Experiments under controlled laboratory conditions showed the moray eel entered the feeding tube containing the fish homogenate significantly (sign test) more often than the tube containing sea water (Fig. 1).

All behaviors observed in the laboratory followed a fixed pattern initiated by the introduction of a food source. The appetitive phase of the moray eel's feeding behavior sequence, during which arousal of activity is observed, can be divided into a detection and identification phase (alert) and a locating phase (search). These are immediately followed by the consummatory phase whereby the moray strikes and consumes its prey.

Figure 2 shows the average time period of the alert versus search acts in a total of 30 sequences using both intact prey and a fish homogenate. Use of the fish homogenate resulted in an extended time period spent performing both these behaviors.

Figure 3 shows the average time interval from the introduction of a food source (fish, shrimp, squid) to arrival at the test-unit. A significant difference in arrival time was shown using a Kruskal-Wallis analysis of variance by ranks. When two food sources were presented simultaneously, no significant difference was observed (Chi-Square analysis).

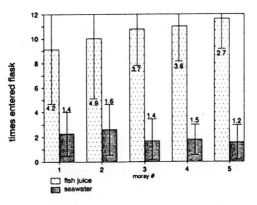

FIGURE NO.1 Average number of entries into flask
(containing either fish homogenate or saline)
in a total of 18 repetitions

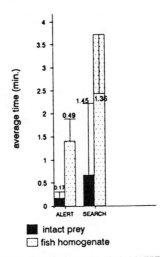

FIGURE NO.2 Average time period of ALERT
versus SEARCH acts in total of 30 sequences

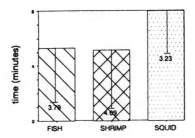

FIGURE NO.3 Time lapse from placement of prey to
arrival at test-unit in a total of 30 repetitions

DISCUSSION

Feeding behavior in fish can be aroused by a wide range of chemicals (Hara, 1975). When using a fish slurry instead of intact fish, a higher number of morays, as well as a greater number of species, arrived at the feeding site. This seems to indicate that moray eels depend on their strong sense of smell for food detection and location.

Olfactory alert is often followed by rheotactic orientation and location (Atema 1980). Eels (Anguilla anguilla) can follow scent trails using smell by comparing successive odor intensities along the gradient (Tesch, 1977). Siderea grisea seems to conform to this behavior. The lateral movements of its head in the water column during alert are in accordance with this phenomenon. There are two phases contributing to the feeding sequence of the moray eel. The first phase, i.e. detection, identification and location, relies heavily on the moray's senses of smell and taste. The time differences for the search and alert acts with the intact prey versus that with fish homogenate show that the moray eel relies on its sense of smell for prey detection and that these chemicals sustain interest for an extended period of time, even when no prey is visible. In several species of fish, the sense of taste can be extended to locate prey (Atema 1980). The search behavior exhibited by the moray eel consisted of continuous probing of the ground with its lower jaw while at the same time moving its head to and fro. The taste buds located on the head, sides and lower jaw are probably used at these times.

The experiments done on prey selection showed a significant preference for fish and shrimp over that of squid. The chemical stimulus emanating from the fish and shrimp elicited a quicker response than from the squid. Under natural conditions, the moray eel is likely to come across shrimp and fish hidden among the coral reef and it is less probable for the moray to encounter squid. Gut content analysis indicates that muraenids feed on a wide variety of fish and crustaceans (Chave & Randall 1971). These results also support the notion that moray eels are opportunistic feeders that feed on whatever prey item they initially encounter. Siderea grisea relies on its sense of taste for the pick-up and ingestion phase of its feeding sequence. There was no significant difference between the number of touches applied to the canister containing food. The chemical stimulus alone was enough to elicit the touching behavior which seems to be a reflex reaction.

The refined sense of olfaction in Siderea grisea is suited to seek out prey located in areas of little or no sunlight. The crevices located on the coral reef are such areas and this is where the moray eel spends most of its time foraging for prey.

REFERENCES

Atema, J.E., 1980, Chemical senses, chemical signals, and feeding behavior in fishes, in: "Fish Behavior and its use in the Capture and Culture of Fishes, J.E. Bardach, J.J. Magnuson, R.C. May, J.M. Reinhart, eds., ICLARM Conference Proceedings 5, International Center for Living Aquatic Resources Management, Manilla, Philippines, pp. 57-101.

Bardach, J.E., Winn, H.E., Menzel, D.W., 1959, The role of the senses in the feeding of the nocturnal reef predators (Gymnothorax moringa and Gymnothorax vicinus), Copeia 2:133-139.

Chave, E.H. and Randall, H.A., 1971, Feeding behavior of the moray eel, Gymnothorax pictus, Copeia, 3:570-574.

Hara, T.J., 1975, Olfaction in fish, in: "Progress in Neurobiology V", G.A. Kerkut and J.W. Phyliss, eds., Pergamon Press, Oxford, pp. 271-335.

Tesch, F.W., 1977, "The biology and management of anguillid ells, London: Chapman and Hall, pp. 434.

TRAUMA COMMUNICATION IN CHANNEL CATFISH

(Ictalurus punctatus)

Mehrnaz Jamzadeh

Biology Department
Benedict College
Columbia, S.C.

INTRODUCTION

Among the animals that are adapted to limited light are the
catfishes. The North American channel catfish, which was used as the
subject of this investigation, is an ostariophysan. These fishes are
known to possess club cells, and most members produce alarm substance
(Frisch 1938, 1941, Pfeiffer 1977). Limitation of visual communication
indicated by the catfish's habitat and specific anatomical features of
the channel catfish, such as the presence of chemosensory cells covering
the body and a high concentration of these cells on parts of the body
such as the barbels, is presumably the result of favored selection for
this method of communication (Todd 1971). Catfishes are capable of
producing and recognizing individual specific pheromones. Through these
pheromones a bullhead catfish, I. nebulosus, can identify not only the
species and sex of a conspecific, but also its age or size, reproductive
state, and hierarchical social status (Lowe-McCannell 1975).

Although channel catfish have been produced and raised in an
intensive culture system for many years, little is known about their
behavioral response to different stimuli and their complicated system of
communication. Prather (1959), reported that 65 percent of the total
catch occurred in the first day and thereafter fishing success in
fishing ponds declined rapidly. Another fact that suggests the presence
of a system of interaction among the fish in the pond concerns the
reduction in weight of the leftover fish in catch-out ponds (Bennett
1962). The following study was conducted to further investigate this
interaction. The experiments were designed to observe the effects of
injury by hook on behavior and survival of a catfish and the response of
its tank-mates compared to a group of fish receiving only the effluent
from the tank containing an injured fish.

METHODS AND MATERIALS

The study was conducted in the laboratory using sets of two tanks
connected by siphons and pumps (Figure 1). Four 190 liter (l) aquaria
with only one glass-side facing the observer were used. Throughout the
study the water was circulated between the two adjacent tanks
maintaining a 24 l/min. rate of flow which provided a turn-over time of

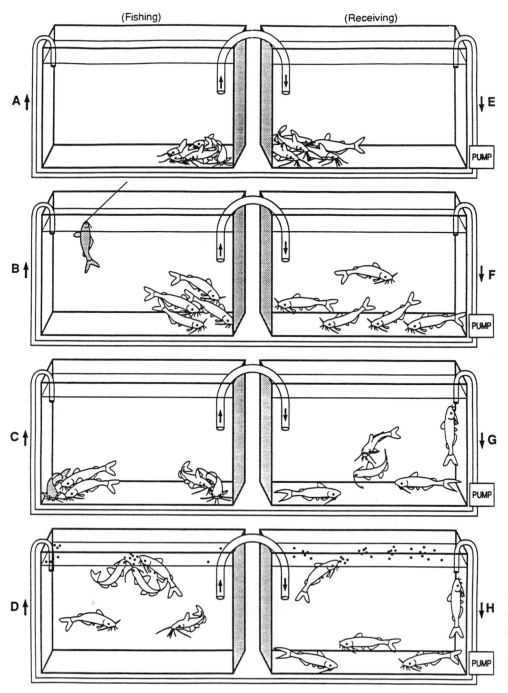

Fig. 1. Sequence of behavioral response of channel catfish to reception of all informational-clues in relation to injury inflicted upon a conspecific (Fishing) tanks versus reception of only chemical clues (Receiving) tanks. See Text.

approximately seven minutes for each tank. To aid in maintaining water quality, a flush of 2 l/min. of fresh water was continuously added to each of the systems. Electrical aquarium heaters were used to keep the water temperature at 25° C. In preventing the fish from jumping or climbing out of the aquaria, wooden framed plastic covers were used over all tanks. The tanks were exposed to a light/dark cycle of 12:12 hours.

In each of the six replications ten (10) different 15 cm catfish from a local fish farm were used. Catfish of this size were chosen since they are physiologically mature but sexually inactive. During the 2-5 days of habituation period and subsequently, the fish were fed 0.5 percent of body weight on 5 mm Purina Trout Chow pellets. These pellets float and keep their consistency more than 20 minutes. The number of pellets consumed in 20 minutes was used to compare the condition of the fish in the different tanks during fishing and no-fishing days. The fish in the tank designated (Fishing) were exposed to fishing by hook and line (Figure 1). The tank adjacent to (Fishing), the receiver tank, was identified as (Receiving) and a system of siphons and pump provided the only connection between the two tanks; i.e., the water circulated from (Fishing) to (Receiving) and through the pump back to (Fishing). Chicken liver and live worms were used as bait throughout the six replications of the study. The fishing process was limited to about an hour during the fishing days, usually resulting in hooking only one fish and occasionally two.

RESULTS

Patterns of activity for the fish in the two connecting tanks (Fishing) and (Receiving) were different during the experimental days. Figure 1 indicates the sequence of events in the two tanks. The response of tank-mates to presence of an injured fish was different from the response of fish receiving only effluent from a tank containing an injured fish.

Bait was introduced on a fishing hook to five fish which aggregated in a pod (Figure 1-A). The first fish to take the bait and consequently be injured was usually the dominant or the largest fish in the tank (Figure 1-B). After a struggle the injured fish would release itself from the hook and seek refuge in the pod which would re-appear in a corner of the tank (Figure 1-C).

The responses of the fish in the receiving tank to a conspecific being injured in the (Fishing) tank were different. While until the time of injury the fish in the (Receiving) tank were attracted to the inlet from the (Fishing) tank, (Figure 1-E), within a few minutes of a fish being hooked in the Fishing tank they moved away from the effluent and took separate positions facing the siphons. This positioning included breaking of the pod and orientation of isolated fish sitting close to or on the bottom of the tank (Figure 1-F). Frequently this positioning was followed by a fight which rapidly involved the whole group (Figure 1-G). The fights continued until all fish except one positioned themselves in the corners or edges of the tank.

In some of the replications following several days of fishing, fighting was noticed prior to initiation of the experiment. On these days, the fights would come to a halt for a period of time after the introduction of the bait but intensify following injury by hooking. In association with vigorous fighting, vertical positioning was observed

(Figures 1-G and H). These fish would take solitary positions in the corners of the tank and not respond to any activity in the tank including feeding or attacks by other fish.

There was a reduction in the number of consumed food pellets (50 pellets/20 minutes) during the experimental days, indicating the existence of special condition in the tanks. To analyze these data statistically, a Chi-square test was conducted comparing days of no fishing to days of fishing and number of days in which fish consumed half or more than half of the pellets in 20 minutes. Two replications from the six indicated significant difference between the two tanks on days of fishing and no fishing ($X^2 = 5.6$, $p = 0.018$; $X^2 = 11.55$, $p = 0.0007$). Figure 2 shows the variation of the rate of consumption of food pellets in two of the six replications of the study. Correlation between injury and occurrence of fighting was also analyzed using a Chi-square test ($X^2 = 7.179$, $p = 0.007$).

There was also a difference in the feeding pattern between the two tanks shortly after the experiments. While fish in the (Fishing) tank would resume normal surface feeding (Figure 1-D), fish in the (Receiving) tank engaged in severe fighting, would not surface feed in groups, and showed limited interest in consuming sunken pellets (Figure 1-H).

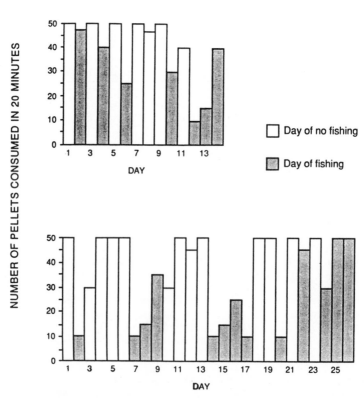

Fig. 2.   Fluctuation of number of food pellets consumed in accordance with days of fishing and no fishing in tanks receiving only chemical cues from a tank containing an injured fish. Two replications are shown.

DISCUSSION

Channel catfish have been the subject of many culture and management studies, but it is apparent from a review of the literature that the question of presence or absence of alarm substance has not been fully investigated. The sequence of events outlined above which resulted in breakage of the pod and the occasional fighting in the (Receiving) tank were evidence for transmission of a chemical which modifies the behavior of the fish. Also, reduction in the speed of food consumption and lack of surface feeding on days when a fish was injured suggested the transfer of trauma-related information between the two tanks. It is unlikely that sound produced by the fish or its activity were transmitted through the siphons, with a pump functioning and as mentioned earlier there were no possibility of visual exchange.

Studies by Bardach, Todd and Crickmen (1967), Todd, Atema and Bardach (1967), Bryant, Elgin and Atema (1978), and Johnsen and Teeter (1980) stressed the importance of chemical communication among catfishes. Although most of these studies were done on brown and yellow bullheads, their results can be generalized to provide information on the characteristics of other catfishes. One such study pertains to territorial fights among yellow bullhead (Todd et al. 1967). The investigators referred to a change in body odor which was recognizable by other members of the species. Todd et al. (1967) suggested that this chemical change in the amino acid composition of the skin mucus was identifiable by chromatographic methods during the 24 hours following the fight, but it was not a lasting phenomenon. This change may very well have been the result of release of the contents of the club cells which do not open directly to the surface of the skin, but injury caused by fighting would be sufficient to release their content. Since catfish have a dominance hierarchy system, information relative to the change of status of any fish is important in reorganization of the hierarchy. Based on the findings by Todd and his colleagues, the sequence of events which was observed in this study could be the result of (a) an alarm reaction (demonstrated by the avoidance of the area around the siphons and also motionless sitting of the fish on the bottom of the tank facing the siphons, and (b) a body odor change of the dominant fish which called for re-establishment of the hierarchy. The latter was supported by the individual positioning of the fish and the occasional fighting which followed the injury.

Another concept important in understanding the chemical communication of alarm among fishes is the distinction between a communication signal and an information signal (Jamzadeh 1991). While a communication signal is directed toward the receiver and contains a specific message an information signal is part of the general existence of the individual or the group. A study by Bryant and Atema (1987) supports this conclusion; their experiments demonstrated that a change of diet caused a change in the individual recognition odor and social interaction of catfish. While release of the alarm substance communicates danger, the individual,s recognition odor is only an information signal identifying one fish from another. With the change in the odor as a result of release of the contents of the club or other cells, status change is communicated.

Although there are some indications that the channel catfish produce an alarm substance, their response to its release indicates a difference between this species and non-predatory schooling ostariophysi. While a study by Anthony (1964) indicates presence of a substance that decreases the possibility of cannibalism, an attribute contributed to production of alarm substance, the niche these large non-

schooling omnivorous fish occupy suggests lack of importance of alarm
substance for their survival. It is feasible to suggest that the
content of the club cells in catfish serves different functions during
its life cycle. Various functions reported in different studies
(Bardach et al. 1967, Bryant et al. 1978, Johnsen and Teeter 1980,
Bradley et al. 1990) may classify the contents of the club cells as
information signals which functions in accordance with the species
needs. I am suggesting the contents of the club cells in a small
schooling prey fish such as the European minnow, Phoxinus phoxinus (L.),
functions as an alarm substance, which may be useful to young schooling
catfish but this function will change to identify the individual
organism and its internalstate in a territorial piscivorous adult
catfish that shows dominance hierarchy.

ACKNOWLEDGMENTS

I wish to thank Drs. B. Bryant, G. Epple, R. Heidinger, and
W. Lewis for their comments and guidance. Thanks are also extended to
the Southern Illinois University Fisheries Research Laboratory, for
providing the facilities and the fish.

LITERATURE

Anthony, M. 1964. Utilization of selected forage organisms by channel
    catfish. Dissertation, Southern Illinois University at
    Carbondale.
Bardach, J.E., Todd, J.H., and Crickmer, R. 1967. Orientation by taste
    in fish of the genus Ictalurus. Science 155:1267-1278.
Bennett, G.W. 1962. Management of artificial lakes and ponds. Reinhold
    Publishing Corporation, Chapman and Hall, Ltd., London.
Bradley, G., Rehnberg, B.G. and Smith, R.J.F. 1990. Behavioral and
    physiological responses to alarm pheromone by ostariophysan fishes
    and a possible modulating role for brain benzodiazepine receptors.
    In: Chemical Signals in Vertebrates 5, (ed. D. MacDonald, D.
    Muller-Schwarze, and S. Natynczuk), pp. 132-138. Plenum, New
    York.
Bryant, B., Elgin, R., and Atema J. 1978. Chemical communication in
    catfish: stress-induced changes in body odor. Biol. Bull
    155:429.
Bryant, B.P. and Atema, J. 1987. Diet manipulation affects social
    behavior of catfish: importance of body odor. J. Chem. Ecol.
    13:1645-1661.
Frisch, K.von. 1938. Zur Psychologie des Fisch-Schwarmes.
    Naturwissenscaften 26:601-606.
Frisch, K.von. 1941. Uber einen Schreckstoff der Fischhaut und seine
    biologische Bedeutung. Z. Vergl. Physiol. 29:44-145.
Jamzadeh, M. 1991. Communication and information signals. (In
    preparation)
Johnsen, B., and Teeter, J.H. 1980. Spatial gradient of chemical cues
    by catfish. J. Comp. Physiol. 140:95-99.
Lowe-McCannel, R.A. 1975. Fish communities in tropical fresh water.
    Longman, London.
Pfeiffer, W.D. 1977. The distribution of fright reaction and alarm
    substance cells in fishes. Copeia 1977, 653-665.
Prather, E.E. 1959. The use of channel catfish as sport fish. Proc.
    13th Ann. Conf. S.E. Assoc. Game and Fish Comm., 331-335.
Todd, J.H. 1971. The chemical languages of fishes. Sci. Amer.,
    224(5):98-108.
Todd, J.H., Atema, J., and Bardach, J.E. 1967. Chemical communication
    in the social behavior of a fish, the yellow bullhead, Ictalurus
    natalis. Science, 158:672-673.

PART 2: Amphibia

CHEMOTESTING MOVEMENTS AND CHEMOSENSORY SENSITIVITY TO

AMINO ACIDS IN THE EUROPEAN POND TURTLE, EMYS ORBICULARIS L.

Yurii Manteifel, Natalia Goncharova, and Vera Boyko

A.N. Severtzov Institute of Evolutionary Animal
Morphology and Ecology, Russian Academy of Sciences
Moscow, Russia

INTRODUCTION

The European pond turtle Emys orbicularis is an amphibious animal detecting and locating food by chemoreception both on land and in water (Honigmann, 1921). This turtle makes regular low-amplitude movements of the lower
jaw that promote the propulsion of water to chemoreceptors within the nasal
and oral cavities which may be regarded as stimulus acquisition behavior
after the terminology of Atema (1987). We term these movements "jaw testing
movements" (JTM). Various changes in the sensory medium and the animal's
emotional state elicit an increase of the JTM frequency which serves as
an indicator of chemosensory perception of stimuli. In this study, we examined the reaction of Emys orbicularis to some L-amino acids. Such agents
are effective stimuli for chemoreceptors of most aquatic animals and are
generally found in the aquatic medium (Bardach, 1975).

CHANGES IN JTM FREQUENCY ASSOCIATED WITH REPRODUCTIVE BEHAVIOR

Ten pairs of E.orbicularis were placed into an aquaterrarium containing
200 liters of water. With insufficient receptivity of turtle females, the
reproductive behavior is accomplished in repeated cycles. Three stages can
be distinguished in each cycle. In stage 1, a male sniffs a female. In stage
2, a male stops near a female with its head, extremities and tail down. In
stage 3, a male, having fixed itself on female's carapax, rubs the female's
nose with his chin.

The average JTM frequency in resting turtles not presented with stimuli was 4.3 per minute in both sexes. In the course of each behavioral
cycle, the JTM frequency gradually increased until stage 3, on average 35
fold in males and 13 fold in females (Fig.1). Bilateral transection of the
olfactory or vomeronasal nerves of males significantly decreased both their
reproductive behavior, especially in stages 2-3, and JTM frequency. No disturbances were observed in males after the unilateral transection of either
nerve or after bilateral transection of the ophthalmic branches of the trigeminal nerve. Transection of both pairs of the olfactory and vomeronasal
nerves in males completely eliminated their reproductive behavior. It seems
probable that JTM plays an important role in chemoreception of the turtle
during reproductive behavior and that the main and accessory olfactory systems are complementary in this regard.

Chemical Signals in Vertebrates VI,   Edited by R.L. Doty and
D. Müller-Schwarze, Plenum Press, New York, 1992

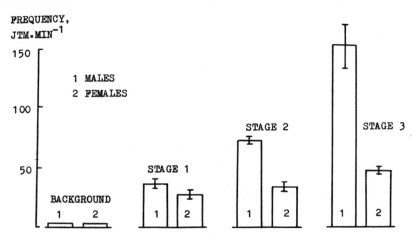

Fig. 1. Frequency of the jaw testing movements in Emys orbicularis (mean + SE) in different stages of the reproductive behavioral cycle.

PERCEPTION AND RELATIVE INFLUENCES OF DIFFERENT AMINO ACIDS

Thirty blinded turtles were tested individually in a plastic aquarium filled with 2 1 of aerated tap water. Amino acid solutions (10^{-4} mol/1, 2 ml) were injected into the water in the vicinity of the turtle's nostrils and the number of JTM was recorded for 1 minute. The water in the aquarium was then changed and another test performed. Two-fold rise of the JTM frequency over the pre-stimulus background level was considered "response". Five of the 6 amino acids tested proved to be significantly stimulatory (Table 1).

SENSITIVITY TO ALANINE, ARGININE AND GLUTAMINE

A study of amino acid sensitivity was carried out on blinded turtles. The water from the aquarium where a tested turtle was placed served as the control stimulus. Test-solutions were prepared with the same water. In each

Table 1. Relative Influences of Amino Acids at a Concentration of 10^{-4} mol/1

Stimulus	Number of Probes	Percent of Responses	Significance (relative to Control), p
Valine	29	59	0.001
Arginine	33	55	0.001
Glutamine	48	54	0.001
Glycine	31	45	0.001
Alanine	29	38	0.002
Histidine	27	22	0.90
Water	115	12	–

series of experiments, 11–12 turtles were used. Solutions of each chemical were presented to individual animals at concentrations gradually increasing by 1–3 orders of magnitude. To quantify the magnitude of the reaction, the background JTM value was subtracted from the JTM value recorded within 30 sec of the presentation of the control or the test stimulus. To evaluate the stimulus effect, we further calculated the difference between the reaction to the test stimulus and the reaction to the previous control stimulus. The threshold concentration of the injected solution for arginine and glutamine was equal to 10^{-18} mol/1 and for alanine to 10^{-17} mol/1 (Fig. 2). At these stimulus concentrations a significant ($p < 0.01$) increase of the mean value of the stimulus effect in comparison with the effects of lower concentrations was present. Measurement of the electric resistance of NaCl solution imitating chemical stimulus has shown that the real threshold does not exceed 0.25 ± 0.01 of the presented values. It has been shown earlier that reception of amino acids at low concentrations is mediated via olfaction (Manteifel and Boyko, 1987).

As seen in Figure 3, at a background glutamine level of 10^{-12} mol/1, the solution of this amino acid at a concentration of 10^{-12} mol/1 was not effective, while at a concentration of 10^{-11} mol/1 it was. The same result was obtained for alanine and arginine (Fig. 3). Thus, for this background the differential threshold is lower than 1000%. More precise evaluations show that this characteristic is between 700 and 800% for alanine and between 600 and 700% for arginine. The data for all 12 turtles were rather similar. If the stimulus dilution in the aquarium is taken into account, it is likely that the differential threshold is close to 200%.

To evaluate the perception specificity, we presented alanine, arginine and glutamine (10^{-12} mol/1) against the artificially created background of another amino acid at 10^{-12} mol/1 concentration (Fig. 3). At the behavioral level, cross-adaptation was absent: the mean stimulus effect did not differ

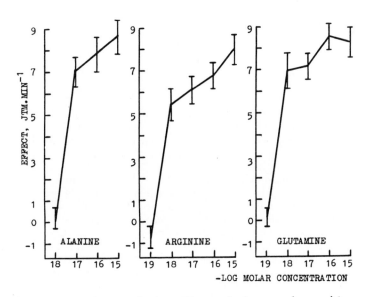

Fig. 2. Dependence of the effect of three amino acids on
the stimulus concentration (mean ± SE).

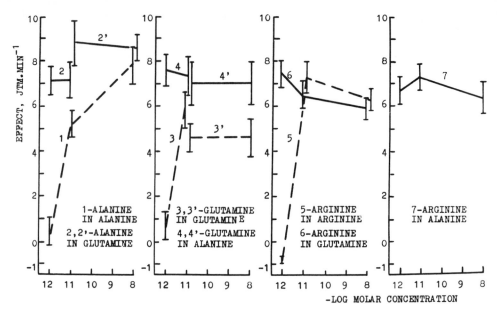

Fig. 3. Concentrational dependence of the effect of three
amino acids at the background level of one of the
chemicals of 10^{-12} mol/l: auto-adaptation and
cross-adaptation (mean $\pm$ SE).

from that of background stimulus presented at a concentration of 10^{-11}
mol/l. These findings indicate considerable specificity of the reception
of each amino acid.

Turtles can distinguish the quality of these three amino acids at the
hedonic level. Unlike glutamine, alanine and arginine often evoke the avoidance reaction. For example, alanine at concentrations of 10^{-17} and 10^{-16}
mol/l induced departure of the turtle in 61 and 58% of cases, respectively.
The subthreshold stimulus (10^{-18} mol/l) and the control never evoked this
reaction ($p < 0.001$). Arginine was less effective but highly significant:
39% of avoidance responses at 10^{-17} mol/l.

CONCLUSIONS

At the stimulus concentration of 2.5×10^{-19} mol/l, only about 15 molecules of the test stimulus enter the turtle's nasal cavity during one jaw
testing movement. Thus, sensitivity is close to the natural limit. The
differential threshold is rather high, and a very low content of amino
acids in tap water (that was confirmed, at least partly, in experiments
with the artificial background as low as 10^{-12} mol/l) must be a necessary
condition for the determination of such a low absolute threshold. Obviously, in natural water bodies with amino acid level of 10^{-8}–10^{-6} mol/l
(Zygmuntova, 1972; Jorgensen, 1986) the absolute sensory sensitivity of
E.orbicularis to amino acids is determined by the stimulus background level
and differential sensitivity. This conclusion may apply to other animals
with high sensory sensitivity to amino acids. High sensitivity and specificity of olfaction for reception of amino acids may be interpreted as
secondary evolutionary adaptation of Emys orbicularis to the aquatic
medium.

REFERENCES

Atema, J., 1987, Distribution of chemical stimuli, in: Sensory Biology of
    Aquatic Animals, J. Atema, R.R. Fay, A.N. Popper and W.H. Tavolga,
    eds, Springer-Verlag, New York.
Bardach, J.E., 1975, Chemoreception of aquatic animals, in: Olfaction and
    Taste, Vol. V, D.A. Denton and J.P. Coghlan, eds., Academic Press,
    New York.
Honigmann, H., 1921, Zur Biologie der Schildkröte, Biol. Zentralblatt,
    41:241.
Jorgensen, N.O.G., 1986, Fluxes of free amino acids in three Danish lakes,
    Freshwater Biol., 16:255.
Manteifel, Yu.B., and Boyko, V.P., 1987, Comparative evaluation of sensi-
    tivity of different chemoreceptor systems of Emys orbicularis, Sensory
    Systems, 1:22 (in Russ.).
Zygmuntova, J., 1972, Occurence of free amino acids in pond water, Acta
    hydrobiol., 14:317.

CORRELATION OF SALAMANDER VOMERONASAL AND MAIN OLFACTORY

SYSTEM ANATOMY WITH HABITAT AND SEX: BEHAVIORAL INTERPRETATIONS

Ellen M. Dawley

Department of Biology
Ursinus College
Collegeville, PA 19426

INTRODUCTION

The salamander family Plethodontidae is an ancient group of vertebrates whose members probably rely on a combination of visual and chemical cues for their perception of the world (Roth, 1986). Chemical cues have been shown to be important in a number of social interactions including territoriality (Jaeger, 1986), mate recognition and choice (Dawley, 1986), and "persuasion" during courtship (Houck and Reagan, 1990). Similar to most other terrestrial vertebrates, plethodontid salamanders have both main olfactory and vomeronasal systems that are physically, and perhaps functionally, separate from one another (Dawley and Bass, 1988; Schmidt, Naujoks-Manteuffel, and Roth, 1988). The vomeronasal system is of particular interest for this group of animals because a unique structure (nasolabial grooves) coupled with a unique behavior (nose-tapping) is used to stimulate the vomeronasal system during many social interactions, including mating. In this paper I present a preliminary survey of variation of vomeronasal and main olfactory system morphometrics for salamanders with different habitat requirements and for both sexes. I attempt to correlate this anatomical variation with ecological and sexually dimorphic behavioral variation.

The majority of living salamanders belong within the family Plethodontidae. Their distinctive derived characters are the absence of lungs and the presence of paired nasolabial grooves. Nasolabial grooves are cutaneous depressions that run from the upper lip to the lateral corner of each nostril. These grooves function as capillary tubes when the bases of these grooves are touched to an object, conveying liquids from the base to the external nares (Whipple, 1906). Using radioactively-labelled proline, a non-volatile liquid, Dawley and Bass (1989) showed that liquids placed at the bases of the nasolabial grooves are preferentially conveyed to the vomeronasal organ and not to the main olfactory epithelium. Blockage of these nasolabial grooves prevented proline from entering through the external nares, although a lesser quantity of proline entered the vomeronasal organs via the internal nares. Plethodontid salamanders nose-tap, touching the bases of their nasolabial grooves to the substrate and objects on the substrate, in a number of different social situations (e.g., Jaeger, 1986). Thus, nose-tapping is analogous to tongue-flicking in snakes, a method by which they physically pick up and convey nonvolatile odorants preferentially to the vomeronasal organs. Only plethodontid salamanders have nasolabial grooves; thus, it is unknown how salamanders in other families stimulate their vomeronasal organs.

The vomeronasal organs of salamanders in nearly all families are found as ventrolateral diverticulae of the main olfactory chambers, remaining confluent with the main olfactory chambers even in adults (Dawley and Bass, 1988). In plethodontids, it is noteworthy that nose-tapping with subsequent nasolabial groove conveyance to the vomeronasal organs results in very little spill-over into the main olfactory epithelium. This is partly due to the anterior placement of the vomeronasal organs so that the posterior end of each nasolabial groove is adjacent to the anterior limit of the vomeronasal organs (Dawley and Bass, 1988). In those other salamander families that have vomeronasal organs in ventrolateral diverticulae (Ambystomatidae and Salamandridae), the vomeronasal epithelium is much more posteriorly placed, so that there is an area of nonsensory epithelium lining the anterior portion of the ventrolateral diverticulae (Jurgens, 1971). In short, odorants that enter through the external nares would have a much longer route to travel before reaching the vomeronasal organs. Other salamander families, including those with fully aquatic species, have reduced lateral diverticulae and vomeronasal organs (Jurgens, 1971).

The vomeronasal system is of interest because it is linked to those areas of the brain that mediate hormonal and behavioral responses of males and females to each other (Simerly, 1990). In addition, the vomeronasal system has been shown to be sexually dimorphic in some mammals (Simerly, 1990). Sexual dimorphism at the sensory receptor, neural, or muscular level may be a possible mechanism for generating sexually dimorphic behavior (Kelley, 1988). Although both male and female plethodontid salamanders have been seen to nose-tap substrates and fecal pellets (Jaeger, 1986), suggesting a role for the vomeronasal system in territory identification and maintenance, only males nose-tap during courtship. Males appear to nose-tap female substrate trails (Gergits and Jaeger, 1990) and to nose-tap females themselves before initiating courtship sequences leading to sperm transfer (Dawley, 1986). In addition, there is sexual dimorphism of the nasolabial grooves; during the breeding season the glands surrounding the nasolabial grooves hypertrophy in males, but not in females (Sever, 1975); these hypertrophied nasolabial grooves are called nasolabial cirri. The degree of this hypertrophy varies with species (see below).

GENERAL METHODS AND RESULTS

Plethodontid salamanders are a diverse family with species inhabiting a diverse set of habitats (Wake, 1966). They include primitive, generalized species living in semiaquatic habitats with extended aquatic larval stages. More derived species do not have an aquatic larval stage but instead undergo direct development. These derived species are semiaquatic or completely terrestrial; some terrestrial species are arboreal or semifossorial. Other derived species show some degree of neoteny, retaining an aquatic larval condition throughout life. The origin and initial radiation of plethodontids probably occurred within the Appalachian Mountains of North America, and eastern North America is still an area of abundance and diversity. Another area of major radiation is Central and South America. Plethodontids are the only group of salamanders to have undergone a major adaptive radiation in the tropics (Wake and Lynch, 1976). This diversity makes plethodontids an ideal group with which to study the evolution of sensory adaptations as they might be correlated with habitat.

For the initial study of general plethodontid vomeronasal and main olfactory organ morphology, I used a species that exhibits a generalized morphology and is completely terrestrial. Plethodon cinereus is a small (32-52mm snout-vent length), slender, woodland salamander of eastern North America that shows minimal sexual dimorphism of the nasolabial grooves. This species is compared to Eurycea wilderae, a species of similar morphology but with a greater degree of nasolabial groove hypertrophy in males of some populations but a complete lack of hypertrophy in males of other populations (referred to as monomorphic males). Adult male and female P. cinereus were collected during the breeding season and immediately sacrificed and processed for histology. Adult E. wilderae were donated from the collection of N. Reagan, University of Chicago, but were judged to be in breeding condition because they possessed packed testes or large oviducal eggs. For habitat diversity comparisons, I collected a series of species from the Appalachian Mountain region which were immediately prepared for histology. These included a completely aquatic species (Leurognathus marmoratus), a number of semiaquatic to terrestrial species that are considered most closely related to L. marmoratus (Desmognathus monticola and D. ochrophaeus), and a large, semiaquatic species with considerable nasolabial groove hypertrophy in males (Eurycea guttolineata). I also looked at five tropical species donated by D. Wake, University of California, all but the first with greatly expanded nasolabial cirri; these included two terrestrial species, Thorius macdougali and Pseudoeurycea cephalica, and three arboreal species, Chiropteroptriton multidentatus, Dendrotriton bromeliacia and Bolitoglossa rufescens.

For all studies, whole heads were processed for plastic embedding as reported by Dawley (1992), cut with glass knives on a rotary microtome at a thickness of $0.5\mu m$, $2.5\mu m$, or $5.0\mu m$, mounted on glass slides, and stained with 1% toluidine blue. Volumes of vomeronasal and main olfactory epithelia were calculated by summing areas of known serial sections. Camera lucida tracings of these serial sections were digitized using a MacIntizer ADB digitizing pad connected to a MacIntosh SE, and areas of each section were calculated by the digitizing program MacMeasure. These areas were summed using the technique outlined by Dornfeld, Slater, and Scheffe (1942) to obtain a volume for vomeronasal and main olfactory organs. Volumes were compared among species and sexes using an analysis of variance followed by Student-Newman-Keuls multiple comparisons tests. When only one value was available per species, this sample was compared to the mean of P. cinereus females, E. wilderae males, or E. wilderae females (Sokol and Rohlf, 1981).

A two-level nested analysis of variance shows that vomeronasal organ volume is significantly different among species ($p < 0.05$) and between sexes ($p < 0.005$). Female P. cinereus have the smallest vomeronasal organs, followed in order of size by male P. cinereus, female E. wilderae, monomorphic male E. wilderae and male E. wilderae. All comparisons are significantly different from each other except female E. wilderae and monomorphic male E. wilderae (Fig. 1). The two species can be easily compared because they have the same range of body sizes (32-52mm snout-vent length). Interestingly, female P. cinereus have the smallest vomeronasal organs despite the fact that they include the largest individuals (51, 52mm). Male E. wilderae with enlarged cirri have the largest vomeronasal organs despite the fact that these males have among the smallest snout-vent lengths (34-38mm). These results show that males have larger vomeronasal organs than do females of the same species and that males of species that have greatly enlarged nasolabial cirri also have even larger vomeronasal organs than do males with minimal hypertrophy of cirri.

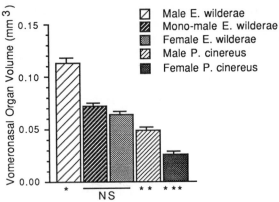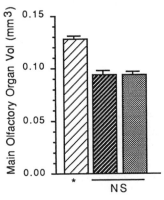

Fig. 1. Volumes of vomeronasal organs (mm³) of male, monomorphic male, and female E. wilderae and of male and female P. cinereus. See Table 1 for sample sizes. *Significantly > all other groups; ** Significantly < preceding 3 groups; *** Significantly < all other groups; NS Not significantly different from each other. Bars represent standard deviations.

Fig. 2. Volumes of main olfactory organs of male, monomorphic male, and female E. wilderae. See Table 1 for sample sizes. Abbreviations as in Fig. 1.

The five Neotropical plethodontid species and the North American E. guttolineata offer an opportunity to compare the positive correlation of nasolabial cirri enlargement with size of vomeronasal organs. All but Thorius macdougali, a miniatured genus, have greatly enlarged nasolabial cirri. The snout-vent lengths of the other four neotropical genera are within the same range as P. cinereus and E. wilderae (32-56mm). Three species, B. rufescens, C. multidentatus, and D. bromeliacia have vomeronasal organs of a similar size as the E. wilderae males with enlarged nasolabial cirri (Table 1). Two other species, P. cephalica and the North American E. guttolineata, have even larger vomeronasal organs (Table 1). These kinds of data confirm a positive correlation between nasolabial cirri enlargement and vomeronasal organ size (Dawley, in prep).

Main olfactory organ sizes show similar correlations (Table 1; Dawley, in prep). Data are unavailable for P. cinereus, but for E. wilderae, males with enlarged nasolabial cirri have significantly larger main olfactory organs than females or monomorphic males (Fig. 2; single classification analysis of variance, $p < 0.025$; Student-Newman-Keuls tests, $p < 0.05$). Female E. wilderae and monomorphic male E. wilderae have similar sized main olfactory organs ($p > 0.5$). In addition, E. guttolineata and the Neotropical plethodontid species, except T. macdougali, have larger main olfactory organs than similarly sized E. wilderae males (Table 1). Thus, there is also a positive correlation between nasolabial cirri enlargement and main olfactory organ size.

Habitat type may also affect vomeronasal and main olfactory organ morphology. In fishes and aquatic salamanders the nasal epithelium is folded (Seydel, 1895); sensory epithelium lines the grooves of the folds and nonsensory epithelium covers the ridges. However, the sensory epithelium of adults of terrestrial species is organized into flat sheets (see description of P. cinereus in Dawley and Bass, 1988). All

406

Table 1. Sample size, vomeronasal organ volume, main olfactory organ volume and snout-vent length for all species

Species	N	VNO Volume[a]	Main Olf Volume[a]	Snout-vent Length[b]
Plethodon cinereus-male	6	0.046mm^3	N/A	32-50mm
P. cinereus-female	4	0.025mm^3	N/A	45-52mm
Eurycea wilderae-male	3	0.111mm^3	0.124mm^3	34-37mm
E. wilderae-male[c]	3	0.069mm^3	0.091mm^3	42-47mm
E. wilderae-female	7	0.061mm^3	0.091mm^3	38-46mm
E. guttolineata-male	1	0.209mm^3**	0.340mm^3**	56mm
Bolitoglossa rufescens-male	1	0.100mm^3†	0.216mm^3**	32mm
Chiropterotriton multidentatus-m	1	0.118mm^3†	0.234mm^3**	37mm
Dendrotriton bromeliacia-male	1	0.176mm^3†	0.310mm^3**	29mm
Pseudoeurycea cephalica-male	1	0.342mm^3**	0.579mm^3**	43mm
Thorius macdougali-male	1	0.025mm^3¢	0.076mm^3¥	16mm
Desmognathus monticola-male	1	0.141mm^3†	0.209mm^3**	43mm
D. ochrophaeus-male	1	0.056mm^3¥	0.124mm^3†	42mm
Leurognathus marmoratus-male	1	0.056mm3¥	0.020mm^3*	54mm

[a]For species with N>1 mean volumes are reported
[b]For species with N>1 ranges of snout-vent lengths are reported
[c]These males do not have enlarged nasolabial cirri; see text
**$p < 0.05$ that single value = average volume of organ of E. wilderae males
† $p > 0.5$ that single value = average volume of organ of E. wilderae males
¥ $p > 0.5$ that single value = average volume of organ of E. wilderae females
¢ $p > 0.5$ that single value = average volume of organ of P. cinereus females
* $p < 0.05$ that single value = average volume of organ of E. wilderae females

salamanders examined in this study had main olfactory organs organized as P. cinereus even if they were semiaquatic or nearly fully aquatic (Dawley, in prep). The only exception was the fully aquatic Leurognathus marmoratus; in that salamander species the olfactory epithelium was organized into flat sheets, but the entire organ was greatly reduced in volume (Table 1). The most closely related group to Leurognathus, the genus Desmognathus, has main olfactory organs similar in size or larger than E. wilderae (Table 1). This suggests that all salamanders examined in this study use their olfactory organs to sense air-borne odors.

The vomeronasal organs of the plethodontid species in this study resembled that of P. cinereus (Dawley and Bass, 1988; Dawley, in prep). Again, the only exception was the fully aquatic L. marmoratus, which has a vomeronasal organ positioned even more anteriorly than in P. cinereus or the closely related Desmognathus species. This may be related to the reduction of the main olfactory organ. The size of the vomeronasal organ of L. marmoratus is proportional to that found in E. wilderae females (Table 1) while Desmognathus species have somewhat larger vomeronasal organs (Table 1).

SUMMARY AND CONCLUSIONS

Male plethodontid salamanders have significantly larger vomeronasal organs and nasolabial cirri than do females of their own species. Males nose-tap females and female trails with these cirri before and during courtship. Females have not been observed to nose-tap in association with mating. Vomeronasal stimulation may be a prerequisite for courtship initiation by males as it is in some other terrestrial vertebrates. Because one possible mechanism for producing sexually dimorphic behavior is through sex differences in sensory receptor, neural, or muscular systems, I suggest that the sexually dimorphic behaviors seen in plethodontid courtship are probably due in part to sexual differences in the vomeronasal system. However, main olfactory organs are also larger in males than in females, indicating a role of the main olfactory system as well.

Vomeronasal and main olfactory organs are significantly larger in species with greatly enlarged nasolabial cirri than those with moderately enlarged cirri. There is no discernible pattern linking development of these chemosensory structures with habitat. All plethodontid species have the olfactory epithelium organized for air breathing. Only the most aquatic species has a reduced main olfactory organ but retains a full sized vomeronasal organ. This last observation supports the idea that the vomeronasal system is prominent in plethodontid mating behavior.

ACKNOWLEDGEMENTS

Parts of this study were funded by Ursinus College through a faculty development grant and a Van Sant grant, by Bowdoin College, and by an NSF ILI Grant USE-9051558. Krishni Patrick, Brian Stein and Alan MacIntyre provided much technical assistance; many thanks for their hard work.

REFERENCES

Dawley, E. M., 1986, Behavioral isolating mechanisms in sympatric terrestrial salamanders, Herpetologica, 42:156.
Dawley, E. M., 1992, Sexual dimorphism in a chemosensory system: The role of the vomeronasal system in salamander reproductive behavior, Copeia, 1992:113-120.
Dawley, E. M. and Bass, A. H., 1988, Organization of the vomeronasal organ in a plethodontid salamander, J. Morphol., 198:243.
Dawley, E. M. and Bass, A. H., 1989, Chemical access to the vomeronasal organ of a plethodontid salamander, J. Morphol., 200:163.
Dornfeld, E. J. Slater, D. W., and Scheffe, H., 1942, A method for accurate determination of volume and cell numbers in small organs. Anat. Rec., 82:255.
Gergits, W. F. and Jaeger, R. G., 1990, Field observations of the behavior of the red-backed salamander (Plethodon cinereus): Courtship and agonistic interactions, J. Herpetol., 93.
Houck, L. D. and Reagan, N. L., 1990, Male courtship pheromones increase female receptivity in a plethodontid salamander, Anim. Behav., 39:729.
Jaeger, R. G., 1986, Pheromonal markers as territorial advertisement by terrestrial salamanders, In: "Chemical Signals in Vertebrates 4," D. Duvall, D. Muller-Schwarze and R. Silverstein, eds., Plenum, New York.
Jurgens, J. D., 1971, The morphology of the nasal region of Amphibia and its bearing on the phylogeny of the group, Ann. Univ. Stell., 46A2:1.

408

Kelley, D. B., 1988, Sexually dimorphic behaviors, <u>Ann. Rev. Neurosci.</u>, 11:119.

Roth G., 1986, Neural mechanisms of prey recognition: An example in amphibians, <u>In:</u> "Predator-prey Relationships," M.E. Feder and G.V. Lauder, eds., Univ. Chicago Press, Chicago.

Schmidt, A., Naujoks-Manteuffel, C., and Roth, G., 1988, Olfactory and vomeronasal projections and the pathway of the nervus terminalis in ten species of salamanders, <u>Cell Tissue Res.</u>, 251:45.

Seydel, O., 1895, Uber die Nasenhohle und das Jacobson'sche Organ der Amphibien <u>Morphol. Jahrb.</u>, 23:453.

Sever, D. M., 1975, Morphology and seasonal variation of the nasolabial glands of <u>Eurycea quadridigitata</u> (Holbrook), <u>J. Herpetologica</u>, 45:322.

Simerly, R. B., 1990, Hormonal control of neuropeptide gene expression in sexually dimorphic olfactory pathways, <u>Trends Neurosci.</u>, 13:104.

Sokol, R. R. and Rohlf, F. J., 1981, "Biometry," Freeman, San Francisco.

Wake, D. B., 1966, Comparative osteology and evolution of the lungless salamanders, family Plethodontidae, <u>Mem. S. Cal. Acad. Sci.</u>, 4:1.

Wake, D. B., and Lynch, J. F., 1976, The distribution, ecology, and evolutionary history of plethodontid salamanders in Tropical America, <u>Nat. Hist. Mus. L.A. Co. Sci. Bull.</u>, 25:1.

Whipple, I. I., 1906, The naso-labial groove of lungless salamanders, <u>Biol. Bull.</u>, 11:1.

PART 3: Reptiles

FORAGING RESPONSES BY THE AMERICAN ALLIGATOR TO MEAT EXTRACTS

Marilyn R. Banta,[1] Ted Joanen,[2] and Paul J. Weldon[1,3]

[1]Department of Biology, Texas A&M University
College Station, TX 77843; [2]Rockefeller Wildlife
Refuge, Grand Chenier, Louisiana 70643; and
[3]Department of Herpetology, National Zoological Park
Smithsonian Institution, Washington, D.C. 20008

INTRODUCTION

Field and laboratory experiments demonstrate that the American alligator (Alligator mississippiensis) uses chemoreception to locate raw meat or animal carcasses in both aquatic and terrestrial environments (Scott and Weldon, 1989; Weldon et al., 1990). Observations during routine feedings indicate that alligators submerge and wave their head from side to side before seizing a morsel of meat in their aquarium (Neill, 1971; Coulson and Hernandez, 1983). More underwater head-waves and mouth-openings were exhibited by alligators to an aqueous extract of beef than to plain water, indicating that these behaviors are elicited by chemical cues (Weldon et al., 1990). The conditions under which free-ranging alligators attend to chemicals from food are unknown, but Weldon et al. (1990) suggest that they respond to the scent of carrion and injured prey.

We report here on foraging behaviors by captive-reared, juvenile A. mississippiensis to extracts of some animals consumed by this species in nature (see Wolfe et al., 1987). In addition, we describe the results of preliminary tests to characterize chemicals eliciting foraging behaviors in alligators.

MATERIALS AND METHODS

Responses to Food Extracts

Fifty alligators (total lengths = 45 - 72 cm), hatched in September from eggs collected in Cameron parish, Louisiana, were tested from June to August. Subjects were maintained in four 2.1 x 0.9-m outdoor enclosures filled 10 cm deep with water. Each enclosure housed up to 13 individuals and contained a platform and shelter. Subjects were maintained for 6 mo on Burris 45% Alligator Food (Burris Pet Food, Franklinton, Louisiana), which consists primarily of fish meat, dried blood meal, corn gluten meal, hydrolyzed feather meal, and dehulled soybean meal. They were then fed ground nutria (Myocastor coypus) meat. Subjects had been presented with nutria extracts during preliminary experiments. Between experiments, they were fed in 49 x 25-cm glass aquaria, the containers in which they were

tested, to reduce habituation of responses to meat extracts in the absence
of food offerings. Subjects were last offered food 5 - 10 days before
experiments began. Water temperature ranged from 27 to 31°C.

Alligators were exposed to extracts of the integument and muscle
tissue of thawed nutria, duck (Anas sp.), turtle (Chrysemys picta), snake
(Nerodia fasciata), frog (Rana catesbeiana), and catfish (Ictalurus sp.);
extracts of crab (Callinectes sapidus) and crayfish (Procambarus sp.) were
prepared from thawed muscle tissue with exoskeletons removed. Extracts were
prepared by adding 600 mL of water to 50 g of meat, blending for 30 sec,
and filtering with cloth and paper. Plain water was used as the control.

Testing was conducted in an area illuminated by four over-head 60 W
bulbs. Subjects were observed from behind a one-way mirror situated 1 m
away from glass aquaria filled 6 cm deep with water and covered with a wire
screen. Plastic tubing, inserted through the screen and taped to the wall
of each aquarium, opened underwater a few centimeters from the aquarium
floor. The other end of the tubing was attached to a funnel situated behind
the mirror.

Each subject was transferred from its home enclosure to an aquarium.
After 10 min, 100 mL of an extract was poured into each aquarium through
the funnel and the following responses were scored for 30 min: submergences
with mouth open (Fig. 1), submergences with mouth closed, underwater head-
waves, and snapping at the water surface.

Fig. 1. Open-mouth submergence by juvenile alligator

Two experiments, involving 12 and 11 different subjects, were
conducted; by presenting an array of extracts over two tests, we reduced
attenuation of foraging responses that may have resulted by withholding
food after repeated exposure to food chemicals. Each subject received a
different condition each day in a Latin square design. Tests were run
between 0830 and 1830 h.

Preliminary Characterization of Foraging Elicitors

Forty alligators of those that had been tested with whole extracts
were used for preliminary characterization of foraging elicitors. Ten
subjects per condition per experiment were selected at random from the pool
of available subjects. Each experiment was conducted over one day. Fifty mL

of nutria extract or plain water were poured directly into each aquarium after subjects had acclimated for 10 min. The number of submergences (open- and closed-mouth) were scored for 10 min trials.

To determine whether foraging elicitors survive heating, 400 g of nutria meat were extracted with 2.4 L of water, and the extract was filtered. Six hundred mL of the extract were boiled (100°C) at atmospheric pressure for 4 h, after which water was added to restore it to its original volume. Another 600 mL of extract were kept at room temperature.

To determine whether foraging elicitors are extractable in chloroform, 400 g of nutria meat were rinsed with 2.4 L of water. The filtered extract was reduced to 300 mL by boiling, cooled to room temperature, and extracted in a separatory funnel. The chloroform layer was decanted and heated for several h at 40°C under 300 mL of water. The resulting extract and the aqueous layer remaining after chloroform extraction each were diluted to 600 mL.

RESULTS AND DISCUSSION

An ANOVA on the results of our first experiment with whole extracts failed to detect significant differences among conditions for any measure ($P \geq 0.8$)(Table 1a). In the second experiment, however, an ANOVA detected significant differences for open-mouth submergences ($P \leq 0.01$) and head-waves ($P \leq 0.005$); closed-mouth submergences ($P \leq 0.07$) and surface snapping ($P \leq 0.09$) approached significance. A Student-Newman-Keuls (SNK) test indicated that the nutria extract elicited more open-mouth submergences and head-waves than did the other materials (Table 1b). This result may indicate the greater efficacy of chemicals from familiar food, which for our subjects was nutria meat, in eliciting foraging behaviors. Interestingly, long-term captive alligators maintained on nutria meat refuse to eat for several days after being offered a new diet, and they

Table 1a. Mean (± standard error) open- (OM) and closed-mouth (CM) submergences, underwater head-waves (HW), and surface snapping (SS) to meat extracts by 12 alligators during 30 min trials

	Crayfish	Catfish	Snake	Duck	Water
OM	0.8 ± 0.7	0.8 ± 0.6	0.8 ± 0.6	1.1 ± 0.6	0.3 ± 0.3
CM	4.3 ± 2.8	6.1 ± 5.0	2.7 ± 2.2	1.6 ± 0.9	3.9 ± 2.2
HW	7.0 ± 3.8	8.2 ± 5.2	5.1 ± 2.7	4.9 ± 2.1	5.6 ± 2.6
SS	1.8 ± 1.3	1.4 ± 0.8	1.5 ± 1.0	1.9 ± 0.8	0.5 ± 0.4

Table 1b. Mean (±SE) foraging responses of 11 alligators to meat extracts during 30 min trials

	Crab	Frog	Turtle	Nutria	Water
OM	2.7 ± 1.5	0.7 ± 0.4	0.7 ± 0.4	8.3 ± 2.8	0.1 ± 0.1
CM	0.8 ± 0.4	1.1 ± 0.8	0.4 ± 0.3	3.4 ± 1.4	0.7 ± 0.3
HW	11.6 ± 5.5	3.4 ± 1.9	3.6 ± 1.9	32.1 ± 9.8	1.2 ± 0.4
SS	2.1 ± 1.1	0.2 ± 0.1	1.6 ± 1.4	4.0 ± 1.7	0.3 ± 0.2

exhibit reduced food intake for up to a week after they resume feeding (Joanen, pers. observ.). A food imprinting hypothesis needs to be tested by presenting alligators reared on different diets with an array of foods and food extracts.

An ANOVA on the results of the experiment with heated nutria extract detected significant differences among conditions ($P \leq 0.05$) (Table 2). A SNK test detected significantly more submergences to the heated extract than to the other conditions ($P \leq 0.05$). Significant overall differences were indicated in the experiment with the chloroform extract of nutria (ANOVA; $P \leq 0.03$), and a SNK test indicated significantly more submergences to the chloroform layer than to the other conditions ($P \leq 0.05$). A separate test indicated that an aqueous extract of nutria and an extract treated with 6N HCl and subsequently neutralized with NaOH were comparable in their ability to elicit submergences. The ability of behaviorally active chemicals to survive heating and acidification suggests that macromolecules are not necessary to elicit a foraging response. The solubility of foraging elicitors in chloroform suggests that alligators respond to lipids.

Table 2. Mean (± SE) underwater submergences by alligators (N = 30 per experiment) to heated and chloroform-treated nutria extracts during 10-min trials

Heated	Unheated	Water
7.5 ± 2.4	4.8 ± 2.1	0.8 ± 0.4

Chloroform layer	Aqueous layer	Water
3.7 ± 1.3	1.1 ± 0.7	1.0 ± 0.4

Our subjects exhibited some open-mouth submergences and other foraging behaviors during control (water) presentations in each of our experiments. We hypothesize that alligators learn to associate meat chemical cues with methods of extract presentation, thus inflating responses to control conditions. Conversely, responses to food chemicals may attenuate without feeding reinforcement. The involvement of learning in the exhibition of foraging responses by crocodilians should be investigated in more rigorous studies.

Acknowledgments. S. Steele helped prepare the manuscript, and D. Marcellini made space available for its preparation. This study was supported by the National Geographic Society.

REFERENCES

Coulson, R.A. and Hernandez, T., 1983, Alligator metabolism: Studies on chemical reactions in vivo, Comp. Biochem. Physiol., 74:1-182.
Neill, W.T., 1971, The Last of the Ruling Reptiles: Alligators, Crocodiles, and their Kin. Columbia University Press, New York.
Scott, T.P. and Weldon, P.J., 1989, Chemoreception in the feeding behaviour of adult American alligators, Alligator mississippiensis. Anim. Behav., 39:398-400.

Weldon, P.J., Swenson, D.J., Olson, J.K., and Brinkmeier, W.G., 1990, The American alligator detects food chemicals in aquatic and terrestrial environments, <u>Ethology</u>, 85:191-198.

Wolfe, J.L., Bradshaw, D.K., and Chabreck, R.H., 1987, Alligator feeding habits: New data and a review. <u>Northeast</u> <u>Gulf</u> <u>Sci.</u>, 9:1-8.

PART 4: Birds

INFORMATION CONTENT OF PREY ODOR PLUMES: WHAT DO FORAGING LEACH'S

STORM PETRELS KNOW?

Larry Clark[1] and Pankaj S. Shah[2]

[1]USDA/APHIS/S&T/Denver Wildlife Research Center, c/o Monell
Chemical Senses Center, 3500 Market Street, Philadelphia, PA
19104; [2]Monell Chemical Senses Center, 3500 Market Street
Philadelphia, PA 19104

INTRODUCTION

Electrophysiological responses to odor have been recorded for concentrations as low as 0.01 ppm for Manx Shearwaters Puffinus puffinus and
Black-footed Albatrosses Diomedea nigripes, indicating that relative to
most birds, procellariiforms have a keen sense of smell (Wenzel and Sieck
1972, cf. Clark 1991; Clark and Smeraski 1990; Clark and Mason 1989). Such
acuity is not unexpected, given the extensive development of the olfactory
anatomy of these species (Bang and Wenzel 1986). Field observations indicate that Procellariiformes use their sense of smell to locate food (Grubb
1972; Hutchison and Wenzel 1980; Lequette, Verheyden and Jouventin 1989).
However, it is not known how far from the source petrels can detect odors.
This information would improve our understanding of procellariiform foraging ecology and engender a broader appreciation of the selective forces
involved in shaping the evolution of the sensory anatomy of this group
(Healy and Guilford 1990). Herein, we report preliminary observations on
the odor sensitivity of Leach's Storm Petrel Oceanodroma leucorhoa to the
major components of natural prey items. The detection data are used to
generate a first order estimate of the odor active space for free ranging
petrels.

SENSITIVITY TO ODORS

If the evaporation rate and threshold sensitivity for odorants are
known, then odor dispersion models can be used to estimate the active space
within which petrels could theoretically detect and use odors to orient
toward prey (Bell and Carde 1984). Towards this end we tested odor responding by Leach's Storm Petrels to krill Euphausia superba by monitoring
changes in cardiac response to volatiles (Shallenberger 1973). The aroma
of krill is composed of a variety of components that have distinct odors
(at least to humans) (Kubota, Uchida, Kurosawa, Komuro and Kobayashi 1989).
Three fractions prepared from a nonpolar extraction of freeze dried krill
were tested: carboxylic acids, phenols and amines. To the human observer
the carboxylic acid and phenol fractions were essentially odorless while
the amine fractions smelled strongly like fish/shrimp. The "fishy" odor of
krill is primarily attributable to pyrazines and N,N-dimethyl-2- phenyleth-

Chemical Signals in Vertebrates VI, Edited by R.L. Doty and
D. Müller-Schwarze, Plenum Press, New York, 1992

yl amine (Kubota and Kobayashi 1988). The relatively less volatile carboxylic acids contain free fatty acids such as linolenic acid (Kubota and
Kobayashi 1988).

Volatiles from each fraction were presented separately to birds via a
dilution olfactometer (Clark and Mason 1989). Responsiveness to an unconditioned odor stimulus was defined as a change in heart rate relative to a
humidified air control (Wenzel and Sieck 1972; Shallenberger 1973; Clark
and Mason 1989). Petrels showed different sensitivity to odors derived
from the three extracts (Figure 1). Petrels were most sensitive to amines
and least sensitive to the carboxylic acid fraction. In general, the
sensitivity probably corresponds to the actual molecular concentration of
components evolving from the mixtures (Mozell, Sheehe, Swieck, Kurtz and
Hornung 1984). Based upon gas chromatographic/mass spectral analysis of
the volatiles derived from krill, the amine fraction contains the more
volatile components, while the carboxylic acid fraction contains the least
volatile components [Clark, unpublished data].

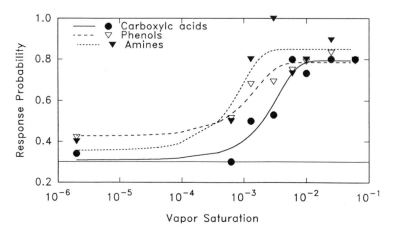

Figure 1. Response probabilities of Leach's Storm Petrels to organically
derived odors. The odors represent mixtures of compounds composed primarily of the three specified classes.

RESPONSE TO ODOR FRACTIONS IN THE FIELD

In the field, petrels orient towards a variety of organically derived
odors, e.g. cod liver oil, fish homogenate (Hutchison and Wenzel 1980;
Lequette et al. 1989). To test how Leach's storm petrels responded to
fractions of krill we used military night vision goggles to observe the
flight behavior over a breeding colony located at the Bowdoin College Field
Station at Kent Island, New Brunswick, Canada.

Leach's storm petrels are only active over land during the night,
presumably because of risk of predation from herring Larus argentatus and
great black-backed L. marinus gulls. The attractiveness of odor fractions
was patterned after studies done at sea (Grubb 1972; Hutchison and Wenzel
1980; Lequette et al. 1989). Briefly, a 1.5 m high platform was set out

over an open field and homogenates of krill or prepared fractions were placed in 80 mm diameter pans set atop a platform by a second individual. Thus, the observer was blind to the identity of the odor source. Prior to testing with the krill fractions, the procedures and methodology were refined by observing flying patterns in response to commercially available cod liver oil. Five stimuli were tested: a water control, homogenate of krill, extracted fractions of amines, phenols and carboxylic acids. These fractions were prepared in identical fashion as that described for the laboratory threshold studies. Observation periods were for 5 min. An observer hidden behind vegetation watched a 90o area downwind from the platform. During that time counts of all birds passing through the observation field, up to a distance of approximately 100 m (the limit of reliable resolution for the night goggles), were recorded. Birds that passed within about 20 m of the platform, or showed a tight circling pattern around the platform [vide Grubb 1972), were counted as having expressed an interest in the odor source. A 10 minute interobservation period followed to allow for odor dispersal and avoid habituation by birds.

There were differences in attractiveness among the stimuli (Figure 2, F=7.77, df=4,30, P=0.0002). Post-hoc tests showed that petrels were equally attracted to all the organic fractions and whole homogenate. However, only the whole homogenate and the carboxylic acid fraction differed from the blank. Subsequent GC/MS analysis of the stimuli did not reveal any obvious fractions that presented themselves as likely candidates for an attractant. Nonetheless, there appear to be subtle but reliable differences in attractiveness of the different components of krill. Interestingly, the component that to humans is perceived as fishy, is not apparently the cue to which petrels are most strongly attracted. Serial dilutions of the original stock carboxylic acid solution showed that attractiveness was linearly related to log concentration of the stimulus (Figure 3).

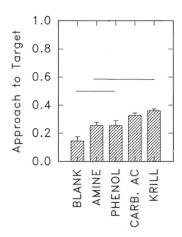

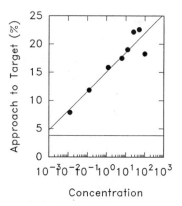

Fig. 2. The proportion of petrels within a visual field that were attracted to the odor target. The horizontal lines indicate homogenous groups based upon a post-hoc test. Vertical bars are + one standard error.

Figure 3. The proportion of petrels within a visual field that were attracted to dilutions of carboxylic acids derived from krill. The horizontal line depicts approaches to the water control.

DISCUSSION

Leach's storm petrels are opportunistic surface feeders. However, the majority of feeding occurs at night (Watanuki 1985). This foraging pattern is consistent with the availability of preferred prey (Brinton 1967; Kawaguchi 1969). For example, E. superba (a major component of petrel diet) generally are near the surface at night, but at a depth of about 100m during the day (Maurano, Marumo, Nemoto and Aizawa 1976; Sekiguchi 1975; Everson 1984). In addition to the temporal variation in availability, prey are distributed patchily and unpredictably, even in areas of uniform physical and chemical oceanographic characteristics (Brown 1980; Everson 1984). Under some conditions petrels may use meteorological anomalies to increase their chances of encountering prey (Brown 1980). Certainly, once in proximity to prey they readily orient using visual cues (Hutchison and Wenzel 1980). However, under low visibility conditions, the only reliable means to survey large tracts of ocean for prey is to use odor cues (Grubb 1972; Hutchison and Wenzel 1980; Lequette et al. 1989; Healy and Guilford 1990). The question remains as to whether this feat is possible on the scale of kilometers (Smith and Paelk 1986; Waldvogel 1987).

We used a three dimensional Gaussian puff model for a continuously generating odor source to model the dispersion pattern of volatile components of krill (Fleischer 1980). For simplicity, we estimated the physical characteristics of a highly volatile amine (pyrazine) and a less volatile fatty acid (linolenic acid) (Reid, Prausnitz and Poling 1986; Kubota et al. 1989). For illustrative purposes, we assumed atmospheric conditions to be neutrally stable, ocean and air temperature to be $10^{\circ}C$, wind speed to be 5 m/s, the inversion layer to be 10,000 m elevation, no cloud cover, and night. The odor source was a small patch of krill about 0.5m2.

The simulation for odor dispersion of pyrazine indicates that petrels may be able to detect this component from 2.5 to 12 kilometers from its source, depending upon which level of detection threshold is assumed (Figure 4). The maximum distance can be achieved at the lowest known sensitivity measured [this study] after a period of 60 minutes. Conservative estimates of 2-3 kilometers are obtained if a detection level of 1 ppm is used (Wenzel and Sieck 1972), and these distances can be achieved in as little as 10 minutes. The predicted active space for the less volatiles fatty acid is no less impressive: 1-5 km after 10 minutes for a threshold of 0.1 ppm and after 60 minutes for a threshold of 0.01 ppm, respectively. These predictions are consistent with anecdotal field observations which report petrels "appearing from nowhere" after jettison of offal from ships (Bent 1963). These distances would also allow petrels to effectively cover large areas while foraging, thereby increasing their efficiency at locating prey. We are currently attempting to determine the physical constants of other components of krill.

Determining the active space for krill is more complex than simulating dispersion of component volatiles. At the extremes, petrels may respond to prey odors in one of two ways: analytically or synthetically. In the latter case, the relative concentrations of components in a mixture is important for behavioral responding. Researchers in insect pheromonal biology have long appreciated this fact (Bell and Carde 1984). Interpretation of odor quality (i.e., identity) may vary as a function of odor concentration. For example, in humans, high concentrations of geosmin takes on the odor of musty basements or soil, while at lower concentrations the same odor is identified as beets. If the synthesis hypothesis of odor

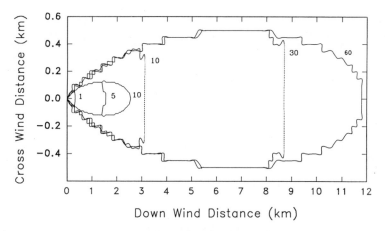

Figure 4. An example of odor dispersion of pyrazine. The odor active space is depicted as an isopleth for a threhsolds of 0.1 ppm (solid lines, smaller dispersion distance) and 0.01 ppm (dotted lines and larger dispersion distance) downwind and crosswind.

quality is correct then, in the case of foraging petrels, the ability to identify the contents of an odor plume will be a function of distance from the source and the evaporation rates of the salient components of the prey odor mixture. Thus, even though individuals components may be above threshold, the distance at which an odor is identified and tracked may be less because the ratios will change as a function of distance from the source, and by implication alter the perception of the odor (Figure 5).

Alternatively, if the odors were interpreted analytically the nature of the odor plume would convey a great deal of information. The trend for the field data suggests that the whole odor fraction is the most attractive. However, rates of attractiveness for carboxylic acid fractions were also high. Carboxylic acids would generally be the least volatile of the fractions and in combination with the higher thresholds, this fraction would have the most restricted active space. When carboxylic acids are encountered, active searching behavior is initiated, presumably because the prey items are reasonably near. This is the behavior most readily detected via the current observation strategy. Orientation to the more volatile amine components would most likely be the best cue for initial orientation because these compounds would disperse most rapidly to great distances at levels within the detection ability of foraging birds. Because this cue may be used as a long distance cue, the circling or target approach criterion used in this study may have underestimated the attractiveness of the cue.

We continue to refine the threshold data and the simulations of odor dispersion. We also are simulating odor dispersion for organic slicks remaining near the surface once the prey have left. Comparing such simulations against steady state odor dispersion simulations may reveal whether it is possible to determine the probability of prey absence from a distance, even though organic volatiles are still present in odor plumes.

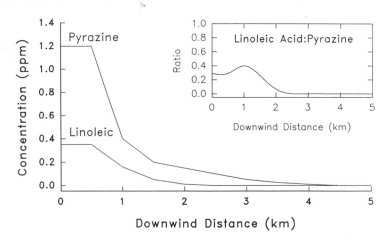

Figure 5. The downwind concentration profiles for pyrazine (solid) and linolenic acid (dotted) as a function of distance from the source. The inset depicts the linolenic acid to pyrazine ratio as a function of distance from the source.

ACKNOWLEDGMENTS

This study was funded by the National Geographic Society grant 4236-89. We also thank C.A. Smeraski, J. Lindsey and B. Schmorelitz for assistance in the field. G. Preti and H. Lawless provided GC/MS analysis of odor fractions. We are especially grateful to the staff of the Bowdoin College Field Station for their logistical support.

REFERENCES

Bang, B.G. and Wenzel, B.M. 1986. Nasal cavity and olfactory system. Pl. 195-225 in Form and function in birds, vol. 3 (A.S. King and J. McLelland, Eds.). London, Academic Press.
Bell, W.J. and Carde, R.T. 1984. Chemical ecology of insects. Sunderland, MA, Sinauer Associates, Inc.
Bent, A.C. 1963. Petrels and pelicans and their allies. New York, Dover.
Brinton, E. 1967. Vertical migration and avoidance capability of euphausiid in the California current. Limnol. Oceanogr. 12:451-483.
Brown, R.G.B. 1980. Seabirds as marine animals. Pp 1-39 in Behavior of marine animals. vol. 4, Marine birds (J. Burger, B.L. Olla and H.E. Winn Eds). New York, Plenum Press.
Clark, L. 1991. Odor detection thresholds in tree swallows and cedar waxwings. Auk 108:177-180.
Clark, L. and Mason, J.R. 1989. Sensitivity of brown-headed cowbirds to volatiles. Condor 91:922-932.
Clark, L. and Smeraski, C.A. 1990. Seasonal shifts in odor acuity by starlings. J. Exp. Zool. 177:673-680.
Everson, I. 1984. Marine Zooplankton. in Pp.463-490. Antarctic Ecology vol 2, R.M. Laws (Ed.) New York, Academic Press.

Fleischer, M.T. 1980. Spills: an evaporation/air dispersion model for chemical spills. U.S. Department of Commerce, National Technical Information Service PB83-109470.

Grubb, T.C. 1972. Smell and foraging in shearwaters and petrels. Nature, Lond. 237:404-405.

Healy, S. and Guilford, T. 1990. Olfactory bulb size and nocturnality in birds. Evolution.

Hutchison, L.V. and Wenzel, B.M. 1980. Olfactory guidance in foraging by Procellariiformes. Condor 82:314-319.

Kawaguchi, K. 1969. [Diurnal vertical migration of macronektonic fish in the western North Pacific]. Bull Plankton Soc. Japan 16:63-66.

Kubtoa, K., Uchida, C. Kurosawa, K., Komuro, A. and Kobayashi, A. 1989. Identification and formation of characteristic volatile compounds from cooked shrimp. Pp 376-385. in Thermal generation of aromas (T.H. Parliment, R.J. McGorrin, and C.-T. Ho, Eds). ACS Sym. Ser. 409, Wash. D.C.

Kubota, K. and Kobayashi, A. 1988. Identification of unknown methyl ketones in volatile flavor components from cooked small shrimp. J. Agric. Food Chem. 36:121-123.

Lequette, B., Verheyden, C. and Jouventin, P. 1989. Olfaction in subantarctic seabirds: its phylogenetic and ecological significance. Condor 91:732-735.

Mozell, M.M., Sheehe, P.R., Swieck, S.W.,Jr., Kurtz, D.B. and Hornung, D.E. 1984. A parametric study of the stimulation variables affecting the magnitude of the olfactory nerve response. J. Gen. Physiol. 83:233-267.

Murano, M. Marumo, R., Nemoto, T. and Aizawa, Y. 1976. Vertical distribution of biomass of plankton and micronekton in Kuroshio water off central Japan. Bull. Plankton Soc. Japan 23:1-12.

Reid, R.C., Prausnitz, J.M. and Poling, B.E. 1986. The properties of gases and liquids. New York. McGraw Hill.

Sekiguchi, H. 1975. Seasonal and ontogenetic vertical migrations in some common copepods in the norther region of the North Pacific. Bull. Fac. Fish., Mie Univ. 2:29-38.

Shallenberger, R.J. 1973. Breeding biology, homing behavior and communication patterns of the wedged-tailed shearwater Puffinus pacificus. Doctoral Dissertation, University of California, Los Angeles.

Smith, S.A. and Paselk, R.A. 1986. Olfactory sensitivity of the turkey vulture (Cathartes aura) to three carrion-associated odorants. Auk 103:586-592.

Waldvogel, J.A. 1987. Olfactory navigation in homing pigeons: are the current models atmospherically realistic? Auk 104:369-379.

Watanuki, Y. 1985. Food of breeding Leach's storm petrels (Oceanodroma leucorhoa). Auk 102:884-886.

Wenzel, B.M. and Sieck, M.H. 1972. Olfactory perception and bulbar electrical activity in several avian species. Physiol. Behav. 9:287-293.

THE OLFACTORY MAP OF HOMING PIGEONS

Silvano Benvenuti, Paolo Ioalè and Floriano Papi

Dipartimento di Scienze del Comportamento animale e dell'
Uomo, Via Volta 6, I-56126 Pisa, Italy, and
Centro di Studio per la Faunistica ed Ecologia Tropicali del
C.N.R., Via Romana 17, I-50125 Firenze, Italy

INTRODUCTION

Until a few decades ago, birds were thought to have a very poor olfactory ability in spite of the fact that many species exhibit well developed olfactory organs, thus raising the question whether they might have come to assume a non-olfactory function. Recent investigations, however, have shown that several bird species rely on olfactory cues for a variety of performances which include the search for food, homing ability, and social and sexual behavior (see Papi, 1990, for references).

As regards homing ability, it has been reported that in a small number of Procellariforms olfactory cues guide the birds in localizing their breeding colony and/or individual nest burrows. Pigeons are a more interesting case since their homing mechanism, unlike that of Procellariforms, makes possible a true navigational performance, in the sense that the birds rely on local odors without needing direct sensory contact with their goal (the home loft). The aim of this paper is to present the most significant findings of our research group on the way in which pigeons acquire and use their navigational abilities.

OLFACTORY DEPRIVATION EXPERIMENTS

The simplest and most basic experiments showing the use of olfactory information for pigeon homing were carried out by depriving birds of their olfactory ability or access to environmental odors during transportation and at release sites. This was achieved by several methods, including mechanical and chemical techniques: a) section of both olfactory nerves; b) unilateral nerve section and nostril plugging at the contralateral side, compared to ipsilaterally treated controls; c) insertion of thin plastic tubes into the nostrils up to the choanae, so as to prevent air from contacting the nasal mucosae; d) temporary anaesthesia of the nasal mucosae by spraying a local anaesthetic in each nostril just before release (the perception of meaningful odors during transportation was obstructed by nasal plugs or by supplying bird containers with filtered or bottled air); e) injection of zing sulfate solution into the choanae - allowing the solution to flow out from the nostrils - one to three days prior to the test release.

Chemical Signals in Vertebrates VI, Edited by R.L. Doty and
D. Müller-Schwarze, Plenum Press, New York, 1992

All these methods consistently produced incorrect initial orientation and impaired homing ability in experimental birds at unfamiliar release sites (see Papi 1990, for references). The impaired homing behavior of anosmic birds is not due to a lack of motivation to fly caused by a non-specific disturbance of the experimental treatment; anosmic birds, in fact, exhibit normal homing behavior - not significantly different from that of unmanipulated controls - when released at familiar sites, where olfactory information can be replaced by familiar landmarks. The demotivation hypothesis, suggested by Keeton (see Papi et al., 1978; Keeton, 1980), is also contradicted by analyses of the sites where pigeons which failed to home were recovered. The geographical distribution of recovery sites shows that anosmic birds fly long distances in wrong directions, not significantly shorter than those flown by controls in the homeward direction (see Wallraff, 1990). Moreover, olfactory deprivation only impairs pigeon homing behavior when birds have not been allowed to smell environmental odors during passive transportation to the release site (Wallraff and Foà, 1981). However, critical importance in determining whether the effects of anosmia are specific or not must be attributed to a series of experiments showing that it is not only possible to eliminate homeward orientation, but also to produce predictably wrong orientation by providing pigeons with false olfactory information. We were, in fact, able to fool pigeons by allowing them to smell ambient air at a "false release site", different from the actual one and situated in the opposite

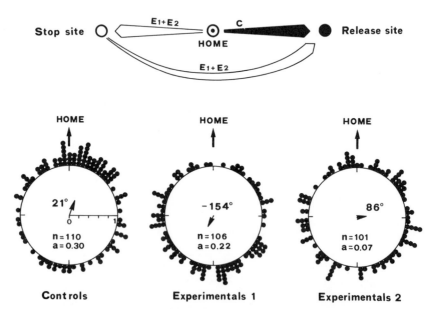

Fig. 1. Scheme of treatment (top) and initial orientation (bottom) in site simulation tests. Experimentals 1 (E1) were transported to, and only allowed to smell atmospheric odors at a stop site located, with respect to the loft, in the opposite direction from that of the release site. Experimentals 2 (E2) transported with E1, were always prevented from smelling atmospheric odors. Controls (C) were transported to, and allowed to smell at the true release site. Dots, which represent birds' bearings, are shown with respect to home direction set to 0°. The length of the mean vectors can be read using the scale in the first diagram. E1 turn out to be oriented in the opposite direction with respect to homeward oriented controls, while E2, which were deprived of meaningful olfactory information, do not differ from random.

direction with respect to home. The initial orientation of the experimental birds turned out to be in the opposite direction with respect to homeward oriented controls (Fig. 1) (see Wallraff 1990, for references).

ACQUISITION OF THE OLFACTORY MAP

The results of olfactory deprivation experiments show that the odors that pigeons perceive at an unfamiliar locality can give them information on the spatial relationship between the release site and the home loft site. This process implies that: (1) different patterns of olfactory stimulation (different odors or different concentrations of the same odorous substances) are present over different areas; and (2) pigeons have an olfactory map of the regions around the loft. The acquisition of navigational abilities is a process which basically develops at the home loft, since pigeons never previously displaced from their loft are able to navigate. The map is acquired by associating the odors carried by the winds with the direction they come from. Then, when unexperienced birds are transported to and released at an unfamiliar site, they can determine the home direction if they remember the direction of the winds which used to carry local odors to the home loft. Once this crucial step has been accomplished, pigeons use a sun or magnetic compass to assume and maintain the homeward direction.

These ideas have been verified by a series of tests in which pigeons were raised in cages equipped with special devices which obstructed or altered the association between the winds and the directions from which they come (see Papi, 1990, for references). The results of these experiments can be summarized as follows: a) pigeons raised in aviaries whose walls were screened from the winds coming from any direction are not able to develop homing ability; b) pigeons exposed to natural winds with their nostrils plugged are unable to navigate; c) if the walls are only screened from the winds coming from two opposite directions, the birds exhibit normal homing behavior only when released in sites along the non-screened directions; d) pigeons raised in cages fitted with deflectors arranged so as to produce a clockwise or counter-clockwise deviation of the winds from any direction show an initial orientation deflected in the same way as the winds to which they have been exposed.

Waldvogel and Phillips (1991), however, do not attribute the deflector loft effect to altered olfactory information, but to the deflection of sunlight, which alters the sun compass of the birds. The role of non-olfactory cues in deflector loft experiments cannot, of course, be excluded, but the main role in the spatial alteration of the wind directions is caused by the deflectors, as is shown by the fact that similar effects on initial orientation are produced by other devices. In fact, pigeons raised in glass corridors where fans produced artificial winds blowing in directions opposite to the natural ones, show an inverted initial orientation with respect to homeward oriented controls (Ioalè et al., 1978).

Foà et al. (1986), using pigeons subjected to anterior commissurotomy, produced the direct evidence that olfactory cues were involved in the deflector loft effect. After surgery, the birds were kept for alternating periods of three days in a clockwise deflector loft, with their right nostril plugged, and in a counter-clockwise deflector loft, with their left nostril plugged. After a treatment of several weeks, the birds showed that they had acquired two different and independent olfactory maps: in test releases carried out with the right or the left nostril plugged, the birds deflected clockwise or counter-clockwise, respectively, with respect to the home direction.

All these experiments show that olfactory cues do not simply trigger the proper motivational conditions for homing, but provide directional information, beside showing that the pigeon map is the outcome of a learning process based on wind-borne olfactory cues. This conclusion paves the way to the reproduction of the process of pigeon map acquisition through the use of artificial winds and odorants. Ioalè et al. (1990), modifying an experimental plan used in a previous experiment (Papi et al., 1974), housed two pigeon groups in two aviaries open to winds from any direction. From time to time the experimentals were exposed to an artificial air current carrying vapors from an odorant: benzoic aldehyde. After a treatment of several weeks, both controls and experimentals turned out to be homeward oriented when released unmanipulated at unfamiliar localities. In subsequent tests both bird groups were released after exposing them to benzoic aldehyde during passive transportation or at the test site: the results show that the experimentals flew off in a direction roughly opposite to that from which they had been used to perceiving the odorant at home, while the homeward orientation of controls was unaffected (Fig. 2).

THE NATURE AND ORIGIN OF OLFACTORY CUES

The evidence that pigeon homing is based on olfactory information raises questions about the nature and origin of wind-borne odorous substances integrated in the map and used in the homing process. Regrettably, we do not know either the chemical structure of the substances which provide pigeons with directional information, or the way in which such substances are generated and dispersed in the atmosphere. It is doubtful whether these questions will be answered soon, considering that nothing is yet known about the cues used by other vertebrates (the salmon, for example) which undoubtedly rely on olfactory information in homing. Apparently, the olfactory system of pigeons is not specialized to detect a narrow band of chemical compounds, as is shown by the fact that some natural and artificial substances (olive oil, turpentine, benzoic aldehyde) may give pigeons directional information (Papi et al., 1974; Ioalè et al., 1990).

THE NEURAL CONTROL OF PIGEON NAVIGATION

Another significant aspect of pigeons' navigational mechanism regards the central control of homing ability. Recent investigation by Papi and Casini (1990) shows that pigeons subjected to removal of the pyriform cortex are capable of correct initial orientation and of homing from familiar sites, whereas they lose their navigational ability in releases at unfamiliar sites.

It is worth noting that the pyriform cortex is reached by neural projection from the olfactory bulbs, so it may be involved in the central processing of olfactory information related to the homing process. Similar indications have been achieved by ablation of the posterodorsolateral neostriatum, an area which is supposed to play a role in spatial orientation and olfactory function (Divac and Gagliardo, in prep.).

OLFACTORY HOMING: A REGIONAL OR UNIVERSAL EXPLANATION OF PIGEON NAVIGATION?

All the experiments quoted previously have been performed in Italy using a stock of pigeons produced in the Po Valley. Results and conclusions which totally agree with the Italian findings have been achieved in other geographical areas, such as Switzerland (Fiaschi and Wagner, 1976), Bavaria (South Germany, see Wallraff, 1990) and Ohio (Bingman, pers. comm.). Conclusions on these findings, however, have been challenged by other authors who do not regard olfactory information as the basic component, but only as one element within a multicue system of navigation. This idea has been generated by the results of experiments carried out in Germany by the Frankfurt and Tübingen

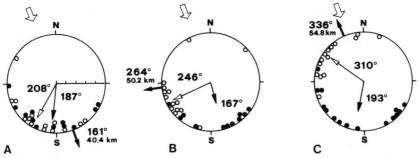

Fig. 2. Initial orientation of pigeons raised in cages exposed to an
artificial wind carrying an odor from a specific direction
(white arrows on top). Test releases were carried out by
exposing the birds to the odor during transportation or at
release sites. Vanishing diagrams A, B and C show the birds'
orientation at three different sites (home direction and distance is given). Controls (open symbols), which had never
experienced the artificial odor at the home site, fly homeward, whereas the experimentals (solid symbols), roughly
orientate in the direction opposite to that from which the
artificial odor was carried to the home site.

research groups and in the United States by the Cornell research group. While
some findings achieved in these areas agree with the Italian findings (Kiepenheuer, 1985; Benvenuti and Brown, 1989), other results have been viewed as
evidence that the olfactory hypothesis is not a "universal" explanation of
pigeons' homing ability (see Wiltschko et al., 1987a).

The current debate on the actual role of olfactory information in different areas cannot be reviewed here. It is worth recalling, however, that
the hypothesis that pigeons raised in different regions may possess genetically differing homing mechanisms is not in agreement with recent findings
showing that olfactory deprivation has the same effect (loss of homing ability) on both Italian and Cornell pigeons raised in Italy (Benvenuti et al.,
1990a). On the other hand, the idea that raising and training procedures may
determine the nature of the cues used in the homing mechanism (Wiltschko et
al., 1987b) contrasts with the results of recent experiments (Papi et al.,
1989; Benvenuti et al., 1990b).

At present, it is hard to account for the fact that the findings reported in different countries have led to such contrasting views. It may, however, be suggested that the discrepancies mentioned have been generated by
inappropriate methods of olfactory deprivation, and by the use of release
sites which were not actually unfamiliar to the pigeons, or not symmetrically
distributed around the home direction. The importance of a correct experimental plan to verify the role of olfactory information on pigeon homing has
been discussed by Wallraff (1990).

In conclusion, a large body of experimental evidence mostly obtained by
our team and that of H.G. Wallraff shows that, at least in the areas where
these experiments were performed, olfactory information plays an irreplace-
able role in homing from unfamiliar sites. Pigeons rely on an olfactory map
which is acquired at the home loft by associating the odors carried by the
winds with the direction from which they come. In some areas other teams
obtained variable results, which have been regarded as evidence that olfactory information may be replaced or supplemented by other kinds of navigational cues. These discrepancies, however, do not seem to be related to

genetic differences or different raising and training procedures. In order to resolve these discrepancies, particularly useful topics for future research will comprise the central mechanisms underlying homing, and the nature and distribution of the substances which constitute the physical basis of olfactory navigation.

REFERENCES

Benvenuti, S., and Brown, A. I., 1989, The influence of olfactory deprivation on experienced and inexperienced American pigeons, Behaviour, 111:113.
Benvenuti, S., Brown, A. I., Gagliardo, A., and Nozzolini, M., 1990a, Are American pigeons genetically different from Italian ones?, J. exp. Biol., 148:235.
Benvenuti, S., Fiaschi, V., Gagliardo, A., and Luschi, P., 1990b, Pigeon homing: a comparison between groups raised under different conditions, Behav. Ecol. Sociobiol., 27:93.
Fiaschi, V., and Wagner G., 1976, Pigeons' homing: some experiments for testing the olfactory hypothesis. Experientia, 32:991.
Foà, A., Bagnoli, P., and Giongo, F., 1986, Homing pigeons subjected to section of the anterior commissure can build up two olfactory maps in the deflector lofts, J. comp. Physiol., 159:465.
Ioalé, P., Nozzolini, M., and Papi, F., 1990, Homing pigeons do extract directional information from olfactory stimuli, Behav. Ecol. Sociobiol., 26:301.
Ioalé, P., Papi, F., Fiaschi, V., and Baldaccini, N. E., 1978, Pigeon navigation: effects upon homing behaviour by reversing wind direction at the loft, J. comp. Physiol., 128:285.
Keeton, W. T., 1980, Avian orientation and navigation: new developments in an old mystery, in: "Acta XVII Congr. Int. Ornithol.", E. Nöring, ed., Dt. Ornithol. Gesel., Berlin.
Kiepenheuer, J., 1985, Can pigeons be fooled about the actual release site position by presenting them information from another site?, Behav. Ecol. Sociobiol., 18:75.
Papi, F., 1990, Olfactory navigation in birds, Experientia, 46:352.
Papi, F., and Casini, G., 1990, Pigeons with ablated pyriform cortex home from familiar but not from unfamiliar sites, Proc. Natl. Acad. Sci. USA, 87:3783.
Papi, F., Gagliardo, A., Fiaschi, V., and Dall'Antonia, P., 1989, Pigeon homing: does early experience determine what cues are used to navigate?, Ethology, 82:208.
Papi, F., Ioalè, P., Fiaschi, V., Benvenuti, S., and Baldaccini, N. E., 1974, Olfactory navigation of pigeons: the effect of treatment with odorous air currents, J. comp. Physiol., 94:187.
Papi, F., Keeton, W. T., Brown, A. I., and Benvenuti, S., 1978, Do American and Italian pigeons rely on different homing mechanisms?, J. comp. Physiol., 128:303.
Waldvogel, J. A., and Phillips J. B., 1991, Olfactory cues perceived at the home loft are not essential for the formation of a navigational map in pigeons, J. exp. Biol., 155:643.
Wallraff, H. G., 1990, Navigation by homing pigeons, Ethology Ecology Evolution, 2:81.
Wallraff, H. G., and Foà, A., 1981, Pigeon navigation: charcoal filter removes relevant information from environmental air, Behav. Ecol. Sociobiol., 9:67.
Wiltschko, W., Wiltschko, R., Gruter, M., and Kowalsky, U., 1987b, Pigeon homing: early experience determines what factors are used for navigation, Naturwissenschaften, 74:196.
Wiltschko, W., Wiltschko, R., and Walcott, C., 1987a, Pigeon homing: different effects of olfactory deprivation in different countries, Behav. Ecol. Sociobiol., 21:333.

PIGEON HOMING:

THE EFFECT OF TEMPORARY ANOSMIA ON ORIENTATION BEHAVIOR

Wolfgang Wiltschko and Roswitha Wiltschko

Fachbereich Biologie der Universität, Zoologie
Siesmayerstraße 70
D W 6000 Frankfurt a.M., Germany

INTRODUCTION

Carrier pigeons, Columba livia, released at unfamiliar sites, do not depart randomly, but fly in directions that are normally not far from the homeward course. This indicates that the birds are aware in what direction their `home' lies, and leads to the question: how do the birds know? Avian orientation has been the subject of intensive research during the last few decades, and some parts of this question can already be answered. Navigation is commonly described as a two-step process: In the first step, the map step, the birds determine their home direction as a compass course, and in the second step, they locate this course with the help of a compass and fly in the respective direction. The second step, the compass step, is rather well understood: Pigeons preferentially use a time-compensating sun compass which is, under overcast, backed up by a magnetic compass. The nature of the map step, however, is still largely unknown.

Among the factors discussed as the physical basis of the navigational map, odors have become the most prominent. When Papi, Fiore, Fiaschi and Benvenuti (1972) proposed the olfactory hypothesis, it met with world-wide attention and stimulated numerous scientists to repeat the respective experiments. However, even today, after almost twenty years of research and considerable effort, the situation concerning the olfactory hypothesis is still unclear. This is true for the theoretical background as well as for experimental evidence. Two models have been proposed, one suggesting that birds might use a mosaic of local odors in the vicinity of their loft (Papi et al., 1972), the other suggesting they use large-scale olfactory gradients (e.g. Wallraff, 1980, 1989). However, meteorologists state that odor distributions of the required kind are not found in nature (e.g. Becker and van Raden 1986). Experimental evidence has also been controversial and divided researchers into two groups: Proponents of the olfactory hypothesis who consistently report positive results of olfactory manipulations (cf. Papi, 1982; Wallraff, 1986), and researchers who remained sceptical, since they found either no or only marginal effects (cf. Keeton, 1980; Schmidt-Koenig, 1987; Waldvogel, 1989). In this paper, we report the results of two series of experiments that may help to explain some of the discrepancies.

Chemical Signals in Vertebrates VI, Edited by R.L. Doty and
D. Müller-Schwarze, Plenum Press, New York, 1992

EFFECTS OF ANOSMIA IN DIFFERENT COUNTRIES

   Since the discrepancies between the findings of various authors
might be due to methodological differences, we began a series of experiments in three different countries using standardized procedures to study
the effects of temporary olfactory deprivation (W. Wiltschko, Wiltschko
and Walcott, 1987a). We compared the response of pigeons from the
following lofts: (1) lofts in Tuscany, Italy, where the importance of
olfaction for orientation had been amply documented (see Papi, 1982), (2)
the Cornell loft in Upstate New York, where inconsistent effects had been
reported (see Papi, Keeton, Brown and Benvenuti, 1978; Keeton, 1980), and
(3) the loft in Frankfurt, Germany, where previous experiments had shown
that at least young inexperienced pigeons were unaffected by olfactory
deprivation (R. Wiltschko and Wiltschko, 1987).

   All test birds were young pigeons born in the same year. They had
some flying experience from training flock tosses, but were unfamiliar
with the release sites used in this study. The methods of depriving
pigeons of olfactory cues were identical in all experiments: During
displacement and while waiting at the release site until being up for
release, the experimental birds had their nostrils plugged with cotton
odorized with citrus oil. The cotton tampons were held in place by cloth
tape. Before release, the plugs were removed and a local anaesthetic was
sprayed into the nostrils, a treatment that heart-rate conditioning
studies suggest renders pigeons anosmic for 4 - 6 h (e.g. Wallraff,
1988). - The control birds wore a band of cloth tape around the upper
mandible which left the nostrils open. Prior to release, they were
treated with a spray that contained only the propellant, but not the
active substance of the anesthetic[1].

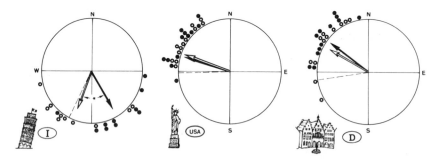

Fig. 1. Examples of the effect of olfactory deprivation in Italy (I),
      Upstate New York (USA) and Frankfurt, Germany (D). The release
      sites were between 30 km and 40 km from home; the home direction
      is indicated by a dashed radius. The symbols at the periphery of
      the circle mark the vanishing bearings of individual pigeons
      (controls: open, anosmic birds: solid), with the arrows repre-
      senting the respective mean vectors (redrawn from W. Wiltschko,
      Wiltschko, Foa and Benvenuti, 1986, W. Wiltschko et al. 1987a, W.
      Wiltschko, Wiltschko and Jahnel 1987b).

---

[1] In some tests a second, completely untreated group of control birds was
released; there are no indications for any difference between the two
control groups (W. Wiltschko et al. 1986, 1987a).

The effect induced by olfactory deprivation differed greatly; examples are presented in Fig. 1. In Italy, we found large effects that were in good agreement with the ones reported by Papi (1982) for Italian pigeons. In Upstate New York, anosmic birds and controls behaved very similarly, which corresponds to the earlier results of Keeton (1980). At Frankfurt, Germany, the treatment was likewise mostly ineffective; in 3 of the 20 the releases, however, a difference was observed.

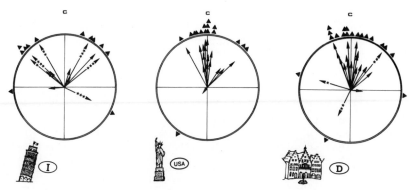

Fig. 2. Vectors of anosmic pigeons plotted relative to the mean of the controls of the same release to illustrate the induced differences. The symbols at the periphery of the circle mark the mean direction. Vectors that deviate significantly from the control direction are marked with askerisks. * = p < 0.05, ** = p < 0.01, *** p < 0.001. (after W. Wiltschko et al. 1987a).

The data on initial orientation are summarized in Fig. 2. - Table 1 shows which of the various parameters of orientation and homing had been affected. In Italy, all data except the vanishing intervals were significantly poorer in the anosmic birds. If parameters were affected in New York and in Frankfurt, the induced differences were significantly smaller than in Italy, with the data from New York and Germany being similar.

Table 1. Medians of Differences Induced by Olfactory Deprivation

Parameter	median difference			significant?		
	Italy	USA	Germany	I/USA	I/D	USA/D
Initial orientation						
Direction	46°**	8°*	18°*	***	***	-
Vector length	-0.16*	0.07	0.02	**	*	-
Vanishing interval (s)	+24*	30*	+11	-	-	-
Homing performance						
Homing speed (km/h)	-22.5**	-8.3**	8.3**	*	*	-
Return rate	-4%	0%	0%	-	-	-

Asterisks at the medians indicate that the parameter in question was significantly affected (Wilcoxon test, matched pairs of data). The last columns indicate differences in the amount of change (Mann Whitney test). Symbols and significance levels as in Fig. 1 and 2.

Because of the standardized procedures, the differences in effectiveness of olfactory deprivation cannot be attributed to methods, but
must reflect true differences in the behavior of birds. This would mean
that olfaction is indeed a very important factor in Italy, while it is
almost negligible in New York. In Germany, the situation appears more
complex: While olfaction seems to be of no importance in most cases, it
significantly affected the behavior at three sites, suggesting that the
importance of olfaction may vary from site to site. Interestingly, the
three sites were all in a region ESE of the loft so that a specific
characteristic of that region might be involved.

EFFECTS OF ANOSMIA IN PIGEONS RAISED AND KEPT UNDER DIFFERENT CONDITIONS

The finding that pigeons from the various lofts do not react in the
same way to olfactory deprivation leads to the question as to why these
differences in orientation behavior occur. Three possible causes must be
considered: (1) genetic differences; (2) regional differences in the
availability and suitability of potential orientation cues; and (3)
differences induced by early experience, i.e. the various ways of raising
pigeons might lead to lasting modifications of the orientational system.
In the course of our comparative study, marked differences in housing and
keeping the birds in the Italian lofts compared to the other lofts had
become obvious. In Italy, the birds had access to large open aviaries
that were largely wind-exposed, and they could always leave their lofts
for free flights, whereas the New York and the Frankfurt lofts were much
more wind-protected, and the birds were allowed to fly freely only once
per day. There were also differences in the training procedures, with the
Italian pigeons normally performing fewer training flights.

We therefore began a second experimental series to test the influence of housing and wind-exposure on orientation by raising two groups of
pigeons, one in our normal, wind-protected garden loft, the other in a
loft of `Italian style'. These birds, siblings of the first whenever
possible, were housed in a newly established loft on the roof of the 6
storied university building which was wind-exposed on all sides; here we
tried to simulate the Italian procedures as well as possible. After some
training releases, which were more numerous for the birds from the garden
loft, the critical tests began; namely, samples from both groups were
made anosmic in the way described above and released together with
controls at the same release site (see R. Wiltschko and Wiltschko, 1989).

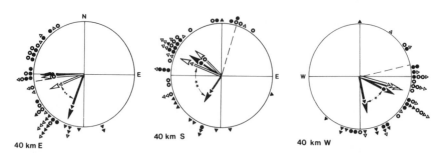

Fig. 3. Examples of orientation behavior of birds from the roof loft
(triangles) and birds from the garden loft (round symbols). Other
symbols as in Fig. 1. (data from R. Wiltschko and Wiltschko,
1989).

Fig. 3 gives three releases as examples. A clear difference between the two groups in their response to olfactory deprivation became obvious: The orientation of the pigeons raised in the normal loft in the garden was not affected, with the exception of two sites in the ESE, so that these findings are in agreement with our previous results (see Fig. 2c). In contrast, the birds that had been raised in the wind-exposed loft on the roof regularly showed a strong reaction to the olfactory treatment which was normally observed in the form of a considerable change of the preferred direction. Fig. 4 summarizes the data on initial orientation, and Table 2 lists the effect on the orientation parameters studied. The difference in the amount of directional change between the two groups was statistically significant.

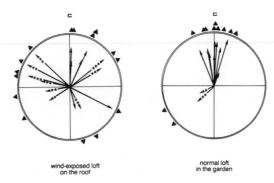

wind-exposed loft
on the roof

normal loft
in the garden

Fig. 4. Vectors of anosmic pigeons plotted relative to the mean of the controls of the same release to illustrate the induced differences. - Symbols as in Fig. 2 (after R. Wiltschko and Wiltschko, 1989).

Table 2. Medians of Differences Induced by Olfactory Deprivation

Parameter	Birds from the Loft on the Roof	Birds from the Loft in the Garden	Significant Difference
Initial orientation			
Direction	63°**	11°*	*
Vector length	-0.04	-0.05	-
Vanishing interval (s)	+17	+48	-
Homing performance			
Homing speed (km/h)	-16.4*	-13.5*	-
Return rate	-16%***	-10%	-

Symbols as in Table 1

These results demonstrate that the conditions at the home loft can indeed be crucial for the pigeons' orientational behavior. The response to olfactory deprivation could be greatly augmented by raising and maintaining the birds in the `Italian style'. Additional tests showed that wind-exposure rather than the amount of training is important for inducing a strong response to anosmia (R. Wiltschko, Schöps and Kowalski, 1989).

GENERAL DISCUSSION

Our findings help to explain why authors reported such great variability in the effects of olfactory manipulations on pigeon homing.
Different loft procedures may induce differences in the relative importance of olfactory input; pigeons that have free, unobstructed access to
wind and air-borne cues depend on olfactory input to a far larger extent.
This is not surprising, since the navigational system is based on learned
orientation mechanism. As pigeons acquire their knowledge of navigational
factors through experience, their "maps" need not be identical in all
parts of the world. Considering the variety of their sensory input (e.g.
Kreithen, 1978), we may expect pigeons to select from the available
potential navigational factors those which provide the most suitable and
most reliable information in their home region, and early experience
might induce lasting preferences for certain cues.

While the proponents of the olfactory hypothesis have looked upon
olfaction as the only navigational cues needed to guarantee oriented
behavior (e.g. Papi, 1982; Wallraff, 1983), the above considerations
suggest that olfactory information is only one of many "map" factors
whose importance varies according to the experience of the birds. The
behavior of our Frankfurt pigeons at the sites ESE of home suggests that
olfactory input might become more important at certain locations,
conceivably because the factors normally used do not give clear navigational information. This would mean that various navigational factors
interact in a complex way, with the role of any one factor depending on
the others.

The idea that olfactory information may be just one of several navigational factors has gained wide acceptance recently (e.g. Walcott and
Lednor, 1983; Able and Bingman, 1987; Schmidt-Koenig, 1987; Waldvogel,
1989). One aspect of our findings, however, lets us hesitate in accepting
this view. In the second experimental series, the pigeons from the garden
loft were not affected by olfactory deprivation, suggesting they did not
use olfactory information. The pigeons from the roof loft were affected
by anosmia, so that we might conclude that the controls used odors while
the anosmic birds relied on other factors. If this is indeed the case, we
would expect that the orientation of the three groups not using olfaction
(roof experimentals, garden experimentals, garden controls) would be
similar. This is, however, definitely not the case. In general, orientation of the garden birds, controls as well as anosmic birds, was similar
to that of controls from the roof loft, the very birds that should have
been using olfactory cues. This would mean that olfactory information
always indicates the same direction for the controls from the roof as
does non-olfactory information for both groups of birds from the garden
loft. In the light of the olfactory hypothesis this is a paradox,
especially when the birds do not depart in the homeward direction, but
show significant deviations from the home course, so-called release site
biases (see Keeton 1973; Fig. 3b,c).

This leads to the question of whether olfactory input really trans-
fers navigational information to the birds, or whether it affects the
behavior in a different way. Olfactory deprivation is known to interfere
with a number of behaviors that are primarily not controlled by olfaction, such as learning a visual task, aggression, and tonic immobility
(e.g. Wenzel, Albritton, Salzmann and Oberjat, 1969; Wenzel and Rausch,
1977). Possibly, we have to refer to these types of interference in order
to understand the effects of ansomia on orientation behavior.

ACKNOWLEDGEMENTS

Our studies were supported by the Deutsche Forschungsgemeinschaft in the program SFB 45. We thank our collegues F. Papi, Pisa and Ch. Walcott, Ithaca, New York, for their cooperation when we conducted the experiments of series 1.

REFERENCES

Able, K. P. and Bingman, V. P., 1987, The development of orientation and navigation behavior in birds, Quart. Rev. Biol., 62:1

Becker, J. and van Raden, H., 1986, Meteorologische Gesichtspunkte zur olfaktorischen Navigationshypothese, J. Ornithol., 127:1

Keeton, W. T., 1973, Release-site bias as a possible guide to the "map" component in pigeon homing, J. Comp. Physiol., 86:1

Keeton, W. T., 1980, Avian orientation and navigation: New developments in an old mystery, in: "Acta XVII Congr. Intern. Ornithol.", Vol. I, R. Nöhring, ed., Deutsche Ornithologen-Gesellschaft, Berlin

Kreithen, M. L., 1978, Sensory mechanisms for animal orientation - can any new ones be discovered?, in: "Animal Migration, Navigation and Homing", K. Schmidt-Koenig and W. T. Keeton, eds., Springer-Verlag, Berlin, Heidelberg, New York

Papi, F., 1982, Olfactory and homing in pigeons: Ten years of experiment, in: "Avian Navigation", F. Papi and H. G. Wallraff, eds., Springer-Verlag, Berlin, Heidelberg, New York

Papi, F., Keeton, W. T., Brown, A. I. and Benvenuti, S., 1978, Do American and Italian pigeons rely on different homing mechanisms?, J. Comp. Physiol., 128:303

Schmidt-Koenig, K., 1987, Bird navigation: Has olfactory orientation solved the problem?, Quart. Rev. Biol., 62:31

Walcott, C. and Lednor, A. J., 1983, Bird navigation, in: "Perspectives in Ornithology", A. H. Brush and C. G. Clark, eds., Cambridge Univ. Press, Cambridge

Waldvogel, J. A., 1989, Olfactory orientation by birds, in: "Current Ornithology", Vol. 6, D. M. Power, ed., Plenum Press, New York

Wallraff, H. G., 1980, Olfactory and homing in pigeons: Nerve-section experiments, critique, hypotheses, J. Comp. Physiol., 139:209

Wallraff, H. G., 1986, Relevance of olfaction and atmospheric odours to pigeon homing, in: "Orientation on Space", G. Beugnon, ed., Privat, I.E.C., Toulouse

Wallraff, H. G., 1988, Olfactory deprivation in pigeons: Examination of methods applied in homing experiments, Comp. Biochem. Physiol. A, 89:621

Wallraff, H.G., 1989a, Simulated navigation based on unreliable sources of information (Models on pigeon homing, Part 1), J. Theor. Biol., 137:1

Wallraff, H.G., 1989b, Simulated navigation based on assumed gradients of atmospheric trace gases (Models on pigeon homing, Part 2), J. Theor. Biol., 138:511

Wenzel, B. M. and Rausch, L. J., 1977, Does the olfactory system modulate affective behavior in the pigeon?, Ann. N.Y. Acad. Sci., 290:314

Wenzel, B. M., Albritton, P. F., Salzmann, A. and Oberjat, T. E., 1969, Behavioral changes on pigeons following olfactory nerve section or bulb ablation, in: "Olfaction and Taste", Vol. 3, C. Pfaffman, ed., Rockefeller Univ. Press, New York

Wiltschko, R. and Wiltschko, W., 1987, Pigeon homing: Olfactory experiments with inexperienced birds, <u>Naturwissenschaften</u>, 74:94

Wiltschko, R. and Wiltschko, W., 1989, Pigeon homing: Olfactory orientation - a paradox, <u>Behav. Ecol. Sociobiol.</u>, 24:163

Wiltschko, R., Schöps, M. and Kowalski, U., 1989, Pigon homing: wind exposition determines the importance of olfactory input, <u>Naturwissenschaften</u> 76:229-231

Wiltschko, W., Wiltschko, R., Foa, A. and Benvenuti, S., 1986, Orientation behaviour of pigeons deprived of olfactory information during the outward journey and at the release site, <u>Monit. Zool. Ital. (N.S.)</u>, 20:183

Wiltschko, W., Wiltschko, R. and Walcott, C., 1987a, Pigoen homing: Different effects of olfactory deprivation in different countries, <u>Behav. Ecol. Sociobiol.</u>, 21:333

Wiltschko, W., Wilschtko, R. and Jahnel, M., 1987b, The orientation behavior of anosmic pigoens in Frankfurt a. M., Germany, <u>Anim. Behav.</u>, 35:1324

THE PUZZLE OF OLFACTORY SENSITIVITY IN BIRDS

Bernice M. Wenzel

Department of Physiology
UCLA School of Medicine
Los Angeles, CA 90024-1751

Serious, if sporadic, research on olfaction in birds has been underway for at least three decades. This current era began with Michelsen's (1959) report of success in training pigeons to discriminate between two odors, and soon burgeoned with Bang's (1960,1971) landmark analyses of olfactory epithelial and bulbar sizes in all avian orders. Inspiration for these investigators came directly from Cobb (1960), who had called attention to the large variation in olfactory bulb prominence throughout Aves and urged consideration of its behavioral significance. Reports of olfactory nerve (Tucker, 1965) and bulb (Wenzel, 1966) responses, as well as respiratory and cardiac changes (Neuhaus, 1963; Wenzel, 1966), in several species to odor stimuli, and of odor tracking by Turkey Vultures (*Cathartes aura*) in the field (Stager, 1964), completed the key evidence for a functional avian olfactory system.

Table 1. Avian species for which odor reliance has been reported

Species	Behavior
Procellariiform sp.	Fly toward food-related odors
Turkey Vulture (*Cathartes aura*)	" " " " " , locate carcasses
Brown Kiwi (*Apteryx australis*)	Find buried food
Common Raven (*Corvus corax*)	" " "
Black-billed Magpie (*Pica pica*)	" " "
Great Tit (*Parus major*)	Avoid bitter food
Graylag Goose (*Anser domesticus*)	Goslings avoid unwanted greens (dill, lavender, etc.)
Domestic goose (*Anser & Branta* sp.)	Goslings choose home box with familiar odor
Leach's Storm-Petrels (*Oceanodroma leuchoroa*)	Home to islands and burrows
European Starling (*Sturnus vulgaris*)	Select foliage for nest lining, orient toward home
Pigeon (*Columba livia*)	Orient in homeward direction
Common Swift (*Apus apus*)	" " " "

Chemical Signals in Vertebrates VI, Edited by R.L. Doty and
D. Müller-Schwarze, Plenum Press, New York, 1992

The subsequent documentation of odor perception or use has now included many different species (see Wenzel, in press, for summary) with olfactory anatomy ranging from massive to minuscule. Table 1 lists these species and characterizes the result for each one. Some of the supporting research is much more extensive than the rest but there is general agreement that most, perhaps all, birds receive useful information from certain odors.

At the same time that the list of odor-using birds has been growing, there has been a lack of progress in studying the basic mechanisms underlying such perception. While endless questions can be pursued about more species, more odors, more conditions, ecological significance and interaction, etc., the primary gaps in our information at present concern the nature of neural activity in the olfactory pathway, the limits of olfactory sensitivity, and the characterization of effective stimuli. In reviewing data on olfactory orientation in birds, Waldvogel (1989) has already pointed out this lack. It is one that should be strongly emphasized and made the focus of the next phase of investigation on avian olfaction. Studies of sensory behavior under field conditions must be complemented with laboratory evidence of appropriate neural activity and suitable acuity to account for the field observations before the conclusion can finally be drawn that the sensory route in question is critical for the behavior. Without such supporting data, alternative explanations are always possible and skeptics will remain unconvinced.

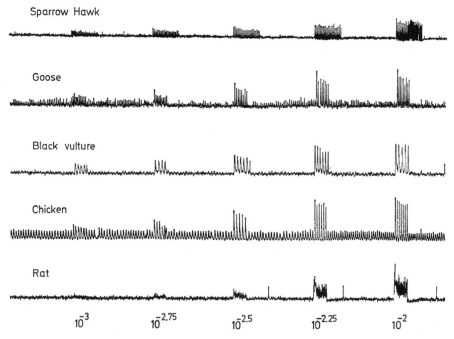

Fig. 1. Primary olfactory nerve responses to amyl acetate breathed in air at concentrations expressed as exponential fractions of vapor saturation at 20° C. Odor was presented every 3 min for durations of 1 min in the first trace and 0.5 in the others. (From Tucker, 1965, with permission of the publisher.)

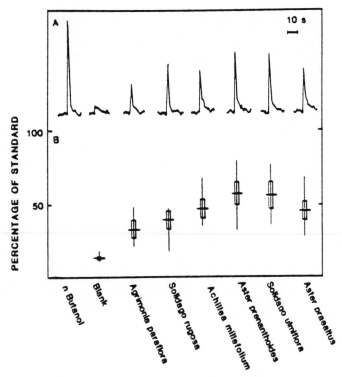

Fig. 2.   A. Representative multi-unit olfactory nerve twig
recordings from an adult male starling measured in
arbitrary units.   B.   Mean (horizontal bar) multi-unit
responses to odorants derived from plant material
relative to n-butanol for five adult starlings (two
males, three females).   Vertical box depicts 1 s.d.;
vertical line depicts range.   (From Clark and Mason,
1987, with permission of the publisher.)

In attempting to validate field observations with laboratory data
collected under more controlled conditions, the stimuli of choice would
include the same naturally occurring odors studied in the field.   Here,
at the outset, a serious problem can be met.   Many such odors are
complex mixtures, of which only one or a few components might be critical for the behavior in question.   Whether the total mixture or a set of
components is more convenient to use, it is necessary to specify odor
intensity in standard chemical units and this can sometimes be very
difficult.   The earliest report of olfactory nerve action potentials in
birds (Tucker, 1965) shows responses of four avian species and the rat
to five concentrations of amyl acetate (Fig. 1).   This odorant was not
selected for its ecological significance but because of its frequent use
in olfactory experiments.   In this case, because the purpose was to
establish the basic fact of avian olfactory responsiveness, a standard
stimulus known to be effective with nonavian species was desirable.   A
recent report (Clark and Mason, 1987), by contrast, shows nerve responses of European starlings (*Sturnus vulgaris*) to the volatiles of several
plant species, some of which the starlings use as nest material (Fig.
2).   In the latter case, no concentrations could be stated and no direct
comparisons could be made with other data.   Nonetheless, the procedure
was appropriate for correlating physiology and behavior.   What it could

not do was measure the sensory threshold in standard units.
Recording can also be done from central olfactory structures, such as
the olfactory bulb, where some processing of the incoming neural
information takes place and the responses of individual units can be
studied. In species with especially large bulbs, like the Northern
Fulmar (*Fulmarus glacialis*), access is relatively easy and a stable
preparation can be made. In one limited study (Hutchison and Wenzel,
1980), the effects of two natural prey-related odors shown to be effective attractants for these birds at sea on the activity of single cells
in the bulb were compared to the effects of a standard odor not associated with the marine environment. Spontaneous background activity was
inhibited by the former and enhanced by the latter. This is another
useful type of physiological validation for an olfactory basis of field
behavior.

Confirming that the olfactory system is involved in those avian
behavior patterns where it is implicated by field observations is obviously feasible, therefore, and should become customary. What is more
problematic is to correlate acuity with specific behavior. Waldvogel's
(1989) summary of absolute olfactory threshold measurements on birds
covered all published reports, which included only five species. All of
the stimuli were standard chemicals so that concentrations could be controlled but some of the odors, such as ethanethiol used with the Turkey
Vulture, were components of the naturally occurring mixture to which the
birds reacted in nature. The species represented covered the size range
for olfactory mucosa and bulb, i.e., from the small system of the Black-
billed Magpie (*Pica pica*) to the large one of the Turkey Vulture. All
of the testing techniques were behavioral; none involved direct recording of neural responses. Regardless of the stimulus or the species,
sensitivity in these studies was about 10^{-6} M, considerably poorer than
that of many other vertebrates including man. Such a high threshold by
laboratory measurements is surprising in light of certain field performance and such low variability seems inconsistent with the large variability in amount of olfactory tissue. The latter puzzle is not directly
relevant to the topic of this meeting; the former is central but has
been addressed in only one study as yet (Smith and Paselk, 1986).

The ultimate verification of natural behavior interpreted as
relying on olfactory perception is to show a correspondence between odor
levels in the natural situation and the bird's olfactory threshold for
the odor(s) involved. Smith and Paselk measured the Turkey Vulture's
absolute threshold for three component odors of decaying meat and also
estimated the distances from a carcass at which such levels could be expected to occur, using standard formulas for dispersion. The results
were markedly discrepant, viz., thresholds of 10^{-5} - 10^{-7} M limited detection to distances very much closer than normal foraging altitudes.

Stager's original study of odor tracking by Turkey Vultures (1964),
in which ethyl mercaptan was released in the field, was confirmed by
Houston (1986) using chicken carcasses. He reported the best performance with one-day old carcasses, compared to older or fresh-killed ones.
The vulture's sensitivity for the chemicals emitted by such a range of
carcasses has not been measured but the differential field performance,
if shown to be reliable, should point to the most detectible odor or
odorant mix. Smith and Paselk's threshold measurements were based on
the first significant change in heart rate as threshold concentration
increased. Although heart rate change is a reliable indicator of perception, it is probably not the most sensitive indicator of acuity. A
comparison with other methods is needed, especially with nerve recording.

Several species of procellariiforms have been attracted to such odors as cod liver oil, krill, and warm bacon fat in experiments conducted in temperate northern oceans and cold Antarctic waters, sometimes under conditions of very poor visibility and with long distances involved (Wenzel, 1980; in press). No threshold measurements have been reported for any of these birds but some of the behavior observed seems to call for detection of very small amounts of odorous material. The Northern Fulmar's olfactory bulb has a complex anatomy with a great many mitral cells (Wenzel and Meisami, in press) and has evidently evolved for some important function(s). Other procellariiform bulbs are comparably sized (Wenzel, 1986) and presumably are comparably organized.

In certain situations, successful use of odor cues could depend as much on relative sensitivity as on absolute. Field studies have not dealt with this function although it is implied by some of the tasks presented. The ability to discriminate among several odors has been studied even less with birds than their ability to detect odors. Clark and Mason (1989) have reported that the discriminative capacity of the Brown-headed Cowbird (*Molothrus ater*) is much better than its small olfactory bulb might suggest and is comparable to human sensitivity.

Perhaps the most urgent task in studying the levels and mechanisms of birds' olfactory sensitivity is to understand the qualities of their odor worlds. Several workers have pointed out that acuity, perhaps even perception itself, may well depend on stimulus quality. At least some avian olfactory systems might be specialists and show little response to most odors. For the systematic study of bird noses, the early advice of Small (1901) can hardly be bettered - "...experiments must conform to the psycho-biological character of an animal if sane results are to be obtained" (p. 206).

REFERENCES

Bang, B.G., 1960, Anatomical evidence for olfactory function in some species of birds, Nature, 188:547.
Bang, B.G., 1971, Functional anatomy of the olfactory system in 23 orders of birds, Acta anatomica, 79, Suppl. 58:1.
Clark, L. and Mason, J.R., 1987, Olfactory discrimination of plant volatiles by the European starling, Anim. Behav., 35:227.
Clark, L. and Mason, J.R., 1989, Sensitivity of Brown-headed Cowbirds to volatiles, Condor, 91:922.
Cobb, S., 1960, Observations on the comparative anatomy of the avian brain, Perspect. Biol. Med., 3:383.
Houston, D.C., 1986, Scavenging efficiency of Turkey Vultures in tropical forest, Condor, 88:318.
Hutchison, L.V. and Wenzel, B.M., 1980, Olfactory function in the Northern Fulmar (Fulmarus glacialis), in "International Symposium on Olfaction and Taste 7," H. van der Starre, ed., IRL Press, London.
Michelsen, W.J., 1959, Procedure for studying olfactory discrimination in pigeons, Science, 130:630.
Neuhaus, W., 1963, On the olfactory sense of birds, in "Olfaction and Taste," Y. Zotterman, ed., Pergamon Press, Oxford.
Small, W.S., 1901, Experimental study of the mental processes of the rat, Amer. J. Psychol., 12:206.
Smith, S.A. and Paselk, R.A., 1986, Olfactory sensitivity of the Turkey Vulture (Cathartes aura) to three carrion-associated odorants, Auk, 103:586.
Stager, K.E., 1964, The role of olfaction in food location by the turkey vulture (Cathartes aura), L.A. County Mus. Cont. Sci., No. 81.

Tucker, D., 1965, Electrophysiological evidence for olfactory function in birds, *Nature*, 207:34.

Waldvogel, J.C., 1989, Olfactory orientation in birds, *in* "Current Ornithology," D. M. Power, ed., Plenum Press, New York.

Wenzel, B.M., 1966, Olfactory perception in birds, *in* "Olfaction and Taste 2," T. Hayashi, ed., Pergamon Press, Oxford.

Wenzel, B.M., 1980, Chemoreception in seabirds, *in* "Behavior of Marine Animals," J. Burger, W. L. Olla, and H. E. Winn, eds., Plenum Press, New York.

Wenzel, B.M., 1986, The ecological and evolutionary challenges of procellariiform olfaction, *in* "Chemical Signals in Vertebrates 4," D. Duvall, D. Müller-Schwarze, and R. M. Silverstein, eds., Plenum Press, New York.

Wenzel, B.M., 1991, Olfactory abilities of birds, *Proc. XX Int. Cong. Ornithol.*, Vol. 3.

Wenzel, B.M., and Meisami, E., 1990, Quantitative characteristics of the olfactory system of the Northern Fulmar (*Fulmarus glacialis*): A pattern for sensitive odor detection?, *in* "Olfaction and Taste 10," L. Döving, ed., GCS A/S, Oslo.

PART 5: Mammals -- Cetaceans, Rodents, Lagomorphs, Ungulates, and Carnivores

A PRIMER OF OLFACTORY COMMUNICATIONS

ABOUT DISTANT FOODS IN NORWAY RATS

Bennett G. Galef, Jr.

Department of Psychology
McMaster University
Hamilton, Ontario, Canada

INTRODUCTION

During a brief period of social interaction, a naive Norway rat (an observer) can extract information from a recently fed conspecific (a demonstrator) sufficient to allow the observer to identify the diet that its demonstrator ate. In our standard procedure (Galef and Wigmore, 1983), each observer rat first interacted for 15 min with a demonstrator rat that had eaten either cinnamon- or cocoa-flavored diet (Diet Cin or Diet Coc). The observer was then isolated and offered a choice between Diets Cin and Coc for 22 hr. In such experiments, observer rats exhibit enhanced preferences for whatever diet their respective demonstrator ate (Galef et al., 1984).

DURATION AND MAGNITUDE OF SOCIAL EFFECTS ON FOOD PREFERENCE

Socially acquired food preferences are both durable and powerful. Figure 1 presents data describing the feeding behavior of two groups of 12 rats, whose members were each offered a choice between Diet Cin and Diet Coc diets for 23½ hr/day, for 17 consecutive days. During the remaining ½ hr of each of the first 5 days of the experiment, each member of one group of

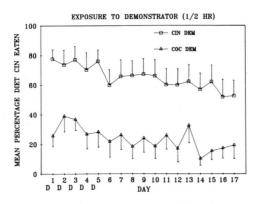

Fig. 1. Mean amount of Diet Cin eaten by observers interacting with demonstrators fed either Diet Cin or Diet Coc, as a percent of total amount ingested (Galef, 1989b).

subjects interacted with a demonstrator rat fed Diet Cin and each member of
the other group of subjects interacted with a demonstrator rat fed Diet Coc.
As can be seen in Figure 1, effects of the diet fed to demonstrator rats on
the food preferences of their observers lasted for 17 days (Galef, 1989b).

The magnitude of social influence on food choice is, perhaps, most
clearly revealed by examining interactions of socially induced diet
preferences with the most powerful known experiential determinant of food
choice, poison-induced learned aversion (Galef, 1986b). I fed observer rats
a palatable, casein-and-cornstarch-based diet (Diet NPT) for 1 hr. After
feeding, I injected each rat with either saline solution or LiCl solution.
After recovering from the effects of injection, each subject interacted
either for ½ hr or for 1 hr with either a bowl containing Diet NPT, a single
demonstrator rat fed Diet NPT or two demonstrator rats each fed Diet NPT.
Finally, each subject was given a choice between Diet NPT and unfamiliar
Diet Coc for 24 hr.

As can be seen in Figure 2, half the observer rats that had learned
aversions to Diet NPT and then interacted with two demonstrator rats, each
of which had eaten Diet NPT (i.e. subjects in Group 2 - DEM, LiCl), totally
abandoned the aversion they had learned to Diet NPT. Social influence
profoundly affected food choices of poisoned rats (Galef, 1986b, 1987,
1989c).

IMPLICATION OF OLFACTORY CUES

Results of a number of experiments indicate that the influence of
demonstrator rats on their observers' food preferences depends on olfactory
cues that pass from demonstrator rats to their observers (Galef and Wigmore,
1983) and that olfactory cues passing from demonstrators to observers and
affecting observers' diet preferences have two components: first, a diet
identifying component (Galef et al., 1985), which permits an observer to
identify the diet that its demonstrator ate; second, a contextual component
that is emitted by a conspecific and makes the diet-identifying component
effective in altering observers' diet preferences (Galef, 1990b; Galef et
al., 1985; Galef and Stein, 1985). As can be seen in Figure 2, simply
exposing observer rats to the smell of a diet (Group Bowl-LiCl) did not
increase the preference of observers for that diet. We observed social
enhancement of observers' diet preferences only when the odor of a diet was
experienced by observer rats in the presence of a demonstrator rat (Groups
1-DEM LiCl and 2-DEM LiCl).

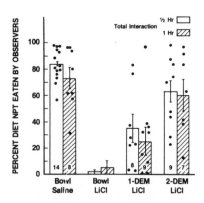

Fig. 2.   Amount of Diet NPT eaten, as a percent of total amount
           ingested by observers during test (from Galef, 1986b).

Results of a series of studies indicate that both diet-identifying olfactory cues and contextual olfactory cues are carried on the breath of demonstrator rats (Galef and Wigmore, 1983; Galef and Stein, 1985). Mass-spectrographic analysis of rat breath revealed that it contains carbon disulfide (CS_2) (Galef et al., 1988) and the results of several experiments (Bean et al., 1988; Galef et al., 1988; Mason et al., 1989) have shown that adding an aqueous solution containing a few parts/million of CS_2 to a food enhances rats' preferences for that food, while adding equal quantities of water to a food does not enhance rats' preferences for it. In sum, our data are consistent with the view that CS_2 in rat breath is an important component of the olfactory context that makes diet-identifying cues, also carried on the breath of demonstrator rats, effective in enhancing the attractiveness of foods to observer rats.

COMPLEXITY AND SOPHISTICATION OF THE COMMUNICATIVE PROCESS

For the remainder of this brief review (for more thorough reviews, see Galef, 1986a; 1988; 1989a; 1990a), I shall focus on the complexity of the messages enhancing food preference that pass between rats. The results of our experiments show that the system for olfactory communication about foods described above can handle more information than one might anticipate.

Information Exchange or Information Parasitism?

The notion of a demonstrator rat providing information to an observer is itself an oversimplification. Naive observer rats do not simply extract information from recently fed demonstrators. Rather, pairs of foraging rats exchange information about foods they have recently eaten (Galef, 1991a).

As illustrated in schematic in Figure 3, I food-deprived pairs of rats for 23 hr, then fed one member of each pair either Diet Cin or Diet Coc and fed the other member of each pair either anise-flavored diet (Diet Ani) or marjoram-flavored diet (Diet Mar). Next, I let the two subjects interact for 30 min. Last, I offered those subjects that had eaten either Diet Cin or Diet Coc a choice between Diets Ani and Mar, one of which had been eaten by each subject's pair mate. Similarly, I offered those subjects that had eaten either Diet Ani or Diet Mar a choice between Diet Cin and Diet Coc, one of which had been eaten by each of these subjects' pair mates.

The 2 x 2 x 2 design of the experiment is easily grasped by examining Figure 4 which shows: (1) the mean percent Diet Ani eaten during the 23-hr test by subjects that ate either Diet Cin or Diet Coc and then interacted with subjects fed either Diet Mar or Diet Ani and (2) the mean percent Diet

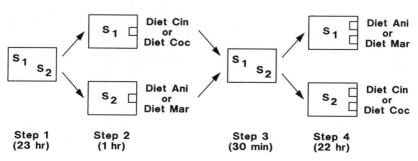

Fig. 3.   Schematic diagram of procedure. S_1 and S_2 = subjects.

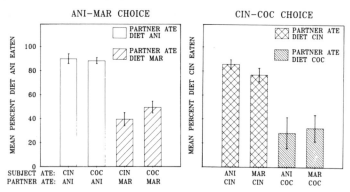

Fig. 4.   Mean percentage Diet Ani or Diet Cin eaten by subjects
offered a choice between either Diets Ani and Mar or
Diets Cin and Coc (from Galef, 1991a).

Cin eaten during the 23-hr test by subjects that ate either Diet Ani or Diet
Mar and then interacted with subjects fed either Diet Cin or Diet Coc.
Subjects in all four conditions exhibited enhanced preferences for foods of
the flavor that their respective partners had eaten. An individual rat is
not either a "demonstrator" or an "observer"; rather, an individual rat can
act both as a demonstrator for and an observer of its fellows.

Information Complexity

By using the olfactory communicative system under discussion, rats are
able to exchange information about rather complex individual histories of
recent food intake (Galef et al., 1990; Galef and Whiskin, in press).
Indeed, if rats were unable to communicate about complex recent feeding
experiences, the system would be of relatively little use. Participants in
an exchange of olfactory information would have to eat foods one at a time,
if they were to provide a decipherable message to those with whom they
interacted.

We (Galef and Whiskin, in press) fed each of 42 rats a food (Combination
A) composed of powdered Purina chow to which we had added (in g/100 g of
chow) 1.0 g cinnamon (Cin), 0.5 g anise (Ani), 0.5 g thyme (Thy), and 0.5 g
cloves (Clo) and a second group of 42 rats a second food (Combination B)
composed of Purina chow to which we had added (in g/100 g of chow) 2.0 g
cocoa (Coc), 1.0 g marjoram (Mar), 0.5 g cumin (Cum) and 0.5 g rosemary
(Ros). We then took one rat that had eaten Combination A and one rat that
had eaten Combination B and let them interact for 30 min. Finally, we
offered each of the 84 rats in the experiment a choice between a pair of
diets: one flavored with an herb or spice from Combination A, the other
flavored with an herb or spice from combination B. Each subject thus chose
between a food containing a flavor that it had eaten itself and a food
containing flavor that its partner had eaten. The experimental design is,
perhaps, most easily grasped by examining the four panels of Figure 5.

Figure 5 shows that, during testing, subjects exhibited a preference for
the flavors that their partners had eaten, not for the flavors that they had
eaten themselves. For example, when, as illustrated in the left-hand panel
of Figure 5, subjects were offered a choice between Diet Cin and Diet Coc,
those subjects that (1) had eaten Combination A (which contained Cin) and
(2) had interacted with partners that had eaten Combination B (which
contained Coc) preferred Diet Coc. This pattern of subjects exhibiting
preferences for flavors in foods that their respective partners had eaten,
rather than for flavors in foods that they had eaten themselves is repeated
in each of the other three panels of Figure 5.

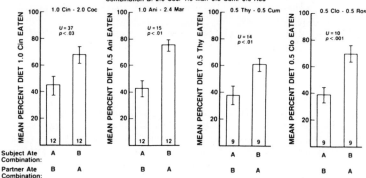

Combination A: 1.0 Cin/ 0.5 Ani/ 0.5 Thy/ 0.5 Clo
Combination B: 2.0 Coc/ 1.0 Mar/ 0.5 Cum/ 0.5 Ros

SUBJECT TREATMENT

Fig. 5.　　　Panels show the percentage of either Diet Cin, Diet
Ani, Diet Thy or Diet Clo eaten by subjects during
testing. Flavorants present in Combinations A and B
are shown above the figure. Choice of diets offered
subjects is indicated at top of panels (from Galef and
Whiskin, in press).

Control subjects (whose data is not presented in Figure 5), simply fed
either Combination A or Combination B and then offered choices between pairs
of foods, one containing a flavor from Combination A and one containing a
flavor from Combination B, exhibited no effect of eating a diet combination
on their subsequent food preferences (Galef and Whiskin, in press).
Consequently, the alterations in experimental subjects' food preferences
(shown in Figure 5) must have been the result of interaction with partners
rather than of subjects' own feeding experiences.

Last, it is worth mentioning that subjects in this experiment surely
experienced more exposure to the flavors in the combination diet which they
ate themselves than to the flavors in the alternative combination diet which
their respective partners ate. Yet, subjects developed preferences for
flavors in the combination diet that their partners ate, not for flavors in
the combination diet that they ate themselves. The results of the present
experiment thus provide additional evidence that simple exposure to foods is
not in itself responsible for the enhanced preference for foods observed
following social interactions between rats that have eaten various foods.

CONCLUSIONS

Taken together, the results described above indicate that aggregations
of rats can act as "information centres" (Ward and Zahavi, 1973) where
successful foragers can exchange information about foods they have eaten and
unsuccessful individuals can extract information allowing them to identify
foods their more successful fellows have eaten. There is reason to expect
that both information exchange among and information parasitism by Norway
rats could prove useful, guiding the unsuccessful to sources of nutriment
(Galef, 1990a; 1991b) and providing the successful with information they
could use if their current sources of nutriment were to fail (Galef, 1991a).

REFERENCES

Bean, N.J., Galef, B.G., Jr., and Mason, J.R., 1989, At biologically
significant concentrations, carbon disulfide both attracts mice and
increases their consumption of bait. J. Wildl. Manage., 52:502.

Galef, B.G., Jr., 1986a, Olfactory communication among rats of information concerning distant diets, in "Chemical Signals in Vertebrates, vol. IV: Ecology, Evolution, and Comparative Biology". D. Duvall, D. Muller-Schwarze and R.M. Silverstein, eds., Plenum Press, New York.

Galef, B.G., Jr., 1986b, Social interaction modifies learned aversions, sodium appetite, and both palatability and handling-time induced dietary preference in rats (R. norvegicus), J. Comp. Psychol., 100:432.

Galef, B.G., Jr., 1987, Social influences on the identification of toxic foods by Norway rats. Anim. Learn. Behav., 15:327.

Galef, B.G., Jr., 1988, Communication of information concerning distant diets in a social, central-place foraging species: Rattus norvegicus, in "Social Learning: Psychological and Biological Perspectives". T.R. Zentall and B.G. Galef, Jr. eds., Lawrence Erlbaum, Hillsdale, NJ.

Galef, B.G, Jr., 1989a, An adaptationist perspective on social learning, social feeding and social foraging in Norway rats, in "Contemporary Issues in Comparative Psychology", D.A. Dewsbury, ed., Sinauer Associates, Sunderland, MA.

Galef, B.G., Jr., 1989b, Enduring social enhancement of rats' preferences for the palatable and the piquant. Appetite, 13:81.

Galef, B.G., Jr., 1989c, Socially-mediated attenuation of taste-aversion learning in Norway rats: Preventing development of "food phobias." Anim. Learn. Behav., 17:468.

Galef, B.G., Jr., 1990a, An historical perspective on recent studies of social learning about foods by Norway rats. Can. J. Psychol., 44:311.

Galef, B.G., Jr., 1990b, Necessary and sufficient conditions for communication of diet preferences by Norway rats, Anim. Learn. Behav., 18:347.

Galef, B.G., Jr., 1991, Innovations in the study of social learning in animals: A developmental perspective, in "Methodological and Conceptual Issues in Developmental Psychobiology", H.N. Shair, G.A. Barr, and M.A. Hofer eds., Oxford University Press, Oxford.

Galef, B.G., Jr., 1991a, Information centres of Norway rats: Sites for information exchange and information parasitism, Anim. Behav., 41:295.

Galef, B.G., Jr., Attenborough, K.S. and Whiskin, E., 1990, Responses of observer rats to complex, diet-related signals emitted by demonstrator rats. J. Comp. Psychol., 104:11.

Galef, B.G., Jr., Kennett, D.J., and Stein, M., 1985, Demonstrator influence on observer diet preference: Effects of simple exposure and the presence of a demonstrator. Anim. Learn. Behav., 13:25.

Galef, B.G., Jr., Kennett, D.J., and Wigmore, S.W., 1984, Transfer of information concerning distant foods in rats: A robust phenomenon. Anim. Learn. Behav., 12:292.

Galef, B.G., Jr., Mason, J.R., Preti, G. and Bean, N.J., 1988, Carbon disulfide:A semiochemical mediating socially-induced diet choice in rats. Physiol. Behav., 42:119.

Galef, B.G., Jr., and Stein, M., 1985, Demonstrator influence on observer diet preference: Analyses of critical social interactions and olfactory signals. Anim. Learn. Behav., 13:31.

Galef, B.G., Jr. and Whiskin, E.E., in press, Social transmission of information about multiflavored foods. J. Comp. Psychol.

Galef, B.G., Jr. and Wigmore, S.W., 1983, Transfer of information concerning distant foods: A laboratory investigation of the "information-centre" hypothesis. Anim. Behav., 31:748.

Mason, J.R., Bean, N.J. and Galef, B.G., Jr., 1989, Attractiveness of carbon disulfide to wild Norway rats, in "Proceedings of the Thirteenth Vertebrate Pest Conference", A.C. Crabb and R.E. Marsh eds., University of California Press, Davis, CA.

Ward, P., and Zahavi, E., 1973, The importance of certain assemblages of birds as "information-centres" for food-finding. Ibis, 115:517.

CASTOREUM OF BEAVER (CASTOR CANADENSIS): FUNCTION, CHEMISTRY AND

BIOLOGICAL ACTIVITY OF ITS COMPONENTS

Dietland Müller-Schwarze

College of Environmental Science and Forestry
State University of New York
Syracuse, NY 13210 USA

INTRODUCTION

Castoreum, the paste found in the paired castor sacs of both sexes
in beaver, Castor canadensis and C. fiber, has been used for medicines
and perfumes since time immemorial. The Romans burned castoreum in lamps
and believed that the fumes caused abortions (McCully, 1969). Trappers
have attracted beaver to castoreum lures for a long time. As for the
natural history of castoreum, Audubon first published a trapper's report
of mud piles topped with strong-smelling castoreum that beaver built at
the banks of their ponds. Two neighbor colonies alternated in marking,
accumulating mud piles up to five feet high (Audubon and Bachman, 1849).
To this date, we don't know the precise role scent mounds play in the
behavior, physiology and population ecology of the beaver. Aleksiuk
(1968) proposed that scent marks may warn transient beaver away from
occupied territories and that scent mounding may be an epideictic display
sensu Wynne Edwards (1962) that communicates population density.

The beaver is one of the few mammals whose chemical communication
can be studied rigorously in the natural habitat. To understand the role
of castoreum in the behavior of beaver, experimenters have presented
whole castoreum to free-ranging beaver families. The experimental scent
mounds (ESM) are typically sniffed, pawed, and scent marked in turn
(Hodgdon, 1978; Bollinger, 1980; Müller-Schwarze and Heckman, 1980;
Svensson, 1980). During the past 6 years we have conducted experiments
to elucidate the function(s) of castoreum and to bioassay its fractions,
single compounds, and their mixtures. In this paper I review our recent
research on the territorial function of castoreum, its chemical composition, and the biological activity of its components.

FUNCTION OF CASTOREUM

The results of four types of field surveys and experiments are
consistent with territorial significance of the beaver's scent mounds.

Number of Scent Mounds

The number of scent mounds in a beaver colony vary directly with the
number of other colonies within 5 km up- or downstream. Beavers in

Chemical Signals in Vertebrates VI, Edited by R.L. Doty and
D. Müller-Schwarze, Plenum Press, New York, 1992

colonies with upstream neighbors did not scent mark more than those with
only downstream neighbors. Thus, if beaver activities provide information by water-borne chemicals, it is not reflected in the number of scent
mounds built by downstream beaver.

Responses to Scent Mounds in Own Territory

Methods

To elicit responses by resident beaver, an ESM is built at the edge
of a pond, upwind from the beavers' activity area. A sample is applied
by pipette to a cork on the mound at about 1700 hrs. The beavers are observed from this time until dark. The observer sits on a slope downwind
from the pond and records the behavior of the beaver on a portable
computer, using a program specifically developed for this project (Houlihan, 1989). The recorded data are edited and printed out in camp on the
same evening. The ESM is checked again the following morning for pawing
or obliteration by the beaver overnight.

Responses

Beavers are attracted to alien scent marks in their territory. Upon
emerging for the evening, they sniff from the water, swim zig-zagging up
the odor plume gradient, go on land, approach the ESM, sniff it, paw it
with their front feet, bring their hindquarters forward onto the mound,
straddle it, and scent mark with their castoreum and/or anal gland
secretion. After marking, they return to the water. They also may pick
up material from the ESM with their front paws and deposit it onto one of
their own mounds. Since beaver generally minimize their time on land,
the strong response is remarkable, and appears to have a vital function.
The response to unfamiliar castoreum even takes precedence over feeding.

Scenting Vacant Colony Sites Reduces Colonization

Scent marks may deter potential competitors without or with reinforcement by an actual encounter with a resident beaver. In the latter
case, the intruder compares the odors of the mark and the encountered
animal ("scent matching", Gosling, 1981). If the beaver's scent marks
serve as deterrent without reinforcement by an encounter with a beaver,
i.e. without scent matching, it should be possible to experimentally
mimic occupancy of suitable sites that presently lack beaver. Indeed,
treating 50 vacant beaver sites with ESMs during 2 seasons significantly
reduced the probability of colonization by beaver (Welsh and Müller-
Schwarze, 1989). Thus scenting of potential beaver sites may serve as an
option, along with other measures, to repel beaver from "problem sites",
at least temporarily.

Population Differences

Responses to castoreum and its constituents varied between New York
beaver populations. The crowded beaver of Allegany State Park (south-
western New York) responded more strongly than those in the Adirondacks
(Huntington Wildlife Forest, Cranberry Lake, and Reinemann Wildlife
Sanctuary) kept low by "nuisance" or fur trapping (Müller-Schwarze,
Houlihan, and Schulte, in prep.). An example is given in Fig. 1.

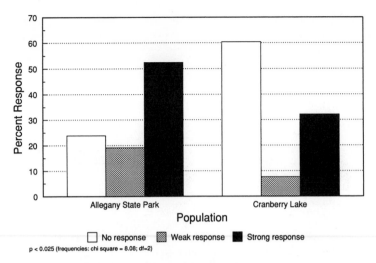

p < 0.025 (frequencies: chi square = 8.08; df=2)

Fig. 1.   Beaver at Allegany State Park responded more strongly to
castoreum than those at Cranberry Lake in the Adirondacks.
"Strong Response" includes flattening and/or marking of
the experiment scent mound.   Mere sniffing is not counted.
Difference between populations: t = 2.42; p < 0.02.

CHEMICAL COMPOSITION OF CASTOREUM

Newly Isolated Compounds

     In a re-investigation of castoreum, 12 new phenolics were identified
(Webster et al., in press) in addition to 10 phenolic compounds described
in the seminal study by Lederer (1946).   Thirteen neutrals, including
oxygenated monoterpenes, were found in castoreum for the first time
(Webster et al., in prep.).   The composition of castoreum varies considerably among samples (unpubl. data).   We are searching for odor patterns
that might be characteristic of individuals, sex and age classes, colonies, or seasons.

BIOLOGICAL ACTIVITY OF CASTOREUM CONSTITUENTS

     A total of 9 castoreum constituents, mostly phenolics (Table 1),
were active in our experiments (Müller-Schwarze and Houlihan, 1991; and
unpublished data).

     So far, there is no evidence that certain compounds control differ-
ent steps of the response, such as sniffing from the water, going on
land, approaching, sniffing, pawing, and marking the scent mound.   The
longer that beaver in the water sniff a sample from a distance, the more
likely they are to obliterate it (Fig. 2).   Note that the mixture of six
compounds loses much of its activity when the two phenols are removed
(sample labeled "4 compounds").

Table 1. Behaviorally Active Castoreum Components

4-ethyl phenol	*1,2-dihydroxybenzene*
4-methyl guaiacol	*4-ethoxyphenol**
4-ethyl guaiacol	borneol
<u>4-n-propyl guaiacol</u>	acetophenone

Italics: compounds that released scent marking.
<u>Underlined</u>: compounds that released land visits.
All compounds released sniffing from the water.
* Not yet described for castoreum.

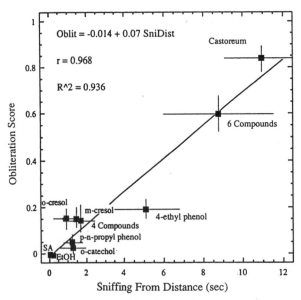

Fig. 2. Samples that initially are sniffed for a longer time from
the water are also more likely to be altered or destroyed by
beaver. Abscissa: average cumulative sniffing time per
beaver present in seconds. Ordinate: number of times an ESM
was flattened, plus 0.5 point for mere scratching or pawing
an ESM. Mixture of 4 compounds lacked 4-ethyl phenol and
p-n-propyl phenol. SA: salicylaldehyde. The solvent ethyl
alcohol (EtOH) served as control. Each sample was tested 8
times. F = 117.69; p = 0.00001 (Allegany State Park, 1989).

Several compounds release the same behaviors, i.e. are redundant in the sense of being interchangeable. Single compounds are less active than whole castoreum. Mixtures are intermediate. For 10 treatments in 1989, single components were sniffed significantly less from a distance than castoreum (ANOVA, $F = 2.222$; $p = 0.0303$; $N = 128$).

DISCUSSION

Possible Signal Value of Variable Castoreum Constituents

To signal species, sex, age or reproductive state, certain compounds, mixtures, or ratios must be consistently present. However, castoreum samples vary drastically in composition, and often some compounds are not detectable at all (unpublished data). We cannot dismiss these compounds as unimportant excretions, until we have excluded some signal value for them. They may signal to other beaver diet, hence quality of the site, and the residents' nutritional state and quality as potential breeding partners, even after they have dispersed from their natal colony. Of the compounds tested in 1989 and shown in Fig. 2, salicylaldehyde, tested by itself, never released a response. It may be simply excreted via castoreum, or function as a component of a mixture.

The Question of Diet-Dependent Castoreum Constituents

Beaver sites differ enormously in the plant species available as diet. As commonly known, beaver prefer aspen and other Populus species. Below these range other softwood species such as willow (Salix ssp.). Conifers are least palatable. And yet, some colonies feed on scotch pine (Pinus sylvestris) or even balsam fir (Abies balsamifera), at least in March/April. We have sampled and analyzed castoreum from beaver on those diets. The results will be published elsewhere.

Many castoreum constituents are also found in trees in the North American beaver's range that are utilized by beaver (own observations). Examples are given in Table 2. Catechol and phenol are also known as degradation products of salicortin and tremulacin in Populus tremuloides (Clausen et al., 1989). Likewise, the plant estrogen genistein and similar isoflavonoids are degraded in the rumen to inactive p-ethyl phenol (Harbourne, 1988).

Table 2. Some Castoreum Compounds That Also Occur in Trees Utilized by Beaver.

Compound	Tree Species
Benzoic acid	Prunus serotina, Pinus sylvestris
Benzyl alcohol	Populus spp.
Borneol	bark-beetle infested pines
Catechol	Populus deltoides
o-Cresol	Thuja occidentalis
4-Ethyl phenol	Picea abies, Populus spp.
Hydroquinone	Pinus resinosa
4-(4'hydroxyphenyl)-2-butanone	Betula spp.
Phenol	Pinus sylvestris
Salicylaldehyde	Salix spp., Populus spp.

From reviews in Rowe, 1989.

Redundancy

Several compounds release the same behavior, albeit each less
strongly than whole castoreum. They are interchangeable, and at the same
time additive in their effects. I suggest that this cue redundancy is
adaptive. As proximate determinants of such "flexibility by design",
several compounds may be very similar chemically and/or belong to the
same physiological pathway. An ultimate cause may be selective pressure
to have available a backup system. This would be particularly useful
when 1) changing vegetation affects diet-dependent pheromone compounds,
2) environmental factors impede propagation of certain odors, or 3) early
olfactory imprinting may rely on variable odor cues from the parent(s).

Redundancy occurs at several levels of chemoemission and chemoreception. In the golden hamster (Mesocricetus auratus), the female
attracts the male by vaginal secretion, flank, Harderian, or ear gland
odors (Johnston, 1986). Priming odors in urine from adult males or
pregnant females accelerate puberty in female mice, but both together are
not more effective. Here, limits of developmental rates may preclude
additive effects (Drickamer, 1982). The level of single compounds
provides other examples of redundancy. Androstenol or androstenone in
the boar's (Sus scrofa) saliva stimulate the mating stance in the sow,
but both together are not more active (Perry et al, 1980). Urine of
crowded female mice, Mus musculus, contains n-pentyl-acetate, cis-2-
penten-1-yl acetate, and 2,5-dimethylpyrazine that delay puberty in other
females. The pyrazine delays by 2.4 days, a mixture of the two acetate
esters by 1.5 days, and all three compounds together by 1.7 days (Novotny
et al., 1986). In tree shrews, Tupaia belangeri, redundant components in
male urine release chinning in both sexes (Stralendorff, 1987).

Turning to chemoreception, the same response can be controlled by
stimuli in different sensory modalities. In domestic sheep, "weaners"
learn to eat wheat from experienced adults. They still perform well with
one or two senses impaired. If vision, hearing, and olfaction are
impaired simultaneously, the sheep perform as poorly as weaners without
"teachers" (Chapple and Wodzicka-Tomaszewska, 1987). Spiny mice, Acomys
cahirinus, find their prey by all senses which are interchangeable, even
though olfaction and hearing may be more important than vision (Langley,
1988). In the domestic pig, visual stimuli attract the sow to the boar.
If they are not available, odors will attract her. Without either,
auditory cues ("chanting") also work (Pearce and Hughes, 1987).

Different chemical senses can be redundant: A male hamster is
attracted to a female using either the vomeronasal organ or the main
olfactory system (MOS)(Powers et al., 1979). The MOS itself is highly
redundant. Ninety percent of the olfactory bulb can be removed, and the
animal still smells very well. By contrast, if only one half of all
hearing is removed, a human loses all highs and lows of music (Graziadei,
1986).

In conclusion, our bioassays of beaver castoreum components lead us
to propose that redundant compounds may be common in vertebrates, and
they may serve important functions. Investigators should be alert to
redundant active compounds.

ACKNOWLEDGMENTS

This work was supported by National Science Foundation grant BNS 88-19982
and by the Adirondack Wildlife Program, New York State Legislature.
Thanks go to Tim Schwender, Brent Speicher, Ken Baginskii, and Chris

Rubeck for extensive field work, and to Dr. Robert M. Silverstein and Dr. Francis X. Webster for chemical advice. I thank Drs. R.M. Silverstein and Stephen Teale, and Bruce Schulte for critical reading of the manuscript.

REFERENCES

Aleksiuk, M., 1968, Scent mound communication, territoriality, and population regulation in the beaver (Castor canadensis KUHL). J. Mammal., 49:759.

Audubon, J.J. and Bachman, J., 1849 (1854), The Quadrupeds of North America," Vol. I, p. 347. Reprinted 1974; New York: Arno Press.

Bollinger, K.S., 1980, Scent marking behavior of beaver (Castor canadensis). M.S. Thesis, Univ. Massachusetts, Amherst.

Chapple, R.S. and Wodzicka-Tomaszewska, M., 1987, The learning behavior of sheep when introduced to wheat. II. Social transmission of wheat feeding and the role of the senses. Appl. Anim. Behav. Sci., 18:163.

Clausen, T.P., Reichardt, P.B., Bryant, J.P. Werner, R.A., Post, K., and Frisby, K., 1989, Chemical model for short-term induction in quaking aspen (Populus tremuloides) foliage against herbivores. J. Chem. Ecol., 15:2335.

Drickamer, L.C., 1982, Acceleration of sexual maturation in female house mice by urinary cues: Dose levels and mixing urine from different sources. Anim. Behav., 30:456.

Gosling, L.M., 1982, A reassessment of the function of scent marking in territories. Z. Tierpsychol., 60:89.

Graziadei, P., 1986, Injury and repair in the olfactory pathway of mammals, pp. 363-372 in: "Biology of Change in Otolaryngology", R.W. Ruben, ed., Elsevier Sci Publ.

Harbourne, J.B., 1988, "Introduction to Ecological Biochemistry." Acad. Press, London.

Hodgdon, H.E., 1978, Social dynamics and behavior within an unexploited beaver (Castor canadensis) population. Ph.D. Dissert., Univ. Massachusetts, Amherst.

Houlihan, P.W., 1989, Scent mounding by beaver (Castor canadensis): Functional and semiochemical aspects. M.S. Thesis, State Univ. New York; Coll. Envir. Sci. & Forestry, Syracuse.

Johnston, R.E., 1986, Effects of female odors on the sexual behavior of male hamsters, Behav. Neur., 46:168.

Langley, W.M., 1988, Spiny mouse's (Acomys cahirinus) use of its distance senses in prey localization, Behav. Processes, 16:67.

Lederer, E., 1946, Chemistry and biochemistry of the scent glands of the beaver, Castor fiber. Nature, 157:231.

McCully, H., 1969, Pliny's pheromonic abortifacients, Science, 165:236.

Müller-Schwarze, D. and Heckman, S., 1980, The social role of scent marking in beaver (Castor canadensis). J. Chem. Ecol. 6:81.

Müller-Schwarze, D. and Houlihan, P.W., 1991, Pheromonal activity of single castoreum constituents in beaver, Castor canadensis. J. Chem. Ecol. 17:715.

Novotny, M., Jemiolo, B., Harvey, S., Wiesler, D. and Marchlewska-Koj, A., 1986, Adrenal-mediated endogenous metabolites inhibit puberty in female mice. Science, 231:722.

Pearce, G.P. and Hughes, P.E., 1987, An investigation of the roles of boar-component stimuli in the expression of proceptivity in the female pig. Appl. Anim. Behav. Sci. 18:287.

Perry, G.C., Patterson, R.L.S., Macfie, H.J.H., and Stinson, C.G., 1980, Pig courtship behaviour: pheromonal property of androstene steroids in male submaxillary secretion. Anim. Prod. 31:191.

Powers, J.B., Fields, R.B. and Winans, S.S., 1979, Olfactory and vomero-nasal system participation in male hamster's attraction to female vaginal secretions, Physiol. Behav., 22:77.

Rowe, J.W., ed., 1989, "Natural Products of Woody Plants," 2 vols., Springer, Berlin.

Stralendorff, F.V., 1987, Partial chemical characterization of urinary signaling pheromone in tree shrews (Tupaia belangeri), J. Chem. Ecol, 13:655.

Svendsen, G.E. 1980, Patterns of scent-mounding in a population of beaver (Castor canadensis). J. Chem. Ecol. 6, 618-620.

Webster, F.X., Rong Tang, and D. Müller-Schwarze, submitted, Isolation and characterization of phenolic constituents of castoreum from the North American beaver (Castor canadensis).

Welsh, G.R. and Müller-Schwarze, D. 1989, Experimental habitat scenting inhibits colonization by beaver, Castor canadensis. J. Chem. Ecol., 15:887.

Wynne-Edwards, V.C., 1962, "Animal Dispersion in Relation to Social Behavior," Hafner, New York.

CHEMICAL IMAGES AND CHEMICAL INFORMATION

Stephan E. Natynczuk and Eric S. Albone

Clifton Scientific, Clifton College, Bristol, BS8 3JH
and School of Chemistry, University of Bristol, Bristol
BS8 ITS, United Kingdom

INTRODUCTION

Much has been written implicating scent profiles, odour fingerprints and scent images in vertebrate chemical communication (Albone,
1984; Gorman et al, 1984; Macdonald et al., 1984; Smith et al., 1985;
Apps, 1990; Natynczuk and Macdonald, this volume), yet we are still
uncertain as to how biological information is organized in a scent or
chemical image. In this brief discussion we wish to further explore
the notion that an animal may present a chemical image to its environment and to discuss the extent to which this complex image, which may
accurately encode an animal's biological condition, is organised.

THE CHEMICAL IMAGE

The chemical image an organism perceives and responds to consists
of a complex of chemical stimuli deriving from the components of a scent
and other materials it encounters in its environment. Because the
animal's sense organs do not sample and respond to ambient chemicals in
the same way as does any particular analytical procedure, it is very
unlikely that the chemical image which the animal perceives will bear a
close relationship to the information profile of a semiochemical source
provided by a particular analytical procedure. It is thus well known
that minor components of mixtures can dominate their perceived odour or
flavour quality (Albone, 1984, p. 34-39), but a full realisation of the
implications of the mismatch between chemical, analytical and semiochemical information has yet to be explored.

CHEMICAL INFORMATION

Stonier (1990) has asserted that information does not depend for
its existence on its perception, but rather reflects the degree of
organisation of a system. The system in our case consists of the ambient chemical environment encountered by an animal, and particularly
those aspects of this environment associated with scent. In as far as
this chemical environment is patterned and organized, it contains chemical information whether or not this information can be read by the
animal encountering it.

Chemical Signals in Vertebrates VI, Edited by R.L. Doty and
D. Müller-Schwarze, Plenum Press, New York, 1992

This chemical information is probed by chemical analysis. The chemical data obtained are bound to be limited by the degree of resolution, sensitivity and selectivity of the analytical techniques used, and by the choices the analyst makes of what compound classes to investigate. Thus, chemical analysis selects from the total chemical information present and it would be fortuitous if the chemical information obtained by the analysis used corresponds to its semiochemical significance. An analogy would be reading a text through an alphabet filter which selectively removed consonants. If this filter could be adjusted to increasingly filter out more consonants, starting say with just b's then the k's, l's, and so on, the text would quickly become quite meaningless. Reconciliation of information obtained by chemical analysis with the chemical image to which the animal responds is the principal dilemma in this field of science.

THE PROBLEM

A major problem confronting mammalian semiochemistry is to relate these two derivatives of the initial chemical information associated with an organism (Albone, 1990). In the following examples we explore a number of investigations of chemical organizations and the biologically useful information obtained.

The chemical information presented by an organism is structured by its biology. The relative concentrations of chemicals recorded in a particular chromatogram may vary and such variations may be assessed statistically. The chemical image hypothesis (Albone, 1984, pp. 5-7) requires one to survey the total chemistry of a scent and assess it for variations which may be related to biologically significant parameters. This approach requires every peak in a sample chromatogram to be included in any statistical analysis. It would be most appropriate to use a statistical technique that summarized all the data and then examine those data for variation in individual components. Recent statistical analyses of gas chromatograms of data of mammalian scent gland material have relied upon analysis of variance (ANOVA) to explore the variation about mean peak areas (see Gorman et al., 1984; Jemiolo et al., 1987; Davies et al., 1988). The functional description of this technique makes it sound like a very attractive method for analysing chromatographic data: "ANOVA is a general technique for partitioning the overall variability in a set of observations into components due to specific influences and to random error (or haphazard variation). The resulting ANOVA table provides a concise summary of the structure of the data and a descriptive picture of the different sources of variation" (Chatfield, 1988, p. 202). ANOVA is, however, a univariate test. It assumes that each peak used in the analysis is, in a qualitative sense, an equivalent variable. So ANOVA is inappropriate for semiochemical gas chromatographic data in view of the differences in sensitivity of organisms to different compounds, as exemplified by odor threshold.

Multivariate methods allow the comparison of samples whose variation may be considered to be due to several causes (Southwood, 1978, p. 435). Such techniques are more appropriate to the analysis of semiochemical material, particularly because this material could have diverse origins; namely, it could be (i) metabolically derived from the mammal itself, (ii) a microbial product, (iii) a microbial modification of host secretion or excretion, or (iv) a mixture derived from the forgoing.

Principal components analysis (PCA) is a multivariate technique for examining relationships between several quantitative variables (Chatfield, 1988, pp. 219-220: Digby and Kempton, 1987, pp. 55-70; Gilbert,

1989). It is one of a family of techniques whose "role is to explore
multivariate data, to provide information rich summaries, to generate
hypotheses (rather than test them) and to help generally in the search
for structure, both between variables and between individuals" (Chatfield 1988, p. 40). Briefly PCA "is concerned with examining the inter-
dependence of variables arising on an equal footing" (Chatfield, 1988,
pp. 219-220). Principal components are chosen in turn to explain as
much of the variation in the sample as possible. Often the first two or
three principal components explain most of the variation and scatter
diagrams of these are helpful in showing relationships in the original
data (chromatograms). Using principal components analysis, chromatograms with similar patterns may be grouped according to important biological criteria such as subspecies (Smith et al., 1985), age, sex and
reproductive condition (Efford, 1985; Natynczuk, 1990; Natynczuk and
Macdonald this volume); such patterning may be taken as indicating
variations in chemical information content.

Natynczuk (1990) successfully used principal components analysis to
group rats' pelage odors, and derivatised preputial gland lipids according to sex and reproductive condition (Figs. 1 and 2) (see Natynczuk and
Macdonald this volume). In this particular case we have two samplings
of the same chemical information presented by the animal, namely (i)
components of the vapor associated with scent glands and (ii) components
extracted with solvents from the scent glands, both of which contain
information about the rats' sex and reproductive condition (Brown,
1985).

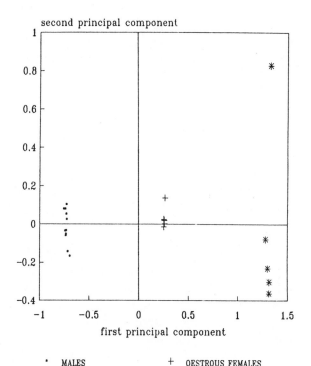

Fig. 1. Principal components analysis of proportional peak areas
obtained using the dynamic solvent effect (headspace mode)
and gas capillary chromatography (Apps, 1990) from male
and female rats' haunch fur.

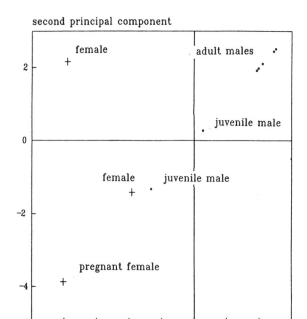

second principal component

Fig. 2. Principal components analysis of proportional peak areas
obtained from gas chromatographic analysis of trimethyl-
silyl derivations of male and female rats' preputial
glands.

However, a study of the inguinal glands of brown hares, Lepus
europaeus; Natynczuk, Coates, Green, and Albone, unpublished data) using
similar headspace and statistical analyses to that of the rat study
provided no clustering and, therefore, no clear explanation of the
chemical data. We expected to find that the scent of the inguinal gland
would reveal the hares' sex and reproductive condition (Mykytowycz
1966), but we found no grouping of chromatograms according to either
criteria (Fig. 3).

Our interpretation of the results was that the hares' inguinal
gland provided a scent peculiar to each individual, much as seems to be
the case of the dogs' anal gland secretion (Natynczuk et al., 1989;
Bradshaw et al., 1990). However, different analytical samplings may
lead to different conclusions.

The divergence between the sampling of chemical information obtained by the animal and by the analytical chemist is unavoidable, so
that although the patterns we obtain through chemical analysis are
unlikely to be identical with the patterns perceived by the animal, they
are both functions of the same chemical information present in the
environment and are practically valuable.

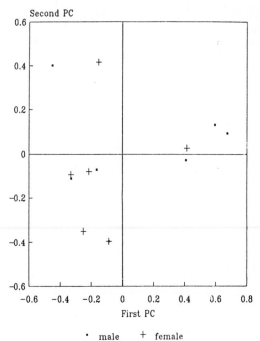

· male + female

Fig. 3. Principal components analysis of proportional peak areas
obtained using the dynamic solvent effect (headspace mode)
and gas capillary chromatography (Apps, 1990) from male
and female brown hares' inguinal gland secretions.

REFERENCES

Albone, E.S., 1984, "Mammalian Semiochemistry", Wiley Interscience,
Chicester.

Albone, E.S., 1990, Mammalian semiochemistry and its application, in:
"Chemical Signals in Vertebrates V", D.W. Macdonald, D. Muller-
Schwarze, and S.E. Natynczuk, eds., Oxford University Press, Oxford
pp. 77-83.

Apps, P.J., 1990, Dynamic solvent effect sampling for quantitative
analysis of semiochemicals, in: "Chemical Signals in Vertebrates
V", D.W. Macdonald, D. Muller-Schwarze, and S.E. Natynczuk, eds.,
Oxford University Press, Oxford, pp. 23-33.

Bradshaw, J.W.S., Natynczuk, S.E., and Macdonald, D.W., 1990, Potential
for applications of anal sac volatiles from domestic dogs, in:
"Chemical Signals in Vertebrates V", D.W. Macdonald, D. Muller-
Schwarze, and S.E. Natynczuk, eds., Oxford University Press, Oxford, pp. 640-644.

Brown, R.E., 1985, The rodents, in: "Social Odors in Mammals", R.E.
Brown and D.W. Macdonald, eds., Oxford University Press, Oxford,
pp. 245-506.

Chatfield, C., 1988, "Problem Solving: A Statistical Guide", Chapman and
Hall, London.

Davies, J.M., Lachno, D.R., and Roper, T.J., 1988, The anal gland secretion of the European badger (Meles meles) and its role in social
communication, J. Zool. Lond., 216:455-463.

Digby, P.G.N., and Kempton, R.A., 1987, "Multivariate Analysis of Eco-
    logical Communities", Chapman and Hall, London.
Efford, M.G., 1985, "The Structure and Dynamics of Water Vole Popula-
    tions", D. Phil. Thesis, Oxford University.
Gilbert, N., 1989, "Biometrical Interpretation", Oxford University
    Press, Oxford.
Gorman, M.L. Kruuk, H., and Leitch, A., 1984, Social functions of the
    subcaudal scent gland secretion of the European badger Meles meles
    (Carnivora: Mustelidae), J. Zool. Lond., 204:549-559.
Jemiolo, B., Andreolini, F., Wiesler, D., and Novotny, M., 1987, Varia-
    tions in mouse (Mus musculus) urinary volatiles during different
    periods of pregnancy and lactation, J. Chem. Ecol., 13:1941-
    1956.
Macdonald, D.W., Krantz, K., and Aplin, R.T., 1984, Behavioral, anatomi-
    cal and chemical aspects of scent markings amongst capybaras
    (Hydrochoerus hydrchaeis) (Rodentia: Caviomormpha), J. Zool. Lond.,
    202:341-360.
Mykytowycz, R., 1966, Observations on odoriferous and other glands in
    the Australian wild rabbit, Oryctolagus cuniculus (L), and the
    hare, Lepus europaeus. II. The inguinal glands. CSIRO Wildl. Res.,
    11:49-64.
Natynczuk, S.E., 1990, Ultrasound and semiochemistry in rat social
    behavior, D.Phil. Thesis, Oxford University.
Natynczuk, S.E., Bradshaw, J.W.S., and Macdonald, D.W., 1989, Chemical
    constituents of the anal sacs of domestic dogs, Biochem. Syst.
    Ecol., 17:83-87.
Natynczuk, S.E. and Macdonald, D.W., 1991, Sex, scent, scent matching
    and the self calibrating rat (in preparation).
Smith, A.B., Belcher, A.M., Epple, G., Jurs, P.C., and Levine, B., 1985,
    Computerized pattern recognition: a new technique for the analysis
    of chemical communication, Science, 228:175-177.
Southwood, T.R.E., 1978, "Ecological Methods", 2nd Edition, Chapman and
    Hall, London.
Stonier, T., 1990, "Information and the Internal Structure of the Uni-
    verse", London, Springer-Verlag.

CHEMICAL SIGNALS IN THE TIGER

R.L. Brahmachary[1], M.P. Sarkar[1] and J. Dutta[2]

[1]Embryology Unit, Indian Statistical Institute
Calcutta 700 035; [2]Department of Chemistry, Bose
Institute, Calcutta 700 009

INTRODUCTION

Like many other mammals, tigers disseminate chemical signals through urine and specialized glandular secretions. The behavior of the tiger, the hunters' prime trophy, was hardly amenable to the rigors of scientific observation and Schaller (1967) first drew attention to a special mode of squirting a milky, lipid-rich fluid upwards and backwards by both sexes of tigers. This has been repeatedly referred to as anal gland secretion, and the consequent confusion in the literature has been described (Brahmachary and Dutta, 1987). This fluid, referred to as Marking Fluid (MF) by this group of authors (Brahmachary and Dutta, 1981, 1987), is ejected through the urinary channel which is apparently unconnected to the anal sac (Hashimoto et al., 1963). Furthermore, Smith et al. (1989), who used the term scent marking (to imply the depositon of MF), have now clearly distinguished MF from anal gland secretion which, on drying, becomes a blackish mass. Numerous observations from very close quarters on a pet tigress has revealed the great innate urge of squirting MF (Choudhury, unpublished; Brahmachary, unpublished).

As described in the present paper, the frequency of ejecting MF and the strategy of marking strongly suggest that MF is a source of chemical signals.

MATERIAL AND METHODS

Eight female and two male adult tigers (8 female, 2 male) have been studied in the open air zoo of Nandan Kanan, India. Two cubs were also reared from the age of about three weeks, specifically, for this study. Collection of MF from adults was possible from three tigresses and one tiger who were familiar with the keepers. Following the movements of tigers behind a chain link mesh, trays or beakers were held up as the animals raised their tails, the prelude to squirting. With practice, part of the ejected MF could be caught in the containers. This was easier when the tigers were let loose in smaller adjoining enclosures which seem to stimulate them to mark this "new" territory. It was noticed that at the age of about ten weeks the tiger cub would usually urinate on a lawn in the morning, after being brought out of the room

where it was confined during the night. This urine was collected daily
in a beaker held near the penis and the cub became habituated to this
routine.

Number and distribution of MF squirting over the "territories"
(enclosures) of three groups of tigers were studied (see Fig. 1). Group
1, consisting of a male tiger (M1) and three tigresses (F1,F2,F3),
roamed in enclosure 1 (En.1) which consisted of a 0.3 acre plot with a
water moat on two sides and having a grassy sward with many trees.
Beyond the common boundary, a chain link fence (1C+W) running along the
moat, there is a replica enclosure (En.2), almost a mirror image of
En.1. Three tigresses (F4,F5,F6) formed group 2 and occupied En.2.
Beyond the second boundary 2C of En.2, there was a vacant area (60 ft.).
Beyond fences A and B (between which lies the masonry shelter (wall)
which also served as the feeding chamber), there were adjoining small
annex enclosures (Anx.) into which the tigers were sometimes allowed to
move. Thus, groups 1 and 2 shared only one common frontier (1C+W).
Group 3 in En.3 contained a male (M2) and two females (F7,F8) whose
immediate neighbor was a lone male (M3). Here fence A is absent and
fence B is present (indicated as 3C in Table 1).

Some of the chemical experiments were carried out in the veterinary
laboratory at the site and many samples were brought to Calcutta under
ice or following chloroform or hexane extraction. GLC was carried out
with a microprocessor-controlled FID instrument (Hewlett Packard 5840 A
dual column).

RESULTS

Squirting behavior

11530 MF squirtings were observed in 1528 hours and 38 minutes.
During the first 9662 MF squirtings, 198 urinations were also recorded.
Table 1 sums up the distribution of MF over the salient features of the
territory. F3 was definitely of "low rank", since very little marking
was performed no data for her have been presented in the table. Further
observation at a later date showed that when F1 was introduced in the
Anx. B, never marked before, was now the second highest site of marking
by M1 and F2 (F3 was no longer present).

Table 1. Marking per unit length (ft.) of the boundaries
          and wall

Tiger	1C+W	2C	B/3C	Wall
F 1	7.7	3.5	0	69.7
F 2	7.7	1.3	0	73.1
M 1	14.4	2.4	0	45.3
F 4	0.7	0.8	0	4.3
F 5	0.3	0.4	0	2.6
F 6	2.1	0.2	0	2.0
F 7	0.2	1.8	0.08	1.5
F 8	0.6	1.3	0.08	1.5
M 2	3.0	0.5	0.06	2.8

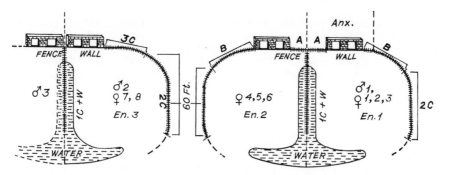

Fig. 1. Diagram of tiger enclosures.

Chemical constituents

All volatiles of MF are potential semiochemicals, and one of the
volatiles has been found to carry the characteristic "musky" aroma of
the tiger and leopard which resembles both civetone and the aroma of
certain varieties of fragrant Indian rice. This aroma was first clearly
perceptible in cub urine at about one year of age. Acidification removes and alkali treatment restores the aroma in both rice and MF,
thereby distinguishing these from civetone. Paper chromatography using
three different solvent systems, n-butanol: acetic acid: water (4:1:1)
(Sol. 1), water saturated butanol (Sol. 2) and ethanol: acetic acid:
water (4:1:1) (Sol. 3) showed that acidified MF aroma and unboiled rice
aroma (extracted in water) have the same R_f in each solvent. To detect
the position of aroma spot, pieces of the chromatogram were made alkaline and allowed to sniff by unbiased persons or blind-folded persons
(Fig. 2). Two Schiff bases contained in MF stained red (Re) and yellow
(Y) with reagent (Feigl, 1966) and were well separated from the aroma
region. The only component of boiled rice aroma identified to date,
(2-acetyl-1-pyrroline (2AP); Buttery et al, 1982) was synthesized using
the method of Schieberle (1988) and co-chromatographed with tiger/un-
boiled rice aroma. On lowering the reaction temperature (from about

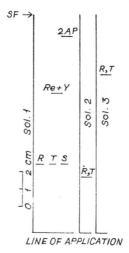

Fig. 2. SF-solvent front; R-rice aroma; T-tiger aroma; S-
synthetic aroma; Re+Y-red and yellow Schiff bases.

170°C) of 2AP synthesis, the yield of another fragrant substance resembling unboiled rice aroma increases. This has the same R_f as that of tiger/leopard aroma (Fig. 2). GLC analysis on squalane and carbowax 20M columns of the head-spaces of water-soaked rice and tiger MF show a peak of the same R_t value of 4.8 and 2.37, respectively (Brahmachary et al., 1990).

TLC and HPTLC (Serva, Heidelberg) plates were used to study the lipid classes of MF. Diglycerides (DG), sterol, free fatty acid, triglycerides (TG), wax esters and sterol esters were detected by suitable staining agents. On GLC chromatograms eleven free fatty acids (FFA) ranging from acetic acid to octanoic acid including some of the isoforms have been identified in the steam distillate of MF of three tigers using GLC chromatography (FFAP column).

DISCUSSION

Tigers spray MF with a frequency that is about 50 times higher than that of urination. This suggests that MF has evolved to play a role in scent marking. The shelter (which may simulate a cave, the sanctum sanctorum of the territory under natural conditions) is the most frequently marked, followed by the common boundary. Also, A, B and 3C beyond which there were no tigers were either not marked at all (A,B) or marked very slightly (3C). However, after a tigress was introduced beyond B of En.1 it became the site of the second highest marking. These facts are consistent with the assumption that MF is a source of chemical signals. The possible candidates for such signals are volatile molecules (disregarding the VNO mediated ones), the most interesting of which has a characteristic, musky aroma, shared by certain strains of unboiled fragrant rice. This is not 2AP; since the compound can now be synthesized, it should be possible to identify it with GCMS and/or FT NMR. The FFA identified in MF are also likely chemical signals, since some of these compounds are already known to play such a role in other species (Gorman, 1976, Curtis et al., 1971, Michael et al, 1974, McClintock, 1978). The quantative ratios of FFA in the MF are being studied and preliminary analysis indicates a much closer relationship between the FFA patterns of the two most closely related individuals among those three tigers. TG, DG, etc., may act as fixatives of the chemical signals so that they can last for quite some time and thus minimize energy expenditure of the animals.

ACKNOWLEDGEMENTS

We thank the staff of Nandan Kanan; H. Chanda, T. Modak, N. Banik, A. Chakraborty, S. Bhattacharya, ISI; Prof. S. Ghose and the staff of the Chemistry Department, Bose Institute; Dr. S.S. Sarkar for preparing the manuscript; Financial support by DOEN, Government of India is acknowledged.

REFERENCES

Brahmachary, R.L. and Dutta, J., 1981, On the pheromones of tigers experiments and theory, Am. Nat., 118:561.
Brahmachary, R.L. and Dutta, J., 1987, Chemical communication in the tiger and leopard, in: "Tigers of the World", R.L. Tilson and U.S. Seal, eds., Noyes Publication, U.S.A.
Brahmachary, R.L., Sarkar, P., Mousumi and Dutta, J., 1990, The aroma of Rice .... and tiger, Nature, 344:26.

Buttery, R. G., Ling, L.C. and Juliano, B.O., 1982, 2-acetyl-1-pyrroline
-- an important aroma component of cooking rice, <u>Chem. Indust.</u>, 4:
958.

Curtis, R.F., Ballantine, J.A., Keverne, E.B., Bonsall, R.W. and Michael, R.P., 1971, Identification of primate sexual pheromones and
the properties of synthetic attractants, <u>Nature</u>, 232:396.

Feigl, F., 1966, "Spot tests in organic analysis", Elsevier Publishing
Co., London.

Gorman, M.L., 1976, A mechanism for individual recognition by odour in
<u>Herpestes auropunctatus</u> (Carnivora: Viverridae), <u>Anim. Behav.</u>,
24:141.

Hashimoto, Y., Eguchi, Y. and Arakawa, A., 1963, Histological observation on the anal sac and its glands of a tiger, <u>Jap. J. Vet. Sci</u>,
25:29.

McClintock, M., 1971, Menstrual synchrony and suppression, <u>Nature</u>,
229:224.

Michael, R.P., Bonsall, R.E. and Warner, P., 1974, Human vaginal secretions: volatile fatty acid content, <u>Science</u>, 186:1217.

Schaller, G., 1967, "The deer and the tiger", Chicago University Press,
Chicago.

Schieberle, P., 1988, Einfluss von Prolinzusaetzen auf die Bildung des
Aromastoffes 2-acetyl-1-pyrrolin in der Weissbrotkruste, <u>Getreide.
Mehl. Brot.</u>, 42:334.

Smith, J.L.D., McDougal, C. and Miquella, D., 1989, Scent marking in
free ranging tigers, <u>Panthera tigris</u>, <u>Anim. Behav.</u>, 37:1.

EFFECTS OF EXOGENOUS TESTOSTERONE ON THE SCENT-MARKING AND AGONISTIC

BEHAVIORS OF WHITE-TAILED DEER

James R. Fudge[1], Karl V. Miller[1], R. Larry Marchinton[1],
Delwood C. Collins[2], and Thomas R. Tice[3]

[1]School of Forest Resources, University of Georgia, Athens
Ga.; [2]Veterans' Administration Medical Center and
University of Kentucky Medical Center, Lexington, Ky.; and
[3]Southern Research Institute, Birmingham, Ala.

INTRODUCTION

Male white-tailed deer (Odocoileus virginianus) exhibit signpost marking behavior during both the sexually active and quiescent periods (Marchinton et al. 1990, Ozoga 1989). During the breeding season, the primary signposts are rubs and scrapes. Rubs consist of small trees on which the buck rubs his antlers and forehead (Moore and Marchinton 1974). Atkeson and Marchinton (1982) identified seasonally active tubular apocrine sudoriferous glands in the forehead tissue of both male and female whitetails. Moore and Marchinton (1974) hypothesized that rubs could serve to delineate dominance areas within a buck's home range, even though such areas are not actively defended against subordinate males.

A scrape is an area of pawed ground located beneath an overhanging branch, which the buck scent-marks with various regions of his head. Bucks often urinate or rub-urinate in or near the scrape. Scrapes may serve to advertise a buck's dominance status to other bucks and does in a specific area (Miller et al. 1987a). Scrapes also may facilitate male-female communication, thereby assisting in mate selection (Marchinton and Hirth 1984). While most scrapes are made by dominant males (Miller et al. 1987a), bucks of all ranks will scent-mark overhead branches during both the reproductive and nonreproductive seasons (Marchinton et al. 1990).

Signpost marking tendencies differ between young, subordinate bucks and mature, dominant bucks. Scraping by yearling bucks is less frequent and occurs later during the rut than that by dominant, prime-age bucks (Miller 1985, Ozoga and Verme 1985). Also, yearlings make 50% fewer total rubs than older bucks (Ozoga and Verme 1985). Only mature dominant males made a substantial number of scrapes during a 4-year study involving 10 different groups of captive deer (Miller et al. 1987a). Rubbing frequency of dominants is higher and peaks earlier in the breeding season than that of lower ranking bucks (Johansen 1987, Marchinton et al. 1990).

Male white-tailed deer exhibit a circannual cycle of testosterone, with low levels during the sexually quiescent period and high concentrations during the fall breeding season (McMillin et al. 1974, Mirarchi et al. 1978). The frequency of agonistic interactions and marking

Chemical Signals in Vertebrates VI, Edited by R.L. Doty and
D. Müller-Schwarze, Plenum Press, New York, 1992

behaviors increase during this time of high testosterone (Hirth 1977, Marchinton et al. 1990) Bubenik and Schams (1986) found peak plasma testosterone concentrations of bucks 1.5-3.5 years of age to be less than those of bucks 4.5-9.5 years of age. In a study of captive whitetails, yearling and 2.5-year-old bucks had significantly lower serum testosterone concentrations than bucks 3.5 years of age and older (Miller et al. 1987a). In a sample of wild bucks, testosterone levels in yearlings did not exceed 1.0 ng/ml and significant differences were found in concentrations between 1.5, 2.5, and 3.5+ year-old bucks (Miller et al. 1987b).

High rates of rubbing, pawing, and scraping appear to be characteristic of dominant bucks (Ozoga and Verme 1985, Johansen 1987, Miller et al. 1987a). Miller et al. (1987a) suggested that testosterone influences aggressive behavior more than hierarchal position. Dominance appears to be affected by antler size, weight, age, and experience, in addition to testosterone levels. Testosterone is related directly to social aggression among red deer (Cervus elaphus) stags but only plays a permissive role in controlling breeding behavior (Lincoln et al. 1972). Mossing and Damber (1981) suggested that testosterone enhanced social aggressiveness in reindeer (Rangifer tarandus) bulls. Conversely, West and Nordan (1976) speculated that high testosterone levels might be brought on by dominance and aggression in male black-tailed deer (O. hemionus columbianus). Therefore, whether testosterone influences social rank and aggressive behaviors or social position influences testosterone levels is not completely understood. In this study, we investigated the effects of exogenous testosterone on agonistic interactions and scent-marking in captive white-tailed deer during the 1987 and 1988 breeding seasons.

METHODS

Behavioral data were obtained from study animals in 3 social groups, maintained in 0.3-ha enclosures at the School of Forest Resources, University of Georgia. Each group contained 3-4 bucks, 1-3 does, and 1-4 fawns. Additional deer, hereafter referred to as hormone profile subjects, were housed individually in 3x6-m stalls at the same facility and were used to monitor testosterone release from hormone implants. Behavioral observations were recorded during 2 breeding seasons, Year 1 (1987) and Year 2 (1988). Agonistic interactions and marking behaviors of adult males were recorded using the "all occurrences" method of Altmann (1974). The behaviors that were recorded were adapted from Johansen (1987). Observations were conducted during the first and last hours of daylight. Each group was observed 3x/week in random order. Weekly sociometric matrices of agonistic encounters were used to determine dominance hierarchies (Altmann 1974).

During Year 1, pre-implant behavioral observations of the 3 social groups began 23 Oct. 1987 and continued for 7 weeks. On 16 Dec., all bucks in each social group were chemically immobilized. Microencapsulated testosterone propionate prepared with poly(DL-lactide-co-glycolide) (Tice and Cowsar 1984) was implanted intramuscularly in the beta buck in each group. Although testosterone production rates in white-tailed deer are unknown, Whitehead and West (1977) reported mid-rut testosterone production in male caribou to be 10-13 mg/day. We assumed production in whitetails to be proportional by body weight to that of caribou (ca. 8 mg/day). Since we estimated an approximate 6 month release period, total testosterone implanted was 1.59 g. With a microcapsule core loading of 62.7%, microcapsule implant weight was 2.53 g. Immediately before implantation, pre-weighed microcapsules were mixed with 10 ml of sterile injection vehicle (2 wt % CMC/1 wt % Tween 80). The resulting slurry was injected intramuscularly into the deer's hindquarter. After implantation, behavioral observations were suspended for 1 week to allow recovery from immobilization and possible early pulsatile implant release. Behavioral sampling continued for 7 weeks and was terminated on 9 Feb. 1988.

Two hormone profile subjects, 1 buck and 1 doe, were implanted with testosterone propionate on the same day as the social groups and monitored to evaluate implant release. The deer were restrained physically and weekly blood samples were collected between 0800-1000 hours for 4 weeks before and 16 weeks after implantation.

In Year 2, observations began on 27 Sept. 1988. On 18 Nov., the alpha and gamma bucks in 2 social groups and the alpha buck in the third group were implanted with microencapsulated testosterone. We replaced testosterone propionate with testosterone in an effort to achieve more consistent release rates. Theoretical release rates were increased to 10 mg/day. Release duration during Year 2 was estimated at 90 days. Placebo microcapsules were implanted in all non-treatment males. A 1-week acclimation period during which no data were recorded followed implantation. Observations continued for 8 weeks until 3 Feb. 1989.

Two hormone profile subjects, 1 buck and 1 doe, were implanted with testosterone 6 weeks later than the social groups. Weekly blood samples were collected for 9 weeks before and 16 weeks after implantation. Concurrent samples were obtained from an additional untreated buck.

Serum testosterone concentrations were determined by radioimmunoassay as described by Mann et al. (1987). Intra-assay coefficients of variation were 4.0% (Year 1) and 8.0% (Year 2). Inter-assay coefficient of variation between years was 8.22%. Sensitivity was 0.21 ng/ml (Year 1) and 0.10 ng/ml (Year 2).

Periods of elevated testosterone for animals in the social groups were inferred from data obtained from the hormone profile subjects. In Year 1, observations were grouped into a pre-implant period (weeks 1-7) and a release period (weeks 8-14). In Year 2, observations were grouped into pre-implant (weeks 1-8), release (weeks 9-12), and post-release (weeks 13-16) periods. Mean occurrence of each behavior was calculated by period for each buck and analyzed in a 2-factor analysis of variance with repeated measures on 1 factor (periods) to determine significant behavioral changes (Winer 1962). Comparisons were made between alpha, beta, and gamma bucks among all periods, and between periods among all bucks. Interactions between period and hierarchical position also were analyzed to determine significant treatment effects. Means with significant F values were tested using the Newman-Keuls' method (Winer 1962).

RESULTS

In Year 1, the hormone profile subjects exhibited increased serum testosterone concentrations for 7 weeks following implantation (Fig. 1a). Peak serum testosterone concentrations in response to the implants were similar in magnitude to the normal rutting testosterone peaks observed prior to implantation. During the second year, the hormone profile subjects exhibited increased concentrations for ca. 4 weeks after implantation (Fig. 1b). There were no changes in dominance hierarchies after implantation, but a beta and gamma buck switched ranks during the post-release period in Year 2. Of the agonistic behaviors recorded, only threats, displacements, and sparring occurred at frequencies sufficient for statistical analysis.

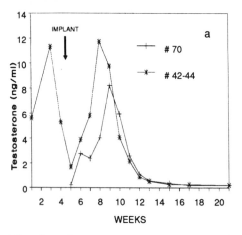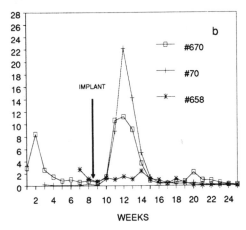

Fig. 1.   Serum testosterone concentrations of: a) a 2.5 year-old male (#42-44) and a 2.5-year-old female (#70) white-tailed deer during Year 1. Microencapsulated testosterone propionate was implanted on 16 Dec.; b) a 1.5-year-old male (#670) and a 3.5 year-old female (#70) white-tailed deer during Year 2. Microencapsulated testosterone was implanted on 3 Jan. Testosterone concentrations of a 2.5-year-old male (#658) that was not implanted are presented for comparison.

In Year 1, significant differences in the number of threats initiated/hour/period (Table 1) occurred among hierarchical positions (F = 9.43, P < 0.05, 2,6 df). Alpha males initiated more threats (P < 0.05, 6 df) than the implanted beta bucks and threats by gamma bucks were almost non-existent. Number of threats did not differ between periods. The number of displacements initiated/hour/period did not differ among hierarchical positions or between periods. Sparring matches initiated/hour/period did not differ among hierarchical positions. However, differences occurred between periods (F = 6.73, P < 0.05, 1,6 df) with the highest number occurring during release period (P < 0.05, 6 df).

Table 1. Mean (SEM) number of observed agonistic acts initiated/hour/period, pooled among hierarchical position, of male white-tailed deer during 2 breeding seasons (1987, 1988).

| | | | Number of acts/hour/period | | | | |
| | | | Year 1[a] | | Year 2[b] | | |
Agonistic Act	Rank	n	Pre-implant	Release	Pre-implant	Release	Post-release
Threat	1	3	0.7 (0.1)	0.7 (0.2)	0.7 (0.2)	1.7 (0.4)	1.8 (0.6)
	2	3	0.5 (0.2)	0.3 (0.1)	0.3 (0.1)	0.2 (0.1)	0.1 (0.1)
	3	3	0.0 (0.0)	0.1 (0.1)	0.0 (0.0)	0.0 (0.0)	0.0 (0.0)
Displacement	1	3	3.7 (1.5)	2.2 (0.3)	1.4 (0.4)	1.7 (0.4)	1.8 (0.7)
	2	3	2.9 (2.1)	1.5 (1.0)	0.7 (0.5)	0.4 (0.3)	0.4 (0.4)
	3	3	0.0 (0.0)	0.1 (0.1)	0.0 (0.0)	0.1 (0.1)	0.0 (0.0)
Spar	1	3	0.9 (0.4)	2.0 (0.9)	0.6 (0.5)	0.5 (0.3)	2.4 (1.1)
	2	3	1.2 (0.2)	2.1 (0.4)	0.7 (0.2)	0.6 (0.2)	0.8 (0.4)
	3	3	0.2 (0.1)	1.5 (0.9)	0.6 (0.3)	0.8 (0.1)	0.9 (0.6)

[a] beta bucks implanted with microencapsulated testosterone proprinate in Year 1
[b] alpha and gamma bucks implanted with microencapsulated testosterone in Year 2

Number of threats initiated/hour/period in Year 2 (Table 1) differed among hierarchical positions (F = 102.72, P < 0.01, 2,6 df) as did number of displacements (F = 6.02, P < 0.05, 2,6 df). Alpha bucks initiated the greatest (P < 0.05, 6 df) number of threats and displacements. Number of threats and displacements did not differ among periods. Sparring matches initiated/hour/period differed among periods (F = 3.87, P < 0.10, 2,12 df) but not among hierarchical position. Highest numbers were initiated during the post-release period (P < 0.05, 12 df).

During Year 1, no significant differences in the frequency of any marking behaviors were observed among hierarchal positions. However, absolute numbers of rubs, pawing, and rub-urination were substantially higher in alpha males (Table 2). Similarly, numbers of the various marking behaviors did not differ between periods except rub-urination which occurred more frequently prior to implantation (F = 4.29, P < 0.10, 1,6 df). Absolute numbers of all marking behaviors except overhead branch marking were higher before implantation.

During Year 2, differences among hierarchical positions occurred in the number of rubs/hour/period (F = 7.99, P < 0.05, 2,6 df), antler thrashing/hour/period (F = 9.33, P < 0.05, 2,6 df), pawing/hour/period (F = 4.71, P < 0.10, 2,6 df), scraping/hour/period (F = 3.88, P < 0.10, 2.6 df), and rub-urination/hour/period (F = 140.64, P < 0.01, 2,6 df). Highest numbers of these behaviors were observed in alpha males (Table 2). Overhead branch marking/hour/period did not differ among hierarchal position. Rates of any of the marking behaviors did not differ among periods although levels of some behaviors tended to be higher in implanted deer during the release period, especially in the gamma bucks.

Table 2. Mean (SEM) number of observed marking acts/hour/period, pooled among hierarchical position, of male white-tailed deer during 2 breeding seasons (1987, 1988).

| Behavior | Rank | n | Number of acts/hour/period | | | | | |
| | | | Year 1[a] | | Year 2[b] | | | |
			Pre-implant	Release	Pre-implant	Release	Post-release
Rubbing	1	3	1.6 (0.7)	0.9 (0.0)	1.0 (0.2)	0.9 (0.2)	1.0 (0.4)
	2	3	1.2 (0.6)	0.8 (0.2)	0.6 (0.2)	0.3 (0.1)	0.0 (0.0)
	3	3	0.4 (0.1)	0.4 (0.3)	0.7 (0.1)	1.0 (0.4)	0.4 (0.2)
Pawing	1	3	2.6 (1.3)	0.2 (0.1)	0.6 (0.4)	0.5 (0.4)	1.4 (0.7)
	2	3	0.1 (0.1)	0.4 (0.4)	0.1 (0.0)	0.0 (0.0)	0.0 (0.0)
	3	3	0.0 (0.0)	0.0 (0.0)	0.0 (0.0)	0.0 (0.0)	0.0 (0.0)
Branch marking	1	3	0.6 (0.3)	0.5 (0.5)	1.2 (0.4)	1.3 (0.5)	0.9 (0.4)
	2	3	0.6 (0.3)	0.6 (0.2)	0.8 (0.3)	0.7 (0.1)	0.4 (0.2)
	3	3	0.2 (0.1)	0.2 (0.2)	0.3 (0.2)	0.4 (0.1)	0.6 (0.2)
Full scrape	1	3	0.3 (0.2)	0.0 (0.0)	1.4 (0.7)	1.2 (0.4)	1.4 (0.9)
	2	3	0.2 (0.1)	0.1 (0.1)	0.1 (0.0)	0.0 (0.0)	0.0 (0.0)
	3	3	0.0 (0.0)	0.0 (0.0)	0.0 (0.0)	0.0 (0.0)	0.0 (0.0)
Rub urination	1	3	0.8 (0.4)	0.1 (0.1)	0.5 (0.2)	1.0 (0.2)	0.4 (0.2)
	2	3	0.1 (0.0)	0.1 (0.1)	0.2 (0.1)	0.2 (0.1)	0.1 (0.1)
	3	3	0.0 (0.0)	0.1 (0.0)	0.0 (0.0)	0.1 (0.0)	0.0 (0.0)
Antler thrash[c]	1	3	--	--	1.5 (0.5)	1.1 (0.2)	2.3 (1.2)
	2	3	--	--	0.0 (0.0)	0.0 (0.0)	0.0 (0.0)
	3	3	--	--	0.0 (0.0)	0.2 (0.0)	0.2 (0.1)

[a] beta bucks implanted with microencapsulated testosterone proprionate in Year 1
[b] alpha and gamma bucks implanted with microencapsulated testosterone in Year 2
[c] not recorded as a separate behavior in Year 1

DISCUSSION

Release periods from the microencapsulated hormone delivery system used in this study were shorter than expected. Nevertheless, implant-induced elevations of serum testosterone concentrations did occur, and appeared to mimic the magnitude and duration of peak testosterone concentrations during the normal rutting period.

Agonistic Behaviors

An interesting trend is evident when the numbers of threats initiated are compared between Year 1 and Year 2. Threat initiation rates of untreated alpha males in Year 1 were similar during both pre-implant and release periods. Similar rates also were exhibited by alpha males in Year 2 during the pre-implant period. However, during the release and post-release periods, the alpha bucks treated with testosterone initiated almost twice as many threats as in the pre-implant period. After implantation, beta bucks, both treated in Year 1 and untreated in Year 2, had a substantial decline in the number of threats initiated. Testosterone supplementation appeared to enhance the expression of this agonistic behavior in dominant bucks, but had little effect on subordinates. Gamma bucks, whether treated or untreated, exhibited low threat initiation rates.

Comparison of the mean number of sparring matches initiated did not reveal any clear relationships with either hierarchical position or testosterone levels. Although recorded as an agonistic behavior in this study, sparring often functions more as a ritualized behavior to assess the fighting ability of other males (Barrette and Vandal 1990). Dominant bucks appear tolerant of younger bucks in sparring matches, allowing subordinates to gain social experience in a relatively low risk agonistic encounter (Koutnik 1983). Thus sparring initiation may be less dependent on testosterone or hierarchal position than threats or displacements.

Marking Behaviors

Bowyer and Kitchen (1987) stated that antler thrashing in cervids is a dominance display. Recorded as a marking behavior in our study, antler thrashing also was performed primarily by dominant bucks. Within the alpha position, exogenous testosterone seemed to enhance the expression of antler thrashing, but had no effect on this behavior in subordinate males.

Testosterone implants had little effect on rubbing or overhead branch marking rates. These marking behaviors appeared to be influenced most by hierarchical position, as dominant males marked at higher rates than subordinates. Overhead branch marking occurs year-round and likely serves as a means of individual recognition among bucks but also may assist in mate selection during the breeding season (Marchinton et al. 1990).

Scraping likely is influenced primarily by dominance rank. However, expression of this behavior in dominant bucks appears to be related to testosterone as implanted dominant bucks exhibited enhanced or prolonged expression of scraping. Scraping by dominant bucks normally peaks in October and November and declines rapidly to very low levels in December and January (Miller et al. 1987a, Marchinton et al. 1990). Implanted alpha bucks in our study continued scraping at high levels through the end of the study in mid-January. Supplemental testosterone had no apparent effect on scraping by beta and gamma bucks. These results support the contention of Miller et al. (1987a) that testosterone levels, social position, and age/experience interact to promote the expression of scraping behaviors.

Both rub-urination and aggressive pawing by untreated dominant bucks in Year 1 decreased during the release period, while rates of these behaviors by treated dominants in Year 2 increased substantially following implantation. Beta and gamma bucks showed little change in rub-urination or pawing rates in either year, regardless of whether they were implanted or not. Thus rub-urination and aggressive pawing apparently are influenced primarily by dominance rank. Expression of these behaviors within the alpha position appears related to testosterone levels. Supplemental testosterone had no apparent effect on rub-urination or aggressive pawing in beta and gamma bucks.

The combination of high testosterone levels and high dominance rank was associated with higher levels of certain agonistic (threats, displacements) and scent-marking (antler thrashing, pawing, rub-urination, and scraping) behaviors. Bucks of lower social rank and correspondingly low testosterone levels appeared to lack this predisposition toward high rates of agonistic initiation and scent marking. Testosterone supplementation in these low-ranking bucks did not influence expression of these behaviors. Therefore, the enhancing effect of testosterone seemed to be restricted to the highest ranking buck in a hierarchy.

Our results are contrary to those of Lincoln et al. (1972) who reported that testosterone supplementation of subordinate red deer stags can result in increased social aggressiveness, although this response may be dose-dependent. Our social groups were much smaller than those observed by Lincoln et al. (1972) and consequently hierarchies were more stable. Strict hierarchies and the forced association of individuals in our enclosures probably resulted in behavioral suppression of the subordinates by the dominants. This suppression may have overridden any potential testosterone-induced aggressiveness in subordinates.

In conclusion, testosterone appears to enhance certain agonistic and scent-marking behaviors in dominant individuals (i.e., threats, displacements, antler thrashing, pawing, rub-urination, and scraping). Supplemental testosterone had no detectable effect on these behaviors in subordinate bucks. Other marking behaviors (antler rubbing and overhead branch marking) were most directly related to hierarchical position and were not affected by supplemental testosterone in dominant or subordinate individuals. Among all deer, scent marking appears to be influenced primarily by hierarchical position and not serum testosterone concentrations. Within a social level, individual variations in marking frequency may be influenced by testosterone concentrations as well as other factors.

LITERATURE

Altmann, J. 1974, Observational study of behavior: sampling methods. Behaviour, 49:227-267.
Atkeson, T. D., and Marchinton, R. L. 1982, Forehead glands in white-tailed deer. J. Mammal., 63:613-617.
Barrette, C., and Vandal, D. 1990, Sparring, relative antler size, and assessment in male caribou. Behav. Ecol. Sociobiol., 26:383-387.
Bowyer, R. T., and Kitchen, D. W. 1987, Significance of scent-marking by Roosevelt elk. J. Mammal., 68:418-423.
Bubenik, G. A., and Schams, D. 1986, Relationship of age to seasonal levels of LH, FSH, prolactin, and testosterone in male white-tailed deer. Comp. Biochem. Physiol., 83:179-183.

Hirth, D. H. 1977, Social behavior of white-tailed deer in relation to habitat. Wildl. Monogr., No. 53. 55 pp.

Johansen, K. L. 1987, "Seasonal Variation in Marking Behavior of White-tailed Deer", M. S. Thesis, Univ. Georgia, Athens. 53 pp.

Koutnik, D. L. 1983, The role of ritualized fighting behaviour in the social system of California mule deer. Biol. Behav., 8:81-93.

Lincoln, G. A., Guinness, F., and Short, R. V. 1972, The way in which testosterone controls the social and sexual behavior of the red deer stag (Cervus elaphus). Horm. Behav., 3:375-396.

Mann, D. R., Free, C., Weloon, C., Scott, C., and Collins, D. C. 1987, Mutually independent effects of ACTH on LH and Testosterone secretion. Endocrinology, 120:1542-1550.

Marchinton, R. L., and Hirth, D. H. 1984, Behavior. Pages 129-168. in: "White-tailed deer: ecology and management". L. K. Halls, ed., Stackpole Books, Harrisburg, Pa.

Marchinton, R. L., Johansen, K. L., and Miller, K. V. 1990, Behavioural components of white-tailed deer scent marking: social and seasonal effects. "Chemical Signals in Vertebrates V", D. W. Macdonald, M. Muller-Schwarze, and S. E. Natynczuk, eds., Oxford Univ. Press, Oxford.

McMillin, J. M., Seal, U. S., Kennlyne, K. D., Erickson, A. W., and Jones, J. E. 1974, Annual testosterone rhythms in the adult white-tailed deer (Odocoileus virginianus borealis). Endocrinology 94:1034-1040.

Miller, K. V. 1985, "Social and Biological Aspects of Signpost Communication in White-tailed Deer", Ph.D. Thesis, University of Georgia, Athens. 114 pp.

Miller, K. V., Marchinton, R. L., Forand, K. J., and Johansen, K. L. 1987a, Dominance, testosterone levels, and scraping activity in a captive herd of white-tailed deer. J. Mammal., 68:812-817.

Miller, K. V., Rhodes, O. E., Jr., Litchfield, T. R., Smith, M. H., and Marchinton, R. L. 1987b, Reproductive characteristics of yearling and adult male white-tailed deer. Proc. Ann. Conf. Southeast. Assoc. Fish Wildl. Agencies, 41:378-384.

Mirarchi, R. E., Howland, B. E., Scanlon, P. F., Kirkpatrick, R. L., and Sanford, L. M. 1978, Seasonal variation in plasma LH, FSH, prolactin, and testosterone concentrations in adult male white-tailed deer. Can. J. Zool., 56:121-127.

Moore, W. G., and Marchinton, R. L. 1974, Marking behavior and its social function in white-tailed deer. Pages 447-456 in "The Behaviour of Ungulates and Its Relation to Management". V. Geist and F. Walther, eds. IUCN Publ. 24. Morges, Switzerland.

Mossing, T., and Damber, J. 1981, Rutting behavior and androgen variation in reindeer (Rangifer tarandus L.). J. Chem. Ecol., 7:377-389.

Ozoga, J. J. 1989, Temporal pattern of white-tailed deer scraping behavior. J. Mammal., 70:633-636.

Ozoga, J. J., and Verme, L. J. 1985, Comparative breeding behavior and performance of yearling vs. prime-age white-tailed bucks. J. Wildl. Manage., 49:364-372.

Tice, T. R., and Cowsar, D. R. 1984, Biodegradable controlled-release parenteral systems. Pharmaceut. Technol., 26-35.

West, N. O., and Nordan, H. C. 1976, Hormonal regulation of reproduction and the antler cycle in the male Columbian black-tailed deer (Odocoileus hemionus columbianus). Part I. Seasonal changes in the histology of the reproductive organs, serum testosterone, sperm production, and the antler cycle. Can. J. Zool., 54:1617-1636.

Whitehead, P. E., and West, N. O. 1977, Metabolic clearance rates of testosterone at different times of the year in male caribou and reindeer. Can. J. Zool., 55:1692-1697.

Winer, B. J. 1962, "Statistical Principles in Experimental Design. McGraw-Hill Inc., New York. 672 pp.

FIELD STUDIES OF CHEMICAL SIGNALLING: DIRECT OBSERVATIONS OF DWARF

HAMSTERS (PHODOPUS) IN SOVIET ASIA

Katherine E. Wynne-Edwards[1], Alexei V. Surov[2] and Alexsandra
Yu. Telitzina[2]

[1]Biology Dept. Queen's Univ., Kingston, Canada K7L 3N6
[2]Institute of Animal Evolutionary Morphology and Ecology
Russian Academy of Sciences of the USSR, Moscow 117071

INTRODUCTION

Chemical signals are vital sources of information in the lives of
most, if not all, vertebrates. Considerable success has been achieved
in experimentally determining the context, regulatory mechanisms and
biochemical bases of vertebrate chemical signalling, particularly in
rodent species. While the microtine and cricetine rodents are amenable
to rearing under laboratory conditions, in the wild the majority are
small, cryptic, nocturnal and virtually impossible to study by direct
observation. What is missing is an integrated picture of the use of
scent marking by free-ranging individuals, and the perception of those
signals by other individuals at later times. This study reports the
results of a direct, observational, field study of scent marking
behavior in wild, free-ranging population of a small, nocturnal rodent -
the Djungarian hamster, Phodopus campbelli.

Several aspects of the ecology and behavior of P. campbelli make
it possible for us to conduct this observational field study. First,
the habitat is very open, so that a small, 20-30 g, hamster is visible.
Only a few habitats such as deserts and arid short grass steppe are open
enough to allow direct observation of small mammals. Second, the
hamsters do not avoid the bright lights necessary to follow them at
night, and will continue to eat, undisturbed, while a human observer is
lying in the grass less than 20 cm away. Third, they are slow. The
maximum running speed of a dwarf hamster is a human's walking stride.
The hind foot in Phodopus has relatively short phalanges, differing from
other genera of the Muridae and contributing to the slow running speeds.
In the same habitat other species, of the same or larger size, are
significantly faster and are impossible to follow continuously by foot.
Fourth, population densities are extremely low. Stable, low population
densities of less than 0.25 individuals/hectare in suitable habitat
(less than 4% change over 15 year trapping records (Flint, 1966)) are
rare in rodent populations (Hayward and Phillipson, 1979; Rogovin,
Shenbrot and Surov, 1991), and provide the opportunity to combine the
techniques of focal individual observations with sampling of the entire
local population.

Chemical Signals in Vertebrates VI,   Edited by R.L. Doty and
D. Müller-Schwarze, Plenum Press, New York, 1992

At least six discrete sources of chemical secretions are involved
in dwarf hamster scent marking behavior in the wild. Adults have a
ventral sebaceous gland (Reasner and Johnston, 1987), a musk gland in
the corner of the mouth which marks items carried in the cheek pouches,
Harderian glands, skin glands behind the ears, urine and feces. In
females, vaginal secretions change in composition, and vaginal scent
marks change in frequency, throughout the female cycle (Wynne-Edwards
and Lisk, 1987a). Laboratory studies have considered the context of
these scent marks during male-female interactions (Wynne-Edwards and
Lisk, 1987a), female-female and male-male interactions (Wynne-Edwards
and Lisk, 1987b), in response to conspecific odors (Reasner and Johnston, 1987) and in multiple-male competitive mating situations (Wynne-
Edwards and Lisk, 1988). Adult male puberty acceleration, sibling
female delay of vaginal opening (Levin and Johnston, 1979) and an
androgen-dependent urinary chemosignal which accelerates reproductive
development in females (Reasner and Johnston, 1988) have also been
demonstrated in P. campbelli. Pre-implantation pregnancy block occurs
in multiple-male mating situations (Wynne-Edwards and Lisk, 1984) and
when the mate is unfamiliar or is removed soon after mating (Wynne-
Edwards et al., 1987). Pregnancy block appears to be a response to any
disruption of the relationship with the male which could decrease
paternal investment and pup survival (Wynne-Edwards, 1987; Wynne-Edwards
and Lisk, 1989). Chemical signalling is intimately involved in social
communication and female reproductive function in Djungarian hamsters.

METHODS

     Studies were conducted at a field site near the town of Erzin
(50.16N, 95.14E) in the Tuva Autonomous region of the Russian Republic
of the Soviet Union. Tuva consists of a relatively arid, high altitude
(approx. 1000m) basin surrounded by mountains. Eight kilometers to the
south of the research area is the international border with the Mongolian People's Republic. Human population densities are very low, and
our only immediate neighbor lives in a traditional, circular, horsehair
yurt. His livelihood and the majority of the ethnic Tuvinian diet
depends on the sheep they herd. The yurts are moved a short distance
each year (about 0.5 km) and there is no human presence in the area from
late August through mid-June. Cattle and Mongolian ponies also range
through the area, but all livestock is penned each night. Our camp is
on the shore of a spring-fed freshwater lake, Terye Xol, which extends
across the international border into the Mongolian People's Republic.
The habitat is characterized by open sand dunes and sand dunes stabil-
ized by Artemisia and Potentilla steppe. Macro-vegetation is dominated
by thorny Caragana bushes ranging in diameter from 0.2 to over 3 m.

     Hamsters were captured through various methods including cylinder
traps, live traps baited with millet soaked in sesame oil and excavation
of burrows. However, the majority of animals in these studies were
captured by hand as they entered or left the burrow of a focal hamster.
Immediately after capture, each individual received an intraperitoneal
implant of a paraffin sealed, L.L. Electronics, radio-transmitter in the
150-151 MHz range. All surgery was conducted under Ketamine (75 mg/kg)
+ Acepromazine (7.5 mg/kg) anesthesia. Animals were kept warm during
recovery and were released either 2 h post-surgery (evening and night
captures) or at dusk the following evening.

     Telemetry signals were used to confirm the presence of an individ-
ual within a burrow, to scan for signals from other hamsters within a
burrow and to immediately relocate individuals which were momentarily
lost from sight in the vegetation. All spatial and behavioral data were
collected by direct observation of the individual

hamster.   Observers worked in teams of two with one individual responsible for following the hamster with a flashlight and recording data
directly into a voice-activated Olympus Pearlcorder with digital
Time/Date encoding.   The second observer was responsible for the
accurate placement of numbered flags, the telemetry receiver/antenna and
spare supplies including batteries etc.   Numbered flags were mapped the
following day.   Observations were made from a distance of less than 2 m
although the observer could approach the focal animal to within 20 cm to
determine the type of food being consumed.   At distances of 1 m or more,
the following observer would remain upright.   In closer proximity, the
observer would crawl or lie down.   Behaviors recorded were four forms of
scent marking, four forms of grooming, five different travelling speeds,
foraging, collecting food into the cheek pouches, eating, social
encounters and mating.

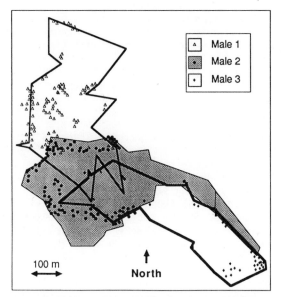

Figure 1.   Schematic representation of the areas visited by each of
            three adult males (polygons) during 1988 showing the approx-
            imate location of ventral gland scent marks (symbols) made
            by each.   Further explanation is in the text.

RESULTS

     Data reported here were collected July 23 to August 24, 1988 (64
hamster nights), July 30 to August 26, 1989 (66 hamster nights) and July
27 to August 12, 1990 (28 hamster nights) and concern the period from
0.5 h before dusk to 0.5 h after dawn.   No scent marking behavior was
observed during daylight hours.

Spatial Distribution of Ventral Gland Marks by Adult Males: Figure 1
shows the total area visited by each of three neighboring adult males
during 1988.   Also shown are the approximate locations of ventral marks
by each male (N=95, 96, and 33, for 10 , 15, and 5 nights of observation

Table 1.   Distribution (% of observed) relative to the substrate of
            scent marks by Raisa and Barbara, two adult females, during
            1990.

Individual	N	Sand	Open Grass	Tussock Grass	*Caragana* Bush	Burrow	Not Recorded
Raisa	445	22.3	15.7	29.4	25.6	5.9	1.1
Barbara	109	11.0	11.9	13.8	52.3	9.2	1.8

respectively). Virtually all foraging and scent marking behaviors were
restricted to non-overlapping ranges, and concentrated near the periphery of those ranges. During mating, males occasionally marked near the
estrous female and outside their usual scent marking boundary. A total
of five females maintained non-overlapping ranges within the map area.
Three had sleeping burrows in the area which all three males visited.
One was visited by just males 6 and 14 and the last was visited only by
male 4. All expeditions by males beyond their own scent marking
boundaries were associated with visits to female burrows.

Context of Adult Female Scent Marks: During 1990, two adult females,
Raisa and Barbara, with non-overlapping home ranges, were followed for
15 and 8 consecutive nights (10,426 and 7,106 min). Raisa and Barbara
were active above ground for 3,633 and 1,258 min respectively during the
observations. A total of 11,490 and 5,002 behaviors were recorded (3.2
and 4.0 behaviors/min) during that time, including 445 (3.9%) and 109
(2.1%) scent marks.

Substrate. Within the study area, the substrate could be classified
into one of five classes ranging from open sand to the area beneath
Caragana bushes (Table 1). The two females differed significantly (X^2 =
36.35, p$\leq$ 0.001) in the use of the substrates chosen for marking,
although each home area was approximately 3.5 ha in size and contained
1,050 - 1,100 bushes. However, for each female, the two substrates
which provided cover (bushes and tussock grass) were used most frequently (55% and 66.1%) as marking sites.

Temporal pattern. Figure 2 shows the temporal distribution of scent
marking by each female on an hourly basis throughout the night.
Observations continued 24 h per day, but neither long distance excur-
sions nor scent marks were observed during daylight hours. Each female
showed a peak in marking frequency between 10 and 11 pm reflecting a
consistent pattern in which the first expedition of the night is the
longest in duration and the most likely to involve distant travel
(unpublished data). When the rate of scent marking is calculated based
on the expedition duration, Raisa marked at a high rate of 0.14 $\pm$ 0.03
marks/min over the first two expeditions each night (N=25) compared with
0.05 $\pm$ 0.01 marks/min (N=23: t_{46} = 2.77, p $\leq$ 0.01) on later expeditions.
With the lower activity level for Barbara, no significant effect on
marking rate was observed.

Behavioral context. Behaviors recorded immediately (approx. 30 sec)
prior to a scent mark, and immediately following a scent mark, were
examined for Raisa and Barbara (Figure 3). For the analysis, behaviors
were grouped into seven categories: Travel (movement at any speed), Eat
(including manipulation of food, collecting food into the cheek pouches
and caching/cache recovery), Groom (including genital grooming, face
washing, ear scratching, flank scratching, etc.), Sniff (of the ground,

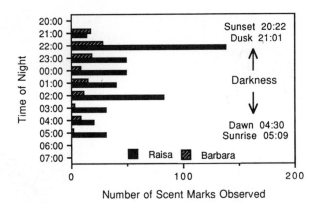

Figure 2. Temporal pattern of scent marking relative to the time of night. Times for sunrise, sunset, dusk and dawn are estimates for the appropriate latitude and longitude on August 1. Data are from Raisa and Barbara during 1990.

vegetation or air), Mark (rolling, vaginal, ventral and urine marks), Alert (with 0, 1 or 2 paws raised) and Dig (usually in the context of collecting food). Figure 3A depicts the frequency distribution of all behaviors recorded during the study. Behavior scores were remarkably similar for the two females with Travel (40.6 vs 39.6%), Eat (19.9 vs 19.8%) and Groom (12.9 vs 11.1%) accounting for over 70% of behaviors. The females differed in some characteristics. Raisa hunted underground beetle larvae while Barbara did not, resulting in significantly higher Dig scores. Barbara reacted to the human observers more often than Raisa, resulting in significantly higher Sniff and Alert scores. Marking behavior tended to occur in bouts, resulting in significant elevation of Prior and Following Mark frequencies compared to be overall sample. Sniff scores Prior and Following Mark for both females were also higher than the overall sample. Raisa frequently marked on departure from a digging site, resulting in high Prior Dig scores and high Following Travel scores. In both females, Groom and Mark had significantly higher scores Prior to Mark than Following Mark (All differences $p \leq 0.01$; Binomial Test).

DISCUSSION

Djungarian hamsters have proven amenable to direct, observational field study. This overcomes the need to infer the behavior of an animal from indirect evidence and provides data which can serve as a reference in the critical design and interpretation of definitive laboratory studies.

The spatial distribution of ventral gland marks by adult males appears to delineate non-overlapping areas of space use with scent marks concentrated near boundaries with conspecific neighbors. The precision with which these boundaries were marked suggests that they are serving a role in male-male communication and probably territorial defense.

Adult females distribute scent marks across all substrates at the site, but appear to concentrate them under bushes and under tussock grass in areas which provide cover. This may increase the effective

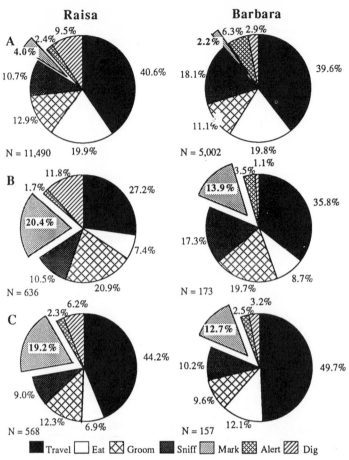

Figure 3. Frequency distributions of behavior by Raisa and Barbara, two adult females, during 1990. A = Overall distribution of behaviors recorded during the study for each individual, B = Behavior frequencies immediately prior to Mark behavior, C = Behavior frequencies immediately following Mark behavior. See text for definitions of behavior categories and results of comparisons.

lifetime of the scent marks by protecting them from dilution by rainfall. The majority of scent marking occurs early in the active period, during the first and second expeditions of each night. This results from increased scent marking rates per minute of activity as well as increased activity during that period of the night.

Sample sizes were large enough to permit preliminary analyses of the behavioral contexts in which scent marking by adult females occurs. Several behaviors, notably grooming and additional marking, occur more frequently immediately surrounding scent marking behavior than in the overall sample. The ritualized pattern of rodent grooming incorporates the majority of chemical signals produced by an individual from the ground while travelling and through rolling. We hypothesize that this ritualized pattern, in combination with a genus (Phodopus) defining character, densely hairy feet, results in the deposition of a unique, individual scent with each step. The elevated incidence of Groom prior to Mark suggests that this perfume may be added to each type of scent mark and may provide additional information useful in navigation or communication. Current experiments are designed to test the contextual perception of naturally occurring scent marks by familiar and unfamiliar conspecifics. Analogous to auditory playbacks used to elucidate the critical components and information content of bird song, these field experiments will present odors from familiar or unfamiliar individuals in familiar or unfamiliar locations and monitor the behavioral responses of resident individuals to them.

ACKNOWLEDGEMENTS

The authors gratefully acknowledge the financial support of Earthwatch - The Center for Field Studies, the Russian Academy of Sciences, the Soros Foundation - Russia, and additional support to K.E. Wynne-Edwards from NSF (DCB 8702298) and NSERC (Canada).

REFERENCES

Flint, W.E., (1966), Die Zwerghamster der Paläarktischen Fauna. Ziemsen-Verlag, Wittenberg Lutherstadt p 38. [In German].

Hayward, G.F. and J. Phillipson, 1979, Community structure and functional role of small mammals in ecosystems, In: Stoddart, D.M. Ed., "Ecology of Small Mammals," Halsted Press, John Wiley & Sons Inc., New York pp 135-211.

Levin, R.N. and R.E. Johnston, 1986, Social mediation of puberty: An adaptive female strategy?, Behav. Neur. Biol., 46:308-324.

Reasner, D.S. and R.E. Johnston, 1987, Scent marking by male dwarf hamsters (Phodopus sungorus campbelli) in response to conspecific odors, Behav. Neur. Biol., 48:43-48.

Reasner, D.S. and R.E. Johnston, 1988, Acceleration of reproductive development in female Djungarian hamsters by adult males, Physiol. Behav., 43:57-64.

Rogovin, K., G. Shenbrot and A. Surov, 1991, Analysis of spatial organization of a desert rodent community in Bolson de Mapimi, Mexico, J. Mamm., 72:347-359.

Wynne-Edwards, K.E., 1987, Evidence for obligate monogamy in the Djungarian hamster, Phodopus campbelli: Pup survival under different parenting conditions, Behav. Ecol. Sociobiol., 20:427-437.

Wynne-Edwards, K.E., U.W. Huck and R.D. Lisk, 1987, The influence of pre- and post-copulatory pair contact on reproductive success in the Djungarian hamster, J. Reprod. Fert., 80:241-249.

Wynne-Edwards, K.E. and R.D. Lisk, 1984, Djungarian hamsters fail to conceive in the presence of multiple males, Anim. Behav., 32:626-629

Wynne-Edwards, K.E. and R.D. Lisk, 1987a, Male-female interactions across the female estrous cycle: A comparison of two species of dwarf hamster (Phodopus campbelli and Phodopus sungorus), J. Comp. Psychol., 101:335-344.

Wynne-Edwards, K.E. and R.D. Lisk, 1987b, Behavioral interactions differentiate Djungarian (Phodopus campbelli) and Siberian hamsters (Phodopus sungorus), Can. J. Zool., 65:2229-2236.

Wynne-Edwards, K.E. and R.D. Lisk, 1988, Differences in behavioral responses to a competitive mating situation in two species of dwarf hamster (Phodopus campbelli and Phodopus sungorus), J. Comp. Psychol., 102:49-55.

Wynne-Edwards, K.E. and R.D. Lisk, 1989, Differential effects of paternal presence on pup survival in two species of dwarf hamster (Phodopus sungorus and Phodopus campbelli), Physiol. Behav., 45:49-53.

"FRAGRANCE ON THE DESERT AIR": THE SEMIOCHEMISTRY OF THE MUSKOX

Peter F. Flood

Department of Veterinary Anatomy, Western College of Veterinary
Medicine, University of Saskatchewan, Saskatoon, Saskatchewan
Canada, S7N 0W0

INTRODUCTION

Muskoxen (_Ovibos moschatus_) are the largest herbivores of the high
Arctic and their range extends from the northern coasts of Greenland and
Ellesmere Island (83°N), the most northerly land area in the world, to as
far south as the Thelon Game Sanctuary (64°N) (Lent, 1988). They are normally inhabitants of the open tundra though there are records of muskoxen
being found south of the tree line (Barr, 1991) and the Pleistocene
muskoxen of Europe apparently occupied forested and steppe environments as
well as the glacial margins (Crégut-Bonnoure, 1984). The genus _Ovibos_
probably first appeared about one million years ago in the Middle
Pleistocene of central Asia though the evidence for this not particularly
secure (Crégut-Bonnoure, 1984).

While muskoxen are taxonomically allied to sheep and goats (the
Caprinae), there is general agreement that their only close relative is the
takin (_Budoras taxicolor_) (Geist, 1984). The confusion that still sur-
rounds the relationship of the muskox and takin to the more typical members
of the Caprinae is exemplified by the fact that Simpson (1945) placed these
two species in the tribe Ovibovini while Haltenorth (1963) gave them their
own subfamily, the Ovibovidae. Despite this uncertainty, it is clear that
the series of Robertsonian translocations that accompanied the evolution of
sheep from its ancestral stock was quite different from the series that occurred in the muskox and takin (Heck, Wurster and Benirschke, 1968).

Known muskox populations were reduced to very low levels in the early
part of this century, probably because of a combination of hunting and ad-
verse climatic conditions (Barr, 1991) but the species is now very successful and roughly 100,000 muskoxen are widely distributed in Greenland and
the Canadian Arctic. Introduced populations are now thriving in several
locations, including their historic range in Alaska (Smith, 1989; Reynolds,
1989), their prehistoric range in Siberia (Yakushkin, 1989), and Northern
Quebec (Le Hénaff and Crête, 1989) where they have never previously ex-
isted.

Large male muskoxen weigh about 350 kg but females are only 60% of
this weight. They are relatively non-selective feeders that eat a wide variety of plants including sedges and dwarf willows, but make no particular
use of mosses or lichens (Hofmann, 1988; Tener, 1965). They seem to have
an excellent ability to digest poor quality forage (Chaplin, 1984).

Muskoxen are not territorial (Tener, 1965) and do not show the elaborate migratory pattern seen in some caribou, though they may have distinct

Chemical Signals in Vertebrates VI, Edited by R.L. Doty and
D. Müller-Schwarze, Plenum Press, New York, 1992

summer and winter ranges (Tener, 1965; Gray, 1987). They are normally found in herds comprising a dominant bull, females of various ages and subordinate males. The herds vary in size but average around ten animals and may be larger in winter when loose associations of over 100 have been reported (Tener, 1965). Mature bulls that have failed to acquire a herd of their own may be found alone or in small groups. Such bulls have been credited with pioneering new territory (Smith, 1989).

A distinct social hierarchy develops at an early age (Reinhardt and Flood, 1983). Dominant bulls attempt to reassert their position during the early part of the rut through characteristic dominance displays and, when this fails, dramatic head clashing (Gray, 1984). Threats from predators, on the other hand, elicit the use of the defense formation, perhaps the best known trait of this species (Gray, 1974). In reality this is an untidy huddle rather than the orderly circle sometimes portrayed.

Muskoxen are seasonal breeders, and under favorable conditions they produce a calf annually (Latour, 1986) but reproduction may be suspended when food is short (Vibe, 1967; Miller, 1989). The rut occurs from mid-July to early October (Gray, 1987) and most females probably conceive in late August and early September. The gestation length of young females in Saskatoon was found to be a little under eight months (Rowell and Flood, 1989). In the wild birth is normally in late April or May. This creates an unusual situation in which the energetically demanding period of early lactation coincides with a time of minimal food availability (Tener, 1965).

There are at least two reasons for curiosity about chemical communication in the muskox. The first lies in the name which is usually seen a doubly inappropriate, the muskox is not an ox and usually it smells of little other than a general ruminant odor; no doubt early explorers most commonly encounter muskoxen during the late summer breeding season so they were not unnaturally impressed by the very obvious rutting odor of the male. The second reason is that muskoxen possess large and sexually dimorphic preorbital glands that can hardly be without a function. Contrary to some accounts (Geist, 1984), muskoxen do not have any specialized pedal glands though ordinary apocrine sweat glands are widely distributed over the body (Flood, Stalker and Rowell, 1989).

Many of the observations to be described here were made on a herd of 13 - 19 muskoxen kept for the last nine years at the Western College of Veterinary Medicine, Saskatoon, Saskatchewan. The herd, which is maintained entirely for research on the ecophysiology of the species, was derived from thirteen calves captured on Banks Island (Flood et al, 1984). The animals are handled extensively and, with the exception of the rutting males, they have remained tractable and tolerate minor experimental interference without undue distress. The muskoxen are kept in a series of interconnecting pens totalling 3 hectares and have free access to water and good quality hay. They are also fed 400 g daily of a pelleted ration consisting primarily of oats; this is unnecessary from a nutritional point of view, but it forms an essential part of the reward system incorporated into the daily handling routine.

A detailed account of muskox behavior can be found in Gray (1987) which includes a distillation of more than 1000 days of observations made at all seasons of the year at Polar Bear Pass, Bathurst Island, Northwest Territories. The summary that follows is largely, but not entirely, derived from two previous papers (Gray, Flood and Rowell, 1989; Flood et al, 1989).

THE RUTTING ODOR

The origin of the rank and pungent odor of the rutting male has never been fully explained, though Teal (1959) recognized that it was associated with urine and was most noticeable in dominant bulls. Our regular observations of captive bulls during the breeding season allowed us to confirm that it is the dominant male who becomes, from the human point of view, malodorous. We were also able to make a detailed study of the superiority display known as "head-tilting" (Gray, 1987) in which a bull walks slowly past his rival with his head tilted to draw attention to the size of the horn boss. While doing this he adopts a curious stiff-legged gait that emphasizes the height of the shoulder. Close inspection during the display revealed that the prepuce was everted to form a pendulous tube about 12 cm in length, tipped with a fringe of matted hair. The preputial tube swung about owing to the staccato rhythm of the walk and simultaneously, urine dribbled from its opening. As a result, the abdominal portion of the skirt (the long, flowing, outer hairs that give the animal its characteristic appearance) became soaked with urine. Subsequent bacterial degradation of the urine may transform and enhance the odor. Subordinate males did not perform the superiority display.

In the quiescent state the preputial tube was retracted to the level of the abdominal wall as in other ruminants and the hair, previously visible at its tip, was drawn inside. Dissection of the prepuce of a 3.5-year-old bull and a 4-year-old castrate showed that about 8 cm of the prepuce was lined with fine, crimped hair that was up to 3 cm in length. It covered the entire circumference of the preputial cavity and was encrusted with pale caseous particles. Histological examination of the prepuce of 3 calves less than 2 weeks old showed that the hairs were present at that age and that their follicles were directed proximally in the more distal part of the prepuce and distally in the deeper part. Sebaceous glands were associated with the hair follicles but no sweat glands could be found.

The nature of the odorants contained in or formed from the urine that is deposited on the coat are not know in detail. An analysis of smegma (Teal, 1961, cited by Tener, 1965) indicated that cinnamaldehyde might be an important constituent and we found large amounts of benzoic acid and pcresol in the preputial flushings of bulls (Flood et al, 1989).

Similar urine sprinkling exercises are performed by other species and in particular by goats, but in this case it is the head, neck and beard that are soaked with urine (Coblentz, 1976). In goats, sprinkling is facilitated by the urethral process of the male which is a delicate, filiform, tubular extension of the urethra beyond the tip of the penis proper. It seems to have evolved as a urine applicator and is not essential for fertility. The muskox, having adopted a different technique for application of urine, has no urethral process.

It is not clear whether the smell of the dominant muskox in rut is entirely attributable to aspects of his behavior or whether there are chemical changes in the urinary constituents. It would appear from assays of androgen concentrations in the feces of muskoxen throughout the year that androgen excretion by the dominant male is much greater than by his subordinates (P.F. Flood, J. Thrush and N.C. Rawlings, in preparation). Androgens, while they may affect behavior, are also likely to be excreted in the urine (Albone, 1984) and may influence the excretion of other volatile products.

THE PREORBITAL GLANDS

The preorbital glands are well developed in muskoxen (Brinkman, 1911, cited by Schaffer, 1940; Lönnberg 1900; Sack and Ballantyne, 1965; Gray et al, 1989). They are pear-shaped structures that weigh around 7 g in females and probably up to 20 g in adult males (Gray et al, 1989). The opening of the gland, which represents the pointed end of the pear, is 2-3 mm in diameter and lies about 6 cm rostral to the medial canthus of the eye. From the opening a skin-lined, tubular invagination penetrates the gland almost to its base. Hairs arising from follicles in the wall of the invagination largely fill the lumen and contribute to the small tuft of hair that lies over the opening on the surface of the face (Fig. 1).

The hair follicles, as usual, give rise to both sebaceous glands and apocrine sweat glands. The sebaceous glands are relatively unspectacular but the peripherally situated sweat glands are huge and constitute the great bulk of the organ. The histological appearance of these secretory tissues does not seem to show any special features but the extent of the apocrine tissue is impressive. Electron microscopy reveals a well developed system of myoepithelial cells (Flood, 1985) that would be effective in expressing secretion from the tissue (Gray et al, 1989).

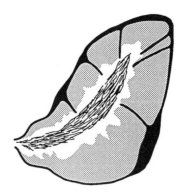

Fig. 1. A mid-line section of the preorbital gland of a muskox. It passes throught the gland opening which is shown on the left. The central hairs are shown by lines, the hair follicles and sebaceous tissue is white and the sweat glands are stippled. The connective tissue is black.

Frequently the facial hair over the opening of the preorbital gland can be seen to be encrusted with dried secretion in both wild and captive animals of either sex. This happens at all times of year though the amount of desiccated material may vary. In captive animals that are distressed, a sticky drop of clear, slightly viscous secretion may appear over the mouths of the glands.

A pooled sample of preorbital gland secretion, collected by manual expression of the glands of several animals, was found to be a complex mixture containing large amounts of cholesterol and benzaldehyde, and a homologous series of straight-chain, saturated γ-lactones ranging from $C_8H_{14}O_2$ to $C_{12}H_{22}O_2$. The ten carbon lactone was the most abundant.

DISCUSSION

Observations in the wild show that the most common behavioral pattern involving the preorbital gland is that usually referred to as "gland rubbing" in which the preorbital area is vigorously rubbed on the outstretched foreleg. This is commonly seen in agonistic situations between contesting bulls but may occur in all ages and either sex during confrontations with predators. Contesting bulls may also rub the preorbital area on rocks or on the sides of pits they have made with their horns. Sometimes these rubbing sites may be overmarked by other bulls that are usually subordinate in rank (Gray et al, 1989). Interestingly, the captive bulls in Saskatoon have never been seen to rub their faces on their forelegs though the superiority display is an almost constant accompaniment of the rut. Perhaps our research animals never learned to do this having been captured as young calves. Or, more likely, they have no need to do so because of all the fence posts that we have provided, the bulls rub their preorbital glands on these very frequently. The effect of rubbing the preorbital gland is mainly to apply the secretion to the face, an extensive and relatively warm surface that promotes evaporation at an appropriate level above the ground.

It would appear that both the principal odor sources of the muskox are behaviorally associated with agonistic situations but one, the urinary odor, is an exclusive and consistent feature of dominant males during the rut and seems to be directed towards other muskoxen, perhaps both rival bulls and cows approaching estrus. The preorbital odor, on the other hand, is apparently used by almost all muskoxen when threatened, but is a particularly prominent feature of the behavioral repertoire of males that are contesting dominance. The nature of the messages is unclear. Do they convey the physiological condition of the individual or do they denote individual identity? Genetically, muskoxen seem to be remarkably homogeneous (Fleischman, 1986) which might make individual olfactory recognition difficult. Whatever the answers to these questions may be, it is clear that muskox odors are for immediate use; they are as evanescent as the tundra wind.

ACKNOWLEDGEMENTS

I would like to express my great gratitude to David Gray, who was indirectly responsible for my own interest in the muskox and has taught us so much about the behavior of muskoxen in the wild. I am also most grateful to graduate students Janice Rowell, Jan Adamczewski and Susan Tedesco for no end of work, stimulus and discussion, and to Colleen Stevens and Ewald Lammerding for their care of the animals. Special thanks are due to Suzanne Abrams and Gillian Muir for their dedication to the chemistry of a sometimes smelly animal.

REFERENCES

Albone, E.S., 1984, Mammalian Semiochemistry, John Wiley and Sons, Chichester.
Barr W., 1991, Back from the brink: the road to conservation of the muskox in the Northwest Territories, in Komatick, No 3, The Arctic Institute of North America, Calgary.
Chaplin R.K., 1984, Nutrition, growth and digestibility in captive muskox calves. Proceedings of the 1st International Muskox Symposium, D.R. Klein, R.G. White, and S. Keller, eds., Biological Papers of the University of Alaska. Special Report No. 4. pp. 195.
Coblentz, B.E., 1976, Functions of scent-urination in ungulates with special reference to feral goats (Capra hircus). Amer. Nat., 110: 549-557.
Crégut-Bonnoure E., 1984, The Pleistocene Ovibovinae of Western Europe: temporo-spacial expansion and paleoecological implications. Proceedings of the 1st International Muskox Symposium, D.R. Klein, R.G. White, and S. Keller, eds., Biological Papers of the University of Alaska. Special Report No. 4. pp. 136-144.
Fleischman, C.L., 1986, "Genetic Variation in Muskoxen (Ovibos moschatus)" M.S. thesis, University of Alaska, Fairbanks.
Flood, P.F., 1985. "Sources of significant smells: the skin and other organs." in Social Odours in Mammals, R.E. Brown and D.W. Macdonald, eds., Clarendon Press, Oxford. pp. 19-36.
Flood P.F., Abrams S.R., Muir G.D., and Rowell J.E., 1989a, Odor of the muskox: a preliminary investigation. J. Chem. Ecology, 15: 2207-2217.
Flood, P.F., Rowell, J.E. Glover, G.J. Chaplin R.K. and Humphreys. S., 1984, The establishment of a small research herd of captive muskoxen. Proceedings of the 1st International Muskox Symposium, D.R. Klein, R.G. White, and S. Keller, eds., Biological Papers of the University of Alaska. Special Report No. 4. pp. 170-172.
Flood, P.F., Stalker, M.J. and Rowell, J.E., 1989b, The hair follicles and seasonal shedding cycle of the muskox (Ovibos moschatus). Can. J. Zool., 67: 1143-1147.

Geist, V., 1984, Goat Antelopes. *in* "The Encyclopedia of Mammals," D.W. Macdonald, ed., Facts on File Publications, New York.

Gray, D.R., 1974, The defence formation of the musk-ox. The Musk-Ox, 14: 25-29.

Gray, D.R., 1987, The Muskoxen of Polar Bear Pass, National Museum of Natural Sciences and Fitzhenry and Whiteside, Toronto, 192 pp.

Gray, D.R., 1984, Dominance fighting in muskoxen of different social status. Proceedings of the 1st International Muskox Symposium, D.R. Klein, R.G. White, and S. Keller, eds., Biological Papers of the University of Alaska. Special Report No. 4. pp. 118-122.

Gray, D.R., Flood, P.F. and Rowell, J.E., 1989, The structure and function of muskox preorbital glands. Can. J. Zool., 67: 1134-1142.

Haltenorth T., 1963, "Handbuch der Zoologie. 8, Klassifikation der Säugetiere", (pp. Artiodactyla 1-167), Walter de Gruyter and Co., Berlin.

Heck, H., Wurster, D. and Benirschke K., 1968, Chromosomes studies on members of the subfamilies Caprinae and Bovinae, the muskox, ibex, aoudad, congo buffalo and gaur. Sägetierk. Mitt., 33: 172-179.

Hofmann R.R., 1988, Anatomy of the gastro-intestinal tract. Church D.C. ed., "The Ruminant Animal: Digestive Physiology and Nutrition", (pp. 14-43)., Prentice Hall, Englewood Cliffs.

Latour P., 1986, Observations on demography, reproduction, and morphology of muskoxen (Ovibos moschatus) on Banks Island, Northwest Territories. Can. J. Zool., 65, 265-269.

Le Hénaff D., and Crête M., 1989, Introduction of muskoxen to northern Quebec: the demographic explosion of a colonizing herbivore. Can. J. Zool., 67: 1102-1105.

Lent P.C., 1988, Ovibos moschatus. Mammalian Species, (No 302): 1-9.

Lönnberg, E., 1900, On the soft anatomy of the musk-ox (Ovibos moschatus). Proc. zool. Soc. Lond. 1900, 142-167.

Miller F.L., 1989, Poor muskox calf representation on Prince Patrick and Eglinton islands. Proceedings of the 2nd International Muskox Symposium, P.F. Flood, ed., National Research Council of Canada, Ottawa, A46.

Reinhardt, V. and Flood, P.F., 1983, Behavioural assessment of captive muskox calves. Behaviour, 87: 1-21.

Reynolds P.E., 1989, Status of a transplanted muskox population in northeastern Alaska. Proceedings of the 2nd International Muskox Symposium, P.F. Flood, ed., National Research Council of Canada, Ottawa, A26-A30.

Rowell J.E., and Flood P.F., 1989, Plasma progesterone concentrations during the estrous cycle and pregnancy in muskoxen. Proceedings of the 2nd International Muskox Symposium, P.F. Flood, ed., National Research Council of Canada, Ottawa, A52-A53.

Sack, W.O. and Ballantyne, J.H., 1965, Anatomical observations on a muskox calf (Ovibos moschatus) with particular reference to thoracic and abdominal topography. Can. J. Zool., 43: 1033-1047.

Schaffer, J., 1940, "Die Hautdrüsenorgane der Säugetiere", Urban and Schwarzenberg, Berlin.

Simpson G.G., 1945, The principles of classification and a classification of mammals. Bulletin of the American Museum of Natural History, 85: 1-350.

Smith T.E., 1989, The role of bulls in pioneering new habitats in an expanding muskox population on the Seward Peninsula, Alaska. Can. J. Zool., 67: 1096-1101.

Teal, J.J. Jr., 1959, Muskox in rut. Polar Notes, 1: 65-71.

Tener, J.S., 1965, Muskoxen in Canada, a biological and taxonomic review. Canadian Wildlife Service. Monograph No. 2. Queen's Printer, Ottawa. 166 pp.

Vibe, C., 1967, Arctic animals in relation to climatic fluctuations, Meddelelser om Grönland, 170: 153-192.

Yakushkin G.R., 1989, The muskox population of the Taymyr Peninsula. Proceedings of the 2nd International Muskox Symposium, P.F. Flood, ed., National Research Council of Canada, Ottawa., A14-A15.

HORMONAL MODULATION OF CHEMOSIGNALS WHICH ELICIT AGGRESSIVE BEHAVIOUR IN THE INDIAN PALM SQUIRREL, FUNAMBULUS PALMARUM

K.M. Alexander and G. Bhaskaran

Department of Zoology, University of Kerala
Kariavattom, Kerala, India

INTRODUCTION

Chemosignals emanating from specialised sebaceous glands located at the oral angles of the Indian palm squirrel, Funambulus palmarum, can elicit aggressive behaviour in consecifics (Bhaskaran and Alexander, 1985). However, the perception of these chemosignals is under the influence of hormones, especially gonadal steroids. Despite the considerable literature on the role of gonadal steroids on aggressive behaviour in both murine and microtine rodents, little attention has been focused on sciurid rodents. Accordingly, this study was undertaken to elucidate the hormonal modulation of aggressive behaviour in one sciurid rodent, the Indian palm squirrel Funambulus palmarum.

MATERIAL AND METHODS

Healthy adult squirrels, weighing 110 gms each, were selected for study. Four weeks after gonadectomy (see Zarrow et al., 1964), each of three groups of gonadectomised squirrels was administered, on a daily basis, 1 mg (in olive oil) of either testosterone propionate, oestradiol benzoate, or hydroxyprogesterone caproate. The hormonal injections were alternated daily from left to right legs and injections were done under ether anaesthesia. Sham operated and olive oil administered animals served as controls. Aggression was studied in the home cage context (HCC) and the neutral cage context (NCC). Normal adult males and females served as stimulus animals. Aggresion in social contexts was evaluatd in the gonad-ectomised and in the hormone-treated animals. The chases and bites were recorded every minute in 10 minute tests. Test animals were kept in isolation for one month. In HCC experiments, 40 castrated males and 40 ovariec-tomised females were studied, of which 30 were hormone treated (Oestrogen, Progesterone and Testosterone for each). Normal male and female animals (ten each) served as controls. Similarly, for NCC the same experimental design was used, but with twice the number of animals, i.e., 80 castrated males and 80 ovariectomised females. In HCC the stimulus animal was introduced into the home cage of the test animal. Concomitant to the higher mortality in aggressive encounters in HCC, a lower number of animals was used. Regarding NCC studies, both the test and stimulus animals were introduced simultaneously into a fresh cage. The data were statistically analysed using Wilcoxon matched pair signed-rank tests and Mann-Whitney U-tests (Siegel, 1956).

Chemical Signals in Vertebrates VI, Edited by R.L. Doty and
D. Müller-Schwarze, Plenum Press, New York, 1992

RESULTS

The aggressive responses of both male and female test animals depended
on whether the subject was a resident or an intruder. In the HCC, males
and females appeared to be equally aggressive.

Gonadectomy significantly curbed aggression in both males and females
in HCC (Fig. 1). Administration of testosterone, oestrogen or progesterone
to the castrated males significantly increased aggression. Among the three
hormones, oestrogen was relatively more effective in elevating aggressive
responses (Fig. 1). Administration of testosterone and progesterone to the
females did not elicit any significant change in aggressive responses over
that of the control ovariectomised females. However, oestrogen caused a
slight increase in aggression over the ovariectomised controls. Gonadectomised females were more aggressive than gonadectomised males.

A comparison of the data obtained for the males and females in HCC and
NCC indicate that both sexes exhibited higher aggressive tendencies in HCC.
Regarding aggressive responses in NCC, normal males were more aggressive
than normal females (Fig. 2). While castration significantly reduced
aggressiveness of the male (Fig. 2), ovariectomy markedly increased the
aggressiveness of the female (Fig. 2). Although testosterone and oestrogen
elevated aggressiveness in castrated males under this conditon, progesterone did not (Fig. 2). Oestrogen is quite effective and elevated the aggressiveness in the castrated male, even beyond the level of that of the
normal male (Fig. 2). Administration of testosterone to ovariectomised

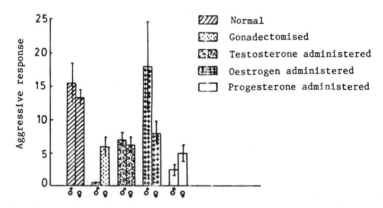

Fig. 1. Aggressive responses of males and females in HCC.

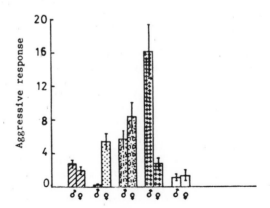

Fig. 2. Aggressive responses of males and females in NCC.

females in NCC produced little change in aggressiveness (Fig. 2). Both ostrogen and progesterone depressed aggression slightly in the ovariectomised female in the NCC (Fig. 2).

DISCUSSION

The test animals were relatively more aggressive in HCC than NCC. The aggression in HCC appeared to be a function of residence in the home cages. Hence, it can be surmised that hyperaggressiveness is correlated with territoriality. Earlier studies on aggression corroborate this observation (De Bold and Miczek, 1984; Rodgers and Randall, 1986). The disparity in the level of aggressiveness, exhibited by the test animals in HCC and NCC, can be attributed to the variation in the motivation for aggression; defence of one's own territory in the former and the establishment of dominance in the latter. This inference is supported by data obtained by Barfield et al. (1972) in rats. Our studies revealed that males in the NCC are more dominance prone than females, exhibiting higher levels of aggression. Investigations by Albert et al., (1986) also support this observation.

The present study indicates that the hormone testosterone is effective in promoting aggression in the male Indian palm squirrel. This corroborates several earlier reports (Gipes, 1982; Line et al, 1985; Albert et al, 1986). However, oestrogen has been noted to be relatively more effective than testosterone in promoting aggression in males. The data obtained in the present study is in conformity with the idea that testosterone may be aromatized to estrogen in the brain (Brain and Bowden, 1976) neural.

The results of the present study reveal the relatively lower impact of progesterone in promoting aggression. The data corroborate earlier reports of Payne and Swanson (1972) and De Jonge et al., (1986). Ovariectomy was associated with a decline in territory-related aggressive behaviour in the females in HCC. As for NCC, ovariectomy resulted in a significant increase of aggressiveness over the normal female level, suggesting that the dominance relationships remained unaltered even after ovariectomy. Based on the observations made by Yahr and Conquelin (1980), the disparity caused by ovariectomy of the female squirrels in HCC and NCC may be due to the interference of social factors. In the HCC, the submissive intruder does not provide enough stimulus for the ovariectomised female to be aggressive, whereas in NCC, the intact normal stimulus animal is aggressive and attacks the test animal, thus providing sufficient provocation for the test animal to fight. It should be noted that ovarian hormones exert an inhibition on aggression based on social dominance, the elimination of which leads to unrestricted expression of the same. Simon and Masters (1987) found that oestradiol benzoate was ineffective in promoting aggression in ovariectomized C57 BL/6J female mice. Administration of this oestrogen seems to have dual effects with respect to the social context of aggression   In HCC, the territorial aggression is positively infuenced by the hormone, whereas in NCC it has a negative effect.

REFERENCES

Alexander, K.M. and Baskaran, G., 1985, Behavioral relevance of secretions of certain specialised integumentary glands of the Indian palm squirrel, Funnambulus plamarum. Ethologie 85. IXXth International Ethological Conference Abstracts., 92.
Albert, D.J., Walsh, M.L., Gorzalka, B.B., Siemens, Y., and Louie, N., 1986, Testosterone removal in rat results in a decrease in social aggression, Physiol. Behav., 36:401-407.

Barfield, R.J., Busch, D.E. and Wallen, K. 1972, Gonadal influence on agonistic behaviour in the male domestic rat, Horm. Behav., 3:247-259.

Brain P.F., 1980, Adaptive aspects of hormonal correlates of attack and defence in laboratory mice. A study in ethology, in: "Multidisciplinary Approaches to Aggression Research", P.F. Brain and D. Benton, eds., 391-413. Elsevier/North Holland, Amsterdam.

Brain, P.F. and Bowden, N.J., 1976, Are androgens converted to oestrogen before they exert their behavioural influences via the brain, Neurosci. Lett., 3:78-79.

De Bold, J. and Miczek, K.A., 1984, Aggression persists after ovariectomy in rats, Horm. Behav, 18:177-190.

De Jonge, F.M., Eerland, E.M.J. and Van De Poll, N.E., 1986, Sex specific interactions between aggressive and sexual behaviour in the rat. Effects of testosterone and progesterone, Horm. Behav, 20:432-444.

Gipes, J.H.W., 1982, The effects of testosterone and scoplamine hydobromide o the aggressive behaviour of male voles, Microtus townsendii. Can. J. Zool., 60:946-950.

Line, S.W., Hart, B.L. and Sanders, L., 1985, Effect of prepubertal castratio on sexual and aggressive behaviour in male horses, J. Am. Vet. Med. Assoc., 186:249-251.

Payne, A.P. and Swanson, H.H., 1972, The effect of sex hormone on the agonistic behaviour of male golden hamster, Mesocricetus auratus, Waterhouse, Anim. Behav., 20:782-787.

Rogers, R.J. and Randall, J.I., 1986, Extended attack from a resident consecific is critical to the development of long lasting analgesics in male intruder mice, Physiol. Behav., 38:427-430.

Siegel, S., 1956, "Nonparametric statistics for the Behavioural Sciences", McGraw Hill, New York.

Simon, N.G. and Masters, D.B., 1987, Activities of male typical aggression by testosterone but not its metabolites in C57 B1/6J female mice, Physiol. Behav, 41:405-408.

Yahr, P. and Coquelin A., 1980, Effect of pre versus pubertal castration on aggression between male gerbils, Behav. Neur. Biol., 28:496-500.

Zarrow, M.X., Yochim, J.M. and McCarthy, J.L., 1964, "Experimental Endocrinology", Academic Press, New York.

INDIVIDUAL ODORTYPES

V.V. Voznessenskaya, V.M. Parfyonova, E.P. Zinkevich

A.N. Severtzov Institute of Evolutionary Animal Morphology
and Ecology, 33 Leninsky Prospekt, Moscow V-71, Russia

INTRODUCTION

Individual recognition and evaluation of individual physiological
state are key issues in chemical communication among populations of
animals. Ultimately, reproductive success and survival depend on the
solution of these problems. The capacity to discriminate between conspecifics as well as the ability to evaluate physiological status on the
basis of individual odor has been demonstrated in a variety of mammals (for
examples, see Halpin, 1980; Bowers and Alexander, 1967; Carr, Krames and
Costanzo, 1970; Johnston, 1983). The occurrence of odor(s) specific to the
individual suggests reliability of individual odortypes. Indeed, some
mammals are known to maintain consistency in odortypes for comparatively
long periods, e.g., for at least two weeks (Kalkowski, 1967). One
principle of such a coding system should be the absence of duplication, at
least in the population of animals that share chemosensory information.

Knowledge about the mechanisms underlying individual odortypes and
their perception is sparse. There are at least two points of view on the
principles of such coding. The first suggests that odortypes are based on
the presence of a small set of identical substances, but the concentration
ratios (patterns) of these substances vary across individuals (Neuhaus,
1956). This is a widely accepted hypothesis, but it fails to explain the
realities of complex mixture discrimination in general, and, in particular,
how dogs might discriminate a target host odortype from among a mixture of
different human odors (Kalmus, 1955). The second hypothesis relies upon
individual-specific substances. In this system, attention to one or a few
members of a subset of a number of highly specific substances (W) provides
perception of an individual odortype. At one end of an extreme, each
individual odortype is coded by a single unique compound (λ). At the
other, myriad compounds are required. In the latter case, and for those
between these extremes, the difference in a single component of the
combinations of highly specific substances is sufficient to encode
individual odortypes, even though individuals may differ in more than one
of these elements. As an initial start to determine how many of these
substances may be necessary to encode individual odortypes, we assume that
the number of such unique substances may approach infinity ($W \to \infty$).

A theoretical analysis of the capacity of the main olfactory system to
discriminate complex mixtures has been published (Minor, 1982). The

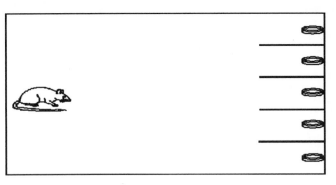

Figure 1.　Schematic of 5-arm maze.　A petri dish containing the urine sample(s) is placed at the end of each arm.

relationships among the number of olfactory receivers, the number of components in a mixture (K) and the probability of error during discrimination (δ_u) have been stated.

The abilities of human beings and other animals to analyze complex mixtures are presently unclear.　In particular the maximum number of components in a mixture that can be discriminated is unknown.　The purpose of our research was to add clarity to these issues and to test the hypotheses mentioned above.

MATERIALS AND METHODS

Test subjects were 45 ♀♀ Wistar rats ("Stolbovaya" nursery, USSR) 4-8 months of age.　Animals were housed under standard vivarium conditions in plastic cages, four per cage.　Lighting conditions were 12 L:12 D (lights off at 0700).　Urine donors were housed individually in the same types of cages.　They all received the same standard diet.　For urine collection, donors were placed into metabolic cages for 12-16 hours during the natural night.　Urine was used during one experimental day only.

A five-arm maze was designed to train rats to detect and respond to the odor of individual rat urine samples (Fig. 1).　A plastic box, 120 cm x 95 cm x 50 cm, was used as an experimental maze.　The front wall consisted of a screen through which the rat's behavior could be observed.　Five arms of the maze, 16 cm x 17 cm x 16 cm each, were located in consecutive order along the back wall of box.　Urine samples (on filter paper inside Petri dishes) were placed in each arm of the maze.　The remaining space within the box formed a starting chamber.　The starting point (S) is located behind the center of the front wall, about 100 cm from each arm.　The rat, which was placed at the starting point, had to make food-motivated run to the urine samples and locate the urine of a chosen target individual in the end of one of the arms of the maze.　Animals were food deprived for 36 hours before the start of the experiment.　Curd rolls were used as an unconditioned food stimulus.

The latencies to correct and incorrect choices, and the rat's movements within the box, were recorded.　Immobility for over 2 minutes after the start of a trial was considered a refusal.　Choices were considered to be correct if the rat ran directly from the starting point to the arm containing the target urine sample wherein it waited or looked for food, or if the rat ran through the maze arms without significant stops, sniffed the samples and then chose the arm containing the target urine sample, wherein it waited for food. Training, per se, ceased when the proportion of correct choices had become stable. Subsequently, the ability to choose a sample of urine from the target individual was tested during 10 trials. The number of correct choices was recorded. Animals that refused to perform the task were culled.

In the first series of experiments animals were trained to respond to individual urine samples. We then tested the rat's ability to discriminate the target urine in a mixture of 2-10 urines versus other mixtures consisting of urine from 2-10 other donors. Each mixture was used only once for each rat, although a mixture may have been used for another rat. Thus, on each trial, the rat must solve a unique problem, i.e., choose the new mixture that contains the target odor from among five new mixtures of urines. In other experiments animals also were tested to discriminate the target urine sample after dilutions from 10-1000 fold.

RESULTS AND DISCUSSION

Wistar rats had little trouble discriminating urine samples from conspecifics individuals ($p < 0.001$; Fig. 2). Moreover, these animals could discriminate the target sample in a mixture from other mixtures containing samples of urine from 2-7 other donors ($p < 0.001$). As the number of urine samples in a mixture increased, the percentage of correct solutions decreased; latencies also increased (Fig. 3). Increased latencies may be explained by an increase in odor information, i.e., perhaps odor images became more complex. This phenomenon is well known in perceptual psychology; the time required to perceive a simple object is less than that required for one having multiple dimensions (Shehter, 1981). Further increasing to 8-10 the number of contributors to the mixture of urines lead to a rapid decline in the ability of the rat to discriminate the mixture containing the target from other mixtures. This decline was accompanied by a decrease in latency, which may reflect the rat's tendency toward chaotic movements in the maze and the absence of attempts to solve the problem.

We did not observe a decrease in the number of correct choices in tests involving dilutions of individual target urine samples by 10-1000 fold. Thus, the decrease in performance when 8-10 samples are mixed cannot be due solely to a decrease in concentration of the target urine.

We also noted few individual differences among the trained rats in their ability to solve the task (Fig. 2). This suggests that a critical increase in odortype information occurs in a mixture of 7-8 individual urine samples. It can be inferred from our experimental data that the individual odortype is a stimulus with certain stable parameters. A single-component difference between complex mixtures might be sufficient to explain the results. In this instance, the simplest way to solve the task would be to evaluate the presence/absence of the single specific component. On the other hand, taking into consideration a limited capacity for simultaneous analysis of discrete information (for human beings, no more than 8 images can be analyzed simultaneously; Miller, 1956) and the single discrete parameter of chemical substances -- a molecule -- it can be concluded that the evaluation of a single specific substance for every individual is not only the simplest solution of this task, but also most likely the only one. This suggestion does not limit the number of substances coding every specific individual. These limitations are concerned only with capacity of a subject to analyze odorous information.

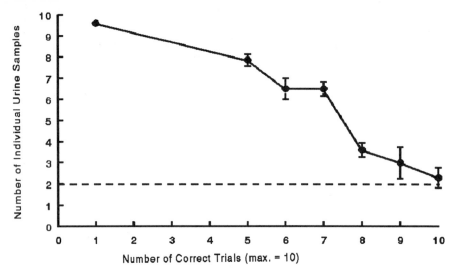

Fig. 2.   Correct choices for 10 trials in the maze for 20 rats.   Task
complexity increased from single urine samples in each petri dish
to 10 samples per dish.   Dotted line indicates chance.

Odortypes may be composed of multiple identifying components, but animals
may learn to focus on single elements to perceive differences in odortypes.

In accordance with the theoretical approach of Minor (1982), the
minimum number of individual-specific coding substances can be evaluated
from the following equations:

$$W_{min} \approx 2Kln\ 1/\sigma_u \qquad and \qquad \lambda_{opt} = W/K,$$

where K is the number of substances in a mixture, λ is the number of
specific substances from W that may be necessary to code a single
indiviudal and σ_u is the error in discrimination. By using data obtained
in our experiments, we can substitute values for K and obtain a solution to
the equation: the minimum number of individual specific substances that
could comprise an odortype (W_{min}) should not be less than 30, and each
individual should optimally have 4-5 (λ_{opt}) of these components. Taking
this into account, we now attempt to evaluate the minimum population size
of a group of animals in which duplicity of some, but not all, components
is allowed while maintaining individual odortypes that could be coded by
different combinations of specific substances. This reduces to C_{30}^4 to C_{30}^5
which approximates 27,500 to 142,500 individuals, i.e., the probability of
meeting two individuals with the same odortype approximates zero.

The population density of Norway rats in natural settings usually
varies from 2 to 113 individuals per 100 sq. m, and dominant males can
sometimes cover an area up to 2000 sq. m (Rylnikov, 1990). In this
instance, the maximum number contacts by the male is 500-2500. Even in
extreme situations the probability of meeting exact odortype copies is
nearly zero.

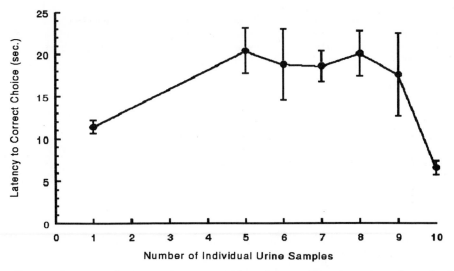

Fig. 3.   Average latency to correctly choose the target urine during
series of 10 trials at each level of complexity by 20 rats.

        Is the actual number of specific substances finite or infinite? Only
an experiment in which many more donors are used can we answer this
question. We suspect that the number must be over 23 -- we used 23 donors
in our experiment and all combinations among the 23 donors could be
discriminated. There is reasonable cause to ask how reception and analysis
of such a considerable number of unique substances can be performed via
olfaction. If we assume the existence of immature (non-specific) receptors, and, subsequent to experience with odorants, biochemical and/or
structural modifications of these receptors, then unique substances that
provide individual odortypes may stimulate the maturation of these non-
specific receptors. If this occurs, then continued experience with odortypes might lead to a decrease in threshold. In prior work, we noted that
a rat's ability to discriminate strangers (in the absence of extensive
training) disappeared after dilutions of the urine by a factor of 10-fold.
In the present work, after extensive training to discriminate the samples,
urine could be diluted at least 1000-fold. A similar change in the ability
to smell another substance after repeated exposures has been demonstrated
(Wysocki, Dorries and Beauchamp, 1989). Under natural conditions, it seems
more probable that rats recognize individuals perhaps first by visual cues
and other simple distinguishing features, including, for example, differences in the concentrations of similar substances. Regularly repeated
contacts may lead to the formation of a more reliable (specific) system of
recognition, i.e., the animal develops the ability to discriminate unique
elements of odortypes. In our experiments we presented rats with a complicated task -- one which is unlikely to be encountered in a natural environment. Although rats may encounter myriad individuals, normally their
odortypes are unlikely to be in the form of a mixture of many individuals
(except perhaps in conditions of very high population density or at com-
munal excretory sites). Rather, the experiment simulated an extreme
condition and perhaps strained the function of the rat olfactory system.
Another interesting direction for investigations is to determine the actual

chemical identification of the specific individual odortypes. The only observation noted thus far is that changes of urine pH (from 1 to 10) do not disrupt performance by trained rats to discriminate individual odortypes. This suggests that odortypes are pH independent.

In summary, we suggest that individual odortype is provided by unique substances. An animal's individual odortype is optimally composed of 4-5 unique substances. Any individual attempting to distinguish among members of a population need not recognize all of these substances -- distinction of individuals may rely upon single constituents when no overlap is present. Finally, the concentrations of these substances in urine, relative to threshold after training, is considerable, apparently at least greater than 1000 times the minimum necessary to distinguish among individuals.

ACKNOWLEDGEMENT

We thank Dr. C.J. Wysocki for his comments, criticisms and assistance in the preparation of this contribution.

REFERENCES

Bowers, T.U., and Alexander, B.K., 1967, Mice: Individual recognition by olfactory cues, Science, 158:1208.

Carr, W.G., Krames, L., and Costanzo, D.G., 1970, Previous sexual experience and olfactory preference for novel versus original sex partners in rats, J. Comp. Physiol. Psychol., 71:216.

Halpin, Z.T., 1980, Individual odors and individual recognition: Review and commentary, Biol. Behav., 5:233.

Johnston, R.E., 1983. Mechanisms of individual discrimination, in: "Chemical Signals in Vertebrates," D. Müller-Schwarze and R.M. Silverstein, ed., Plenum Press, New York-London.

Kalkowski, W. 1967, Olfactory basis of social orientation in white mouse, Folia biol. (PRL), 15:69.

Miller, G.A., 1956, The magic number 7 plus or minus: Some limit in our capacity for porcessing information, Psychol. Rev., 63:81.

Minor, A.V., 1982, Theoretical possibilities of olfactory signals coding, in: "Chemical Signals in Animals," Academician V.E. Sokolov, ed., Nauka Publishers, Moscow (in Russian).

Neuhaus, W., 1956, Die Unterscheidungsfähigkeit des Hundes für Duft-gemische, Ztschr. Vergl. Physiol., 39:25.

Rylnikov, V.A., 1990, Breeding, age composition and mortality, in: "Norway Rat. Systematics, Ecology, Population Control," V.E. Sokolov and E.V. Karasjova, ed., Nauka Publishers, Moscow (in Russian).

Shehter, M., 1981, "Visual Discrimination: Regulation and Mechanisms," Pedagogica Publishers, Moscow (in Russian).

Wysocki, C.J., Dorries, K., and Beauchamp, G.K., 1989, Ability to perceive androstenone can be acquired by ostensibly anosmic people, Proc. Nat. Acad. Sci., USA, 86:7976.

ODOR DISCRIMINATION IN FEMALE MICE AFTER LONG-TERM EXPOSURE

TO MALE ODORS: GENOTYPE-ENVIRONMENT INTERACTION

N. Kenneth Sandnabba

Department of Psychology
Abo Akademi University
SF-20500 Abo, Finland

INTRODUCTION

It has been shown in several studies that female mice are capable of selecting males on the basis of certain odor cues indicative of genetic quality (e.g. Yamazaki et al., 1978, 1979; Lenington and Egid, 1985; Sandnabba 1986a; Novotny et al., 1990). Female mice have also been found to show a preference for male odors of high ranking and aggressive individuals (Scott and Pfaff, 1970; Jones and Nowell, 1974; Sandnabba, 1985; 1986b). An interesting question arising from these observations is whether a dominant and highly aggressive male is always preferred, independent of genotype.

Olfactory learning reportedly plays, in some species, an important role in establishing mate selection (Keverne, 1983; Fillion and Blass, 1986). According to Mainardi (1964), the sexual preferences of female (but not male) mice are, in fact, strongly influenced by precocious learning of the parents' characteristics; when reared in the absence of their fathers, they showed a loss of the normal ability to discriminate between males of different strains. Cross-fostering has also been found to influence the odor discrimination of adult mice (Aldhous, 1990). Thus, it seems that a process of "olfactory imprinting" to a male odor occurs and that it determines in adult mice of at least some strains the choice of objects for sexual behavior.

The present study reports an experiment concerned with the following questions: (1) Do female mice, independent of their own genotype, prefer the odors of highly aggressive males?; and (2) Does pre-exposure from birth onwards to alien male odors change the odor preferences of adult female mice?

METHODS

Subjects

Mice from three different strains were used. The TA (Turku Aggressive) and TNA (Turku Non-Aggressive) mice belonged to the 53d generation of selection for high and low levels of aggressiveness respectively (Lagerspetz, 1964; Sandnabba, 1986a). The mice of intermediate aggressiveness

Chemical Signals in Vertebrates VI, Edited by R.L. Doty and
D. Müller-Schwarze, Plenum Press, New York, 1992

used in this experiment came from the parental Swiss albino strain (here called normal, N) from which the selective breeding originally started in 1959 (Lagerspetz, 1964). Litters were weaned at 21-23 days post-natally, and from that point the experimental females were housed in 22 x 17 x 14 cm clear polycarbonate cages with wire lids. The males were housed individually in similar cages measuring 20 x 10 x 12.5 cm. Water and standard animal food pellets were provided at libitum. The animals were housed in air-conditioned rooms, maintained at 21-23 centigrades, on a twelve-hour alternating light/dark cycle (lights on at 0700 h). The different strains were kept in separated rooms. Altogether, 108 experimental female mice (36 TA, 36 TNA, and 36 N) were used. The soiled sawdust originated from 120 males (40 TA, 40 TNA, and 40 N).

Procedure

Prior to delivery, six pregnant females of each strain were placed into the rooms containing only mice of the same strain. Thus, from birth onwards, they were exposed only to the odors specific for each strain of mouse. Further, approximately 0.5 dl of soiled sawdust taken from a male in the same room was placed into each breeding cage daily. After weaning, two females from each litter were placed together in the same cage and the exposure to the same strain odor was prolonged for approximately five months in the manner described above. The cages were cleaned once every week when the old bedding was discarded.

Within each strain, the experimental females were thus divided into three groups consisting of twelve animals each. The females in groups TA I, TNA I, and N I were exposed to TA male odors. The females in groups TA II, TNA II , and N II were exposed to TNA male odors. The females of the TA III, TNA III, and N III groups were exposed to odors from normal males.

After approximately six months of exposure the odor preferences of the experimental females were tested in an aluminium X - maze, each arm of the maze measuring 15 x 15 x 12 cm. Each of the four arms contained one type of soiled sawdust (TA, TNA, or N) or clean sawdust. The experimental female was placed in the uncovered middle section of the maze with its nose pointing to the arm containing clean sawdust. For a period of five minutes, the total duration of the time spent on the different types of bedding was recorded. The latencey to the first movement from the middle section to another area was also recorded. Sawdust soiled by only one animal was used for each individual testing situation, and the maze was cleaned and the sawdust changed after each trial. Each experimental female was tested only once since earlier findings have shown that the differences are clearest on the first test occasion (Sandnabba, 1986a).

RESULTS

The significance of differences between the different treatment groups was calculated by two-way analysis of variance (Manova), followed by post-hoc t-tests for dependent and independent groups. There was no significant effect of the different types of odor exposures observed, but an interaction between the type of bedding in the maze and the strain of the female was shown ($p < 0.01$).

From Fig. 1. it can be seen that all the TA females, independent of odor exposure, showed a preference for the area containing soiled bedding from the nonaggressive TNA males relative both to the arms containing TA- and normal-soiled beddings (p-values ranged from 0.001 to 0.05). The preference of the normal-exposed TA females in favor of the TNA-soiled area compared to the Normal-soiled area did not reach a significant level. The

TA exposed females showed a preference for the clean sawdust compared both to the TA- and TNA-soiled areas (p<0.001).

As shown in Fig. 2., the TNA females spent the most time in the areas containing normal-soiled and clean sawdust (p<0.001) relative to the time spent in the area containing odors of the males of their own strain. In addition, the TA-soiled area was more popular (p-values ranged from 0.001 to 0.05) than the TNA-soiled area.

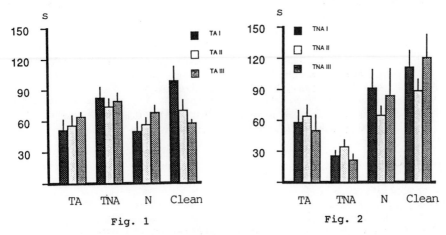

Fig. 1. Mean total time in seconds (SEM) spent by TA females on different types of bedding in the X - maze (I = TA-, II = TNA-, and III = N-exposure)

Fig. 2. Mean total time in seconds (SEM) spent by TNA females on different types of bedding in the X - maze (I = TA-, II = TNA-, and III = N-exposure)

From Fig. 3 it can be seen that the normal females, except for the TNA exposed ones, preferred the areas covered by TNA-soiled and clean sawdust compared with the odors of males from their own strain (p-values ranged from 0.001 to 0.01). The normal females exposed to odors of males from their own strain also preferred the TA soiled area to the normal-soiled area (p<0.001). The TNA exposed normal females showed no significant preferences.

Fig. 4 shows that both the TA and TNA females exposed to TNA odors showed longer latencies to the first movement when compared to the TA and TNA females exposured to TNA and normal odors (p-values ranged from 0.001 to 0.05).

DISCUSSION

The present experiment shows that variations in male odor stimulation from birth onwards did not radically change the odor preferences of the female mice. The females were also generally attracted to males of a genetic status different from their own and not solely to the odors of the males with a high potential for aggressive behavior. The results suggest

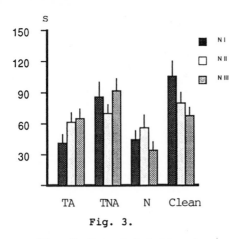
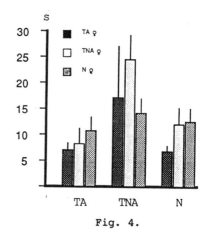

Fig. 3.   Fig. 4.

Fig. 3. Mean total time in seconds (SEM) spent by the N females
on the different types of bedding in the X - maze (I =
TA-, II = TNA-, and III = N-exposure).

Fig. 4. Mean latency in seconds (SEM) to the first movement
from the middle section to another area of the X -
maze.

discrimination on the basis of genetic constitution rather than as a result
of learning in these strains of mice.   The preference choices of the females may serve to select against marked aggressiveness within the gene-
pool.

     The surprisingly small effects of the long-term exposure to the alien
male odors may have served explanations.   Since the three different strains
used were generated from the same parental strain, they are supposed to
differ mainly in aggressiveness, for which they have been selectively bred,
as well as in characteristics dependent upon the same pleiotropic genes as
the aggressiveness (Lagerspetz and Lagerspetz, 1983).   Perhaps differences
only in the aggressive potential of the males are not sufficient to radically change the odor preferences of the females, or conversely, aggressiveness is such an important factor in sexual communication that it cannot
easily be masked through experience.   The preferences for unfamiliar odors
reported by Hepper (1986) were not observed in this study, and it seems
that familiarity is less important than unrelatedness.   The overall longer
mean latency to the first movement of the experimental TNA exposed females
in the maze could be interpreted as a lack of power to stimulate females in
the odor cues of the non-aggressive TNA males, cues which generally have
been reported to be present in the odors of intact males (Ropartz, 1968).

     Although female mice usually prefer the odors of highly aggressive and
dominant males (Scott and Pfaff, 1970; Jones and Nowell, 1974; Sandnabba,
1990), the present results shows that there are no easy pathways in the
study of the correlations between aggressiveness or dominance status of the
male and the mating preferences of the female.   There seems to be a complex
and dynamic network of communication operating not only on the basis of
information about aggressiveness and social status, but also on the genetic
constitution of the animals involved.   A genetic basis for individual
recognition provides scope for selective mating (Yamazaki et al. 1978).

After all, the fitness of a species is hardly enhanced by a forever growing potential for aggressive behavior since the consequense of excesssive fighting could be disastrous for the species, whereas stability from one generation to another in this respect would be favorable for the species. Natural selection seems to favor the moderately aggressive individual which is capable of rapidly coping with changes in the environment.

REFERENCES

Aldhous, P., 1990, Olfactory recognition cues and the development of intra-sexual kin discrimination in male laboratory mice, in: "Chemical Signals in Vertebrates 5," D. W. Macdonald, D. Mller-Schwarze, and S. E. Natynzuk, eds., Oxford University Press, Oxford.

Fillion, T. B., and Blass, E. M., 1986, Infantile experience with suckling odours determines adult sexual behaviour in male rats, Science, 231:729.

Hepper, P. G., 1986, Kin recognition: functions and mechanisms. A review, Biol. Rev., 61:63.

Jones, R. B., and Nowell, N. W., 1974, A comparison of the aversive and female attractant properties of urine from dominant and sub-ordinate male mice, Anim. Learn. Behav., 2:141.

Keverne, E. B., 1983, Pheromonal influences on the endocrine regulation of reproduction, Trends Neurosci., 6:381.

Lagerspetz, K. M. J., 1964, "Studies on the Aggressive Behaviour of Mice," Suomalainen Tiedeakatemia, Helsinki.

Lagerspetz, K. M. J., and Lagerspetz, K. Y. H., 1983, Genes and aggression, in: "Aggressive Behavior: Genetic and Neural Approaches," E. C. Simmel, and M. Hahn, eds, Lawrence Erlbaum Ass., Hillsdale.

Lenington, S., and Egid, K., 1985, Female discrimination of male odors correlated with male genotype at the T-locus: a response to T-locus or H-2 locus variability, Behav. Genet., 15:53.

Mainardi, D., 1964, Relations between early experience and sexual prefer-ences in female mice, Atti Associazione Genetica Italiana, 9:141.

Novotny, M., Jemiolo, B., and Harvey, S., 1990, Chemistry of rodent phero-mones: molecular insights into chemical signalling in mammals, in: "Chemical Signals in Vertebrates 5," D. W. Macdonald, D. Muller-Schwarze, and S. E. Natynczuk, eds., Oxford University Press, Oxford.

Sandnabba, N. K., 1985, Differences in the capacity of male odours to affect investigatory behaviour and different urinary marking patterns in two strains of mice, selectively bred for high and low aggressive-ness, Behav. Process., 11:257.

Sandnabba, N. K., 1986a, "Heredity, Fighting Experience and Odour Cues: Factors Determining the Aggressive Interaction in Mice," Department of Psychology at Abo Akademi, Turku.

Sandnabba, N. K., 1986b, Effects of selective breeding for high and low aggressiveness and of fighting experience on odor discrimination in mice, Aggressive Behav., 12:359.

Sandnabba, N. K., Differences between aggressive and non-aggressive mice in odour signals and marking behaviour, in: "Chemical Signals in Verte-brates 5," D. W. Macdonald, D. Mller-Schwarze, and S. E. Natynczuk, eds., Oxford University Press, Oxford.

Scott, J. W., and Pfaff, D. W., 1970, Behavioral and electrophysiological responses of female mice to male urine odors, Physiol. Behav., 5:407.

Yamazaki, K., Yamaguchi, J., Andrews, P. W., Peake, V., and Boyse, E. A., 1978, Mating preferences of F2 segregants of crosses between MHC-congenic mouse strains, Immunogenetics, 6:253.

Yamazaki, K., Yamaguchi, J., Andrews, P. W., Peake, V., Boyse, E. A., and Thomas, L., 1979, Recognition among mice: evidence from the use of Y-maze differentially scented by congenic mice of different major histocompatibility types, J. Exp. Med., 150:755.

OLFACTORY AND VOMERONASAL MECHANISMS OF SOCIAL COMMUNICATION

IN GOLDEN HAMSTERS

Robert E. Johnston

Department of Psychology
Cornell University
Ithaca, NY   14853

INTRODUCTION

A number of hypotheses have been put forward to characterize the differences in function between the olfactory and vomeronasal sensory systems
in vertebrates; these include the suggestions that (1) the vomeronasal system is especially important for reproductive function (Wysocki, 1979, 1989);
(2) the vomeronasal system is primarily responsible for responses to relatively large, non-volatile compounds (Ladewig and Hart, 1980; Halpern and
Kubie, 1980; Wysocki, Wellington and Beauchamp, 1980), (3) the vomeronasal
system may be especially important for relatively hard-wired responses to
specialized chemical signals whereas the olfactory system may be more concerned with olfactory cues that need to be learned (Meredith, 1983), (4)
the vomeronasal system mediates rewarding properties of scent (Halpern,
1988), (5) the olfactory system is especially important for maintenance
functions such as feeding (Wysocki, 1979), (6) the olfactory system is
primarily involved in functions that involve pattern (mixture) recognition,
such as might occur in individual, kin or species recognition (Johnston,
1985b). All of these hypotheses have merit, but at present they do not
provide a general theory or even an adequate account of all of the phenomena that have been described.

There may be clear taxonomic differences in the functions of the two
systems. For example, snakes seem to use the vomeronasal system in prey
trailing and identification whereas no similar functions have so far been
demonstrated for mammals (Halpern, 1987). There also may be differences
between species, even within closely related taxonomic groups, but since
we do not yet thoroughly understand the functions of these two systems in
even one species it is difficult to discuss the reasons for possible species
differences.

In my laboratory we have begun a series of experiments in which we
are attempting to take a systematic approach to olfactory communication in
golden hamsters by examining a number of different social behaviors and
endocrine responses that are influenced by chemical signals and determining
the roles of the vomeronasal and olfactory systems in each one. I think it
is important to investigate relatively simple, discrete responses whenever
possible; in this paper I report on: (1) hormonal responses of male hamsters to females and their odors, (2) scent (flank) marking by male hamsters,

Chemical Signals in Vertebrates VI, Edited by R.L. Doty and
D. Müller-Schwarze, Plenum Press, New York, 1992

(3) scent (flank and vaginal) marking by female hamsters, and (4) ultra-sonic calling by female hamsters.

ANDROGEN RESPONSES TO FEMALES OR THEIR ODORS

It is known that male hamsters, like males in many other species of mammals, produce surges in circulating androgens shortly after exposure to females. In hamsters one stimulus that is sufficient to cause such re-sponses is the scent of vaginal secretions (Macrides, et al., 1974). In this first set of experiments, undertaken with Cheryl Pfeiffer, we examined the roles of the vomeronasal and olfactory systems in mediating androgen responses in males to female vaginal secretions.

Lesions of the vomeronasal system were produced by surgical removal of the vomeronasal organ; lesions of the olfactory system were produced by irri-gation of the nasal cavity with a 0.5% solution of zinc sulfate (Johnston and Mueller, 1990, Johnston, 1992). Histology of the nasal cavity was per-formed to determine the success of these procedures. These same methods were used in all experiments reported in this paper.

Males were tested in their home cages in a testing room containing no other hamsters. Vaginal secretions were collected from estrous female donors and stored at -20°C until use. An amount of secretion approximately equal to the amount collected from one estrous female was placed on a glass microscope slide and placed in the male's cage for 10 min. The latency to investigate the slide and the duration of sniffing and licking at the slide were recorded. Thirty minutes after the introduction of the stimulus the animals were lightly anesthetized, a blood sample obtained, and plasma was stored in the freezer (-20°C) until radioimmunoassay.

Lesions of the olfactory mucosa produced deficits in the ability of animals to find and investigate the slide with vaginal secretions. Those animals with olfactory lesions took significantly longer to find the slide than the control animals. The total time spent investigating the secretions was primarily determined by prior sexual experience of the males -- sexually experienced lesioned males investigated significantly longer than both the sexually naive lesioned males and the control males (which were also sexually naive).

Lesions of the olfactory mucosa did not, however, have any effect on the androgen responses of males to vaginal secretions. All groups of males exhibited significant increases in circulating androgen levels 30 minutes after investigating this scent (Johnston, 1990; Pfeiffer, 1988).

Removal of the vomeronasal organ also changed the males' behavioral responses to vaginal secretions. The latency to investigate the secretion was much longer in the lesioned, sexually experienced group than in the control group and the duration of investigation was much shorter in the lesioned, sexually naive group than in the sham control group. There were, however, interactions with sexual experience that we could not confidently interpret since we did not have a sham lesioned, sexually experienced group.

Most interestingly, vomeronasal lesions eliminated the rise in circu-lating androgen levels that normally follow investigation of vaginal secre-tions (Fig. 1). Neither the sexually experienced nor the sexually naive males that had their vomeronasal organs removed showed increased androgen levels after exposure to vaginal secretions, while the sham control animals did. Thus, the vomeronasal system seems to be essential for androgen surges caused by this scent (Johnston, 1990; Pfeiffer, 1988).

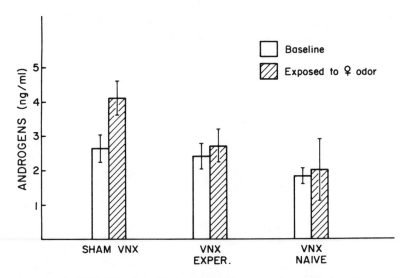

Fig. 1   Mean (±SEM) levels of plasma androgens in male hamsters during
baseline conditions and 30 min. after exposure to vaginal secretions: the three groups of males were sexually naive, sham
lesioned males (SHAM VNX, N=11), sexually experienced males with
vomeronasal organs removed (VNX EXPER, N=12) and sexually naive
males with vomeronasal lesions (VNX NAIVE, N=12). Only the SHAM
VNX males showed significant increases in androgens after exposure
to vaginal secretions (t=2.13, p < .025).

    This result led us to ask if the vomeronasal system was necessary for
the androgen responses that follow actual interactions with estrous females.
Using groups of sexually experienced and sexually naive males that were
either lesioned or not lesioned, we discovered that males with vomeronasal
organs removed did show increases in androgen levels after interacting with
estrous females. Similarly, males with lesions of the olfactory mucosa
also showed androgen responses after interactions with such females. Thus,
lesions of neither the vomeronasal nor the olfactory system alone were
sufficient to eliminate androgen surges in males after interactions with
females. It is also worth noting that neither of these lesions caused
significant changes in the level of male sexual performance (Johnston, 1990;
Pfeiffer, 1988).

    These results raised the question of whether any nasal chemosensory
information is necessary for androgen surges in males in response to interactions with estrous females. To examine this question we tested sexually
naive and sexually experienced males with estrous females after the males
had lesions of both the vomeronasal organ and the olfactory mucosa. Such
males did not mate with females, as one would expect from previous work
(Murphy and Schneider, 1970; Meredith, 1980). Nonetheless, the sexually
experienced males still produced androgen surges after such interactions
(Fig. 2). This demonstrates that for such males cues perceived through
the olfactory and vomeronasal system are not necessary for androgen surges
and also that sexual behavior is not necessary for such surges (Johnston,
1990; Pfeiffer, 1988).

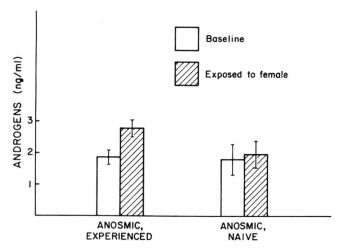

Fig. 2, Mean (±SEM) androgen levels of males after interactions with an
estrous female; males were either sexually experienced (ANOSMIC
EXPERIENCED, N=18) or sexually naive (ANOSMIC NAIVE, N=13); these
males had lesions of the vomeronasal organs and lesions of the olfactory mucosa. Only the experienced males responded with an
androgen surge (t=1.82, p < .05).

Perhaps even more interesting, however, is that in sexually naive
males lesions of both olfactory and vomeronasal systems did eliminate androgen surges following interactions with females. This suggests that
prior to sexual experience chemosensory input is necessary for such androgen responses. One could hypothesize that what is necessary in particular
is the vomeronasal system, and that it may have some specialized mechanism
that is genetically and/or developmentally specified to respond to particular
components of the vaginal secretion. Once a male has sexual experience with
a female he may learn an association between this scent, sexual activity and
other chemical and non-chemical cues. After such learning activity,
the more hard-wired mechanism is then no longer necessary. Whether this
hypothesis is correct or not, it is clear that sexual experience does
effect the mechanisms responsible for androgen surges in males caused by
interactions with females.

In summary, we see that the vomeronasal system is essential for androgen responses to at least one scent from females, that in sexually naive
males chemosensory input is essential for androgen responses to females
themselves, and that sexual experience lessens the importance of chemosensory information for this type of response.

COMMUNICATIVE BEHAVIORS: SCENT MARKING AND ULTRASONIC CALLING

Three communicative behaviors of hamsters that are influenced by conspecific scents are flank marking, vaginal marking, and ultrasonic calling
(for review see Johnston, 1985a). Both males and females flank mark, de-
positing the scent of the flank gland and perhaps other secretions from the
head region and side of the body. Particularly frequent when agonistic
motivation is high, flank marking also occurs at lower frequencies associ-

ated with maintenance activities and grooming (Johnston, 1975a). In males this behavior is to some extent dependent on testicular androgens (Johnston, 1981). The scent that is most important in stimulating marking, at least in males, is the flank gland scent of another hamster (1975b). Vaginal marking deposits the copious and strong smelling vaginal secretions; this behavior is particularly frequent in response to males or their odors and it shows striking correlations with the female's reproductive condition, peaking in frequency a day prior to estrus (Johnston, 1977). It functions, at least in part, as an advertisement of imminent sexual receptivity. Both marking behaviors are more frequent in response to odors of conspecifics as opposed to odors of closely related species of hamsters (Johnston and Brenner, 1982). Ultrasonic calling is performed by both males and females, primarily in sexual contexts. It is stimulated by calls of other hamsters and by odors or contact with opposite-sexed conspecifics; such calls seem to be primarily used as a means of finding or maintaining contact with sexual partners (Floody, 1979).

We found that flank marking that was stimulated by scent of other hamsters was mediated via the olfactory system but not the vomeronasal system in both male and female hamsters. Vomeronasal lesions of male hamsters tested in the presence of scents from another male did not affect flank marking frequency, whereas lesions of the olfactory system resulted in striking reductions in marking frequency. In this latter experiment the mean (±SEM) number of marks per male per trial for a group of 11 males was 12.1±2.0 over the six days following saline treatment but only 2.7±1.0 for six days following zinc sulfate treatment (t=4.066, p<.005; Johnston and Mueller, 1990). Similar experiments with female hamsters exposed to odors of other females showed the same effect of lesions of the olfactory mucosa (Fig.3; Johnston, 1902). In another experiment, females were tested in the presence of both male and female odors; the effects of olfactory lesions were similar, whereas vomeronasal lesions had no effect on flank marking frequency (Johnston, 1992).

Female hamsters were also examined for the effects of lesions on the two sensory systems on vaginal scent marking and ultrasonic calling in response to the combined scents of males and females (Johnston 1992). Vaginal marking was not influenced by vomeronasal organ removal, but was decreased by lesions of the olfactory mucosa (Fig. 3). In contrast, ultrasonic calling was decreased by lesions of the vomeronasal organ and by lesions of the olfactory mucosa. In the vomeronasal experiment the number of calls per female per trial before surgery was 2.7 ± 0.4 and after surgery it was 1.0 ± 0.2 (t=4.343, p<.005). In the olfactory lesion experiment calls before treatment were 2.0 ± 0.1 per female per trial, whereas after treatment they were 0.9 ± 0.3 (t= 4.711, p .0005).

Thus the olfactory system mediates scent-stimulated flank and vaginal marking. Ultrasonic calling in response to conspecific scent is influenced by input through both the olfactory and vomeronasal systems.

CONCLUSION

We have described three functions that seem to be mediated specifically by either the vomeronasal or the olfactory system: flank marking, vaginal marking, and androgen responses of males to the vaginal secretion of females. The two marking behaviors are mediated by the olfactory system whereas the androgen responses are mediated by the vomeronasal system. These findings are consistent with some recent hypotheses about the functions of the two systems, notably that the vomeronasal system has special impor-

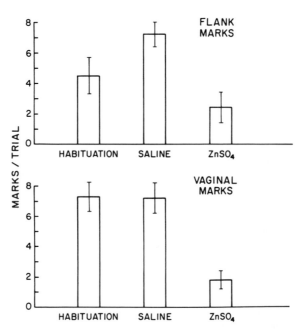

Fig.3,   Mean (±SEM) number of scent marks by females (N=8) in the presence
of scent from other females (marks per female per 5 min. trial).
Females were tested first for an 18 day habituation period, then
for six days following saline treatment and finally for another
six days following zinc sulfate treatment. The decrease in flank
marking frequency after zinc sulfate treatment compared to the
saline control condition was highly significant (t=6.094, p <
.0005), as was the decrease in vaginal marking (t= 9.799, p <
.0005).

tance for neuroendocrine function related to reproduction (Wysocki, 1979, 1989). Also, since scent stimulation of both scent marking behaviors requires species recognition (Johnston and Brenner, 1981), olfactory mediation of these behaviors is consistent with the hypothesis that the olfactory wywtem may be primarily (although not uniquely) responsible for behaviors that require social recognition (Johnston, 1985b).

REFERENCES

Floody, O.R., 1979. Behavioral and physiological analysis of ultrasound production by female hamsters (Mesocricetus auratus). American Zoologist, 19, 443-455.

Halpern, M., 1988. Vomeronasal system functions: Role in mediating the reinforcing properties of chemical stimuli, in: "The Forebrain of Reptiles", W. K. Schwerdfeger and W.J.A.J. Smeets, eds., S. Karger, Basel.

Halpern, M., 1987. The organization and function of the vomeronasal system. Annual Review of Neuroscience, 10, 325-362.

Halpern, M. and Kubie, J.L., 1980. Chemical access to the vomeronasal organs of garter snakes. Physiology and Behavior, 24, 367-371.

Johnston, R.E., 1975a. Scent marking by male hamsters (Mesocricetus auratus) I. Effects of odors and social encounters. Zeitschrift für Tierpsychologie, 37, 75-98.

Johnston, R.E., 1975b. Scent marking by male golden hamsters (Mesocricetus auratus) II. The role of the flank gland scent in the causation of marking. Zeitschrift für Tierpsychologie, 37, 138-144.

Johnston, R.E., 1977a. The causation of two scent marking behaviors in female golden hamsters. Animal Behaviour, 25, 317-327.

Johnston, R.E., 1981. Testosterone dependence of scent marking by male hamsters (Mesocricetus auratus). Behavioral and Neural Biology, 31, 96-99.

Johnston, R.E., 1985a. Communication, in: "The Hamster: Reproduction and Behavior", Harold I. Siegel, ed., Plenum Press, New York.

Johnston, R.E., 1985b. Olfactory and vomeronasal mechanisms of communication, in: "Taste, Olfaction and the Central Nervous System", D. W. Pfaff, ed. Rockefeller University Press, New York.

Johnston, R.E., 1990. Chemical communication in golden hamsters: From behavior to molecules and neural mechanisms, in: "Contemporary Issues in Comparative Psychology", D.A. Dewsbury, ed. Sinauer, Sunderland, MA.

Johnston, R.E., 1992. Vomeronasal and/or olfactory mediation of ultrasonic calling and scent marking by female golden hamsters. Physiology and Behavior, 51,

Johnston, R.E. & Brenner, D., 1982. Species specificity of scent marking in hamsters. Behavioral and Neural Biology, 35, 46-55.

Johnston, R.E. & Mueller, U.G., 1990. Olfactory but not vomeronasal mediation of scent marking by male golden hamsters. Physiology and Behavior, 48, 701-706.

Ladewig, J. & Hart, B.L., 1980. Flehmen and vomeronasal organ function in male goats. Physiology and Behavior, 24, 1067-1071.

Macrides, F., Bartke, A., Fernandez, F. & D'Angelo, W., 1974. Effects of exposure to vaginal odor and receptive females on plasma testosterone in the male hamster. Neuroendocrinology, 15, 355-364.

Meredith, M., 1980. The vomeronasal organ and accessory olfactory system in the hamster, in: "Chemical Signals in Vertebrates and Aquatic Invertibrates", D. Müller-Schwarze and R. Silverstein, eds. Plenum Press, New York.

Meredith, M., 1983a. Sensory physiology of pheromone communication, in: "Pheromones and Reproduction in Mammals", J.G. Vandenbergh, ed., Academic Press, New York.

Murphy, M.R. & Schneider, G.E., 1970. Olfactory bulb removal eliminates mating behavior in the male golden hamster. Science, 167, 302-304.

Pfeiffer, C.A., 1988. Factors contributing to acute androgen surges in male golden hamsters: The role of olfaction and female cues. Ph.D. Dissertation, Cornell University.

Wysocki, C.J., 1979. Neurobehavioral evidence for the involvement of the vomeronasal system in mammalian reproduction. Neuroscience and Biobehavioral Reviews, 3, 301-341.

Wysocki, C.J., 1989. Vomeronasal chemoreception: Its role in reproductive fitness and physiology, in: "Neural Control of Reproductive Function", J. M. Lakoski, J.R. Perez-Polo and D.K. Rassin, eds., A.R. Liss, New York.

Wysocki, C.J., Wellington, J.L. and Beauchamp, G.K., 1980. Access of urinary nonvolatiles to the mammalian vomeronasal organ. Science, 207, 781-783.

OLFACTORY BIOLOGY OF THE MARSUPIAL SUGAR GLIDER - A PRELIMINARY STUDY

D. Michael Stoddart, A. J. Bradley and K. L. Hynes

Zoology Department, University of Tasmania, Australia 7001

INTRODUCTION

The marsupial sugar glider, *Petaurus breviceps* Waterhouse, is a small gliding possum which inhabits many forest types along Australia's east and northern coasts, from southern Tasmania to Darwin, and extends northwards into Papua New Guinea (Strahan 1983). It is abundant and widespread in Tasmania, following a successful introduction in the mid-1850s. Sugar gliders are almost exclusively arboreal and seldom venture to the forest floor. Their diet is mainly invertebrates, though the sugary sap from trees of the genus *Eucalyptus* and the nectar and pollen from *Banksia* are eagerly sought during winter. Highly social mammals, sugar gliders nest and breed in tree holes in small colonies of about 6 to 10 individuals (Suckling 1984). While group structure is not fully understood it is usual for more than one fully mature male to be present. In Tasmania they breed once each year, mating in the winter (July to early September) and give birth after a gestation of about 16 days. A single young, or occasionally twins, remain in the pouch attached to a teat for a further two months. It is possible for females to carry a second litter but this usually occurs only when the first has been lost (Stoddart and Bradley, in press).

The dermal scent glands of the sugar glider were first investigated by Schultze-Westrum (1965). A frontal gland, found only in males, lies between the eyes and was described by Schultze-Westrum as being rubbed on the sternum of cohabiting females, and in this way establishing a community odour. A gland lying at the base of the throat, called a sternal gland by Schultze-Westrum (1965) and again occurring only in males, was observed to be rubbed on the substratum and occasionally on nest partners and was assumed to have a territorial demarcation function. The glands lining the receptive pouch were thought to be involved in facilitating the newborn to find the pouch and to provide information on the sexual status of the female. Paracloacal glands were thought to be involved in territory marking as well as having a role in the establishment of community odour. All of Schultze-Westrum's (1965) work was conducted on a group of 9 sugar gliders from New Guinea (subspecies *papuanus*), held under laboratory conditions in Munich.

Chemical Signals in Vertebrates VI, Edited by R.L. Doty and
D. Müller-Schwarze, Plenum Press, New York, 1992

This paper reports on a continuing study of the olfactory biology of the sugar glider being undertaken in our laboratory in Hobart, Australia, and under natural conditions in the Tasmanian bush, and it attempts to expand Schultze-Westrum's (1965) observations under more natural conditions.

THE SCENT GLANDS

This section will discuss the structure of the gular gland (= sternal gland of Schultze-Westrum (1965)), and the frontal gland.

The gular gland. The gular gland of a fully adult male sugar glider measures about 7 mm in length and 4 mm in width. It lies in the midline of the ventrum at the posterior margin of the throat. It is composed of two layers of secretory tissue - a superficial layer of sebaceous and a profound layer of apocrine tissue - sandwiched between a thin epidermis and the dermal muscle layer. The organ may measure up to 2 mm in thickness.

The frontal gland. The frontal gland straddles the midline of the front of the head, measuring up to 8 mm in length and 6 mm in width. Its structure is similar to that of the gular gland but never exceeds 1.8 mm in thickness.

THE ANNUAL CYCLE OF GLANDULAR DEVELOPMENT

We studied the annual cycle of gular and frontal gland development on a population of free-living sugar gliders inhabiting an open *E.delegatensis* forest in central Tasmania. Individuals were caught in custom built live-traps (Mawbey 1989) and released at the site of capture following the removal of tissue biopsy samples. Full details of the methodology may be found in Stoddart and Bradley (1991), Bradley and Stoddart (1991), and Stoddart and Bradley (unpublished). We developed a system for taking repeated biopsies from both frontal and gular glands; this enabled us to use each individual as its own control, as well as to examine the mean change in gland biology. TABLE 1 shows the mean changes in a) gross glandular depth, b) sebaceous tissue depth, c) apocrine tissue depth d) mean nuclear diameter of sebaceous and e) apocrine tissue together with f) the mean level of testosterone in the blood at four key points in the year. The data show that during the breeding season, and immediately afterwards, the scent glands are most highly developed, although the differences were rarely statistically significant. Our methodology allows us to examine the annual cycle in *individual* sugar gliders and this will be examined elsewhere; the data in Table 1 suggest that substantial differences occur between individual males. We do not know what governs these differences but a preliminary analysis of the consequences of social dominance in the laboratory reveals that some individuals appear to suffer from an inhibition of testicular activity and as a consequence exhibit lowered levels of plasma testosterone (Stoddart and Bradley unpublished). We are currently investigating this phenomenon further to ascertain what roles the odour of frontal and gular secretions play in the inhibition.

SCENT MARKING

In this section we examine frontal and gular gland scent marking within two separate contexts, *viz.* arena marking, and following disruption of the social group.

524

Table 1

Selected frontal and gular gland dimensions and mean plasma testosterone for four seasons of the year (upper value frontal, lower gular)

	Pre-breeding May-June	Breeding July-Sept	Post-breeding Oct-Dec	Interbreeding Jan-April
a) Gross gland depth (μm)	1193.1±527(11) 1687.0±384(12)	1619.0±428(25) 1980.0±409(20)	1790.0±373(17) 1980.0±357(19)	1571±436(20) 1623.0±464(21)
b) sebaceous tissue depth (μm)	445.1±253.3(11) 684.2±238.7(12)	543.0±269.2(25) 826.4±237.9(20)	552.5±208.6(17) 890.7±184.8(19)	444.2±97.8(20) 633.2±289.8(21)
c) apocrine tissue depth (μm)	810.4±412.8(11) 1003.3±216.9(12)	1033.7±339.7(25) 1154.3±309.1(20)	1244.5±303.2(17) 1090.1±240.5(19)	1085.8±447.1(20) 963.4±315.7(21)
d) mean nuclear diameter, sebaceous (μm)	6.4±0.7(11) 6.4±0.5(120)	7.6±0.6(25) 7.8±0.7(20)	7.4±0.7(17) 7.4±1.9(19)	6.6±0.7(20) 6.5±0.5(21)
e) mean nuclear diameter, apocrine (μm)	6.4±1.0(11) 6.5±0.6(12)	6.4±1.0(11) 6.5±0.6(12)	7.5±0.7(17) 7.8±1.0(19)	6.8±0.6(20) 6.6±0.6(21)
f) mean plasma testosterone (nM)	5.6±4.8(26)	5.6±4.8(26)	9.5±14.5(29)	3.9±2.5(27)

All data are means plus/minus 1SD

n = number in parenthesis

525

Arena marking

Scent marking behaviour in a clean arena measuring 2 m x 2 m x 2 m
and constructed of breeze block (clinker brick) was observed in both a
clean arena and in one previously occupied by sugar gliders. To clean
the walls and door of the arena, warm water and detergent was applied
with a scrubbing brush. Detergent was removed with copious quantities
of cold water. The arena was allowed to dry before observations began.
Data were recorded by direct observation (Pienaar, 1988).

Three males and one female sugar glider were introduced into a
clean arena and 16 occurrences of scent marking were observed in three,
three-hour observation periods. Only the socially dominant male was
observed to mark and in 69% of all occasions the frontal gland was
rubbed against an object or the substratum (the ceiling, the upper part
of the walls, the interior of the nest box, the vegetation, and the
floor). On a further 31% of occasions, the gular gland was applied
against an object or the substratum. The upper parts of the walls were
marked the most frequently (50% of all marks), followed by the ceiling
(18%), interior of the nest box (18%), and branches and vegetation
(14%).

In the previously scented arena the frequency of scent-marking
increased sharply. A total of 69 scenting episodes were recorded in
three, three-hour periods and while the dominant male was responsible
for 62% of them, the female was responsible for 18%, the second-ranking
male for 13% and the subordinate for 7%. Sixteen of these episodes
involved the gular gland and the dominant male was responsible for 14
(88%) of them, the remaining 12% was the responsibility of the other
males. The interior of the nest box was marked the most frequently (45%
of all marks), followed by the upper part of the arena walls (43%), the
ceiling (7.5%), branches and vegetation (2.5%) and the floor (2%).

Disruption of the social group

In order to examine whether the dermal glands of male sugar gliders
were used within the social context of the sudden presence of an unknown
interloper (such as a transient would be in the natural environment
Stoddart and Bradley unpubl.), strangers were introduced into colonies
containing either one or two males. In colony 1, with two males one of
which was markedly dominant to the other, a socially subordinate
'intruder' elicited no increase in gular scent marking by the resident
dominant; neither did a castrate male. A dominant male 'intruder',
however, elicited a substantial increase in marking (Fig. 1a). The
resident male in colony 2, reacted to a dominant 'intruder' in the
identical manner to the dominant in colony 1, with a substantial
enhancement of marking (Fig. 1b). All marks, with a single exception,
were made with the gular gland. The data in Fig. 1a and, to a lesser
extent Fig. 1b, indicate that colony members are marked far more
frequently than the physical environment, and that maximum rates were
induced by the presence of an 'intruder'.

DISCUSSION

The data presented above suggest that the two dermal scent gland
complexes may be used differentially in different contexts. When only

scent was encountered, and the social complexion of the group was not under threat, all scent marking was directed heavily towards the nest site; no instance of marking conspecifics was observed, even when activity inside the nest box was studied by video filming. The gular gland was used significantly less often than the frontal gland (p<0.05). Fadem and Cole (1985) noted that grey short-tailed possums (*Monodelphis domesticus*) scent-marked more readily in an arena which had been previously occupied and still retained the scent of its previous occupants, and similar

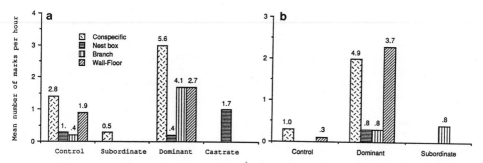

FIG 1 The mean number of scent marks per hour made by the resident males in two colonies of sugar gliders subjected to socially subordinate and dominant 'intruders'. No castrated male was available for use in Fig 1b.

observations have been reported by a number of workers (e.g. Harris and Murie, 1982 for *Spermophilus columbianus*, Mykytowycz and Hesterman, 1970 for *Oryctolagus cuniculus*, Rasa (1973) for *Helogale undulata*, von Holst and Buergel-Goodwin (1975) for *Tupaia belangeri*) In this respect the sugar glider appears to follow the mammalian pattern.

When the cohesion of the group was threatened, however, only the socially dominant male scent marked and almost exclusively he used the gular gland. Marking was directed mainly towards conspecifics though the physical environment was not ignored. Our data extend those of Schultze-Westrum (1965) and, in one major respect, differ from them. Schultze-Westrum (1965) observed that the frontal gland was rubbed on conspecifics. It has already been noted that conspecific marking was only performed under conditions of social unrest and then only with the gular gland of the dominant resident male. The frontal gland moreover was repeatedly used to mark objects - not conspecifics - within the context of arena marking; this use for the frontal gland was not reported by Schultze-Westrum (1965). The preliminary nature of these data are recognised and we admit that there may be subspecific differences between the southern and northern subspecies of sugar glider, yet the scent marking behaviours we have observed have been very visible and easy to record and we are satisfied that no marking passed unrecorded during our observation periods.

This study indicates that the two scent gland complexes in the sugar glider are used for scent marking under very different social circumstances. Our endocrinological work (Stoddart and Bradley unpubl.) has indicated that the hormonal underpinning of the two systems is not identical, lending support to the above reported behavioural observations.

REFERENCES

Bradley, A.J and Stoddart, D.M. 1991, A non-destructive small scale biopsy method for obtaining repetitive tissue samples from skin glands in a marsupial. Aust. Mammal. 14: 151

Fadem, B.H. and Cole, E.A. 1985, Scent-marking in the grey short-tailed opossum (*Monodelphis domestica*). Anim. Behav. 33: 730

Harris, M.A., and Murie, J.O. 1982, Responses to oral gland scents from different males in Columbian ground squirrels. Anim. Behav. 30: 140

Mawbey, R 1989 A new trap design for capture of sugar gliders, *Petaurus breviceps*. Aust. Wildl. Res. 16: 425

Mykytowycz, R and Hesterman, E.R. 1970, The behaviour of captive wild rabbits, *Oryctolagus cuniculus* (L) in response to strange dunghills. Form. et Funct. 2: 1

Pienaar, V 1988, The behaviour of *Petaurus breviceps*. Unpublished Honours thesis, University of Tasmania

Rasa, O.E.A. 1973, Marking behaviour and its social significance in the African dwarf mongoose, *Helogale undulata rufula*. Z. für Tierpsychol. 32: 293

Schultze-Westrum, T 1965, Innerartliche Verstandigung durch Dufte beim Gleitbeutler *Petaurus breviceps papuanus* Thomas (Marsupialia: Phalangeridae). Z. für vergleich. Physiol. 50: 151

Stoddart, D.M. and Bradley, A.J. 1991, The frontal and gular dermal scent organs of the marsupial sugar glider (*Petaurus breviceps* Waterhouse). J. Zool. (Lond) 225: 1

Stoddart, D.M. and Bradley, A.J. (unpubl) The relationship between free cortisol, male body mass, testosterone, and relative density in a free- living population of the sugar glider (*Petaurus breviceps* Waterhouse 1939; Marsupialia: Petauridae). J. Anim. Ecol. - submitted

Strahan, R., (ed) 1983, Complete Book of Australian Mammals. Angus & Robertson, Sydney

Suckling, G.C., 1984, Population ecology of the sugar glider, *Petaurus breviceps*, in a system of fragmented habitats. Aust. Wildl. Res. 11: 49

von Holst, D and Buergel-Goodwin, U 1975, Chinning by male *Tupaia belangeri*: the effects of scent marks of conspecifics and of other species. J.Comp. Physiol. 103: 153

PREDATOR-ODOR ANALGESIA IN DEER MICE:

NEUROMODULATORY MECHANISMS AND SEX DIFFERENCES

Martin Kavaliers, Duncan Innes and Klaus-Peter Ossenkopp

Division of Oral Biology, Faculty of Dentistry
and Department of Psychology
University of Western Ontario
London, Ontario, Canada   N6A 5C1

INTRODUCTION

Animals respond to the threat of predation with a number of
defensive behaviors, including: flight, immobilization or 'freezing',
risk assessment, increased wariness and the suppression of non-defensive
behaviors (Blanchard et al. 1990). It has become evident that a
reduction in nociceptive and pain sensitivity (antinociception or
analgesia, respectively) is a also a major correlate of predator
exposure. Activation of endogenous analgesic mechanisms has been
demonstrated in laboratory mice and rats exposed to a cat (Lester and
Fanselow, 1985; Lichtman and Fanselow, 1990; Kavaliers and Colwell, 1991,
1992), deer mice and white-footed mice exposed to a weasel (Kavaliers,
1988, 1990), and laboratory mice exposed to the calls of avian predators
(Hendrie, 1991). There is mounting evidence that these analgesic
responses are an important component of an animal's defense repertoire.
Defensive systems that are activated by either innate or learned danger
stimuli, such as that of predators, may inhibit nociceptive and pain
sensitivity (i.e. induce analgesia) associated with either the perception
of, defense against, and/or recuperation from the danger stimuli (Bolles
and Fanselow, 1980). As such, analgesia is advantageous in predator
exposure, in which responding to noxious stimulation might compete with
and/or disrupt effective defensive behaviors.

Results of studies with laboratory stressors have revealed that,
depending on the characteristics of the stimuli, either endogenous opioid
(i.e. endorphin, enkephalin) or non-opioid mediated antinociceptive
mechanisms are activated (Amit and Galina, 1986). A similar response
pattern is evident for predator-induced analgesia. Relatively long (15
min) predator exposures can induce opioid analgesia while brief (30 sec
to 5 min) exposures are suggested to induce non-opioid analgesia,
involving serotonergic-1A (5-HT$_{1A}$), and to a lesser extent,
benzodiazepine mechanisms (Kavaliers, 1988, 1990; Lichtman and Fanselow,
1990; Hendrie, 1991; Kavaliers and Colwell, 1991, 1992). In addition,
there are indications of sex differences in opioid and non-opioid
mediated predator analgesia. Male mice exposed to a cat display greater
opioid analgesia than females, while females exhibit larger non-opioid
mediated responses (Kavaliers and Colwell 1991).

Chemical Signals in Vertebrates VI, Edited by R.L. Doty and
D. Müller-Schwarze, Plenum Press, New York, 1992

In this paper, we consider opioid and non-opioid involvement in the mediation of predator odor-induced analgesia in male and female deer mice. We describe the effects of brief (30 sec) and prolonged (15 min) durations of exposure to a synthetic weasel odor (Crump, 1980; Crump and Moors, 1985) on the nociceptive responses of deer mice. We also examined the effects of the opiate antagonist, naloxone, and the 5-HT$_{1A}$ agonist, 8-hydroxy-2-(din-n-propylamino) tetralin (8-OH-DPAT) (Dourish et al. 1986), on the predator induced responses.

METHODS

Subjects. Sexually mature deer mice, <u>Peromyscus maniculatus artemisae</u>, (20-30 g; 2-months -1 year of age) were housed as mixed- sex pairs, in polyethylene cages provided with cotton nesting material and Beta chip bedding at 21 ± 1°C under a 20L:8D cycle. Food (Purina Rat Chow) and water were provided ad lib. The laboratory bred mice were derived from a population present near Kamploops, British Columbia where weasels are present and are predators of deer mice (Hall, 1981). The mice used in this study had not previously encountered weasel odor.

<u>Experimental Procedures</u>. During the mid-photoperiod eight male and female mice were exposed, while in their home cages, for either 30 sec or 15 min to synthetic weasel odor (2-propylthietane and 3-propyl-1, 2-dithiolane (Crump, 1980, Crump and Moors 1985). Nociceptive responses of the mice were determined 15 min prior to exposure and immediately (15-30 sec) and 15 min after exposure. Nociception was measured as the latency of a foot-licking response to an aversive thermal stimulus (50 + 0.5 C, hot-plate, Omni-Tech, OH). Either immediately, or 15 min before being exposed for 15 min and 30 sec, respectively, to the weasel odor, male and female mice received intraperitoneal injections of either naloxone hydrochloride (1.0 mg/kg/10 ml saline), 8-OH-DPAT (0.50 mg/kg/10 ml saline) or isotonic saline vehicle (10 ml/kg). Thermal response latencies of individual mice were recorded prior to injection and predator exposure and immediately after exposure. These doses of naloxone and 8-OH-DPAT have no evident effects on the locomotor activity and nociceptive responses of mice (Kavaliers and Colwell, 1991, 1992; Rodgers and Shepherd, 1989). Data were analyzed by repeated measures analysis of variance and post-hoc comparisons were done with Newman-Keuls tests with the significance level set at 0.05.

RESULTS

Male and female mice exposed to the weasel odor for either 30 sec or 15 min were analgesic, displaying significantly (p<0.01) greater thermal response latencies than either prior to exposure (Fig. 1). Such response latencies were also markedly longer than those exhibited by mice exposed to peppermint as a control for novelty (Kavaliers, 1990). Maximum analgesia was evident directly after exposure with a decline to basal levels by 15 and 60 min for the 30 sec and 15 min exposures, respectively. Mice exposed to the weasel odor for 15 min displayed significantly greater amplitude (p<0.01) and duration (p<0.01) of analgesia than did mice that were exposed for 30 sec (Fig. 1). There were also sex differences in the predator-induced analgesia. The analgesia induced by 15 min of exposure to the weasel odor was significantly (p<0.05, for 0 and 15 min post exposure) greater in male than female mice (Fig. 1). In contrast, the analgesia induced by the 30

sec exposure was significantly greater (p<0.05 at 0 min post exposure) in
the female than male mice (Fig. 1). In both sexes naloxone blocked
(p<0.01, for both sexes) the analgesia induced by 15 min of exposure to
the weasel odor, with the nociceptive responses of the naloxone-injected
mice being equivalent to those of unexposed mice (Fig. 1 A, B). Neither
8-OH-DPAT nor saline had any significant effects on the analgesia arising
from 15 min of exposure to the weasel odor. The analgesia induced by 30
sec of exposure to the weasel odor was blocked (p<0.01) by 8-OH-DPAT in
both male and female mice (Fig. 1 C,D). Neither naloxone nor saline had
any significant effects on the analgesia arising from the 30 sec
exposure.

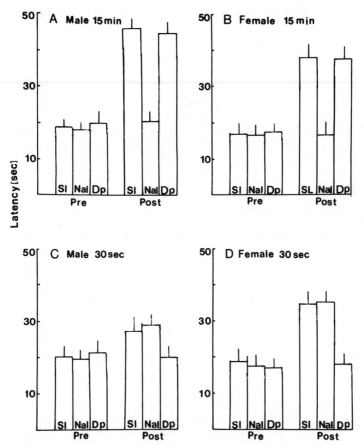

Figure 1. Effects of intraperitoneal injections of
either the opiate antagonist naloxone (Nal: 1.0
mg/kg), the 5-HT$_{1A}$ agonist, 8-OH-DPAt (Dp; 0.5 mg/kg)
or the saline vehicle (S1, 10 ml/kg) on the thermal
(50 C) nociceptive response latencies of male and
female deer mice exposed to weasel odor. A, B = 15
min male and female, and C, D = 30 sec male and female
exposure groups, respectively. Nociceptive responses
were determined 15 min after injection (Post), as well
as 15 min prior (Pre) to predator exposure and
injection. N=8, in all cases. Vertical lines denote
standard errors of the mean.

DISCUSSION

Opioid and Non-opioid Predator Analgesia

These observations with weasel odor indicate that predator associated olfactory cues have an important role in activating endogenous opioid and non-opioid mediated analgesic mechanisms. This is reinforced by the fact that brief exposures to either a novel odor or a non-predator have minimal analgesic effects (Kavaliers 1990, Kavaliers and Colwell 1991a). These findings also reveal that actual physical presence of or attack by a predator is not an essential prerequisite for the activation of analgesic mechanisms.

Exposure of deer mice for 15 min to just the odor of a weasel induces prolonged (60 min) opioid mediated antinociceptive responses that are blocked by naloxone. Likewise, adult meadow voles, which have been shown to avoid weasel odor in the field (Sullivan et al. 1988), also display a marked opioid mediated analgesic response after 15 min of exposure to weasel odor (in preparation). These antinociceptive responses are of a similar duration and magnitude as the analgesia observed after peripheral administration of opiate agonists such as morphine (Innes and Kavaliers, 1987). Furthermore, in both voles and deer mice these analgesic responses are affected by opiate antagonists acting at a number of different specific opioid receptor types (mu, delta, and kappa). This suggests that a 15 min predator exposure leads to the activation of multiple central endogenous opioid systems that are involved in the modulation of a variety of behavioral and physiological functions and may contribute to other behavioral modifications (e.g. locomotor activity, feeding levels) observed after extended predator exposures.

Brief (30 sec) exposures of deer mice and meadow voles (in preparation) to weasel odor also elicits analgesia that is of short duration (15 min) and insensitive to naloxone and other more specific opiate antagonists. This non-opioid analgesia is, however, blocked by compounds with agonist activity at $5\text{-}HT_{1A}$ sites, including, 8-OH-DPAT, buspirone and ipsapirone. The antagonistic effect of these compounds may be related to the inhibition of $5\text{-}HT_{1A}$ transmission by an agonist action on somatodendritic autoreceptors (Dourish et al., 1986). Since such activity may also underlie the anxiolytic effects of these compounds (Dourish et al. 1986), these findings are consistent with the involvement of anxiety in the initiation of this particular form of adaptive antinociception. This does not, however, preclude a role for other mechanisms (e.g. benzodiazepine, GABA, neurosteroid) in the mediation of other predator odor-induced responses.

A similar dichotomy in analgesic responses and control mechanisms is also evident under conditions of social conflict. Brief exposure of subordinate male mice to an aggressive, dominant conspecific (or olfactory signals associated with the dominant) leads to 5-HT-sensitive analgesia in the subordinate mouse, while extended interactions induce opioid-mediated responses (Rodgers and Randall, 1987; Rodgers et al. 1991). As with predator exposure, the opioid mediated analgesia can be related to stress and passive defense responses, while the non-opiuoid mediated analgesia is associated with anxiety, fear, and accompanying more active defensive behaviors.

Non-opioid analgesic consequences of brief predator exposure have

not always been reported. The extent of opioid and non-opioid mediation
of predator-induced analgesia is suggested to be sensitive to the
nociceptive test (Lichtman and Fanselow, 1990). Genetic factors and a
prey population's history with a predator can also affect predator
recognition (Hirsch and Bolles, 1980) and the expression of opioid and
non-opioid mediated predator analgesia. Deer mice derived from a
mainland population with a history of weasel exposure display both opioid
and non-opioid mediated analgesia, whereas, mice from an island
population with no weasel exposure history display primarily opioid
mediated responses (Kavaliers, 1990).

Sex Differences in Predator Analgesia

The present observations with deer mice, as well as findings from
laboratory mice and voles (Kavaliers and Colwell, 1991; in preparation),
suggest that there are also sex differences in predator-induced
analgesia. After 15 min of exposure to weasel odor, male deer mice and
voles display significantly greater naloxone sensitive, opioid analgesia
than females, whereas after a 30 sec exposure female mice display
significantly greater non-opioid, $5-HT_{1A}$ mediated, 8-OH-DPAT sensitive,
analgesia than male mice. These sex differences are also consistent with
the sex differences in 5-HT mediated antipredator defensive reactions and
ultrasound production in rats exposed to a cat (Blanchard et al., 1991).

The present findings are consistent with the suggestions of greater
opioid mediated behavioral stress responses in males than females, and
greater serotonergic sensitive anxiety in females than males. Sex
differences have been demonstrated in opioid mediated antinociception
with males, including that of deer mice, displaying significantly greater
levels of exogenous opiate-induced analgesia than females (Innes and
Kavaliers, 1987; Bodnar et al., 1988). There are also sex differences in
laboratory stress-induced analgesia, with females, including that of deer
mice, displaying lower levels of endogenous opioid mediated
antinociception than males (Bodnar, et al., 1988). These sex differences
in opiate effects, whereby males display greater analgesic responses than
females, have been proposed to be modulated by gonadal steroids, with
there being a sexual dimorphism in the density and pattern of central
opiate binding sites (Simmmerley et al., 1988; Hammer, 1990). Whether or
not estrous stage influences the level of predator-induced analgesia
remains to be determined, though, there is evidence that the
susceptibility to predation in deer mice varies across the estrous cycle
(Cushing, 1985).

The observations of larger 5-HT mediated antinociceptive responses
in female than male mice agree with evidence for greater anxiety related,
5-HT mediated responses in females than males (Kennet et al., 1986,
Johnson and File, 1991). These sex differences are also consistent with
the greater behavioral anxiety in female than male rats that have been
exposed to cat odor (Blanchard et al., 1990). These sex differences in
5-HT mediated anxiogenic related responses also agree with the apparent
greater susceptibility of female deer mice to detection by weasels
(Cushing, 1985).

ACKNOWLEDGEMENTS

We thank Phero Tech Inc. and Dr. D. Crump for providing the weasel
odor. We also thank Susan Lipa, Donna Tysdale and Norman Yu for their

technical assistance. The research described here was supported by Natural Sciences and Engineering Research Council of Canada grants to M.K. and K.- P.O.

REFERENCES

Amit, N., and Galina, Z.H., 1986, Stress-induced analgesia: adaptive pain suppression, Physiol. Rev., 30:561-602.

Blanchard, D. C., Shepherd, J. K., Rodgers, R. J., Agulla, R., Flores, T., and Blanchard, R. J., 1991, Sex differences in antipredator defensive reactions and the effects of anxiolytic drugs, Soc. Neurosci. Abstr., 17:877.

Blanchard, R.J., Blanchard, C.D., Rodgers, J., and Weiss, S.M., 1990, The characterization and modelling of antipredator defensive behavior, Neurosci. Biobehav. Rev., 14:463-472.

Bodnar, R.J., Romero, M.T., and Kramer, E., 1988, Organismic variables and pain inhibition: roles of gender and aging, Brain Res. Bull., 21:947-953.

Bolles, R.C., and Fanselow, M.S., 1980. A perceptual-recuperative model of fear and pain, Behav. Brain Sci., 3:291-323.

Crump, D.R., 1980, Thietanes and dithiolanes from the anal gland of stoat (Mustela erminea), J. Chem. Ecol., 6:341-347.

Crump, D.R., and Moors, P.J., 1985, Anal gland secretions from the stoat (Mustela erminea), and the ferret (Mustela putorius forma furo): Some additional thietane components, J. Chem. Ecol., 11:1037-1043.

Cushing, B.S., Estrous mice and vulnerability to weasel predation, Ecology, 66:1976-1978.

Dourish, C.T., Huston, P.M., and Curson, G., 1986. Putative anxiolytics 8-OH-DPAT, buspirone and TVX Q are agonists at $5-HT_{1A}$ autoreceptors on the raphe nuclei, Trends Pharmacol. Sci., 7:212-214.

Hall, E. R., 1981, "The Mammals of North America, Vol. 2, " Wiley-Interscience, New York.

Hammer, R.P., Jr., 1990, μ-Opiate receptor binding in the medial preoptic area is cyclical and sexually dimorphic, Brain Res., 515:187-192.

Hendrie, C.A., 1991, The calls of murine predators activate endogenous analgesia mechanisms in laboratory mice, Physiol. Behav., 49:569-573.

Hirsch, S.M., and Bolles, R.C., 1980, On the ability of prey to recognize predators, Z. Tierpsychol., 54:71-84.

Innes, D.G.L., and Kavaliers, M., 1987, Opiates and deer mouse behaviour: differences between island and mainland populations, Can. J. Zool., 65:2504-2512.

Johnston, A.L., and File, S.E., 1991, Sex differences in animal tests of anxiety, Physiol. Behav., 49:245-250.

Kavaliers, M., 1988, Brief exposure to a natural predator, the short-tailed weasel, induces benzodiazepine-sensitive analgesia in white-footed mice, Physiol. Behav., 43:187-193.

Kavaliers, M., 1990, Responsiveness of deer mice to a predator, the short-tailed weasel: population differences and neuromodulatory mechanisms, Physiol. Zool., 63:388-407.

Kavaliers, M., and Colwell, D.D., 1991, Sex differences in opioid and non-opioid mediated predator-induced analgesia in mice, Brain Res., submitted.

Kavaliers, M., and Colwell, D.D., 1992, parasite modification of predator responses in mice: nociceptive models and neuromodulatory mechanisms, Anim. Behav., in press.

Kennet, G.A., Chaouloff, F., Marcou, M., and Curzon, G., 1986. Female rats are more vulnerable than males in animal models of depression: the possible role of serotonin. Brain Res., 382:416-421.

Lester, L.S., and Fanselow, F.S., 1985, Exposure to a cat produces opioid analgesia in rats, Behav. Neurosci., 99:756-759.

Lichtman, A.H., and Fanselow, M.S., 1990, Cats produce analgesia in rats on the tail-flick test: naltrexone sensitivity is determined by the nociceptive test stimulus, Brain Res., 533:91-94.

Rodgers, R. J., and Randall, J. I., 1987, Defensive analgesia in rats and mice. Psychol. Rec., 37:335-347.

Rodgers, R. J., and Shepherd, J. K., 1989, 5-HT agonist, 8-hydroxy-2-(di-n-propylamino)tetralin (8-OH-DPAT) inhibits non-opioid analgesia in defeated mice: influence of route of injection. Psychopharmacology, 97: 163-165.

Rodgers, R. J., Shepherd, J. K., and Donat, P., 1991, Differential effects of novel ligands for 5-HT receptor subtypes on non-opioid defensive analgesia in mice. Neurosci. Biobehav. Rev., in press.

Simerly, R. B., McCall, L. D., and Watson, S. J., 1988. Distribution of opioid peptides in the pre-optic region: immunohistochemical evidence for a steroid-sensitive enkephalin sexual dimorphism. J. Comp. Neurol., 276:442-459.

Sullivan, T.D., Crump, D.R., and Sullivan, D.S., 1988, Use of predator odors as repellants to reduce feeding damage by herbivores. III Montane and meadow voles (Microtus montanus and Microtus pennsylvanicus), J. Chem. Ecol., 14:363-377.

SCENT COMMUNICATION IN THE RAT

Stephan E. Natynczuk[1,2] and David W. Macdonald[2]

[1]Clifton Scientific, Clifton College, Bristol, BS8 3JH
and School of Chemistry, Univ. of Bristol, Bristol BS8
ITS, U.K.; [2]Wildlife Conservation Research Unit, Dept. of
Zoology, South Parks Road, Oxford, OX1 3PS, U.K.

INTRODUCTION

The use of scent in mammalian social behavior has been studied in a
wide range of species (Brown and Macdonald, 1985), and there is a consensus amongst workers in semiochemistry that some of an animal's social
odors may reflect its hormonal profile (Aron, 1979; Milligan, 1980;
Quay, 1982; Yahr and Commins, 1982; Albone, 1984; Krebs and Dawkins,
1984; Jemiolo et al., 1987). In this paper we will consider briefly the
links between scent, biochemistry and certain aspects of evolutionary
theory with regard to the Norway rat, Rattus norvegicus. For reasons of
space, we will confine ourselves to a discussion of the males' olfactory
investigation of females.

OBSERVING RATS TO DETERMINE SOCIALLY IMPORTANT SCENT SOURCES

Rats are endowed with a number of odor sources (Albone, 1984;
Brown, 1985; Natynczuk, 1990b). An important step towards assessing the
social importance of these odors is to determine how rats use their
various scent glands during social behavior. Once an odor source has
been identified as socially important, its information content can be
assessed chemically and hypotheses concerning chemical activity can be
tested experimentally.

The sebaceous glands of the rat, compared to such glands of some
other small mammals, have received little attention (Brown and Macdonald, 1985). Sebaceous signalling is especially interesting because of
the links between endocrinology and lipid metabolism. Sebum has many of
the qualities one would expect of a good source for a chemical signal.
It is responsive in terms of quantity and quality to endocrinological
variations and it offers a mechanism for signalling individuality, as
well as hormonal status. For example, oestradiol has a marked effect in
suppressing the secretion of sebaceous glands in the dorsal region of
female rats (Ebling and Skinner, 1967, Ebling, 1977). Changes in the
rate of sebum secretion due to changes in the concentration of circulating female sex hormones could be a simple and straightforward mechanism
for chemical signalling, either directly or via modification of sebum by
micro-organisms resident on the skin.

Chemical Signals in Vertebrates VI, Edited by R.L. Doty and
D. Müller-Schwarze, Plenum Press, New York, 1992

In contrast to the many potential scent sources of the genital region (Albone, 1984,) the haunch provides essentially only one potential source of scent, the skin sebaceous glands, which in the case of rats have been largly overlooked (Brown, 1985). Large sebaceous glands do, however, occur in the genital region in the form of preputial glands in the male rat and clitoral glands in the female (see below). The function of sebum and, indeed, of most natural waxes is open to speculation (Kolattukudy, 1976). The inhibition of some micro-organisms by surface lipids (Koidsumi, 1957; Martin, 1964; Nicholaides, 1974) facilitates colonization by others, so that quantitative changes in the sebum components could control the ecological balance of skin organisms, and the generation of semiochemicals by selected micro-organisms (Albone et al., 1977).

In this phase of our research, male and female hybrids (F1) of wild caught and laboratory rat strains (Wistar and DA) of rats were observed in a semi-natural habitat (Natynczuk, 1990a; Natynczuk and Macdonald, 1992a). The importance of each of the rats' scent sources (Harderian, Meibomian, skin sebaceous, genital, etc.) was assessed by determining the proportion of sniffs each scent source received. A sniff was defined as a rat placing its nose in contact with the body of another rat. The rats' bodies were divided into a number of zones at which sniffs could be directed (Natynczuk, 1990b). Male rats were either brothers of the females they were tested against or not related at all.

The results showed that the females' haunches received the largest proportion of sniffs; the genitals were sniffed almost as often (Natynczuk, 1990 a,b). Male rats directed significantly more sniffs towards the genital region of unrelated than related oestrous females. The males' sniffing behavior at the genitals of oestrous and non-oestrous related females did not differ significantly when assessed in terms of the proportion of sniffs they received, although the exact provenance of the odors being investigated there could not be determined.

CHEMICAL AND ANATOMICAL ANALYSIS OF BEHAVIORALLY IMPORTANT SEMIOCHEMICAL REGIONS

Yahr and Commins (1982) discussed the central role of the endocrine system in synchronizing and controlling social behavior and scent marking activity, thus increasing the accuracy with which an animal is able to signal its hormonal status. An animal able to interpret such signals is at great selective advantage because it is thus better able to make predictions about that individual's behavior (Krebs and Dawkins, 1984).

Natynczuk (1990a) and Natynczuk and Macdonald (1992b) demonstrated significant pre- and post-pubertal effects on preputial gland mass in males and significant differences in the clitoral gland masses of oestrous and non-oestrous females. In addition such measures were also altered during pregnancy. The preputial and clitoral glands are known to be a rich source of lipids of varying complexity (Clevedon Brown and Williams, 1972; Albone, 1984) and differences in sex hormone profiles between sexes are reflected in the quantitative chemistry of the glands (Natynczuk, 1990a; Natynczuk and Macdonald, 1992b). Thus, information pertinent to an individual's hormonal status is potentially available to other rats in the form of sebum-derived chemical signals.

To test whether sebaceous glands of females' haunches differ significantly in output as a funtion of the oestrous cycle, histological sections were made from the shoulders and haunches of female rats obtained from a pest control operation at an Oxfordshire farm. The female rats' phase of oestrus was determined at the time of capture by vaginal epithelial cell counts (Zarrow et al., 1964). Natynczuk (1990a) found significant differences in the surface area to volume ratio of sebaceous alveoli in the shoulder and haunches between oestrous and non-oestrous female rats. The mean surface area to volume ratios per rat for the haunch were significantly smaller than those of the shoulder in oestrous females, with fewer significant differences in the case of non-oestrous females. These data, which are in broad agreement with Ebling's findings (Ebling and Skinner, 1967, 1975; Ebling, 1974), strongly imply a fluctuating production of sebum in the haunch of wild female rats in synchrony with the oestrous cycle.

Do rats' body scents represent chemical images containing biologically important information? Albone's (1984 p.6) chemical image hypothesis (see also Natynczuk and Albone in this volume) is an "holistic one in which nature of the entire chemical image presented by one animal to another is surveyed". Quantitative changes in this chemical spectrum being a signal of change in certan physiological variables, and a reflection of underlying endocrinological changes such as those associated with the oestrous cycle.

We used the Dynamic Solvent Effect (DSE) (Apps, 1990) and capillary gas chromatography to compare the generalized body odors of the haunch of sexually mature male rats with those of oestrous and dioestrous female rats. Female rats with regular oestrous cycles were sampled in dioestrus and again at their following oestrus. Each male rat was sampled twice over a period equivalent in time to a female's oestrus and dioestrus. Peaks were matched (by retention time) by eye from each chromatogram. The results showed enormous individual variation in the size and ratio of peak areas, and no one peak was a reliable indicator of either sex or reproductive condition. The same variation meant that simply comparing the chromatograms by eye did not reveal three obvious classes of rats. However, post hoc superimpositon of the headspace chromatograms suggested that the males' scent profiles were more constant in their pattern and size of peaks than were those of the females.

Twenty-two peaks common to all chromatograms were located and their proportional peak areas used in a Principal Component Analysis (PCA) of male and oestrous and non-oestrous female rats' chromatograms. PCA is a multivariate technique for examining relationships among several quantitative variables in the search for structure, both between variables and between individuals (Gilbert, 1989). The statistical analysis showed that the sexual status of the rats, i.e., the first principal component, accounted for 79.5% of the variance in proportional peak areas for the 3 classes of experimental animals. The PCA revealed a close grouping of scent profiles (see Natynczuk and Albone, this volume) and suggests that the general body odor of the rat provides a passive signal of the animal's hormone profiles (Natynczuk, 1990a).

SCENT MATCHING AND SEBACEOUS SKIN GLANDS IN SIGNALLING OESTRUS

The rat's oestrous cycle period is four to seven days and the duration of oestrus, per se, is about 12 hours (Meehan, 1984, p.41). During this time profound metabolic changes occur which affect behavior (Rubin and Barfield, 1983) and scent (Albone, 1984). Increased levels of oestrogen limit the rate of lipogenesis (Ebling, 1977) and the resultant quantitative change in sebum biosynthesis could produce a change in scent. One or both of the following factors might significantly alter the scent profile to reflect a physiological change: (i) the reduced output of sebum may be associated with a change in the proportions of its component lipids, which in turn changes the substrate available for modification to scent by the host's skin micro-organisms; (ii) alteration of the ecological balance of skin micro-organisms through changes in the proportions of sebum lipids. Ebling and Skinner (1967) did identify changes in secretion rates in terms of changes in the ratios of certain sebum components: fatty acid and ratios nc_{16}/nc_{18} and $nc_{18:1} {}_9/nc_{18}$ increase in androgen treated sebaceous tissue. This discussion of sebum also applies to the preputial and clitoral glands, with those of oestrous females being significantly smaller than those of non-oestrous females (Natynczuk, 1990a; Natynczuk and Macdonald, 1992b).

If changes in the females' scent profiles with the oestrous cycle can be explained by changes in sebum output, and a difference in scent quality occurs between the shoulder and the haunch, then the female rats' oestrous scent is self-calibrating, i.e., the male would not have to cross-refer to a learned or innate reference odor, but simply match the scent of the same female's shoulder and haunch. The difference in scent would be the measure of oestrus.

Frequent grooming would be required to clean away old sebaceous signals, and rats tend to keep themselves scrupulously clean. Meehan (1984, p.310) reports that rodents spend up to 20% of their waking time grooming. There may also be an enhancement of scent via the action of saliva on the sebum, not necessarily due to any chemical reaction between the saliva and the sebum, but due simply to wetting the fur.

The sniffing behavior of male rats towards females is compatible with the self-calibrating hypothesis. The rat haunch has a relatively large surface area and accordingly receives a large number of sniffs, although statistically it receives no more per unit area than any other region of the rat's trunk. However, the sequence of sniffing is important. Contingency tables leading to the calculation of transitional probabilities (Bakeman, 1986) of one sniff following another were constructed to allow a simple comparison of the temporal patterns of olfactory behavior. These indicated that males frequently sniff the forequarters and haunch of a female in quick succession (Natynczuk, 1990a; Natynczuk and Macdonald, 1992a).

DISCUSSION

Chemical signalling must be the oldest form of communication (Macdonald and Brown, 1985). It is surely the only plausible means whereby primitive organisms might originally have communicated, and probably played a crucial role in the evolution of the eucaryotes (Broda, 1975). Within an animal, communication is via electro-chemical and chemical channels. There is feed-back between the tissues and the genetic machinery which drives the survival and evolution of the organism (Dawkins, 1986). If the genes of conspecifics could (chemically) communicate directly with each other or, failing that, via the products of biochemical processes under direct genetic control, then communication between survival machines (Dawkins, 1986) would be at its most efficient. This notion is easily applied to the workings of primer pheromones which can be considered as extra-organism hormones (Novotny et al., 1990) targeted at receptive conspecifics.

The effectiveness with which sebum chemistry mirrors endocrinology and broadcasts an animals's hormone status has profound implications for the role of social odors (Krebs and Dawkins, 1984). The rats sebaceous skin glands are easily accessible by other rats and contain information about an individual's personal identity, sex, and reproductive condi-

tion. General body odors may contain additional information of special
social value, for example, high levels of testosterone are correlated
with aggression and dominance in male mammals (Barfield, 1984). Apps,
Rasa, and Viljoen (1988) have shown that the body odor of male mice
provides a good prediction of the individual's tendencies to initiate
aggressive interactions.

REFERENCES

Albone, E.S., 1984, "Mammalian Semiochemistry", Wiley Interscience,
    Chicester.
Albone, E.S., Gosden, P.E., and Ware, G.C., 1977, Bacteria as a source
    of chemical signals in mammals, in, "Chemical Signals in Verte-
    brates", D. Muller-Schwarze and M.M. Mozell (eds.), Plenum Press,
    New York, pp. 35-43.
Apps, P.J., 1990, Dynamic solvent effect sampling for quantitative
    analysis of semiochemicals, in:, "Chemical Signals in Vertebrates
    V", D.W. Macdonald, D. Muller-Schwarze, and S.E. Natynczuk, (eds.),
    Oxford University Press, Oxford.
Apps, P.J., Rasa, A., and Viljoen, H.W., 1988, Quantitative chromato-
    graphic profiling of odors associated with dominance in laboratory
    mice, Aggressive Behavior, 14:451-461.
Aron, C., 1979, Mechanisms of control of the reproductive function by
    olfactory stimuli in female mammals, Physiol. Reviews, 59:229-284.
Bakeman, R., 1986, "Observing Interaction: An Introduction to Sequential
    Analysis", Cambridge University Press, Cambridge.
Barfield, R.J., 1984, Reproductive hormones and aggressive behavior,
    Biol. Perspect. Aggress., pp. 105-134.
Broda, E., 1975, "The Evolution of the Bioenergetic Process", Pergamon
    Press, Oxford.
Brown, R.E., 1977, Odor preference and urine-marking scales in male and
    female rats: effects of gonadectomy and sexual experience on
    responses to conspecific odors, J. Comp. Physiol. Psychol., 5:1190-
    1206.
Brown, R.E., 1985, The rodents, in, "Social Odours in Mammals",
    R.E. Brown and D.W. Macdonald (eds), Oxford University Press,
    Oxford, pp. 245-506.
Brown,R.E. and Macdonald, D.W. (eds), 1985, Social odours in mammals,
    Clarendon Press, Oxford.
Clevedon Brown, J., and Williams, J.D., 1972, The rodent preputial
    gland, Mammal Review, 2:105-147.
Dawkins, R., 1986, "The Blind Watchmaker", Longman Scientific and Tech-
    nical, Harlow
Ebling, F.J., 1974, Hormonal control and methods for measuring sebaceous
    gland activity, J. Invest. Derm., 62, 161-171.
Ebling, F.J., 1977, Hormonal control of mammalian skin gland, in:
    "Chemical Signals in Vertebrates", D. Muller-Schwarze and M.M.
    Mozell (eds.), Plenum, New York, pp. 17-33.
Ebling, F.J., and Skinner, J., 1967, The measurement of sebum production
    in rats treated with testosterone and oestradiol. Br. J. Derm.,
    79:386-393.
Ebling, F.J., and Skinner, J., 1975, The removal and restitution of hair
    fat in the rat, Br. J. Derm., 92:321-324.
Gilbert, N., 1989, Biometrical Interpretation, Oxford University Press,
    Oxford.
Jemiolo, B., Andreolini, F., Wiesler, D., and Novotny, M., 1987, Varia-
    tions in mouse (Mus musculus) urinary volatiles during different
    periods of pregnancy and lactation, J. Chem. Ecol., 13:1941-1956.

Koidsumi, K., 1957, Antifungal properties of cuticular lipids in insects, J. Insect Physiol., 1:40-50.

Kolattukudy, P.E., 1976, "The Chemistry and Biochemistry of the Natural Waxes",P.E. Kolattukudy (ed), Elsevier, Amsterdam.

Krebs, J.R., and Dawkins, R., 1984, Animal signals: mind reading and manipulation, in: "Behavioural Ecology: An Evolutionary Approach", J.R. Krebs, and N.R. Davies (eds.), 2nd edition, Blackwell Scientific, Oxford.

Macdonald, D.W. and Brown, R.E., 1985, The smell of success, New Scientist, 106:10-14.

Martin, J.T., 1964, Role of cuticle in the defense against plant disease, Ann. Review Phytopathol., 2:81-100.

Meehan, A.P., 1984, "Rats and Mice: Their Biology and Control", Rentokil Ltd., East Grinstead.

Milligan, S.R., 1980, Pheromones and rodent reproductive physiology, Science, 45:251-275.

Nicholaides, N., 1974, Skin lipids: Their biochemical uniqueness, Science, 186:19-26.

Natynczuk, S.E., 1990a, "Ultrasound and Semiochemistry in Rat Social Behavior", D. Phil. Thesis, Oxford University.

Natynczuk, S.E., 1990b, Behavioural cues to semiochemically important body regions of Rattus norvegicus, in: "Chemical Signals in Vertebrates V, D.W. Macdonald, D. Muller-Schwarze, and S.E. Natynczuk (eds.) Oxford University Press, Oxford, pp. 445-450.

Natynczuk, S.E., and Macdonald, D.W., 1992a, Sex, scent, scent matching and the self calibrating rat, in preparation.

Natynczuk, S.E., and Macdonald, D.W., 1992b, The rat preputial gland: ontogeny and chemistry, in preparation.

Novotny, M.V., Jemiolo, B., and Harvey, S., 1990, Chemistry of rodent pheromones: Molecular insights into chemical signalling in mammals, in: "Chemical Signals in Vertebrates V", D.W.Macdonald, D. Muller-Schwarze, and S.E. Natynczuk (eds.), Oxford University Press, Oxford, pp. 1-22.

Quay, W.B., 1982, Olfaction in central neural and neuroendocrine systems: Integrative review of olfactory representations and interactions, in: "Chemical Signals in Vertebrates", D. Muller-Schwarze, and R.M. Silverstein (eds.), Plenum, New York. pp. 115-117.

Rubin, B.S. and Barfield, R.J., 1983, Progesterone in the ventromedial hypothalamus facilitates oestrous behaviour in the ovariectomized oestrogen-primed rat, Endocrinol., 113:797-804.

Yahr, P., and Commins, D., 1982, The neuroendocrinology of scent marking, in: "Chemical Signals in Vertebrates", D. Muller-Schwarze and R.M. Silverstein (eds.), Plenum, New York, pp. 115-117.

Zarrow, M.X., Yochim, J.M. and McCarthy, J.L., 1964, Experimental Endocrinology: a source book of basic techniques, Academic Press, New York and London.

QUASI-OLFACTION OF DOLPHINS

Vitaly B. Kuznetsov

Department of Vertebrate Zoology
Faculty of Biology
Moscow State University
119899 Moscow, Russia

INTRODUCTION

Since olfactory bulbs regress in toothed whales long before the moment of birth (Sinclair, 1966; Oelshläger, Buhl, 1985), they possess no olfaction. Taste buds are found in neonates and young animals, but they are reduced to such a degree that the adults appear to possess no gustatory perception (Yamasaki et al., 1978; Li Yemin, 1983). Morphological data have made it possible to conclude that the toothed whales do not possess the typical vertebrate senses of olfaction and taste.

This unique situation is one major reason why the experimental investigation of chemical perception in dolphins is a very important issue for the solution of a number of general vertebrate chemoreception problems. It may also be important for the solution of such general neurology problems as structure-function relationships in the vertebrate brain. In particular, it is not clear why all of the olfactory brain structures (with the exception of the anterior olfactory nuclei) are present in dolphins when, at the same time, olfactory bulbs and tracts are absent (Filimonov, 1965; Jacobs et al., 1971).

The question of chemoreceptivity in these animals brings together the interests of many different biological fields. The study of this question represents the main focus of the author's 20 year investigation of dolphin chemoreceptivity.

DOLPHIN SENSITIVITY TO CHEMICAL STIMULI

Conditioning Experiments

Conditioning experiments were performed at the Outrish Station on the Black Sea with adult bottlenose dolphins (Tursiops truncatus). The animals were trained to stay in a vertical posture with the head above the water and mouth opened. All test solutions (40 ml of a chemical stimulus in sea or fresh water) were compared to 40 ml of sea or fresh water alone. These liquids were poured into the oral cavity and the dolphins were trained to perform different instrumental reactions to them.

Chemical Signals in Vertebrates VI, Edited by R.L. Doty and
D. Müller-Schwarze, Plenum Press, New York, 1992

Table 1.   The Sensitivity of 2 Black Sea Bottlenose Dolphins to Fresh
Water Solutions of Chemical Stimuli

Chemical Stimuli	Threshold Concentration (mM) Dolphin 1		Dolphin 2		Level of Significance
Sucrose	600*	300*	600*	300*	
Glucose	1000*	500*	1000*	500*	
NaCl	300		-		<0.01
Quinine HCl	0.03		0.01		<0.01
Picric acid	0.05		0.02		<0.01
HCl	100		-		<0.01
Citric acid	50		50		<0.001
Oxalic acid	30		-		<0.01
Capronic acid	0.1		0.05		<0.01
Valeric acid	0.01		0.01		<0.001
Indole	0.001		0.001		<0.001

* = The proportion of correct choices is at chance level (about 50%).

The dolphins were found to be insensitive to sucrose. Only the
sensitivity to the bitter substance picric acid (0.043 mM) was similar to
that of most other mammals. Fresh water could be discriminated from sea
water, but sea water could not be distinguished from twice diluted sea
water. Dolphins were insensitive to 50 mM citric acid in (pH 2.5), but
were capable of sensing 0.1 mM capronic acid at the pH of sea water.

The dolphins' sensitivity to fresh water solutions of standard taste
chemicals is presented in Table 1. The animals showed low sensitivity to
sodium chloride and acids. At the same time their thresholds to odorous
substances - carboxylic acids and indole - were several hundred times
lower than that of most other mammals.

The thresholds of these dolphins to citric acid, quinine and sodium
chloride are similar to threshold values reported by Friedl et al. (1990)
in the Pacific bottlenose dolphin (respective values: 17 mM; 0.012 mM;
and 100 mM). However, the threshold values we observed for sucrose were
much higher than the value of 90 mM reported in the Pacific bottlenose,
implying differences among species may be present.

It is possible that similar taste sensitivity occurs in other forms
of sea mammals. To investigate this hypothesis, conditioning experiments
with pinnipeds were performed. The Northern fur seal's (Callorhinus
ursinus) sensitivity to citric and valeric acids was tested. The taste
sensitivity of this animal to valeric acid (10 mM) was 5 times lower than
that to citric acid (2 mM). In the Steller sea lion (Eumetopias jubatus)
and Caspian seals (Phoca caspica) the threshold concentration of citric
acid was 2 mM. This suggests that the taste sensitivity of the dolphin
is significantly different from that of these two other sea mammals
(Kuznetsov, 1982; 1986).

Autonomic Reaction Experiments

The chemical sensitivity of other dolphin species was also evaluated. The autonomic reactions in response to odorous chemical presentations were recorded in harbor porpoises (Phocoena phocoena) and common

dolphins (_Delphinus delphis_). The bottlenose dolphins were also tested to the solutions used in the conditioning experiments. This series served as a control series.

The dolphin was placed into a basin and a tube was inserted into its oral cavity. A steady flow of sea water was provided to wash the oral cavity. The chemical solutions were introduced into the tube variable intervals. Pure water was given as a control. The reactions to pure sea water and the reactions to chemicals were compared.

The results are shown in Table 2. The data show that the order of chemical sensitivity in all three dolphin species is similar and rather high (0.1 - 0.00085 mM). We conclude that the sensitivity of this type of chemical sense resembles olfaction (Kuznetsov, 1979).

Table 2. The Sensitivity of Black Sea Dolphins (Bottlenose Dolphins, Common Dolphins and Harbor Porpoise) to Sea Water Solutions.

Dolphin species	Number of subjects	Chemical Stimuli	Concentration	Level of Significance
Bottlenose dolphin	3	Valeric acid	0,1 mM	<0.001
	3	Quinine HCl	0.1 mM	<0.005
	2	HCl	150 mM	<0.01
	1	Urine	1%	<0.01
Common dolphin	2	Feces suspension	0.00001%	<0.001
Harbor porpoise	4	Scatole	0.0017 mM	<0.01
	4	Camphor	0.003 mM	<0.01
	4	Trimethylamine	0.0085 mM	<0.001
	2	Trimethylamine	0.00085 mM	<0.05

CHEMORECEPTION AND BEHAVIOR IN DOLPHINS

We may conclude that the dolphin chemical sense is dissimilar to that of other vertebrates, and does not represent olfactory or gustatory sensation, since these senses are regressed. The question arises as to whether this type of sensitivity influences the dolphin's behavior or whether it is a rudimentary type of perception. This series of experiments was aimed to address this point.

Seven dolphins were conditioned to perform rather difficult instrumental actions. Each animal had to perform one type of action in response to one chemical and to perform a second type of action in response to another chemical. Correct choices were reinforced by giving fish. The learning criterion was nine out of ten correct responses.

The dolphins discriminated the chemical stimuli successfully, all reaching criterion within 30-100 trials. Thus, they were capable of associating chemical and visual, as well as echolocation, information. The capacity to associate a chemical with a definite spatial location indicates that they have the capacity to use chemoreception for orientation.

In these studies, dolphins proved to be capable of solving very difficult tasks using cues from chemical stimuli, including compounds of six chemical classes (carboxylic acids, amines, aromatic compounds, terpenes, 5-and 6-membred-heterocyclic compounds). This suggests that dolphins possess a well-developed system of chemoreception which is functionally connected with brain systems that control certain behaviors.

This notion is further supported by the fact that male dolphins have a perianal gland with mechanisms for discharging its secretion and can perform complicated social actions on the basis of chemical information (Kuznetsov, 1989).

COMPARISON OF TONGUES IN DOLPHINS, PINNIPEDS AND UNGULATES

We investigated the histological tongues of bottlenose dolphins (Tursiops truncatus) of different ages, harbor porpoises (Phocoena phocoena), Amazon river dolphins (Inia geoffrensis) and beluga (Delphinapterus leucas), and also the tongues of pennipeds: the Caspian seal (Phoca caspica) and the Northern fur seal (Callorhinus ursinus). In addition, we evaluated the tongues of cows and pigs because dolphins are supposed to have evolved from primitive ungulates.

In contrast to the tongues of other animals, the adult dolphin tongue is devoid of papillae. However, bottlenose dolphin fetuses and neonates possess fungiform and circumvallate papillae in the region of tongue root (Fig. 1A). These structures resemble those seen on the cow tongue.

Histological investigation of the papillae demonstrates typical taste buds with taste pores and innervation (Fig. 1B). The taste pores are numerous. There are about 1600 taste pores in the bottlenose dolphin neonate's tongue. Thus, taste reception appears to well-developed in the neonatal dolphin and may provide cues for feeding during suckling.

In 2- to 3-year old dolphins the circumvallate papillae have already been transformed into pits -- specialized structures filled with mucous secreted by special glands. At the bottom of each pit one can see irregularly formed papillae (Fig. 1C). The taste buds are reduced. By adulthood, no taste buds are present.

In the tongues of the ungulates and pinnipeds, the circumvallate papillae were covered by cornified epithelial cells. In contrast, the surface structure of the pit papillae on the dolphin tongue root is drastically different. The surface is covered by microvilli (Fig. 1D). In some regions, the microvillar structure is variable. Nine types of cells which contain microvilli could be detected in the proximity of the pit papillae. Their average diameter was 0.1 μm and their length varied from 0.2 to 2.0 μm.

The papillae cells in the harbor porpoise tongue also are covered by microvilli, as are those of the Amazon river dolphin. Preliminary data suggest that in the beluga tongue a similar microvillar system exists. It is possible that such peculiarities are characteristic for all toothed whales. To our knowledge, no structures comparable to these have been observed in tongues of other mammals.

CHEMICAL SENSE OF DOLPHINS - QUASI-OLFACTION

Vertebrates possess two main types of chemoreception -- olfaction and gustation. We suggest that the main function of gustation is homeosta-

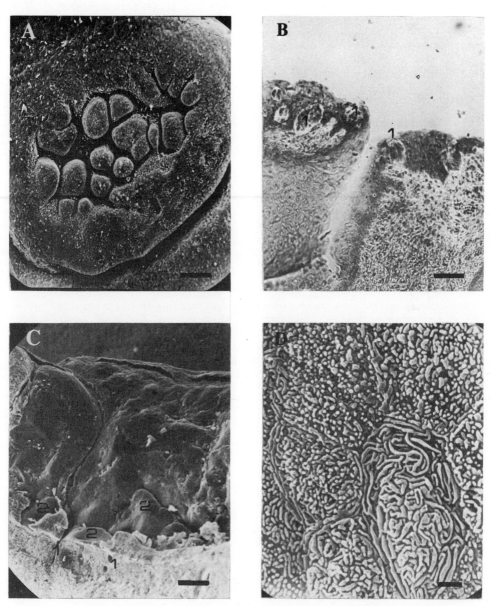

Fig. 1. A: Scanning electron micrograph (SEM) of the circumvallate
papillae of a neonatal bottlenose dolphin 1100 mm in body
length. Bar = 330 μm. B: Photomicrograph of the epithelial structure of the neonatal circumvallate papilla. 1 =
taste bud. Bar = 50 μm. C: SEMs of the vertical section
of the pit of the adult harbor porpoise, 1 = the opening
of the gland duct, 2 = the irregularly formed papillae at
the bottom of the pit. Bar = 166 μm. D: SEMs of the
microvillar cells in the pit's papillae of the bottlenose
dolphin. Bar = 2 μm.

sis; namely, to provide stability of the internal medium of the organism. Taste is crucial for acceptance or rejection of different foods. Animals can be easily conditioned to a taste aversion- attraction reaction, while other types of reflexes are much more difficult to establish.

The main function of olfaction is a communicative one. It is used during reproductive, territorial, and social interactions. Olfaction also serves for individual recognition of chemicals.

Atema (1980) demonstrated that fishes solve complex problems using only the sense of olfaction, although their taste and smell capabilities are equally well-developed. Fishes use taste sensation when they look for dead prey or when they are presented with the most simple attraction-aversion learning paradigm. The chemical communication signals and other chemical signals of living prey are perceived by means of olfaction.

Dolphins demonstrate a relatively high sensitivity to odorous substances, and the morphology of receptors is radically different from that of taste buds. They possess chemical communication and can solve complex tasks using the chemical sense. Therefore we conclude that the dolphin chemical sense substitutes for olfaction. At the same time it is not olfaction. This is why we call it "quasi- olfaction".

We suppose that the development of echolocation in ancestors of toothed whales led to a reduction in the olfaction (Kuznetsov, 1988). The olfactory epithelium could not function in conditions of large and rapid changes of pressure (0 to more than 80 kPa) arise in bony nares upon generation of echolocation clicks (frequency more than 100 kHz). The process of evolution resulted in total spatial separation of two functions -- pulse generation and chemical sensitivity. Instead of an olfactory epithelium, quasi-olfactory receptors developed in the dolphin tongue.

CONCLUSIONS

Toothed whales were considered by Turner (1891) to be the only vertebrates unable to perceive smells, i.e. anosmatics. However, the presence of quasi-olfaction in dolphins suggests that there are no anosmatics among vertebrates.

The development of quasi-olfaction in dolphins makes it doubtful, as currently assumed, that dolphin olfactory brain structures are not involved in processing of chemical information. Therefore, theories which suggest that the olfactory brain region of the dolphin may be functionally reorganized may need to be revised (e.g., Filimonov, 1965).

Trigeminal nerve chemoreception is mediated by fibers which inner- vate the olfactory epithelium and oral cavity. It is believed that this type of chemoreception is not specific and carries out accessory func- tions. Until now, trigeminal nerve chemoreception has been viewed as part of the general chemical sense (Parker, 1922). However, the trigemi- nal system is sensitive to odorous substances. In fish, the sensitivity of trigeminal chemoreception is 0.01 - 0.001 mM for various odorous substances (Belousova et al., 1983). Among vertebrates, only the toothed whales appear to have no olfaction. In dolphins, trigeminal chemorecep- tion is likely one basis for the mediation of the quasi-olfaction dis- cussed in this paper.

548

REFERENCES

Atema, J., 1980, Chemical senses, chemical signals, and feeding behavior in fishes, in: "Fish behavior and its use in the capture and culture of fishes", J.E. Bardach, J.J. Magnuson, R.C. May, J.M. Reinhart, eds., Manila.

Belousova, T.A., Devitsina, G.V., and Malyukina, G.A., 1983, Functional peculiarities of fish trigeminal system, Chemical Senses, 8:121.

Filimonov, I.N., 1965, On the so-called rhinencephalon in the dolphin, J. Hirnforsch., 8:1.

Friedl, W.A., Nachtigall, P.E., Moore, P.W.B., Chun, N.K.W., Haun, J.E., Hall, R.W. and Richards, J.L., 1990, Taste reception in the Pacific bottlenose dolphin (Tursiops truncatus Gilli) and the California sea lion (Zalophus californianus), in: "Sensory Abilities of Cetaceans," J.A. Thomas and R.A. Kastelein, eds., Plenum Press, New York.

Jacobs, M.S., Morgane, J., and McFarland, W., 1971, The anatomy of the brain of the bottlenose dolphin (Tursiops truncatus), Rhinic Lobe (Rhinencephalon) 1. The paleocortex, J. Comp. Neurol,. 141:205.

Kuznetsov, V.B., 1979, Chemoreception in dolphins of the Black Sea: bottlenose dolphin (Tursiops truncatus), common dolphin (Delphinus delphis) and common porpoise (Phocoena phocoena), Dokl. Acad. Nauk SSSR, 49:1498 (in Russian)

Kuznetsov, V.B., 1982, Taste perception of sea lion, in: "Abstracts of Reports of the Eighth All-Union Conference, Astrakhan". (in Russian)

Kuznetsov, V.B., 1986, Chemical sense in dolphins, in: "Chemical Communication of Animals. Theory and Experience", Nauka, Moscow. (in Russian)

Kuznetsov, V.B., 1988, Problems of olfaction reduction in toothed whales (Odontocety), Zn. Obsch. Biol., 49:129. (in Russian)

Kuznetsov, V.B., 1989, Chemoreception and communication in dolphins, in: Abstracts, Fifth International Theriological Congress, 22-29 August 1989, Rome, V.1:371.

Li Yuemin, 1983, The tongue of Baiji, Acta Zool. Sin., 29:35.

Oelschlager, H., and Buhl, E.H., 1985, Development and rudimentation of the peripheral olfactory system in the harbor porpoise, Phocoena phocoena (Mammalia: Cetacea), J. Morphol., 184:351.

Parker, G.H., 1922, "Smell, Taste and Allied Senses in Vertebrates", Lippincott, Philadelphia.

Sinclair, J.G., 1966, The olfactory complex of dolphin embryos, Texas Rep. Biol. Med., 24:426.

Turner W., 1891, The convolutions of the brain: a study in comparative anatomy, J. Anat. Physiol., 25:105.

Yamasaki, F., Komatsu, S., and Kamiya, T., 1978, Taste buds in the pits at the posterior dorsum of the tongue of Stenella coeruleoalba, Sci. Rep. Whales Res. Inst. Tokyo, 30:285.

VOMERONASAL ORGAN SUSTAINS PUPS' ANOGENITAL LICKING IN PRIMIPAROUS RATS

I. Brouette-Lahlou, E. Vernet-Maury, F. Godinot, and J. Chanel

Université Claude Bernard/Lyon-Physiologie neurosensorielle
F-69622 Villeurbanne cedex - France

INTRODUCTION

In the rat, anogenital licking of pups by the dam is vital during the first 3 weeks. A volatile constituent from pups preputial gland secretion has been found to direct and regulate the pups' anogenital licking. Dodecyl propionate was identified as the main active compound in the pups' preputial gland (Brouette-Lahlou et al., 1991 a,b). It seems that the main olfactory system is specialized in the reception of volatile compounds, whereas the vomeronasal system is more effective in detecting compounds of extremely low volatility. Access of large and non-volatile molecules to the vomeronasal organ was demonstrated by Wysocki et al. (1985) in studies of the social and feeding behavior patterns in mammals. Dodecyl propionate, a high molecular weight compound (MW: 242), provides an opportunity to address this hypothesis concerning the vomeronasal function in the reception of relatively non-volatile constituents used in pheromonal communication.

Fleming et al. (1979) demonstrated that maternal behavior in virgin female rats is modulated by both the vomeronasal and olfactory systems. Thus, completely cutting the vomeronasal nerve or partially transecting the olfactory bulb did not interfere with the development of maternal behavior in multiparous rats. In the same year, Marques investigated the relative roles of the main olfactory and vomeronasal systems in the response of female golden hamster to pups; after vomeronasal nerve cuts, most "killers" became "carriers" and built better nests than before deafferentation. $ZnSO_4$ treatment, which impairs the main olfactory system alone, had little effect, but when combined with transection of vomeronasal nerves, converted killers into carriers. Jirik-Babb et al. (1984) observed that adequate maternal behavior develops in primiparous rats submitted to vomeronasal nerve section or to $ZnSO_4$ nasal irrigation, but not the two treatments combined. In mice, normal maternal processes are also expressed in the absence of an intact accessory olfactory system (Lepri et al., 1985).

In the present study, we analyzed the respective roles of the main and accessory olfactory systems in the maternal behavior of the primiparous rat, as indexed by the specific licking of pups' anogenital areas.

Chemical Signals in Vertebrates VI, Edited by R.L. Doty and
D. Müller-Schwarze, Plenum Press, New York, 1992

METHODS

Primiparous dams underwent one of the following treatments : vomero-nasal organ removal (VN), irrigation of nasal cavities with 5% $ZnSO_4$ solution (ZN), both treatments (VN + ZN), surgical control (SC), saline irrigation control (SA), and normal control (N).

Vomeronasal lesions were undertaken on intraperitoneally anaesthetized dams placed in a decubito supine position on a cork slab with the caudal regions slightly elevated to prevent blood from interfering with breathing. After retracting the jaws, an incition was made in the palate mucosa and the edges of the wound separated. Following the checking of the cavities with a needle, the two vomeronasal organs were extracted with the aid of a microaspirator via two small bilateral incisions made with a fine electric drill. Animals assigned to the control group underwent all the surgery with the exception of VN extraction and checking. Peripheral anosmia was performed on slightly etherized rats placed on a cork slab with the head lower than the rest of the body. 5% $ZnSO_4$ solution was introduced into the nasal cavities via external naris using a standard syringe which had been blunted; while irrigating one naris, $ZnSO_4$ solution drained from the opposite naris. The operation was repeated in the opposite naris to insure bilateral $ZnSO_4$ nasal irrigation. Sham lesions involved anaesthesia and infusion of saline through the nasal cavities.

The dams and their litters were observed from day of birth until weaning. Anogenital licking time was recorded every two days.

Two behavioral experiments were conducted to determine whether VN, ZN, VN + ZN, SC, SA, SC + SA, N rats might detect Dodecyl propionate on pups' heads (Experiment 1) or on impregnated filter paper in glass dishes (Experiment 2). Dams were tested once before surgery or $ZnSO_4$ irrigation when pups were about 4–day–old and again a day after treatment.

In Experiment 1, a solution of Dodecyl propionate (20 ng dissolved in dichloromethane) was dripped onto the heads of 5–day–old pups and their anogenital area was cleaned. A stimulus or control pup was placed in the opposite position to that of the dam, in the test cage (identical to the dam's home cage). A 5 min. observation took place during which any licking of pups' heads or anogenital regions was quantified after the successive introduction of 3 pups (one DP head pup – one control pup – again one DP head pup. Experiment 2 took place two days after experiment 1. It was designed to determine whether DP remains attractive to different treated dams, independently to pups' body cues. In this phase, treated dams had access to glass dishes (0.8 cm depth, 2.8 cm diameter) containing filter paper (0.5 cm^2) impregnated with DP (10 ng) dissolved in dichloromethane versus filter paper impregnated with 0.5 ml dichloromethane alone (control). Each prepared filter paper was dried at ambient temperature. Sniffing and licking behavior of dams towards the glass dishes was recorded in a 3 min. observation schedule.

In the dam in Experiment 1, the completeness of vomeronasal organ destruction was verified histologically after sacrificing all dams at the end of experiments. After removing the mandible and all soft tissues with the exception of the mucosa and oral surface of the hard palate, the heads were immersed in Bouin's fixative and then decalcified differently for specimens with nasal cavity damage vs nasal cavity intact and for specimens with vomeronasal and nasal epithelium both destroyed. Heads were then paraffin embedded and serial sections ($\approx 10\mu$) stained with 0.05% cresyl violet.

RESULTS

Histological examination showed that, with the exception of 3 rats, the vomeronasal organ of the VN dams had been completely lesioned. In contrast, ZnSO$_4$ irrigation did not seem to affect the vomeronasal epithelium, although the olfactory epithelium appeared to have been completely in both the ZN and the ZN + VN dams.

In order to test for the remaining olfactory sensitivity after ZnSO$_4$ irrigation, we buried a small piece of chocolate under 1cm of rat bedding. All VN dams found the chocolate bits, confirming that they had not suffered from olfactory epithelium impairment. None of the ZN and VN + ZN dams similarly found the chocolate bits.

Maternal anogenital licking (MAGL) time was significantly increased after treatment both in VN and VN + ZN dams while in ZN animals as well as in the control groups (SA, SC, SA + SC), MAGL time was equivalent to baseline levels (Fig. 1).

In Experiment 1 (pups' heads odorized with DP) significant less time was spent by VN and VN + ZN dams licking pups' heads as compared to pretreatment values.

DISCUSSION

Three primary results were obtained : (a) Peripheral anosmia after ZnSO$_4$ irrigation of olfactory mucosa had no effect on DP detection. (b)

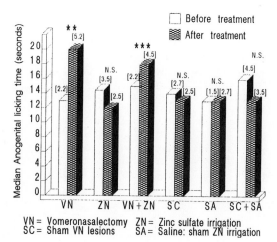

Fig. 1. Median (semi-interquartile range) anogenital licking time (seconds) in experimental groups analyzed by the Wilcoxon test, before and after treatment. N.S. : non significant, ** : p < 0.01, *** : p < 0.001.

Table 1. Number of pups licked on the head odorized with Dodecyl Propionate and median time (semi-interquartile range) spent by
dams licking them before (PRE) and after (POST) treatment dams.
Data were analyzed by χ^2, Wilcoxon (T) and Mann–Whitney (U)
tests.

Rat treatment	Number of pups licked / Total number of pups presented to dam			Median time (in sec) spent licking pups' heads			
	PRE	χ^2 test	POST	PRE	T test	POST	U test
VN	13/20	4.9*	5/20	6.0 [5.5]	0*	2 [1.5]	1.78*
ZN	14/20	0.NS	13/20	3.5 [1.5]	57 NS	3 [1.5]	2.5*** / 3.1***
VN+ZN	10/20	10.8***	0/20	4.0 [3.0]	0*	0 [0]	0.8
SC	10/16	0.NS	9/16	3.0 [2.5]	9.5 NS	5 [3]	NS
SA	10/16	1.32 NS	6/16	4.0 [2.25]	15 NS	5 [1.75]	3.2***
SC+SA	9/16	0.5 NS	6/16	2.0 [1.5]	10 NS	2 [2]	

Experiment 2 (glass dishes test): Vomeronasalectomy prevented VN dams from
detecting DP.

Table 2. Median (semi-interquartile range) time (in sec.) spent by
dams sniffing glass dishes containing CH_2Cl_2 + DP impregnated filter paper vs CH_2CL_2 impregnated filter paper
(not indicated) before and after treatment.

	Pre-treatment	Post-treatment	Wilcoxon test
VN	6 (3.5)	2 (2.5)	T = 0; p = 0.05
ZN	9 (3.5)	4 (4)	T = 3; NS
VN + ZN	7 (2)	2 (3)	T = 0; p = 0.05
SC	8.5 (4.5)	11.5 (3)	T = 9; NS
SA	5 (1.5)	6 (1.5)	T = 9; NS
SC + SA	8 (1)	8.5 (3)	T = 9; NS

VN removal enhanced pups' anogenital licking time. (c) Similar results were obtained with VN removal plus ZnSO$_4$ irrigation. The vomeronasal organ seems to be the chemosensory detector of Dodecyl Propionate; its location - near the opening and its anatomical relationships with the oral cavity - might explain the phenomenon - and suggest its implication in pheromone mediated behavior.

REFERENCES

Brouette-Lahlou, I., Vernet-Maury, E., and Chanel, J., 1991a, Is rat-dam licking behavior regulated by pups' preputial gland secretion? Anim. Learn. Beh., 19:177.

Brouette-Lahlou, I., Amouroux, R., Chastrette, F., Cosnier, J., Stoffelsma, J., and Vernet-Maury, E., 1991b, Dodecyl Propionate, the attractant from rat pup preputial gland. Characterization and Identification, J. Chem. Ecol., 17:1343.

Fleming, A., Vaccarino, F., Tambosso, L., and Chee, P., 1979, Vomeronasal and olfactory system modulation of maternal behavior in the Rat. Science, 203:372.

Jirik-Babb, P., Manaker, S., Tucker, A.M., and Hofer, M.A., 1984, The role of the accessory and main olfactory systems in maternal behavior of the primiparous rat, Beh. Neur. Biol., 40:170.

Lepri, J.J., Wysocki, Ch. J., and Vandenbergh, J.G., 1985, Mouse vomero-nasal organ : effects on chemosignal production and maternal behavior, Phys. Beh., 35:809.

Marques, M.D., 1979, Roles of the stain olfactory and vomeronasal systems in the response of the female hamster to young, Beh. Neur. Biol., 26:311.

Wysocki, C.J., Beauchamp, G.K., Redinger, R.F., and Wellington, J.L., 1985, Access of large and non-volatile molecules to the vomeronasal organ of mammals during social and feeding behaviors, J. Chem. Ecol., 11:1147.

PART 6: Mammals -- Humans

BILATERAL AND UNILATERAL OLFACTORY SENSITIVITY:

RELATIONSHIP TO HANDEDNESS AND GENDER

Richard E. Frye,[1,2] Richard L. Doty,[1,2] and Paul Shaman,[1,3]

[1]Smell and Taste Center, [2]Department of Otorhinolaryngology --
Head and Neck Surgery, School of Medicine, and [3]Department of
Statistics, The Wharton School, University of Pennsylvania,
Philadelphia, Pennsylvania 19104.

INTRODUCTION

Unlike most major sensory systems, the majority of olfactory projections are ipsilateral. Both hemispheres can process olfactory information in a manner analogous to what is seen in other sensory systems, although, as noted below, they may do so differently. Gordon and Sperry (1969) found that patients whose corpus callosum and other forebrain commissures were surgically sectioned could only names odors presented to the left nostril; odors presented to the right nostril could be identified by pointing to an object associated with the smell.

A number of recent studies have suggested that suprathreshold olfactory tasks such as odor matching and memory are processed by the right hemisphere (Eskenazi et al., 1988; Zatorre & Jones-Gotman, 1990). Evidence for lateralization of olfactory sensitivity is less clear. Two studies have found left-right threshold differences related to handedness, although these studies do not agree on the direction of the difference (Toulouse and Vaschide, 1900; Youngentaub et al., 1982). Other studies have been unable to find any left:right differences in olfactory sensitivity (Koelega, 1979; Zatorre & Jones-Gotman, 1990).

The present paper summarizes the salient findings of a recent study in which we sought to determine if one side of the nose is better than the other in detecting the odor of 2-butanone. A determination was made as to the influence of handedness, gender, and blocking of the contralateral nostril on unilateral olfactory sensitivity.

METHODS

Subjects

Sixteen right handed men, 21 right handed women, 17 left handed men, and 21 left handed women served as subjects (mean age = 21.85, SD = 5.99). Handedness was determined using the revision by Briggs and Nebes (1975) of Annett's (1967) handedness inventory. None reported having any problems smelling odorants and all were in good health. Thirty six of the subjects performed the olfactory tests with their contralateral nostril blocked by a piece of 3M Microfoam[TM] tape, and 39 performed the test with it open.

Chemical Signals in Vertebrates VI, Edited by R.L. Doty and
D. Müller-Schwarze, Plenum Press, New York, 1992

General Experimental Design

The design of this study was straight forward. A subject first completed a permission form and the Handedness Inventory. Nasal resistance was then determined in each nostril, followed by the detection threshold tests of the right and left sides of the nose. Nasal resistance was again determined, followed by a suprathreshold olfactory function test (not described in this paper). Subsequently, the subject left the laboratory for a break of several hours. Upon return, if the relative nasal resistance had not shifted to the opposite nostril, we attempted to induce a shift by applying pressure under the armpit, in the palm of the hand, or by auditory stimulation on the more patent side of the nose (Rao & Potdar, 1970) and the subject was retested for smell function.

Olfactory Threshold Measurement

The concentration series of 2-butanone was produced by a 10-stage air-dilution dynamic olfactometer (Doty et al., 1988). The following half-log step odorant concentrations were used (in ppm): 0.6, 1.0, 1.8, 3.2, 5.6, 10.0, 17.8, 31.6, 56.2, and 100. Olfactory thresholds were determined using a single staircase forced-choice procedure described in detail elsewhere (Doty et al., 1984). Within the threshold test sessions, the order in which each side of the nose received the stimulus was randomized (with the exception that a given side received no more than three consecutive trials).

Data Analysis

We sought to determine whether gender, handedness, or blockage of the contralateral nostril during testing influenced the magnitude of either threshold measure or the difference between the right and left thresholds. We subjected the data to a multivariate (left nostril, right nostril) repeated measures (session 1, session 2) weighted analysis of covariance. The threshold measures and the nasal resistance covariates (for each nostril and each session) were logarithmically transformed. Since the threshold measurement was calculated from trials occurring in approximately the last third of the test session, the post-test maximum radius airflow measurement was used in this analysis.

RESULTS

Table 1 lists the F-values and their respective p values for each factor in our analysis. The column for combined thresholds indicates if the factor influenced both the right and left side of the nose. The threshold difference column shows if any factor influenced the difference between the left and right threshold measures. Essentially, this difference value represents the degree and direction of left:right superiority in olfactory sensitivity. Although both left and right thresholds were lower for the second test session than the first, the difference between the left and right thresholds were the same in both sessions; therefore, the effects of the first and second sessions were combined in the analysis.

A significant difference between the left and right threshold values was found, with the left side exhibiting a slightly lower average threshold

Table 1. Statistical tests for the influence of gender, handedness, block-age of the contralateral nostril during testing, session, and airflow for the combined left and right olfactory threshold values and the difference between the left and right threshold values.

Factors	Combined Thresholds			Threshold Difference		
	df	F Ratio	p Value	df	F Ratio	p Value
All Factors				11/278	2.29	0.011
Gender	8/278	2.75	0.007	4/278	3.85	0.005
Handedness	8/278	2.52	0.016	4/278	3.87	0.004
Blockage	8/278	3.38	0.001	4/278	4.79	0.001
Gender x Hand	4/278	4.45	0.003	2/278	6.55	0.002
Gender x Block	4/278	3.14	0.015	2/278	6.04	0.003
Hand x Block	4/278	2.95	0.020	2/278	5.72	0.004
Gender x Hand x Block	2/278	5.39	0.005	1/278	10.47	0.001
Session	2/278	7.31	0.001	1/278	0.40	0.530

than the right [Mean in log ppm (SE), Combined Left = 2.139 (0.090), Right = 2.145 (0.081)]. This is illustrated in Figure 1a, which is a scatterplot of the left vs the right threshold value for each subject. The diagonal line separates those cases with a higher threshold value for the right side of the nose (those below the line) from those cases with a higher value for the left side of the nose (those above the line). The majority of these cases are below this line indicating that, in general, the left side of the nose was more sensitive.

In general, females evidenced lower olfactory thresholds than males (Figure 1b). On average, females were slightly more sensitive on the right than the left side of the nose, whereas the opposite was true for males (Figure 1b).

Both male and female subjects who had their contralateral naris blocked during testing typically showed higher thresholds than those who did not have their contralateral naris blocked; this difference was greater for females than for males (Figures 1c & d). Contralateral naris blockage also influenced which side of the nose was more sensitive. Greater sensi-tivity on the left than the right side of the nose was observed when the contralateral nostril was open during testing, whereas the opposite was true when the contralateral naris was closed (Figures 1c & d).

In general, right handers were more sensitive to the test odor than were left handers. However, handedness influenced left and right threshold values for males and females differently. As can be seen in Figure 2, left and right threshold values were higher for male right handers than male left handers and lower for female right handers than female left handers. However, the relationship between handedness and threshold was more pro-nounced in the left than the right nostril, especially for females. Thus, left handed females were more sensitive in the right than the left nostril and right handed females were more sensitive in the left than the right nostril. The opposite relationship occurred for the male subjects. This effect was increased by blockage of the contralateral naris during testing, with a larger increase occurring for females than for males.

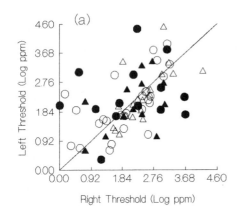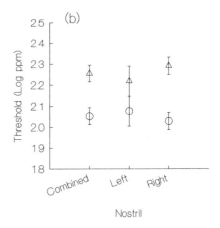

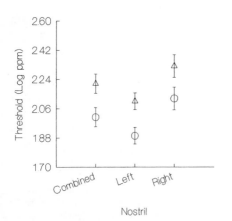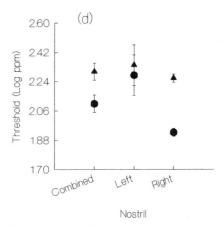

Figure 1. (a) Average left and right threshold values for each subject.
Mean (± SEM) thresholds for (b) males (triangles) and females
(circles), and subjects tested with thier contralateral nostril
(c) opened (empty symbols) and (d) blocked (filled symbols).

DISCUSSION

   The literature concerning unilateral olfactory sensitivity and odor
identification is unclear. In the present study, we found the average
difference between the right and left thresholds to be very small (when
untransformed this difference is less than 1 ppm). We have shown that the
difference between the right and left thresholds is a complex function of
handedness and gender. Overall, our findings suggest that right handed
subjects are more sensitive on the left side of the nose and left handed
subjects are more sensitive on right side of the nose. These results are
in accord with Toulouse and Vaschide (1900), but contradict those of
Youngentaub et al. (1982) who reported that left handed subject were more
sensitive on the left side of the nose and right handed subjects were more
sensitive on the right side of the nose.

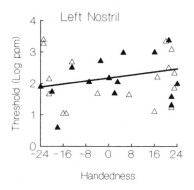

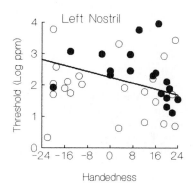

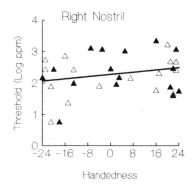

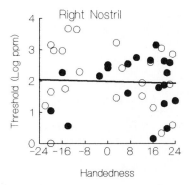

Figure 2. Relationship between handedness and olfactory threshold for male
          (triangles) and female (circles) subjects. Positive handedness
          values correspond to right handed subjects and negative values
          correspond to left handed subjects. Filled symbols represent
          subjects who blocked thier contralateral nostril during testing
          and empty symbols represent subjects who did not block their
          contralateral nostril during testing.

ACKNOWLEDGEMENTS

Supported, in part, by Grant DC 00161 from the National Institute on Deafness and Other Communication Disorders and Grant HL 37911 from the National Heart, Lung, and Blood Institute. We thank Ronald Hartika, Bruce Londa, David Marshall, and Richard Meinel for technical assistance during the study.

REFERENCES

Annett, M., 1967, The binomial distribution of right, mixed and left handedness, Quart. J. Exp. Psych., 19:327.
Briggs, G.G. and Nebes, R.D., 1975, Patterns of hand preference in a student population, Cortex, 11:230.
Doty, R.L., Deems, D.A., Frye, R.E., Pelberg, R. and Shapiro, A., 1988, Olfactory sensitivity, nasal resistance and autonomic function in patients with multiple chemical sensitivities. Arch. Otolaryngol. Head. Neck. Surg., 114:1422.
Doty, R.L., Shaman, P., Applebaum, S.L., Giberson, R. and Sikorsky, L., Rosenberg, L., 1984, Smell identification ability: Changes with age, Science, 226:1441.
Eskenazi, B., Cain, W.S., Lipsitt, E.D. and Novelly, R.A., 1988, Olfactory function and callosomy: A report of two cases, Yale J. Bio. Med., 61:447.
Gordon, H.W. and Sperry, R.W., 1969, Lateralization of olfactory perception in the surgically separated hemispheres of man, Neuropsychol., 7:111.
Koelega, H.S., 1979, Olfaction and sensory asymmetry, Chem. Senses. Flav., 4:89.
Rao, S. and Potdar, A., 1970, Nasal airflow with body in various positions, J. Appl. Physiol., 28:162.
Toulouse, E. and Vaschide, N., 1900, L'asymetrie sensorielle olfactive, Revue Philosophique, 49:176.
Youngentob, S.L., Kurtz, D.B., Leopold, D.A., Mozell, M.M. and Hornung, D.E., 1982, Olfactory sensitivity: Is there laterality?, Chem. Senses, 7:11.
Zatorre, R.J. and Jones-Gotman, M., 1990, Right-Nostril advantage for discrimination of odors, Percept. Psychophys., 47:526.

CHEMOSENSORY EVOKED POTENTIALS

Thomas Hummel and Gerd Kobal

Department of Pharmacology and Toxicology
University of Erlangen-Nurnberg
Universitatsstr. 22, W-8520 Erlangen, Germany

INTRODUCTION

Research on human olfaction is complicated by the fact that, at
sufficiently high concentrations, many odorants produce trigeminally-
mediated sensations. Therefore, most patients with complete anosmia
still are able to respond to a great variety of substances. For example, menthol and eucalyptol produce cooling or even stinging sensations
at higher concentrations. Doty et al. (1978) found that in 15 anosmic
patients, only vanillin and phenyl ethyl alcohol were not discernible by
irritation or other somatosensory sensations (out of a total of 47 odorants). Our own studies in patients with Kallman's syndrome (Hummel,
Pietsch and Kobal, 1991) revealed that they also were unable to perceive
hydrogen sulfide. The function of the trigeminal nerve may be assessed,
in the absence of olfactory stimulation, by means of the non-odorous
stimulus carbon dioxide (CO_2), which reportedly produces only painful
sensations (Thurauf et al., 1991).

Although recent achievements have been made in the clinical testing
of the sense of smell (Doty et al., 1984; Cain and Rabin, 1988), a major
difficulty in assessing the olfactory function of some patients relates
to the nature of subjective responses. Thus, patients may be unwilling
to report the perception of an odorant or, for physical reasons, are
unable to take a sniff. In addition, as in the case of comatose patients and those with advanced Alzheimer's disease, cognitive factors
can limit the value of the investigations. That is, if cognitive deficits are severe enough, testing based on the patient's responses may be
precluded. Such problems could be overcome by measures less dependent
upon subjective responses. Kobal (1981) introduced a technique which
made it possible to obtain chemosensory evoked potentials (CSEP) in man.
Late near-field evoked potentials are extracted out of the electroen-
cephalograph (EEG) and represent stimulus-related neocortical activity
(Huttunen et al., 1986). Characteristics of the CSEP such as the
relation to stimulus concentration and flow rate have been thoroughly
investigated (for review, see Kobal and Hummel, 1991a).

In this paper, results from patients suffering from various diseases are discussed and preliminary results of an investigation concerned with the effects of benzodiazepines on olfactory sensation is
reported.

Chemical Signals in Vertebrates VI, Edited by R.L. Doty and
D. Müller-Schwarze, Plenum Press, New York, 1992

METHODS

For chemosensory stimulation, an apparatus was employed which
delivered chemical stimulants without altering the mechanical or thermal
conditions at the mucosa (Kobal, 1981). This monomodal chemical stimulation was achieved by mixing pulses of the stimulants in a constantly
flowing air stream with controlled temperature and humidity (36.5 °C;
80% relative humidity; total flow rate 140 ml/s). Stimulus duration was
200 ms, with a rise time below 20 ms. Stimuli were applied nonsynchro-
nously to breathing. Subjects were comfortably seated in an acoustical-
ly and electrically shielded air-conditioned chamber. White noise was
used to mask switching clicks of the stimulator. The three substances,
carbon dixoide (52% v/v), hydrogen sulfide (0.78 ppm), and vanillin
(2.06 ppm), were tested in each session. All three stimulants were
applied 15 times to both the left and right nostrils. The interstimulus
interval was approximately 40 s.

The EEG (bandpass 0.2-70 Hz) was recorded from 5 positions of the
international 10/20 system referenced to A1. EEG-records of 2048 ms
duration were digitized (sampling frequency 250 Hz), averaged, and
evaluated by OFFLAB programs (Kobal, 1981). EEG-records contaminated by
eye blinks (Fp2/A1) or motor artifacts were discarded from the average.

In addition to the recording of CSEP, a simple odor identification
test was employed. Eleven odorants stored in brown glass bottles were
presented in a randomized order to either the left or right side while
the experimenter gently closed the other nostril with the tip of his
finger. The verbal descriptions of the odorants were recorded and
later grouped into 4 categories (Kobal and Hummel, 1991a).

RESULTS AND DISCUSSION

Anosmic Patients

During the last three years, 45 anosmic patients (ranging from 17
to 69 years of age) were investigated in our laboratory (Kobal and
Hummel, 1991b). In most of the cases, anosmia was caused by trauma
(58%) or viral infection of the upper airways (16%). None of the anosmic patients responded to stimulation with vanillin or hydrogen sulfide.
However, in all of them CSEP could be obtained after trigeminal stimulation with CO_2 (Fig. 1). Thus, it is possible to independently investigate the two sensory systems primarily responsible for the perception of
odorants. This technique may be of advantage in many clinical situations, such as the monitoring of olfactory disorders. Specifically, the
testing of malingering patients complaining of anosmia no longer needs
to be the difficult task ENT-clinicians are presently confronted with.

Parkinson's Disease

It has been demonstrated that Parkinson's disease is accompanied by
olfactory deficits (Doty et al., 1988). Eight patients suffering from
Parkinson's disease were tested against 8 healthy, age and sex matched
subjects (Kobal and Hummel, 1991a). The healthy subjects scored better
on the odor identification test. Significant effects were obtained for
phenyl ethyl alcohol, anethol, coffee, isoamyl acetate, vanillin, and
benzaldehyde. However, for acetic acid, menthol, eucalyptol, limonene,
and linalool, no significantly different responses could be observed
when the two groups were compared. Thus, the largest effects between
the two groups were observed for those substances, which at the given
concentration, possessed no, or minimal, trigeminal stimulation.

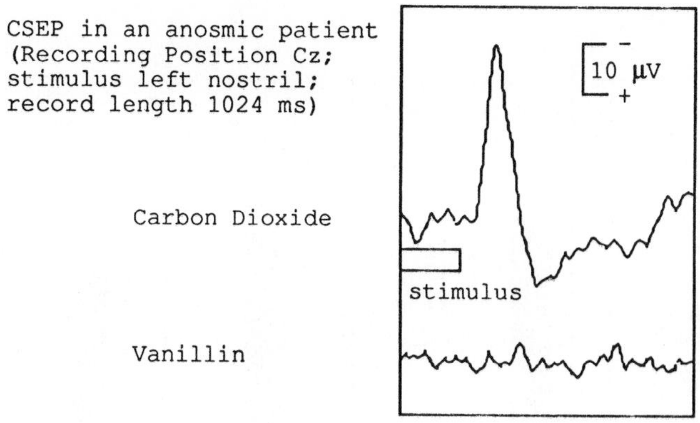

CSEP in an anosmic patient
(Recording Position Cz;
stimulus left nostril;
record length 1024 ms)

Carbon Dioxide

Vanillin

Fig. 1.   CSEP in a patient suffering from ansomia.

In a comparable manner, CSEP to vanillin and hydrogen sulfide revealed significantly longer latencies in the Parkinsonian group.   In contrast to stimulation with these two odorants, there were no differences in the responses to the trigeminal stimulant CO_2.

Temporal Lobe Epilepsy

In a preliminary study (Kobal et al., 1992), 14 patients were investigated, 8 of whom suffered from left temporal lobe epilepsy.   The remaining 6 patients had a right-sided, temporal epileptical focus. CSEP latencies following left side stimulation were prolonged in patients with left-sided epileptical foci.   In contrast, in patients with a right-sided focus, CSEP latencies were prolonged when the right nostril had been stimulated.   These findings were less pronounced after stimulation of the trigeminal system.   It was concluded that olfactory sensations may be processed mainly ipsilaterally to the stimulated nostril.

Effects of Benzodiazepines

Receptors of benzodiazepines appear to be highly concentrated in the olfactory bulb and the limbic system (Braestrup, Albrechtsen, and Squires, 1977).   Therefore, it is of interest to gain information about the physiological role of benzodiazepines in the perception of odorants. In an initial study of 4 volunteers, the effect of diazepam (10 mg) was compared to that of a placebo.   Measurements were taken before and 60 minutes after oral administration of the drug.   The responses of two odorants, the pleasant-smelling phenyl ethyl alcohol (40 ng/ml) and the unpleasant-smelling hydrogen sulfide (0.78 ppm) were compared.   When diazepam had been administered, hydrogen sulfide tended to be perceived as less intense when compared to placebo, whereas there seemed to be no change in the perceived intensity of phenyl ethyl alcohol.   Amplitudes of CSEP after stimulation with hydrogen sulfide revealed a tendency to decrease when the tranquilizer had been administered (Fig. 2).   However, CSEP to stimulation with phenyl ethyl alcohol did not change.   These preliminary results indicated the possible influence of benzodiazepines on the processing of olfactory stimuli.

In summary, chemosensory evoked potentials have found their way into a number of fields of research.   It is hoped that this technique will help us to better understand human olfactory function.

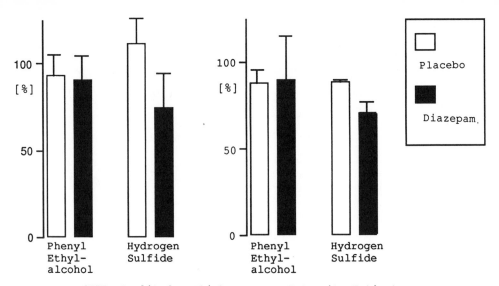

CSEP: Amplitudes N1/P2   Intensity Estimates
(Recording Position Cz)

Fig. 2.   Effects of diazepam on CSEP amplitudes and intensity
estimates compared to placebo (means, SEM, n=4).
Data obtained 60 minutes after drug administration
are related to baseline values.

ACKNOWLEDGEMENTS

This work was supported by DFG grant Ko812/2-2/3.

REFERENCES

Braestrup, C., Albrechtsen, R., and Squires, R.F., 1977, High densities
of benzodiazepine receptors in human cortical areas, Nature, 269:
702.

Cain, W.S., and Rabin, M.D., 1989, Comparability of two tests of olfactory functioning, Chem. Senses, 14:479.

Doty, R.L., Brugger, W.E., Jurs, P.C., Orndorff, M.A., Synder, P.J., and
Lowry, L.D., 1978, Intranasal trigeminal stimulation from odorous
volatiles: psychometric responses from anosmics and normal humans,
Physiol. Behav., 20:175.

Doty, R.L., Shaman, P., and Dann, M., 1984, Development of the University of Pennsylvania Smell Identification Test: a standardized
microencapsulated test of olfactory function, Physiol. Behav.,
32:489.

Doty, R.L., Deems, D.A., and Stellar, S., 1988, Olfactory dysfunction in
parkinsonism: a general deficit unrelated to neurological signs,
disease stage, or disease duration, Neurology, 38:1237.

Hummel, T., Pietsch, H., and Kobal, G., 1991, Kallmann's syndrome and
chemosensory evoked potentials, Eur. Arch. Oto-Rhino-Laryngol.,
248:311.

Huttunen, J., Kobal, G., Kaukoronta, E., and Hari, R., 1986, Cortical
responses to painful CO_2-stimulation of nasal mucosa: a magnet-
encephalographic study in man, Electroenceph. Clin. Neurophysiol.,
64:347.

Kobal, G., 1981, "Elektrophysiologische Untersuchungen des menschlichen Geruchssinnes", Thieme, Stuttgart.

Kobal, G., and Hummel, T., 1991a, Olfactory evoked potentials in humans, in: "Smell and Taste in Health and Disease", T.V. Getchell, R.L. Doty, L.M. Bartoshuk, and J.B. Snow, Jr., eds., Raven Press, New York, pp. 255-275.

Kobal, G., and Hummel, T., 1991b, Chemosensory evoked potentials in anosmic patients, Rhinology, submitted.

Kobal, G., Hummel, T., Schuler, P., and Stefan, H., 1992, Chemosensory evoked potentials in patients with temporal lobe epilepsy (in preparation).

Thurauf, N., Friedel, I., Hummel, C., and Kobal, G., 1991, The mucosal potential elicited by noxious chemical stimuli: is it a peripheral nociceptive event?, Neurosci. Lett., 128:297.

INDIVIDUAL DIFFERENCES IN PREFERENCES: SENSORY SEGMENTATION AS AN

ORGANIZING PRINCIPLE

Howard R. Moskowitz

Moskowitz Jacobs Inc.
Valhalla, New York

INTRODUCTION

Individual differences in hedonics and preference pervade the chemical senses (Ekman & Akesson, 1964; Moncrieff, 1966; Pangborn, 1970), evidencing themselves both qualitatively and quantitatively. The pattern relating stimulus intensity to degree of liking differs from subject to subject. The difference may be one of degree (so that one subject always likes the stimulus set more than another subject does), or one of pattern (so that the two subjects differ in terms of the specific stimulus level at which liking peaks). Individual differences are not simply artifacts of and emergent phenomena from an "artificial" test environment. Well stocked supermarkets offer the same general product in several flavors, flavor intensities, etc.

The Inverted U-Shaped Hedonic Curve

The organizing principle for hedonics is straight-forward and appealing. More than a century ago the German psychologist Wilhelm Wundt (see Beebe-Center, 1932) speculated that there might exist a "Hedonic Law". For any stimulus continuum, overall liking follows an inverted U shaped curve, when plotted against stimulus intensity. As the physical intensity increases liking would first increase, then peak at an optimum level, and then decrease as the stimulus intensity passes the optimum.

Sensory Preference Segmentation as a Second Organizing Principle

Using the hedonic function as an organizing principle researchers discovered that the shape of function differed across subjects. The average hedonic function resulted from the aggregate of disparate patterns. It is at this point that the organizing principle of sensory segmentation enters.

Sensory segmentation posits that there exist a limited number of basic or "fundamental" patterns of preference. Individuals can be classified as belonging to one of these segments, based upon the pattern relating their liking ratings to sensory intensity. These sensory segments may either be opposing groups, nodal points or centers of gravity of a distribution. It remains for experimenters to decide whether these segments are primaries or nodal points.

Chemical Signals in Vertebrates VI, Edited by R.L. Doty and
D. Müller-Schwarze, Plenum Press, New York, 1992

Moskowitz (1985) developed an algorithm using the empirical relation between each subject's own liking rating and a common reference scale of sensory intensity. The algorithm searches for common patterns of hedonic functions and then clusters together subjects showing the same patterns.

EXPERIMENTAL EVIDENCE - THREE FOOD CATEGORIES

Much of the experimental evidence for sensory segmentation comes from consumer evaluations of actual marketed foods, along with prototypes created by research and development. This paper presents three sets of data; for roast and ground coffee, for carbonated beverage, and for cooking sauce, respectively.

Test Procedures

The three studies were run separately. In each study the consumer panelists were chosen to be users of the product. The coffee study was run in one evening, the carbonated beverage and sauce studies were run over two evenings. Each session lasted 4 hours.

Each subject evaluated the products in a randomized order to remove potential bias. The subject rated each product on a wide variety of sensory and liking attributes, using an anchored 0-100 scale (Moskowitz, 1985).

The coffee study involved 10 different products, and a base of 140 subjects. The carbonated beverage study involved 29 products, and a base of 144 subjects. For both studies the subjects evaluated every one of the samples. The cooking sauce study involved 31 products, and a base of 120 subjects. Each subject evaluated 16 of the 31.

Data Analysis

Data analysis followed procedures outlined by Moskowitz (1985). These were the steps:

1) Calculation of a basic data matrix (product x attribute)
2) On a subject by subject basis, relate that subject's liking ratings to the consensus sensory levels (step 1), by means of regression analysis. Use a quadratic function for each subject's liking function
3) Determine the sensory level at which a subject's liking peaks
4) Develop a vector of these optimal levels, one vector per subject, with the numbers in the vector corresponding to different attributes
5) Reduce the matrix of optimal levels by factor analysis
6) Cluster the subjects. Subjects in the same cluster show similar patterns of optimal levels. They are in the same segment.

RESULTS

Roast And Ground Coffee

Two sensory segments emerged from segmentation. Figure 1 shows how liking varies with perceived coffee bitterness. The two segments show different patterns. Similar, but not as dramatic segment differences occur for liking vs darkness.

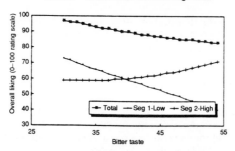

Coffee - Total Panel Versus Segments

Fig. 1.   Overall Liking Versus the Perceived Bitter
Taste of Roast and Ground Coffee

Carbonated Beverage

Carbonated beverages are more complex than coffees. Carbonation adds
trigeminal inputs. The color of the beverage itself can also vary as a
function of the flavoring level, among other factors. Additionally, the
beverage varies in sweetness, flavor intensity, etc.

Two segments emerged from carbonated beverage -- again, low versus
high impact. The dynamics differ across these two segments. The high
impact segment wants more sweetness, whereas the low impact segment wants
an intermediate level of sweetness (Figure 2).

Cooking Sauce

In contrast to coffee and carbonated beverages, cooking sauces are
highly complex, creating complicated texture and flavor profiles. Segmentation revealed two segments - a segment that wants redder products and is
indifferent to spicy flavor, and a segment that wants a darker, milder, and
thicker product. Figure 3 shows the dynamics of overall liking versus
perceived spiciness.

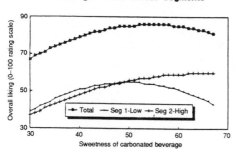

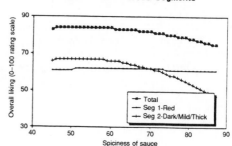

Fig.2 - How Overall
Liking Varies With The
Perceived Sweetness Of
Carbonated Beverage

Fig. 3 - How Overall
Liking Varies With The
Perceived Spiciness Of
Cooking Sauce

DISCUSSION

Is Sensory Segmentation Real, Or Just An Artifact Of Data Analysis?

Over the years researchers have developed a variety of clustering methods to segment the consumer population. Most of these methods use psychological batteries or purchase behavior patterns. Sensory segmentation, in contrast, uses the actual patterns which relate overall liking to sensory attributes.

In defense of the underlying "reality" of sensory segmentation, we find that sensory segmentation reveals dramatic differences in the pattern of liking ratings. Segmentation by purchase behavior, gender, income, etc. shows little differentiation between the segments in terms of the liking ratings. Segmentation by sensory preferences shows a much wider range of differences (see Figures 1 - 3).

A further argument comes from the segmentation method itself. The sensory segmentation was never based upon the level of liking per se. It cannot be argued that the sensory segmentation divides consumers on the basis of "lovers" versus "haters" of a product. The segmentation algorithm uses individual sensory optima. The algorithm is specifically designed to be independent of level of liking. Hence, we must conclude that sensory segmentation taps into meaningful, rather than artifactual, differences among consumers.

AN OVERVIEW

The organizing principle is powerful. It provides a mechanism by which one can examine different cultures, different ages of individuals, different disease states, etc. Sensory segmentation has already proven itself viable as a procedure in the business community by which to create new foods and fragrances. The time now has come to apply the procedures in the scientific realm to build knowledge as well.

REFERENCES

Beebe-Center, J.G., 1932, The Psychology Of Pleasantness And Unpleasantness. New York, Van Nostrand Reinhold.
Ekman, G., and Akesson, C.A., 1964, Saltiness, sweetness and preference: A study of quantitative relations in individual subjects. Report 177, Psychological Laboratories, University Of Stockholm, Sweden.
Moncrieff, R.W., 1966, Odour Preferences. London, Leonard Hill.
Moskowitz, H.R., 1985, New Directions for Product Testing and Sensory Analysis of Foods. Westport, Food and Nutrition Press.
Pangborn, R.M., 1970, Individual variations in affective responses to taste stimuli. Psychonomic Science, 21, 125-128.

INFLUENCE OF ANDROSTENOL AND ANDROSTERONE ON THE EVALULATION OF MEN OF

VARYING ATTRACTIVENESS LEVELS

R.E. Maiworm[1], W.U. Langthaler[2]

[1]Department of Psychology, University of Olshausenstr.
40/60 2300 Kiel, Germany
[2]Department of Psychology, University of Munster
Fliednerstr. 21 4400 Munster, Germany

INTRODUCTION

The components of human axillary secretions, especially odoriferous steroids, may take part in human olfactory communication in a manner analogous to the odours from specialized skin glands in animals (Labows et al., 1982). Filsinger & Fabes (1985), for expample, suggest that some of them (especially androstenol, androstenone) may be putative human pheromones. With regard to androstenol and androstenone, evidence has been found for a connection between these odoriferous substances and the covert or open behavior of the receivers (e.g., assessment of person, self-ratings, choice of locations: cf. Kirk-Smith, 1978; Filsinger & Fabes, 1985; Benton & Wastell, 1986). However, it has not been shown that such substances function as human pheromones and induce a definite behavior or physiological response.

In contrast to androstenol, androsterone is a weakly androgenic steroid. As a part of steroid metabolism in the skin, androsterone is metabolized from dihydroepiandrosterone (Sharp et al., 1976; Kaufmann et al., 1990). Androsterone is found in lipids coating the axillary hair and may be secreted by sebaceous glands (Toth, 1983). Labows et al. (1979) found no free androsterone in fresh apocrine secretion, although free androsterone was found after the secretion was acted on by bacteria. Androsterone is also found in human urine. This steriod has a musky odor and olfactory properties (e.g., subjective intensity) similar to those of androstenol (Kloek, 1961).

Little information is available about the influence of androsterone on the open or covert behavior of humans. Gustavson et al. (1987) used androsterone as a control odor to androstenol when testing its influence on human choice behavior (rest room stalls). They did not find any effects of androsterone on the behavior of either men or women.

In a previous study, androstenol, as well as androstenone, was found to influence the self-ratings of the mood of female subjects. Furthermore, Maiworm and Langthaler (1990) found an effect of androstenone on women's assessment of photographs of moderately attractive men. This effect was related to the stage of the menstrual cycle.

Chemical Signals in Vertebrates VI, Edited by R.L. Doty and
D. Müller-Schwarze, Plenum Press, New York, 1992

The present study examines the potential influence of 5-alpha-an-drost-16-en-alpha-ol and 3-alpha-hydroxy-5-alpha-androsten-17-one on the assessment of the attractiveness and other attributes of men. We hypothe-sized that, compared to two control groups (namely solvent and no treat-ment), androstenol and androsterone will influence women's subjective ratings of the attractiveness and other attributes of photographed men. The effects of the substances were expected to vary with the degree of the attractiveness of the men, as determined from peer ratings.

METHODS

The double-blind study, in which 102 female students (mean age = 23.3 years) participated, was conducted by two female experimenters. The sub-jects were paid a moderate amount of money to participate.

In an earlier study photographs of 60 men and women had been taken under standardized conditions. The attractiveness had been assessed by 118 female and male students. For the present study, we choose 5 men from this sample whose mean attractiveness scores covered the range of the attrac-tivenss scores of the sample; namely, a scarely attractive man (value 24), a less attractive man (value 48), a moderately attractive man (value 53), and two highly attractive men (value 65, 72).

At the beginning of each session, a standardized questionnaire was administered to assess each subject's state of mood (EWL; Jahnke & Debus, 1978). One of the three test substances was applied to the upper lip of each of ten subjects. The olfactory stimuli were stored in air-tight, 2.5 ml plastic containers. Each container held 1.5 ml of either androstenol, androstenol, androsterone (1 mg/ml), or the control stimulus ethanol (99%, research grade). In one control condition, no test substance was applied. The subject subsequently rated five randomly presented photographs of men, one by one, on 30 bipolar adjective scales developed in the course of a former study (Maiworm & Langthaler, 1990), and on a 100-point attractive-ness scale which ranged from "1" (indicating "not attractive") to "100" (indicating "highly attractive"). After the assessment of the photographs, the questionnaire about the subject's mood state was again administered and the subjects were interviewed (e.g., about menstrual cycle phase and oral contracptive use). A multifactorial two (experimental) by two (control) experimental design was used. The data were analyzed by a MANOVA with adjusted alpha levels (SPSS; Statistical Package for Social Science; two factorial model and contrasts).

RESULTS

The test substances and the phases of the menstrual cycle had an effect on the assessments (interaction, $p < 0.006$). The use of oral con-traceptives also influenced the evaluations (main effect, $p < 0.034$) and the phase of the menstrual cycle (main effect $p < 0.001$). In accord with a previous study, the data showed that the assessment of men under the influ-ence of androstenol is related to menstrual cycle phase ($p < 0.005$; main effect). Relative to the third phase (days 20-29), the evaluations con-cerning "open" ($p < 0.05$) were more positive during the first phase (days 1-9) and the evaluations of "attractive" ($p < 0.05$) were more positive during the second phase (days 10-19). A relationship between the influence of androsterone and menstrual cycle phase was present for the assessment "expressive" ($p < 0.05$ between the first and third and the second and third cycle phases). Androsterone was found to have a significant impact on the subjects who rated the men as "warm" ($p < 0.01$), "good" ($p < 0.01$), "mascu-line" ($p < 0.05$) and "black" ($p < 0.05$), whereas androstenol significantly

Table 1. Significant mean (SEM) ratings under androsterone and androstenol in comparison to control (contrasts, t-tests, adjusted alpha).

Scarcely Attractive Man, Value 14

	Androsterone		Ethanol		No Treatment	
Sensitive	3.18	(0.33)	3.68	(0.32)	3.98	(0.44)*
Aggressive	4.27	(0.37)	3.70	(0.37)	2.87	(0.36)*
Intelligent	3.23	(0.34)	3.75	(0.34)	4.35	(0.33)*
Hard	4.81	(0.31)	4.61	(0.31)	3.91	(0.21)*
Attractiveness	9.20	(2.70)	14.95	(2.70)	22.50	(3.43)*
	Androstenol		Ethanol		No Treatment	
Sensitive	3.36	(0.36)	3.68	(0.32)	3.98	(0.44)*
Intelligent	3.30	(0.17)	3.75	(0.34)	4.35	(0.33)

Little Attractive Man, Value 38

	Androsterone		Ethanol		No Treatment	
Attractive	1.95	(0.25)	2.91	(0.31)*	3.00	(0.34)*
Rash	2.33	(0.22)	2.83	(0.24)	3.08	(0.27)*
Black	2.71	(0.31)	3.43	(0.25)	3.88	(0.28)**
Nonchalant	3.24	(0.40)	4.00	(0.27)	4.40	(0.27)**
Attactiveness	25.14	(4.41)	37.35	(4.85)	37.30	(4.85)*
	Androstenol		Ethanol		No treatment	
Macho	4.43	(0.26)	3.85	(0.25)	3.14	(0.24)

Moderately Attractive Man, Value 48

	Androsterone		Ethanol		No Treatment	
Warm	5.00	(0.22)	4.00	(0.23)**	4.20	(0.25)*
Intelligent	5.59	(0.23)	4.62	(0.28)*	5.01	(0.16)
Loveable	5.14	(0.24)	4.36	(0.26)*	4.30	(0.18)*
Good	4.96	(0.21)	4.00	(0.34)*	4.50	(0.18)
	Androstenol		Ethanol		No Treatment	
Lovable	5.10	(0.23)	4.36	(0.26)	4.30	(0.18)*

Higher Attractive Man, Value 59

	Androsterone		Ethanol		No Treatment	
Attractive	5.13	(0.22)	4.41	(0.31)	4.08	(0.20)*
Intelligent	5.86	(0.14)	4.91	(0.28)**	5.00	(0.22)*
Like to meet him	5.08	(0.32)	4.75	(0.36)	3.50	(0.40)*
Sure	4.91	(0.36)	3.95	(0.26)	3.94	(0.34)*
Open	4.58	(0.36)	3.87	(0.33)	3.87	(0.35)*
Erotic	3.58	(0.35)	4.25	(0.34)	3.46	(0.35)*
Expressive	5.26	(0.16)	4.58	(0.27)	4.44	(0.28)*
I like him	5.61	(0.21)	5.25	(0.25)	4.62	(0.25)**
Sexy	4.13	(0.35)	4.25	(0.33)	3.65	(0.35)*
Attractiveness	66.61	(3.30)	58.87	(4.69)	51.83	(4.82)*

Higher Attractive Man, Value 60

	Androsterone		Ethanol		No Treatment	
Sure	3.32	(0.40)	5.28	(0.26)****	4.77	(0.36)**
Open	4.00	(0.30)	5.52	(0.38)***	5.06	(0.33)*
Stimulates me	2.86	(0.31)	4.06	(0.33)*	3.58	(0.27)
Rash	2.86	(0.22)	4.18	(0.25)****	4.12	(0.27)****
Erotic	3.09	(0.35)	4.32	(0.31)*	3.42	(0.24)
Sexy	3.32	(0.36)	4.38	(0.32)*	3.87	(0.30)
	Androstenol		Ethanol		No Treatment	
Sexy	3.20	(0.24)	4.38	(0.32)*	3.87	(0.30)
Sure	4.08	(0.35)	5.28	(0.26)*	4.77	(0.36)

---

* = p < 0.05, ** p < 0.01, *** p < 0.005, **** p < 0.001

influenced the ratings of "black" (p. < 0.05). Within one of the control groups (no treatment), the assessment of the attributes "macho" was affected by the phase of the menstrual cycle (p < 0.05). Androsterone raised the attractiveness ratings for the more attractive men, and lowered the attractiveness ratings for the less attractive men. The ratings of the higher attractive men (photographs) were more strongly influenced by the test substances than those of the scarcely or moderately attractive man (Table 1; contrasts by MANOVA, t-tests, alpha level adjusted).

The subjects' ratings concerning the personality profile and their mood states were within the range of values of the standard sample. A comparison (paired t-tests) of the pre/post-test scores indicated that the women described themselves as less tired (p < 0.05), less sensitive (p < 0.05), and less depressed (p < 0.05) under the androstenol condition at the end of the session than at the beginning. Within the androsterone condition, they also described themselves as less sensitivie (p < 0.05). At the end of the test session the women described themselves as less-assured and more anxious under the androsterone condition than under the control condition (p < 0.05).

DISCUSSION

The hypothesis concerning the influence of androstenol and androsterone was confirmed. Both substances had an effect on the womens' subjective perception of the attractiveness and other attributes of the photographed men, and this effect varied with the degree of peer-rated attractivenss of the men. With respect to androsterone, this result was unexpected, since to our knowledge this there is no mention of this phenomenon in the literature. However, we did not find a specific pattern; factor analyses did not yield any components that explained a satisfactorily high percentage of the variance. Since there was no significant difference in the ratings of the men between the control conditions (ethanol, no treatment), androstenol and androsterone appear to have an impact on the assessments. Somehow, both substances function as "levellers" by raising the evaluation of moderately attractive men and by lowering the evaluation of the higher attractive man (tendency towards the mean). The effects of androstenol on the estimation of other attributes were also found to be influenced by the menstrual cycle, which, in a previous study (Maiworm & Langthaler, 1990), turned out to be more evident for androstenone. The men were evaluated much more negatively in the third part of the menstrual cycle. Under the influence of both substances, the assessments of erotic attributes (sexy, erotic) tended to be more negative, whereas the assessments of non-sexual attributes (e.g., emotional, warm, sensitive, nice) tended to be more positive. Within the androstenol condition, the women described themselves as being in a good mood (less depressed, active), whereas in the androsterone condition, they described themselves as being less self-assured and courageous.

REFERENCES

Benton, D., and Wastell, V., 1986, Effects of androstenol on human sexual arousal, Biological Psychology, 22:141-147.
Filsinger, E.E., Braun, J.J., and Monte, W.C., 1985, An examination of the effects of putative pheromones on human judgements, Ethology and Sociobiology, 6, 227-236.
Filsinger, E.E. and Fabes, R.A., 1985, Odor communication, pheromones and human families, Journal of Marriage and the Family, 47, 3   49-359.
Gustavson, A.R., Dawson, ME., and Bonnet, D.G., 1987, Androstenol, a putative human pheromone, affects human (Homo sapiens) male choice performance. Journal of Comparative Psychology, 101, 210-212.

Jahnke, W., and Debus, G., 1978, Die Eigenschaftsworterliste. EWL. Hogrefe. Gottingen.

Kaufmann, F.R., Stancyk, F.Z., and Gentzschein, E, 1990, DHT and DHTS metabolism in human genital skin. Fertility and Sterility, 54, 251-254.

Kirk-Smith, M., Booth, D.A., Caroll, D., and Davies, P., 1978, Human social attitudes affected by androstenol. Research Communications in Psychology, Psychiatry and Behavior, 3, 379-384.

Kloek, J., 1961, The smell of some sex-hormones and their metabolites. Reflections and experiments concerning the significance of smell for the mutual relation of the sexes. Psychiatria, Neurologia, Neurochirurgia, 64, 106-344.

Labows, J.N., Preti, G., Hoelzle, E., Leyden, J., and Kligman, A., 1979, Steriod analysis of human apocrine secretion. Steriods, 34, 249-258.

Labows, J.N., McGinley, K.J., and Kligman, A., 1982, Perspectives on axillary odor. Journal of the Society of Cosmetic Chemists., 34, 193-202.

Maiworm, R.E. & Langthaler, 1990, Influences of androstenol, androsterone, menstrual cycle andoral contraceptives on the attractivity ratings of female probands. Ninth Congress of the ECGO. The Netherlands, 2-7.9, 1990.

Sharp, F., Hay, J.B., Hodgins, M.B., 1976, Journal of Endocrinology, 70, 491.

Toth, I., and Faredin, L., 1983, Steriods excreted by human skin. Acta Medica Hungaria, 40, 139-145.

INFLUENCE OF ODORS ON HUMAN MENTAL STRESS AND FATIGUE

Masashi Nakagawa, Hajime Nagai, Miyuki Nakamura,
Wataru Fujii and Takako Inui

Suntory Ltd., Institute for Fundamental Research
Mishimagun, Osaka 618, Japan

INTRODUCTION

There have been few objective studies of the psychological effects
of odors (Dodd et al., 1983). Torii and his colleagues (1988) studied
the effect of odors on brain activity. They reported that the odor of
jasmine increased the amplitude of contingent negative variat'ion (CNV)
due to a stimulating effect, while the fragrance of lavender, which is
said to have a sedative effect, decreased CNB amplitude. Sugano (1990)
reported that several fragrances increased cerebral blood flow, as
measured by single photon emission computed tomography. In this study,
we evaluated the effect of odoriferous compounds on mental stress and
fatigue, and measured the subjects' heart rate and the pupillary light
reflex. The Profile of Mood State (POMS) questionnaire was also administered (McNaire et al., 1964).

METHOD

Subjects

Nine women and three men, all in good health, were recruited from
local universities and institute staff. They ranged in age from 18 to 22
years. Caffeine intake and vigorous exercise were prohibited for 5 hours
prior to the experiment, and all subjects reported having slept well on
the night preceding the test day.

Design and Procedures

Each subject came to the laboratory for two 100-minute sessions. In
one session odor was pumped into the room (see below), whereas in another
no odor was presented. Sessions were separated by at least a week
interval and were counterbalanced in order, with half of the subjects
receiving the odor condition the first week and half the non-odor condition the second week. Each subject came into the electromagnetically
shielded room 30 minutes before a test session for adaptation. After
initial setup procedures, a ten-minute resting baseline measurement
period was initiated, followed by another ten-minute resting period just
before the mental workload session. Odor was generated by blowing 5
liter/min of air through 1.3% sweet fennel of essential oil in liquid

Chemical Signals in Vertebrates VI, Edited by R.L. Doty and
D. Müller-Schwarze, Plenum Press, New York, 1992

paraffin, a process that was started 10 minutes before the mental task in the odor condition. Using a personal computer, the subject performed the mental arithmetic task (two figures additions) for 60 minutes. The subject was asked to work out twenty arithmetic problems in one scroll of screen within 40 seconds. Ninety scrolls had to be completed in 60 minutes. After the arithmetic session, a recovery period was recorded for 20 minutes. Heart rate was recorded throughout the entire period. The pupillary light reflex was measured at the end of the first ten-minute baseline period, before the mental arithmetic period, after 40 minutes of work load, at the end of work, and at the end of session. A subjective symptom questionnaire was filled out before the session, and the POMS was used to assess subjective state during the experiment.

Physiological Measure Processing

A 7T18 signal processor (Nippon Denki Sanei, Ltd.) was used for on-line data acquisition and storage. The raw electrocardiogram activity data (ECG) were fed into an analytical system for heart rate change developed in our laboratory (Sayer, 1973, Akselrod et al., 1981). The respiratory sinus arrhythmia (RSA) components were analyzed by a FFT power spectrum. Total power from 0.15 Hz to 0.50 Hz was evaluated as physiological effects of task. A binocular iriscorder C2515 (Hamamatsu Photonics) was used to measure the pupillary light reflex. Photo-stimulation to the right eye for 0.25 seconds induced a construction of the pupil. The pupillary area and constriction velocity were analyzed by pupil-lography (Utumi, 1988) (Fig. 1). Initial pupillary area and con-striction ratio were used as the index of the physiological effects of task. Data were averaged across three trials.

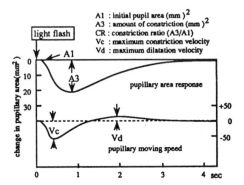

Fig. 1. Reaction graphs of pupillary light reflex

Subjective Mood State Measures

A Japanese version of POMS was used to assess subjective states during the experiment. The questionnaire consisted of 65 visual analog scale ratings anchored by mood-related adjective words, e.g., tension, depression, anger, vigor, fatigue, confusion and no relationship with mood state.

Results

Respiratory sinus arrhythmia (RSA) components were analyzed from ten subjects, excluding two subjects with unstable heart rate values. A significant decrease (df=9, t=2.28, p<0.05; before, df=9, t=2.97, p<0.05; after) was found between values obtained before and after the arithmetic task in the odor condition, while no significant difference was found in the non-odor condition. There was no significant difference on mental workload between conditions. (Fig. 2)

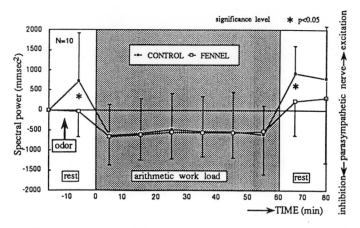

Fig. 2. Time-course data of Respiratory Sinus Arrhythmia (RSA)
Bars represent standard deviations which showed the complementary half of integral whole.

Pupillary Light Reflex

Pupillary light reflex data from ten subjects was compared between
odor and non-odor conditions.   There was no significant difference in
initial pupillary area (A1) between conditions.   A significant difference
(DF=9, t=3.62, p<0.01) was indicated in pupillary constriction during the
mental arithmetic workload period (Fig. 3).   In the non-odor condition,
the constriction ratio (CR) of the pupil to a flash of light was reduced
by about 25% compared to the baseline rest period.   The diminution of
pupillary CR, probably caused by a fatigue during mental arithmetic
workload, was relieved by the odor presentation.

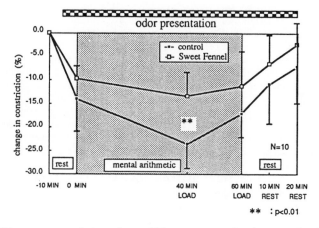

Fig. 3. Time-course data of pupillary constriction ratio (CR) Bars
represent standard deviations which showed the complementary half of integral whole.

Subjective Mood State (POMS)

A change of mood state from the baseline rest condition to after the workload task was evaluated. Due to the mental arithmetic for 60 minutes, the mood state scores for tension, vigor and confusion decreased, while the scores for depression and fatigue increased in the non-odor condition (Fig. 4). Significantly lower scores for depression (DF=11, t=2.52, p<0.05) and fatigue (DF=11, t=2.08, p<0.10) were found in the odor presentation condition compared to the non-odor condition.

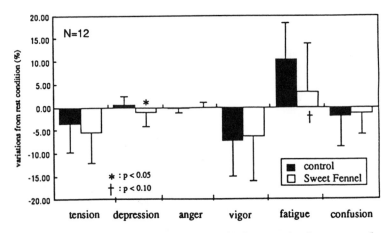

Fig. 4. Change of Mood State (POMS) before and after mental work loads. Bars represent standard deviations which showed the complementary half of integral whole.

DISCUSSION

The odor of sweet fennel suppressed the total power of the respiratory sinus arrhythmia (RSA) components and the diminution of the pupillary light reflex. There was no effect of odor on heart rate. Components of RSA increased due to excitation of the parasympathetic nervous system. The diminution of RSA components before and after workload in the odor presentation condition suggests that sweet fennel suppresses the parasympathetic nervous system. This idea is also supported by the effect of the odor on the constriction ratio (CR) of the pupillary light reflex during the workload task. Light-induced pupillary constriction is caused by excitation of the parasympathetic nervous system. Further support is provided by an inhibition test of the stress ulcer induced by restraint and immersion in the mouse (Nagai et al., 1990). The formation of a stress ulcer is induced by the over-secretion of gastrins, which is promoted by the excitation of parasympathetic nerves. In the POMS, self-reported fatigue and depression were lower during the odor presentation than when no odor was presented during the period of time when subjects were exposed to mental stress. These results indicate that some relationship exists between symptoms and the excitation of the autonomic nervous system.

REFERENCES

Akselrod, S., 1981, Power spectrum analysis of heart rate fluctuation, Science, 213:220.
Dodd, G.H. and Van Toller, C., 1983, The biology and psychology of perfumery, Perfumer and Flavorist, 8:1.
McNaire, D.M. and Lorr, M., 1964, An analysis of mood in neurotics, J. Abnormal and Social Psychology, 69:620.
Nagai, H., Nakagawa, M., Nakamura, M., Fujii, W. and Inui, T., 1990, Proceeding of the 24th Japanese Symposium on Taste and Smell, pp 195.
Sayers, B. McA., 1973, Analysis of heart rate variability, Ergonomics, 16:17.
Sugano, H., 1990, Effects of odors on mental function, Proceedings of the 23rd Japanese Symposium on Taste and Smell, pp 301.
Torii, S., Fukuda, H., Kanemoto, M., Miyauchi, R., Hamauzu, Y. and Kawasaki, M., 1988, Contingent negative variation (CNV) and the psychological effects of odor in perfumery, in "The psychology and biology of fragrance", Chapman and Hall, New York.
Utumi, T., 1988, Open-loop infrared video pupillography, Nichigankaishi, 83:315.

PERCEPTUAL ANALYSIS OF COMPLEX CHEMICAL SIGNALS BY HUMANS

D.G. Laing[1] and B.A. Livermore[2]

[1]CSIRO, Sensory Research Centre, North Ryde NSW 2113
Australia, and [2]Macquarie University, School of Behavioural
Sciences, North Ryde NSW 2111, Australia

INTRODUCTION

Odors encountered during the daily activities of humans and animals,
including those from flowers, sweat, excrement and cooking, are almost
always complex arrays consisting of dozens, often hundreds, of odorous
constituents. Rarely is an odor due to a single chemical. Nevertheless,
despite their chemical complexity, commonly encountered odorants are
usually identified within a second or two with the aid of a few sniffs.
The sense of smell, therefore, can detect, analyze, discriminate and
identify a complex odorant in a very short space of time, an action which
as yet cannot be matched by any instrument.

However, in view of the short amount of time taken to identify
complex odorants such as the pleasant aroma of coffee or the repugnant
odor of feces, it seems unlikely that the sense of smell attempts to
analyse all the odorous constituents to achieve identification.

What information, therefore, does the sense use to identify a
complex aroma? A solution to this question could provide important
insight as to the features of a complex odorant that humans and animals
use in assessing the acceptability of food, and by animals, insects and
fishes when interpreting chemical signals related to alarm, reproduction
or territory.

Recently, a series of studies aimed at resolving this question were
conducted with humans in our laboratories. These studies investigated
the influence of training and experience, odor type, test method, and
composition of complex odorants on the ability of humans to discriminate
and identify the constituents of odor mixtures.

GENERAL METHODOLOGY

A computer-controlled air dilution olfactometer with eight channels
(Laing and Francis, 1989) was used to produce a single concentration of
each odorant or mixtures containing up to eight odorants. Subjects in
all but the selective attention experiment (see below) were seated before
a computer monitor, a pressure sensitive Apple graphics tablet, and an
odor outlet. Sampling instructions were displayed on the monitor.

Chemical Signals in Vertebrates VI, Edited by R.L. Doty and
D. Müller-Schwarze, Plenum Press, New York, 1992

Located on the graphics tablet was a column of seven labels, each with a single odor description, e.g., 'almond' for benzaldehyde.

To initiate a trial, a subject touched an age and sex label, located on the tablet, with the electronic pen. These actions triggered delivery of the stimulus. Five seconds later an instruction on the monitor indicated to the subject to commence sniffing the odor outlet and the subject indicated which odor(s) were present by touching the appropriate label with the pen. A trial ended when the subject touched a 'Selection Finished' label or after 50 seconds. The subject then waited 50 seconds before commencing the next trial, which was automatically initiated by the computer. Each time a subject touched a label signifying the choice of odor, the word on the label appeared on the monitor, e.g., 'spearmint'. Once the 'Selection Finished' label had been touched, the correct answer(s) appeared on the monitor so that the subject had immediate feedback regarding his or her performance.

Odorants in each experiment were of approximately equal perceived intensity and moderate strength. Adjustment of intensity was achieved using panels of approximately 20 subjects who rated the perceived intensity of each odorant by placing a mark on a 130 mm graphic rating scale. The scale had the words 'No Odor' and 'Extremely Strong' at the ends. The presentation order of the stimuli was randomized and differed for each subject. Specific numbers of trials per subject are indicated below within the description of each experiment.

Subjects in the selective attention experiment received odors from a previously described olfactometer (Laing, 1988) and recorded their responses on score sheets. Feedback was given orally by the experimenter after each trial; sampling times and intertrial intervals were similar to those described above.

EXPERIMENTS AND RESULTS

In the first series of studies, the responses of untrained, trained and expert subjects were compared. The subjects were (i) 123 untrained members of the public who were visitors to CSIRO over 3 days, (ii) 10 psychology students trained for 1 week to identify the individual test odorants, and (iii) 8 expert flavorists and perfumers. After several minutes of familiarisation with the pool of 7 test odorants, untrained subjects were given 5-10 test trials with stimuli consisting of between 1 and 5 odorants. Trained subjects were presented with single odorants daily for one week with feedback as to the correctness of their responses after each trial, and were required to achieve > 6/7 correct responses over the two final sessions before participating in the study. Subjects in the trained and expert groups were both required to sniff the single test odorants twice daily for 1 week at home before beginning the test sessions, which were conducted in exactly the same manner for both groups. During a test session, each subject from the trained and expert groups received a different random series of 17 stimuli which consisted of one or up to five odorants from the pool of seven over 10 and 3 sessions, respectively.

The odorants and descriptions used were acetic acid (vinegar), benzaldehyde (almond), (-)-carvone (spearmint), ethyl caprylate (fruity), eugenol (cloves), (-)-limonene (orange) and alpha-pinene (camphor).

The results from the three test groups are shown in Figure 1 and indicate that subjects had great difficulty in identifying all the constituents of 5-component mixtures; indeed very few subjects from any group could identify all the odorants in 4-component mixtures. Clearly, the results indicate that humans have very great difficulty in discriminating and identifying more than 3 components in mixtures. This conclusion is supported by the result that the most common number of odorants selected with 3-, 4- and 5-component mixtures was 3, regardless of whether the choice was correct or incorrect, suggesting that mixtures never appeared to contain more than three components.

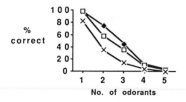

Fig. 1. Percent correct identifications by expert (filled diamonds), trained (open squares) and untrained subjects (crosses) when identifying constituents of stimuli containing up to 5 components.

In an attempt to determine whether the type of odorants or the method of testing had limited the ability of subjects to identify mixture constituents, two further studies were undertaken. In the first, subjects were tested with a group of 8 'good' blending odorants and a group of 8 'poor' blenders (both sets were selected by a perfumer and flavorist) in two separate experiments and had to identify the constituents of stimuli having between 1 and 8 odorants. 'Good' blenders were phenethyl alcohol (rose), Exaltolide (musky), cinnamic aldehyde (cinnamon), gamma-nonalactone (coconut), ethyl butyrate (fruity), limonene (orange), furaniol (burnt caramel) and benzaldehyde (almond). 'Poor' blenders were skatole (bad breath), Champignol (mushroom), cis-3-hexenol (cut grass), methyl salicylate (Dencorub), 2,5-dimethylphenol (antiseptic), diallyl sulphide (garlic), styrene (aeroplane glue), and mandarin aldehyde (mandarin). In the second study, a similar group of odorants to those used in the first three studies was chosen and a selective attention procedure was employed to identify the constituents of stimuli containing between one and six odorants. Briefly, subjects in the latter study had either to identify all the constituents presented as before, or only one target odorant, e.g., spearmint, at each test session. Over several weeks all the test odorants were the target odorant. In both of these studies subjects were given one week of training prior to testing and had to achieve a similar criterion of correctness, as described above, before qualifying for the study.

As shown in Figures 2 and 3, few subjects could identify more than 4 odorants regardless of the type of odorant or the test method, with little advantage being gained using 'poor' blending odorants, or when subjects had only to identify a single target odorant. Again, as described above, the most common number of odorants selected, regardless of the correctness of the choice, was three in both studies.

The results from all five studies described above support the conclusion that humans have a limited ability to discriminate and identify the constituents of mixtures, the capacity being limited to about 3 or 4. For a sensory system which is renowned for its ability to

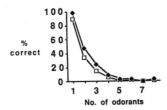

Fig. 2. Percent correct identifications of stimuli consisting of
         up to 8 'good' blenders odorants (open squares) or 'poor'
         blending odorants (filled diamonds).

discriminate among thousands of odorants, and appears to almost always
have to operate in an environment containing mixtures of odorants, the
capacity found here is unexpected and poor.

But does the sense of smell need to identify many or indeed any of
the individual constituents to identify a mixture?

The names of many odors, for example, coffee, rose, feces, are the
names of objects from which these odors emanate. Furthermore, identification of the odor of these objects is rapid, occurring within a
second -- clearly insufficient time to discriminate and identify even a
few of the constituents. Accordingly, it is proposed here that a complex

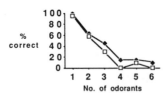

Fig. 3. Percent correct identifications using a multiple iden-
         tification procedure (open squares) or selective attention
         procedure (filled diamonds).

aroma such as that of coffee, which contains hundreds of odorous
constituents, is processed and stored in olfactory memory as a single
entity. Furthermore, it is proposed that the engram for coffee has
spatial and temporal characteristics arising from spatial and temporal
'filtering' at the periphery and/or other olfactory centers.

Spatial filtering may occur at the periphery through competition for
receptor sites and receptor cells by the many mixture constituents,
resulting in the normal input of many individual constituents being
suppressed. The normal pattern of receptor cells activated by an individual constituent of a mixture may be dramatically altered and reduced
through such inhibition, with perhaps relatively few cells being activated by a particular constituent. The resulting pattern of activated

receptor cells following stimulation by an aroma such as that of coffee, therefore, is likely to be very different to the patterns of activation produced when each of the constituents is presented individually. Further alteration of this pattern is possible through lateral inhibition in the glomerular layer or between mitral cells in the olfactory bulb (Shepherd, 1972), and perhaps by other unknown mechanisms at other olfactory centres before the pattern of activation is stored in memory. The spatial pattern resulting from a complex aroma may therefore be no more complex than that of a single chemical and this suggestion is supported by limited studies with the cell metabolic marker 2-deoxyglucose which show that activation patterns at the level of the olfactory bulb for the odor from rat cages (Stewart et al., 1979) or scent glands (Skeen, 1977) are no more complex than for single chemicals such as limonene (Bell et al., 1987) or amyl acetate (Skeen, 1977).

Identification of an odor from the spatial pattern in memory, however, is likely to have a second dimension involving a temporal component. Electrophysiological studies, for example, indicate that the time taken to activate a receptor cell differs between odors by as much as several hundred milliseconds (Getchell et al., 1984). 'Fast' odorants would therefore be in a favoured position to occupy receptor sites and depolarise cells, or simply act as antagonists and block receptor sites at the periphery, thereby preventing activation by 'slow' odorants. Thus, the final coded message for coffee aroma in memory may have spatial and temporal features.

To test the hypothesis that complex odorants may be processed and stored in olfactory memory as a single unique item, subjects were presented, in a further experiment, with stimuli consisting of between one and eight odorants, where each odorant consisted of dozens, probably hundreds, of odorous constituents. The complex odorants used were smokey, strawberry, lavender, kerosene, rose, honey, cheese and chocolate, and were commercial preparations donated by Dragoco Australia (except kerosene). The two most likely outcomes of the study were that (i) a maximum of 3-4 complex odorants would be identified, i.e., complex odors would be processed as single entities, or (ii) the massive number of odorants delivered to the receptors when mixtures of the complex odorants were the stimuli would produce new patterns of activation at the periphery, bulb, etc., not represented in memory, resulting in perhaps none of the 'individual' complex odorants being identified.

The result was clear-cut. A maximum of four complex odors were identified, supporting the hypothesis that complex aromas are stored in memory as a single unique entity (Fig. 4).

SUMMARY AND CONCLUSIONS

The results of the six studies clearly show that only about four odorants, be they complex (object) odors or single chemicals, can be discriminated and identified in mixtures. These results also suggest that stimuli which are complex mixtures of odorants, the so-called "object" odors, are processed and stored in memory as a single entity. Thus the neural input from each of the individual mixture constituents of "object" odors is likely to be small and the pattern of cells activated by an individual constituent bears little resemblance to the one produced when the odorant is presented as the sole stimulus. Hence discrimination and identification of individual constituents in such complex arrays is extremely difficult.

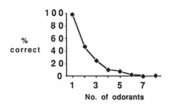

Fig. 4. Percent correct identifications of stimuli consisting of up to 8 complex odorants.

Nevertheless, the evidence is clear that up to four odorants can be identified in mixtures. This is not quite the paradox it may at first seem. In many mixtures obtained from natural sources it is not uncommon to identify several 'impact' components (Nursten, 1979). Chemical and olfactory analysis usually show that these 'impact' components are present in quantities far in excess of their olfactory threshold and despite the suppressing and modulating effects of other odors in the mixture, the few 'impact' components stand-out from the nondescript but characteristic background aroma.

Clearly, further studies are required to resolve why humans can only identify up to four components of mixtures and whether complex 'object' odors are processed similarly to individual chemical odors.

How humans utilize the information from mixtures has particular relevance to the perception of pheromone mixtures by their animal counterparts. For example, it has been reported that less than three or four individual chemicals in a complex mixture are important for initiating specific behaviors in insects (Rumbo, 1983). In contrast, in studies with monkeys, fractions of glandular extracts containing large numbers of constituents have been found necessary for behavior induction (Smith et al., 1985). These examples indicate that the utilization of information from a limited number of mixture components or from the mixture as a whole occurs in nature in species other than humans, and suggest common peripheral strategies and coding mechanisms across species.

REFERENCES

Bell, G.A., Laing, D.G., and Panhuber, H., 1987, Odor mixture suppression: Evidence for a peripheral mechanism in human and rat, Brain Res., 426:8-18.

Getchell, T.V., Margolis F.L., and Getchell, M.L., 1984, Perireceptor and receptor events in vertebrate olfaction, Prog. Neurobiol., 23:317-345.

Laing, D.G., 1988, Relationship between the differential adsorption of odorants by the olfactory mucus and their perception in mixtures, Chem. Senses, 13:463-471.

Laing, D.G., and Francis, G.W., 1989, The capacity of humans to identify odors in mixtures, Physiol. Behav., 46:809-814.

Nursten, H.E., 1979, Why flavour research? How far have we come since 1975? in: "Progress in Flavour Research." D.G. Land and H.E. Nursten, eds., Applied Science Publishers, London.

Rumbo, E., 1983, Differences between single cell responses to different components of the sex pheromone in males of the lightbrown apple moth (Epiphyas postuittana), Physiol. Entomol., 8:195-201.

Shepherd, G.M., 1972, Synaptic organization of the mammalian olfactory bulb, Physiol. Rev. 52:864-917.

Skeen, L.C., 1977, Odor-induced patterns of deoxyglucose consumption in the olfactory bulb of the tree shrew, Tupaia glis, Brain Res., 124:147-153.

Smith, A.B., Belcher, A.M., Epple, G., Jurs, P.C. and Lavine, B., 1985, Computerised pattern recognition: A new technique for the analysis of chemical communication, Science, 228:175-177.

Stewart, W.B., Kauer, J.S. and Shepherd, G.M., 1979, Functional organization of rat olfactory bulb analysed by the 2-deoxyglucose technique, J. Comp. Neurol., 185:715-734.

THE INFLUENCE OF ULTRADIAN AUTONOMIC RHYTHMS, AS INDEXED BY

THE NASAL CYCLE, ON UNILATERAL OLFACTORY THRESHOLDS

Richard E. Frye and Richard L. Doty

Smell and Taste Center and Department of Otorhinolaryngology --
Head and Neck Surgery, School of Medicine
University of Pennsylvania, Philadelphia, PA

INTRODUCTION

Cyclic side-to-side variations in the autonomic tone of the nasal
mucosae produce corresponding changes in nasal patency. This 90 minute to
4 hour "nasal cycle" is correlated with a number of ultradian rhythms,
including asymmetries in left:right cerebral EEG activity and differential
performance on visual/spatial psychological tasks (Eccles, 1978; Klein et
al., 1986; Werntz et al., 1984). Several authors have proposed that the
nasal cycle is part of an overall physiological rhythm known as the Basic
Rest-Activity Cycle (BRAC). Thus, relatively greater airflow through the
left nasal chamber is associated with the 'REST' phase of the BRAC and
relatively greater airflow through the right nasal chamber is associated
with the 'ACTIVITY' phase of the BRAC. During the 'REST' phase of the BRAC
there is a preponderance of right hemispheric EEG activity, a spatial
cognitive mode, and a parasympathetic predominance in the stomach, intes-
tines, and other unpaired body organs. During the 'ACTIVITY' phase of the
BRAC, greater left EEG hemispheric activity, a verbal cognitive mode, and
sympathetic predominance in unpaired organs occurs (Kennedy et al., 1986).

This paper briefly describes studies which (a) support the notion that
the nasal cycle is associated with several rhythms of the BRAC and (b)
suggest that the nasal cycle is related to olfactory sensitivity.

THE NASAL CYCLE AND SLEEP STAGE

Goldstein, Stoltzfus and Gardocki (1972) found that left hemispheric
total integrated EEG activity was of greater magnitude during REM sleep
periods than during non-REM sleep periods (and visa versa). Given the
relationship between hemispheric EEG activity and the nasal cycle, relative
airflow should be greater on the right than on the left side of the nose
during REM episodes. The data presented below support this hypothesis.

Three male and three female subjects [mean age (SD) = 29.9 years
(12.9)] participated in the twelve-hour recording sessions. Recording
electrodes were attached to the face and scalp of each subject, and nasal
thermistors were uniformly placed below the outer edge of each naris.
Since asymmetric peripheral stimulation such as lateral recumbency can
influence the expression of the nasal cycle (Haight and Cole, 1984), body
movements were viewed by a video camera with infrared lighting.

Chemical Signals in Vertebrates VI, Edited by R.L. Doty and
D. Müller-Schwarze, Plenum Press, New York, 1992

The amplitude of one representative nasal airwave from each nostril was measured from the recording every two minutes. These values were then logarithmically transformed and a value which represented the relative phase of the nasal cycle was derived by calculating the percentage of total nasal airflow through the right naris (Rao and Potdar, 1970).

A significantly greater percentage of total nasal airflow occurred through the right nasal chamber during periods of REM sleep than during periods of non-REM sleep, regardless of body position ($p < 0.001$; Figure 1a).

FOOD INTAKE AND THE NASAL CYCLE

Oscillations in oral activity, gastric contractility, plasma noradrenaline levels, spontaneous vigilance, and napping behavior are all believed to be associated with the BRAC and presumably reflect relative shifts in the systemic predominance of sympathetic and parasympathetic autonomic nervous system activity (Hiatt and Kripke, 1975; Levin et al., 1979; Oswald et al., 1970). If greater airflow occurs in the left nasal chamber during periods of relative parasympathetic dominance and a causal association between stimulation of the visceral organs by ingestion of food and activation of parasympathetic activity exists, then eating a meal should lead to proportionately more airflow in the left than in the right nasal chamber.

Eight female and seven male subjects participated in six one-hour consecutive daily test sessions. Each subject was tested before and after a 15-minute period during which either eating or no eating occurred. In order to establish the degree of nasal airflow laterality, the nasal resistance of each subject was measured by anterior rhinomanometry before and after the food/no-food period. Nasal resistance was calculated at the first Broms radius and subsequently logarithmically transformed. A percent right nostril resistance measure was then derived (Rao and Potdar, 1970).

To determine whether food intake resulted in a change in the laterality of nasal resistance, a three-way repeated measures analysis of variance with factors of measurement time (before vs. after the 15-minute food/no-food period), food condition (food vs. no-food), and meal time (i.e., breakfast, lunch, dinner) was applied to the percent right nostril resistance data. This analysis revealed a significant measurement time by food condition interaction ($p = 0.002$). This interaction reflects the fact that the amount of airflow shunted through the left nasal chamber increased across the food condition and decreased across the no-food condition (Figure 1b).

UNILATERAL OLFACTORY SENSITIVITY AND NASAL RESISTANCE

The nasal cycle may influence olfactory sensitivity in several ways. First, systemic fluctuations in autonomic activity could influence central arousal centers. Second, such fluctuations might influence odorant access to receptors by altering not only airflow patterns to the olfactory epithelium, but by differentially influencing the thickness and consistency of the nasal mucosal secretions on each side of the nose. Third, variations in hemispheric integrated EEG activity might influence smell function by modulating the activity of a given hemisphere, assuming hemispheric specialization for olfactory processing exists.

Thirty three men and 44 women were tested in two sessions separated by approximately four hours. During each session, 2-butanone unilateral olfactory thresholds were determined using a single staircase forced-choice procedure described in detail elsewhere (Doty, et al., 1988). Approximately half of the subjects were required to block airflow through the nostril not sampling the odorant. Nasal resistance was assessed by anterior rhinomanometry before and after each threshold test. At the beginning of the second session, if the phase of the nasal cycle had not changed from its phase during the first session, we attempted to induce a phase shift by applying pressure under the armpit, in the palm of the hand, or by auditory stimulation on the more patent side of the nose (Rao and Potdar, 1970).

High resistance in the right nasal chamber was associated with elevated olfactory thresholds on both sides of the nose, whereas high resistance in the left was associated with comparatively depressed

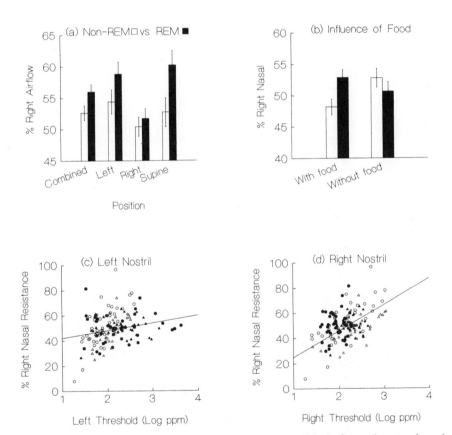

Figure 1. The nasal cycle (a) during sleep and (b) before (empty bars) and after (filled bars) the food or no-food period. The relationship between the nasal cycle and (c) left and (d) right olfactory thresholds.

olfactory thresholds on both sides of the nose for those subjects who did not block the non-odor sampling nostril (p = 0.05). Figures 1c and d show the influence of the phase of the nasal cycle on left and right olfactory thresholds of these subjects. Extending the associations between the BRAC rhythms and the nasal cycle to our findings, greater airflow through the left nasal chamber is correlated with parasympathetic predominance, greater right hemispheric integrated EEG activity, and decreased olfactory sensitivity. In contrast, greater airflow through the right nasal chamber is associated with sympathetic predominance, greater left hemispheric integrated EEG activity, and heightened olfactory sensitivity.

CONCLUSION

The present set of studies demonstrates intimate and complex associations between measures of olfactory sensitivity and autonomic processes associated with the nasal cycle.

ACKNOWLEDGEMENTS

This research was supported by the Eleanor Dana Center for Clinical Sleep Research, Grant DC00161 from the NIDCD, and Grant SBIR HL37911 from the NHLBI. The authors would like to thank Drs. Nancy Kribs, Andrew Mester, Adrian R. Morrison, Deborah Rosin, Jeanne Schuett and Paul Shaman, and Mr. Adrian Goldszmidt for their assistance.

REFERENCES

Doty, R.L., Deems, D.A., Frye, R.E., Pelberg, R., and Shapiro, A., 1988, Olfactory sensitivity, nasal resistance, and autonomic function in patients with multiple chemical sensitivities, Arch. Otolarynol. Head. Neck. Surg., 114:1422.

Eccles, R.,1978, The central rhythm of the nasal cycle, Acta. Otolaryngol., 86:464.

Goldstein, L. Stoltzfus, N.W., and Gardocki, J.F., 1972, Changes in inter-hemispheric amplitude relationships in the EEG during sleep, Physiol. Behav., 8:811.

Haight, J.S.J. and Cole, P., 1984, Reciprocating nasal airflow resistances, Acta Otolaryngol., 97:93.

Hiatt, J.F. and Kripke, D.F., 1975, Ultradian rhythms in waking gastric activity, Psychosomatic Med., 37:320.

Kennedy, B., Ziegler, M.G., and Shannahoff-Khalsa, D.S., 1986, Alternating lateralization of plasma catecholimines and nasal patency in humans, Life Sci., 38:1203.

Klein, R., Pilon, D., Prosser, S., and Shannahoff-Khalsa, D., 1986, Nasal airflow asymmetries and human performance, Biol. Psych., 23:127.

Levin, B.E., Rappaport, M., and Natelson, B.H., 1979, Ultradian variations of plasma noradrenaline in humans, Life Sci., 25:621.

Oswald, I., Merrington., J, and Lewis, H., 1970, Cyclical "On Demand" oral intake by adults, Nature, 225:959.

Rao, S. and Potdar, A., 1970, Nasal airflow with body in various positions, J. App. Physiol., 28:162.

Werntz, D.A., Bickford, R.G., Bloom, F.E., and Shannahoff-Khalsa, D.S., 1983, Alternating cerebral hemispheric activity and the lateralization of autonomic nervous function, Human Neurobiol., 2:39.

TOPOGRAPHICAL EEG MAPS OF HUMAN RESPONSES TO ODORANTS --

A PRELIMINARY REPORT

W.R. Klemm[1] and Stephen Warrenburg[2]

[1]Department of Veterinary Anatomy & Public Health,
Texas A&M University, College Station, TX 77843
[2]International Flavors and Fragrances, Inc.
Union Beach, N.J. 07735

ABSTRACT

EEG recordings from 19 scalp loci from 10 young adult females were used to assess the physiological response to seven odorants (birch tar, galbanum, heliotropine, jasmine, lavender, lemon, and peppermint). The odorants were randomly delivered to a face mask via valve-selectable tubing pathways that diverted air flow through one of seven sample vials or through an empty vial to serve as a control. Subjects first scored the odors on a scale of 1-9 along continua for sleep-arousal, intensity, and pleasantness. Topographic maps were constructed from the amplitude spectra in the frequency bands of delta (1-4 Hz), theta (4-8 Hz), alpha (8-13 Hz), and beta (13-30 Hz). Eight seconds of representative and artifact-free EEG were selected for FFT analysis before onset of stimulus delivery, and at three times after stimulus onset. EEG was also quantified at 30 seconds after stimulus termination.

Subjects differed in their subjective responses to the various odorants, with the most consistently arousing and strong odors being galbanum, lavender, and lemon. Heliotropine was notably weak. The most pleasant odors were lemon and peppermint, while lavender was consistently unpleasant. All subjects showed EEG map changes for several odorants, although the odorants giving the most pronounced EEG map changes differed across subjects. EEG map responses typically involved several-to-many scalp loci in one or more frequency bands. EEG map changes were seen even to weak odors and even in some cases when the subject was not consciously aware of stimulus presentation. These methods seem appropriate for evaluating odors. They have the potential for testing many hypotheses.

INTRODUCTION

Humans have a very sensitive sense of smell, being able to resolve only a few molecules of certain odorants (Engen, 1987). Despite that sensitivity, olfaction is poorly perceived in the human consciousness. People react variably to odors and fragrances and commonly have difficulty in describing the quality of a given smell. Some investigators believe that olfaction is akin to "cortical blindness," wherein the subcortical processing can yield behavioral responses even when conscious awareness of the stimuli is missing (Hecean and Albert, 1978).

Because olfactory input is widely distributed in the amygdala and phylogenetically primitive cortex, without direct projections to neocortex, it is not surprising that people have poor ability to recognize and describe odors. We might also expect that the EEG, which is most directly derived from neocortex, is poorly responsive to odorant stimulation. Perhaps that assumption accounts for the sparse literature on odorant effects on the EEG.

Yet, there is evidence that the EEG is responsive to odorants (see Discussion). In recent years, a new technique for EEG analysis has been introduced, topographical mapping (Maurer, 1989; Wong, 1991). These procedures have numerous advantages over older EEG approaches:

1. The methods not only involve computer analysis of digitized data, but the EEG is obtained simultaneously over the entire scalp, typically requiring 19 or more channels of data.

2. Data are represented as a two-dimensional spatial signature that is reflective of the brain state that generates it.

3. The display is color-coded, which makes it easier to interpret visually. Color juxtaposition emphasizes spatial distinctions and a sense of gradient in the field.

4. Such mapping facilitates the detection of hemispheric asymmetries and other regionally specific changes.

5. One can display the topography of the numerous features of the same signal in the same way. Such features include, for example, the actual voltages of each channel at a given point in time or various mathematical values (such as variance, frequency spectrum, cross- and autocorrelation coefficients, coherence, and fractal dimension) derived over a given epoch.

6. Significance probabilities can be mapped to show population differences between treatments or to compare a single unknown data set to a control or reference data set.

7. Mapping is the most convenient way to convey three-dimensional information in an immediately understandable form. A sequence of spatial maps can be viewed side by side in temporal order. Images can be cartooned dynamically.

8. Dipole sources and their spatio-temporal changes can be displayed.

Of the reported EEG studies of olfaction, most have used only a few sites on the scalp. Because it is not clear in advance which neocortical sites are affected by odors, it seems important to map the EEG from many scalp sites and the topographical features of the EEG. Only recently, for example, have multiple electrodes been used, and even then the four to nine electrode sites that were monitored were insufficient for an accurate interpolation (Lorig et al., 1990; Lorig and Roberts, 1990).

METHODS

Subjects were 10 female college students who were recruited by advertisement and paid to participate in the study. They were not tested
within one hour since the last meal or the last drink of cola or coffee
and were screened to eliminate those who were knowingly allergic to odors.
All subjects passed a screening test of identifying five common odorants.

Affective response to odors was scored along two continua: sleepi-
ness-arousal and pleasure-displeasure. We modified the scoring system of
Russell et al. (1989). This test seemed to have adequate reliability,
convergent validity, and discriminant validity in four studies. Our subjects rated each dimension on a scale of 1 to 9 (a score of 1 indicated
extreme sleepiness and 9 indicated hyperexcitability). Subjects also
scored odors similarly for strength. Subjects were asked if they recognized any of the odors or what they resembled. We presented odors in
counterbalanced sequence for both rating and EEG testing.

Odorants were delivered via face mask from a pump system that permitted selective routing of air flow (2 L/minute) through one of eight sample
vials containing a water blank, birch tar, galbanum, heliotropine, jasmine, lavendar, lemon, or peppermint. The flow line leading from the blank
sample was led through an activated charcoal filter to remove residual
odors from previous stimuli.

EEGs were recorded from 19 electrodes, permanently mounted in standard 10-20 configuration in a commercial EEG cap (Electro-Cap International, Inc.). Linked earlobes served as the reference. Impedances of all
electrodes were less than 10K ohms. Eye movements were monitored by an
electrode mounted between the eyes. Subjects sat in a comfortable reclin-
ing chair, with eyes closed, and were told to think whatever they wanted
to. Subjects were told to press a signal-marker button once each time
they detected a new odor. Visual monitoring of the EEG indicated that
sleepiness was not a problem, presumably because of the stimulus conditions. Each odor was preceded by a 1 minute pre-exposure recording period, followed by two minutes of odor delivery, followed by another minute
of no odor.

Data were collected on a 19 channel computerized topographical EEG
analyzer (Brain Atlas III, Biologic Corporation). Amplifier settings had
frequency cut-offs at 1 and 30 Hz and a gain of 20,000. Signal was digitized at 128 Hz. For each channel, we computed frequency spectra on four
representative and artifact-free 2-sec epochs before odorant onset, at
approximately 10 secs after odorant onset, at about 1 minute after odorant
onset, and at the end of the odorant presentation. Additionally, a
"recovery" sample was taken at 30 seconds after the odorant was turned
off. Data were presented as color-coded topographical maps of the frequency spectra in the four standard bands of delta, theta, alpha, and
beta.

RESULTS

Subjective ratings (Table 1) revealed that arousal and strength
ratings were correlated. This was not true for pleasure: galbanum and
lavender were scored relatively high for arousal/strength but were considered distinctly unpleasant. Heliotropine was considered very weak and
near neutral for arousal and pleasantness.

Table 1. Subjective ratings of the odors (Mean ± S.E.). Scale = 1 to 9,
n = 10.

AFFECT	Bir.	Gal.	Hel.	Jas.	Lav.	Lem.	Pep.
arousal	4.9 ±0.8	7.3 ±0.5	4.6 ±0.6	5.8 ±0.6	7.0 ±0.7	6.8 ±0.3	6.0 ±0.8
strength	5.5 ±0.7	7.1 ±0.5	2.2 ±0.5	6.1 ±0.8	7.4 ±0.6	6.8 ±0.4	6.3 ±0.4
pleasure	3.2 ±0.8	3.2 ±0.5	5.6 ±0.7	5.4 ±0.7	2.3 ±0.3	7.1 ±0.2	7.2 ±0.4

The possible effects of odor were determined by comparing FFT maps
after odor onset with the map just before odor onset (Fig. 1). In addition, it was helpful to examine the control trial in which no odor was
presented throughout a one-minute pre-exposure period, the two minute
blank odor period and the one minute post-blank period.

Most of the subjects had control maps like those previously reported
(Wong, 1991) for a large population studied under comparable recording
conditions. That is, the maps were generally symmetrical throughout,
alpha activity was the largest and had occipital dominance, delta and
theta activity had medium amplitudes and were more widely spread into
central and frontal areas, while beta activity was of low amplitude and
ill defined throughout the map. Two subjects had significant amounts of
widespread delta activity and four subjects had very little alpha. In any
case, emphasis was placed on the change produced after odor onset, relative to the pre-onset baseline and to whatever changes occurred in the
blank odor trial.

EEG map responses typically could involve several-to-many scalp loci
and one or more frequency bands (Fig. 1). In some situations EEG map
changes would habituate in the first minute of stimulation, while in
others the response persisted even after the odor was turned off. EEG map
changes could be seen even to weak odors that were ranked weak on the
strength scale and even in some cases when the subject was not consciously
aware of stimulus presentation. Most odors in most subjects were consciously detected, with a latency of two to eight seconds after the flow
valve was activated (transit time from sample vial to face mask was on the
order of two secs). All subjects showed EEG map changes for at least
several odors. Subject 017, for example, responded distinctly to heliotropine, lavender (Fig. 1), and lemon. Odors giving the most pronounced
EEG map changes were not always the same across subjects. For example,
subject 011 had distinct responses to birch tar and peppermint and subject
013 had distinct responses to jasmine.

DISCUSSION

These results seem to justify further study. In particular, we plan
to include more subjects and to develop formal analysis of variance models
for testing of main effects and the various interactions on the average
amplitude within each frequency band.

There are many hypotheses that can be tested with quantitative topographical EEG measures. Habituation of EEG response needs to be

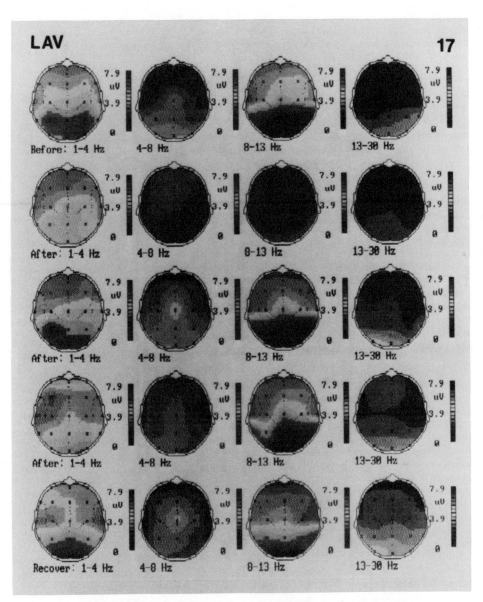

Fig. 1. Responses of subject 017 to lavender. During "pre-stimulus" (first row), there was widespread delta and occipital
theta, along with marked occipital alpha. At 10 seconds
after odor onset, delta, theta, and alpha decreased initially (second row). As odor continued, alpha habituated
(third and fourth row), while theta intensified. Beta
activity increased. After the odor was turned off (last
row), the widespread theta and beta persisted. Affect
scores were 6, 5, and 4, for arousal, strength, and
pleasure, respectively.

determined, as well as the possibility that EEG responsivity can be enhanced by response to a given odor is consistent across subjects or specific to individuals. Concentration effects on topographical change are also readily examined. Recent spectral data from Lorig et al. (1990), taken from two frontal and two temporal leads of 18 subjects, showed that the topography of theta activity was concentration-dependent for the two odors studied, lavender and spiced apple.

We can evaluate EEG maps for chaotic attractors, which are tendencies for a sequence of EEG voltages to be drawn to similar values. Reportedly, the olfactory EEG "burst" that carries the perceptual information from the bulb to the pyriform cortex of the rabbit reflects a chaotic attractor (Freeman and Skarda, 1990). A comparable olfactory effect in cortical EEG has not been looked for, particularly in humans.

We can test to see if conscious awareness of odor presence is necessary to produce a topographical change. Our preliminary data indicate that undetected odors can cause EEG changes. Lorig et al. (1990) report that beta activity during lavender stimulation was increased at temporal leads in 7 of their subjects who failed to detect stimulation with low concentrations of lavender. Even when asleep, humans react behaviorally, autonomically, and centrally to odorant stimulation. Peppermint odor, for example, activates (desynchronizes) the human EEG during sleep. Sometimes behavioral awakening would occur (Badia et al., 1990). It seems that EEG responses can be dissociated from self-reported perception (Lorig and Schwartz, 1988). Even simple breathing of room air through the nose causes different EEG alpha and beta activity, compared to effects of mouth breathing, and effects can vary with inhalation vs. expiration (Lorig et al., 1988).

We can test for correlation of the EEG response with subjective ratings or with attentional variables or the ability to verbalize odor identity or with the concurrent conscious association with a visual representation of the odor (e.g., seeing a lemon in the mind's eye while smelling lemon). Other cognitive influences may be operative. For example, the frontal EEG evoked response called contingent negative variation (CNV) can be differentially altered by certain odors (Torii et al., 1988; Lorig and Roberts, 1990). But, the effect may be cognitive rather than physiological because CNV amplitude changed according to the subjects' experiences about the odors' identity, even when in conflict with the actual odor composition (Lorig and Roberts, 1990).

We conclude that topographical EEG mapping may prove useful in the study of human olfaction.

ACKNOWLEDGMENTS

This research was sponsored by International Flavors & Fragrances, Inc., Union Beach, New Jersey.

REFERENCES

Badia, P., Wesensten, N., Lammers, W., Culpepper, J. and Harsh, J., 1990, Responsiveness to olfactory stimuli presented in sleep, Physiol. Behav., 48: 87-90.
Engen, T., 1987, Remembering odors and their names, Amer. Scientist, 75: 497-502.

Freeman, W. J., and Skarda, C. A., 1990, Chaotic dynamics versus represen-
     tationalism. Behav. Brain. Sci., 13:167-168.
Hecean, H., and Albert, M. L., 1978, "Human Neuropsychology," Wiley &
     Sons, New York.
Lorig, T. S., and Schwartz, G. E., 1988, Brain and odor: I. Alteration of
     human EEG by odor administration, Psychobiology, 16:281-284.
Lorig, T. S., Schwartz, G. E., Herman, K. B., and Lane, R. D., 1988, Brain
     and odor: II. EEG activity during nose and mouth breathing. Psycho-
     biology, 16:285-287.
Lorig, T. S., Herman, K. B., Schwartz, G. E., and Cain, W. S., 1990, EEG
     activity during administration of low-concentration odors. Bull.
     Psychonom. Soc., 28:405-408.
Lorig, T.S. and Roberts, M., 1990, Odor and cognitive alteration of the
     contingent negative variation, Chem. Senses, 15:537-545.
Maurer, K. (Ed.), 1989, "Topographic brain mapping of EEG and evoked
     potentials," Springer-Verlag, New York.
Russell, J. A., Weiss, A., and Mendelsohn, G. A., 1989, Affect grid: a
     single-item scale of pleasure and arousal. J. Personality. Soc.
     Psychol., 57:493-502.
Torii, S., Fukuda, H., Kanemoto, H., Miyanchi, R., Hamauzu, Y., and Kawa-
     saki, M., 1988, Contingent negative variation and the psychological
     effects of odor. In: "Perfumery: the Psychology and Biology of
     Fragrance", S. van Toller and G. Dodd, Eds., Chapman and Hall, Lon-
     don.
Wong, P. K. H., 1991, "Introduction to Brain Topography," Plenum, New
     York.

AUTHOR INDEX

Gilmartin, M., 120, 124

Gilmore, D., 264, 266

Ginetz, R.M., 376, 381

Ginsberg, B.E., 197, 202

Ginzberg, A.S., 347, 348

Gionfriddo, J.P., 313, 315

Giongo, F., 431, 434

Gipes, J.H.W., 501, 502

Gjessing, L.R., 177, 180

Gladysheva, O.S., 43, 47

Glahn, J.F., 312, 313, 316

Glenn, E.M., 130, 132

Glosser, G., 59, 60, 64

Glover, G.J., 494, 497

Glowinski, J., 75, 76

Godinot, F., 551

Goethe, F., 155, 159

Goetz, F.W., 345, 348, 358, 359, 364

Gold, H., 89, 95

Goldman, B.D., 123

Goldman, P., 177, 180

Goldstein, L., 595, 598

Goncharova, N., 397

Good, A., 205, 210

Good, I.J., 176, 180

Gordon, H.W., 559, 564

Gorman, M.L., 281, 284, 465, 466, 470, 474, 475

Gorski, R.A., 120, 122

Gorzalka, B.B., 501

Gosden, P.E., 538, 541

Gosling, L.M., 458, 463

Gossow, H., 155, 159

Goz, H., 375

Granno-Reisfeld, N., 149, 153

Gray, D.R., 493-496, 498

Gray, J., 130, 132

Gray, W.R., 108, 110, 114

Graziadei, P., 21, 23, 25, 26, 462, 463

Graziadei, P.P.C., 25, 149, 150, 153

Green, B.G., 311, 316

Green, M., 219, 223

Green, W.J., 335, 342

Greene, H.W., 335, 341

Greenfield, K.L., 121, 122

Greenstein, Y.J., 73, 76

Greenwald, G.S., 264, 266

Gregorova, S., 200, 202

Gregson, R.A.M., 59, 60, 64

Greig-Smith, P.W., 313, 316

Gresik, E.W., 121, 123

Grijalva, C.V., 67, 75

Grill, H.J., 298, 302

Grillo, M., 149, 150, 153

Groot, C., 349, 355

Grubb, T.C., 421-424, 427

Gruber, S.H., 335, 336, 340-342

Grudman, M., 73, 76

Gruter, M., 433, 434

Guarino, J.L., 312, 315, 316, 330

Gubernick, D.J., 195, 196

Gubits, R.M., 121, 123

Guengerich, F., 25

Guevara-Aguilar, R., 297, 298, 302

Guilford, T., 421, 424, 427

Guinness, F., 478, 483, 484

Gunnet, J.W., 51, 53

Gustafson, A.W., 32, 34

Gustafsson, B.E., 177, 180

Gustavson, A.R., 575, 578

Gutknecht, J., 229, 234

Guyenet, P.G., 50, 53

Haberly, L.B., 39, 40

Haegele, C.W., 343, 348

Haga, T., 11, 14

Hagg, T., 299-303

Haight, J.S., 21, 25

Haight, J.S.J., 595, 598

Hainey, S., 121, 122

Halasz, N., 52, 53

Hall, D.H., 312, 316

Hall, E.R., 534

Hall, R.W., 544, 549

Hallworth, G.W., 25

Halpern, M., 107-109, 113, 114, 255, 257, 515, 521

Halpin, Z.T., 503, 508

Haltenorth, T., 493, 498

Hama, K., 149, 153

Hamann, U., 169, 172

Hamauzu, Y., 581, 585, 604, 605

Hamilton, G.D., 263, 266

Hamilton, J.K., 108, 114

Hamilton, W.D., 179, 180

Hammer, R.P., Jr., 533, 534

Hampl, R., 199, 200, 202

Hamza, M., 330

Hanada, T., 57, 89, 92

Hands, C., 200, 202

Hanokoglu, I., 23, 26

Hansen, T.R., 118

Hanski, E., 89, 95

Hara, T.J., 285-287, 345, 348-350, 354, 355, 358, 359, 363, 364, 375, 381, 387

Harbourne, J.B., 461, 463

Harding, J., 23, 25

Hari, R., 565, 569

Harpaz, S., 383

Harris, M.A., 527, 528

Harris, Z.L., 279, 280

Harsh, J., 604

Hart, B.L., 129, 132, 501, 502, 515, 521

Hart, G.H., 129, 132

Hartikka, J., 301, 302

Hartman, G., 349, 355

Harvey, S., 201, 203, 312, 317, 462, 463, 509, 513, 540, 542

Hashimoto, Y., 471, 475

Hastie, N.D., 121, 123, 124

Hatanaka, T., 27

Haun, J.E., 544, 549

Hawes, M.L., 99, 105

Hawk, H.H., 129, 132

Hawkins, G.N., 115, 118

Hay, D.E., 343, 348

Hay, J.B., 575, 579

Hayaishi, O., 120, 123

Hayward, G.F., 485, 491

Hazenberg, M.P., 176, 181

Head, R., 149, 153

Healy, S., 421, 424, 427

Heap, P.F., 269, 270, 283, 284

Hebel, N., 60, 64

Hecean, H., 600, 604

Heck, G.L., 82, 83

Heck, H., 493, 498

Heckman, S., 457, 463

Hefti, F., 301, 302

Held, W.A., 120-124

Heldman, J., 23, 26

Heller, J., 120, 122

Helti, F., 299, 303

Hemley, R., 120, 123

Hempstead, J., 149, 153

Hendrie, C.A., 529, 534

Henegar, J., 25

TAXONOMIC INDEX

SNK test, *see* Student-Newman-Keuls test

Snowshoe hares, 281

Social behavior
  flank gland secretions and, 99-105
  MHC and, 175-179, 197-201

Social communication, 515-521, 537-541

Sockeye salmon (Oncorhynchus nerka), 376

Sodium chloride, 90, 91, 136, 137, 283, 284, 399, 544
  IC lesions and, 298, 300
  sucrose and, 3-6
  trigeminal nerve system and, 87

Sodium dodecyl-sulfate, 339

Sodium o-vanadate, 45, 46

Sour taste, 137, 138

Sparrows, 17

Spawning pheromones, 343-347

Spiny mice, 462

Squid, 384, 386

Squirrels, 499-501

Standard opponents, *see* Standard stimulus test

Standard stimulus test, 198-201

Starlings, 17, 312, 325, 326, 443, 445

Stellate sturgeon, 135-137

Steller sea lion, 544

Stereobilin, 339

Sterilization, 239

Sterlet, 161

Steroidal pheromones, 285-287, 344-345, 371

Sterols, 474

Stickleback fish, 179

Stone sculpin, 371-372

Stress
  adrenal function and, 271, 273, 275
  odors and, 581-584

Stria terminalis, 51

Student-Newman-Keuls (SNK) test, 405, 415, 416

Sturgeon, 135-138, 161

Styrene, 589

Substance P, 149

Substantia nigra, 50

Succinic acid, 339

Succinimide, 339

Sucker fish, 360

Sucrose, 136, 137, 544
  IC lesions and, 298
  salty taste response and, 3-6
  trigeminal nerve system and, 87

Sulfatase, 345

17,20βP-Sulphate (17,20βP-S), 358, 360

Superovulation, 115, 117, 264

Swallows, 17

Sweat, MHC in, 205, 206, 210

Sweet taste, 9, 13, 137, 138

Swifts, 17, 443

Takins, 493

Tannin, 329

Taste, 135-138, *see also* Taste buds
  amino acids in, 365-368
  α-gustducin in, 9-13
  IC lesions and, 297-301
  sucrose effect on salty, 3-6
  trigeminal nerve system and, 85-88

Taste buds, 15-19

Taurocholic acid, 344, 345

Temporal lobe epilepsy, 567

Terminalis systems, 141-146, *see also* Nervus terminalis

Testosterone, *see also* specific types
  aggression and, 499-501, 541
  functional properties of, 285, 286, 287
  MHC and, 199-201
  as pheromone, 344, 357, 360
  scent glands and, 524, 525

Testosterone glucuronide, 357, 360

Testosterone propionate
  aggression and, 499
  posterolateral glands and, 125-128
  scent marking and, 477-483

Tetracycline, 177

TH, *see* Tyrosine hydroxylase

Thalamocortical mechanisms, 59-64

Theta-bursting, 73-75

Thiamine deficiency, 60-63

Thin-layer chromatography (TLC), 474

Threonine, 136-138

TI dopaminergic neurons, *see* Tuberoinfundibular dopaminergic neurons

Tigers, 471-474

Time-averaged models, 79

TLC, *see* Thin-layer chromatography

TMT, *see* Trimethyl thiazoline

Tonic immobility, 335-340

Toothed whales, 143, 144-146, 543, 546, 548

Topographical EEG maps, 599-604

Tortoises, 30

α-Transducins, *see* α-Gustducin

Transfer of training, 208-209, 226-227

Trans-3-hexenol, 93

Trauma communication, 389-394

Tree shrews, 462

Tremulacin, 461

Trigeminal nerve system, 55, 91, 548-549
  bird repellents and, 311, 319-322
  CSEP and, 565, 566
  olfaction and taste interactions of, 85-88
  sensory segmentation and, 573

Triglycerides, 474

17α,20β,21-Trihydroxyprogesterone, 344

Trimethylamine, 545

Trimethyl thiazoline (TMT), 305-310
Tris-borate, 108
Tris-chloride, 90
Tris-glycine, 108
Trout, 349-354
Tryptophane, 136, 138
Tsetse flies, 79
T-tests, 278
Tuberculosis, 313
Tuberoinfundibular (TI) dopaminergic
    neurons, 49-53
Turbulence, 79-80, 82
Turkey vultures, 443, 446
Turpentine, 432
Turtles, 414
  amino acid sensitivity in, 397-400
  olfaction in, 90-94
  olfactory epithelium of, 55-58
  trigeminal nerve of, 55, 91
  vomeronasal responses in, 89-91
Twin studies
  on bacterial flora, 176
  on odor discrimination, 219-222
Tyrosine, 136-138
Tyrosine hydroxylase (TH), 149-152

Ultradian autonomic rhythms, 595-598
Ultrasonic vocalizations, 305, 516,
    518-519
Ultraviolet/visible absorption spec-
    trometry, 336, 339
Ungulates, 546
Urea, 107, 339
Urinary chemosignals, 261, 335
  discrimination of human, 219-222
  major proteins in, 120-121
  marking fluid in, 471-472, 474
  MHC and, 167-168, 170-172, 175-
    179, 191, 193-195, 197-199,
    205-207, 225, 237-240
  odortypes in, 504-508
  pheromones in, see Pheromones
  reproduction and, 245-250
  rutting odors and, 494, 495, 497
  scent marking and, 477, 481, 483

Vaginal markings, 518-519, 520
Vaginal smearing, 247, 249, 250, 271-
    272, 275, 281
Valeric acid, 56, 57, 92, 544, 545
Valine, 136-138, 398
Vanillin, 81, 295, 565-567

Vanillyl alcohol, 295
Verapamil, 91
Veratryl alcohol, 295
Vision
  bird repellents not enhanced by,
    323-329
  α-gustducin and, 10-13
Voles, 255, 256
  mating in, 231
  pheromones and, 121, 265, 277
  posterolateral glands of, 125
  predator odors and, 281, 532, 533
  reproduction and metabolism in,
    277-280
Vomeronasal system, 33, 89-91, 253-
    256, 462
  absence of, 143, 144
  adrenal function and, 271, 272, 274
  anogenital licking sustained by,
    551-555
  AOB and, 52, 53
  habitat and gender correlated
    with, 403-408
  pheromones and, 27-30, 119, 120,
    277
  phosphatase activity of, 43-47
  predator odors and, 281-284
  prey odors and, 107-113
  social communication and, 515-521
Vulval skin gland secretion, 115-118

Wax esters, 474
Weasel odors, 529-530, 532, 533
Whales, 141-146, 543, 546, 548
White-footed mice, 529
White Sea cod, 85-88
White-tailed deer, 477-483
Wilcoxon matched-pairs signed-ranks
    test, 308, 377, 499, 554
Working memory, 60, 63-64

X chromosomes, 195, 201
Xenobiotics, 21, 23
X-maze test, 510-512

Y chromosomes, 195, 199, 200
Yellowfin sculpin, 285-287, 371
Y-mazes, 156, 191, 192, 214, 237,
    238, 240, 372

Zebra danio, 357
Zingerone, 315